Siegfried Haussühl
Physical Properties of Crystals

1807–2007 Knowledge for Generations

Each generation has its unique needs and aspirations. When Charles Wiley first opened his small printing shop in lower Manhattan in 1807, it was a generation of boundless potential searching for an identity. And we were there, helping to define a new American literary tradition. Over half a century later, in the midst of the Second Industrial Revolution, it was a generation focused on building the future. Once again, we were there, supplying the critical scientific, technical, and engineering knowledge that helped frame the world. Throughout the 20th Century, and into the new millennium, nations began to reach out beyond their own borders and a new international community was born. Wiley was there, expanding its operations around the world to enable a global exchange of ideas, opinions, and know-how.

For 200 years, Wiley has been an integral part of each generation's journey, enabling the flow of information and understanding necessary to meet their needs and fulfill their aspirations. Today, bold new technologies are changing the way we live and learn. Wiley will be there, providing you the must-have knowledge you need to imagine new worlds, new possibilities, and new opportunities.

Generations come and go, but you can always count on Wiley to provide you the knowledge you need, when and where you need it!

William J. Pesce
President and Chief Executive Officer

Peter Booth Wiley
Chairman of the Board

Siegfried Haussühl

Physical Properties of Crystals

An Introduction

WILEY-VCH Verlag GmbH & Co. KGaA

The Authors

Prof. Dr. Siegfried Haussühl
Institute of Crystallography
University of Cologne
Zülpicher Str. 49b
50674 Cologne
Germany

Translation
Peter Roman, Germany

Library of Congress Card No.:
applied for

British Library Cataloguing-in-Publication Data
A catalogue record for this book is available from the British Library.

Bibliographic information published by Die Deutsche Nationalbibliothek
Die Deutsche Bibliothek lists this publication in the Deutsche Nationalbibliografie; detailed bibliographic data is available in the Internet at http://dnb.d-nb.de.

Printed in the Federal Republic of Germany

Printed on acid-free paper

Printing: Strauss GmbH, Mörlenbach
Bookbinding: Litges & Dopf Buchbinderei GmbH, Heppenheim
Wiley Bicentennial Logo: Richard J. Pacifico

ISBN: 978-3-527-40543-5

Contents

1 Fundamentals *1*

1.1 Ideal Crystals, Real Crystals *1*

1.2 The First Basic Law of Crystallography (Angular Constancy) *3*

1.3 Graphical Methods, Stereographic Projection *4*

1.4 The Second Basic Law of Crystallography (Law of Rational Indices) *8*

1.5 Vectors *10*

1.5.1 Vector Addition *10*

1.5.2 Scalar Product *13*

1.5.3 Vector Product *14*

1.5.4 Vector Triple Product *17*

1.6 Transformations *18*

1.7 Symmetry Properties *19*

1.7.1 Symmetry Operations *19*

1.7.2 Point Symmetry Groups *24*

1.7.3 Theory of Forms *32*

1.7.4 Morphological Symmetry, Determining the Point Symmetry Group *42*

1.7.5 Symmetry of Space Lattices (Space Groups) *42*

1.7.5.1 Bravais Types *42*

1.7.5.2 Screw Axes and Glide Mirror Planes *45*

1.7.5.3 The 230 Space Groups *46*

1.8 Supplements to Crystal Geometry *47*

1.9 The Determination of Orientation with Diffraction Methods *48*

2 Sample Preparation *51*

2.1 Crystal Preparation *51*

2.2 Orientation *54*

Physical Properties of Crystals. Siegfried Haussühl.

ISBN: 978-3-527-40543-5

3 **Definitions** *57*
3.1 Properties *57*
3.2 Reference Surfaces and Reference Curves *59*
3.3 Neumann's Principle *60*
3.4 Theorem on Extreme Values *61*
3.5 Tensors *62*
3.6 Theorem on Tensor Operations *65*
3.7 Pseudo Tensors (Axial Tensors) *70*
3.8 Symmetry Properties of Tensors *72*
3.8.1 Mathematical and Physical Arguments: Inherent Symmetry *72*
3.8.2 Symmetry of the Medium *74*
3.9 Derived Tensors and Tensor Invariants *78*
3.10 Longitudinal and Transverse Effects *80*

4 **Special Tensors** *83*
4.1 Zero-Rank Tensors *83*
4.2 First-Rank Tensors *85*
4.2.1 Symmetry Reduction *85*
4.2.2 Pyroelectric and Related Effects *86*
4.3 Second-Rank Tensors *89*
4.3.1 Symmetry Reduction *89*
4.3.2 Tensor Quadric, Poinsots Construction, Longitudinal Effects, Principal Axes' Transformation *93*
4.3.3 Dielectric Properties *99*
4.3.4 Ferroelectricity *106*
4.3.5 Magnetic Permeability *108*
4.3.6 Optical Properties: Basic Laws of Crystal Optics *112*
4.3.6.1 Reflection and Refraction *118*
4.3.6.2 Determining Refractive Indices *127*
4.3.6.3 Plane-Parallel Plate between Polarizers at Perpendicular Incidence *130*
4.3.6.4 Directions of Optic Isotropy: Optic Axes, Optic Character *133*
4.3.6.5 Sénarmont Compensator for the Analysis of Elliptically Polarized Light *136*
4.3.6.6 Absorption *139*
4.3.6.7 Optical Activity *141*
4.3.6.8 Double refracting, optically active, and absorbing crystals *148*
4.3.6.9 Dispersion *148*
4.3.7 Electrical Conductivity *150*
4.3.8 Thermal Conductivity *152*
4.3.9 Mass Conductivity *153*
4.3.10 Deformation Tensor *154*

4.3.11 Thermal Expansion *159*
4.3.12 Linear Compressibility at Hydrostatic Pressure *164*
4.3.13 Mechanical Stress Tensor *164*
4.4 Third-Rank Tensors *168*
4.4.1 Piezoelectric Tensor *173*
4.4.1.1 Static and Quasistatic Methods of Measurement *174*
4.4.1.2 Extreme Values *180*
4.4.1.3 Converse Piezoelectric Effect (First-Order Electrostriction) *182*
4.4.2 First-Order Electro-Optical Tensor *184*
4.4.3 First-Order Nonlinear Electrical Conductivity (Deviation from Ohm's Law) *194*
4.4.4 Nonlinear Dielectric Susceptibilty *195*
4.4.5 Faraday Effect *204*
4.4.6 Hall Effect *205*
4.5 Fourth-Rank Tensors *207*
4.5.1 Elasticity Tensor *214*
4.5.2 Elastostatics *217*
4.5.3 Linear Compressibility Under Hydrostatic Pressure *220*
4.5.4 Torsion Modulus *221*
4.5.5 Elastodynamic *222*
4.5.6 Dynamic Measurement Methods *231*
4.5.7 Strategy for the Measurement of Elastic Constants *266*
4.5.7.1 General Elastic Properties; Stability *267*
4.5.8 The Dependence of Elastic Properties on Scalar Parameters (Temperature, Pressure) *270*
4.5.9 Piezooptical and Elastooptical Tensors *271*
4.5.9.1 Piezooptical Measurements *272*
4.5.9.2 Elastooptical Measurements *273*
4.5.10 Second-Order Electrostrictive and Electrooptical Effects *285*
4.5.11 Electrogyration *286*
4.5.12 Piezoconductivity *288*
4.6 Higher Rank Tensors *288*
4.6.1 Electroacoustical Effects *288*
4.6.2 Acoustical Activity *289*
4.6.3 Nonlinear Elasticity: Piezoacoustical Effects *290*

5 Thermodynamic Relationships *297*
5.1 Equations of State *297*
5.2 Tensor Components Under Different Auxiliary Conditions *301*
5.3 Time Reversal *305*
5.4 Thermoelectrical Effect *307*

6 Non-Tensorial Properties *309*
6.1 Strength Properties *309*
6.1.1 Hardness (Resistance Against Plastic Deformation) *310*
6.1.2 Indentation Hardness *315*
6.1.3 Strength *317*
6.1.4 Abrasive Hardness *318*
6.2 Dissolution Speed *323*
6.3 Sawing Velocity *324*
6.4 Spectroscopic Properties *326*

7 Structure and Properties *329*
7.1 Interpretation and Correlation of Properties *329*
7.1.1 Quasiadditive Properties *331*
7.1.2 Nonadditive Properties *338*
7.1.2.1 Thermal Expansion *339*
7.1.2.2 Elastic Properties, Empirical Rules *341*
7.1.2.3 Thermoelastic and Piezoelastic Properties *344*
7.2 Phase Transformations *347*

8 Group Theoretical Methods *357*
8.1 Basics of Group Theory *357*
8.2 Construction of Irreducible Representations *364*
8.3 Tensor Representations *370*
8.4 Decomposition of the Linear Vector Space into Invariant Subspaces *376*
8.5 Symmetry Matched Functions *378*

9 Group Algebra; Projection Operators *385*

10 Concluding Remarks *393*

11 Exercises *395*

12 Appendix *407*
12.1 List of Common Symbols *407*
12.2 Systems of Units, Units, Symbols and Conversion Factors *409*
12.3 Determination of the Point Space Group of a Crystal From Its Physical Properties *410*
12.4 Electric and Magnetic Effects *412*
12.5 Tables of Standard Values *414*

12.6 Bibliography *421*
12.6.1 Books *421*
12.6.2 Articles *427*
12.6.3 Data Sources *431*
12.6.4 Journals *433*

Preface

With the discovery of the directional dependence of elastic and optical phenomena in the early 19th century, the special nature of the physical behavior of crystalline bodies entered the consciousness of the natural scientist. The beauty and elegance, especially of the crystal-optical laws, fascinated all outstanding physicists for over a century. For the founders of theoretical physics, such as, for example, Franz Neumann (1798-1895), the observations on crystals opened the door to a hidden world of multifaceted phenomena. F. Pockels (1906) and W. Voigt (1910) created, with their works *Lehrbuch der Kristalloptik* (Textbook of Crystal Optics) and *Lehrbuch der Kristallphysik* (Textbook of Crystal Physics), respectively, the foundation for theoretical and experimental crystal physics. The development of lattice theory by M. Born, presented with other outstanding contributions in Volume XXIV of *Handbuch der Physik* (Handbook of Physics, 1933), gave the impetus for the atomistic and quantum theoretical interpretation of crystal-physical properties. In the shadow of the magnificent success of spectroscopy and structural analysis, further development of crystal physics took place without any major new highlights. The application of tensor calculus and group theory in fields characterized by symmetry properties brought about new ideas and concepts. A certain completion in the theoretical representation of the optical and elastic properties was achieved relatively early. However, a quantitative interpretation from atomistic and structural details is, even today, only realized to a satisfactory extent for crystals with simple structures. The technological application and the further development of crystal physics in this century received decisive impulses through the following three important discoveries: 1. High-frequency techniques with the use of piezoelectric crystals for the construction of frequency determining devices and in ultrasound technology. 2. Semiconductor techniques with the development of transistors and integrated circuits based on crystalline devices with broad applications in high-frequency technology and in the fields of information transmission as well as computer technology. 3. Laser techniques with its many applications, in particular, in the fields of optical measurement techniques, chemical analysis, materials processing, surgery,

Physical Properties of Crystals. Siegfried Haussühl.

ISBN: 978-3-527-40543-5

and, not least, the miniaturization of information transmission with optical equipment.

In many other areas, revolutionary advances were made by using crystals, for example, in radiation detectors through the utilization of the pyroelectric effect, in fully automatic chemical analysis based on X-ray fluorescence spectroscopy, in hard materials applications, and in the construction of optical and electronic devices to provide time-delayed signals with the help of surface acoustic waves. Of current interest is the application of crystals for the various possibilities of converting solar energy into electrical energy. It is no wonder that such a spectrum of applications has broken the predominance of pure science in our physics institutes in favor of an engineering-type and practical-oriented research and teaching over the last 20 years. While even up to the middle of the century the field of crystallography-apart from the research centers of metal physics-mainly resided in mineralogical institutes, we now have the situation where crystallographic disciplines have been largely consumed by physics, chemistry, and physical chemistry. In conjunction with this was a tumultuous upsurge in crystal physics on a scale which had not been seen before. With an over 100-fold growth potential in personnel and equipment, crystal physics today, compared with the situation around 1950, has an entirely different status in scientific research and also in the economic importance of the technological advances arising from it. What is the current state of knowledge, and what do the future possibilities of crystal physics hold? First of all some numerical facts: of the approximately 45,000 currently known crystallized substances with defined chemical constituents and known structure, we only have a very small number (a few hundred) of crystal types whose physical properties may largely be considered as completely known. Many properties, such as, for example, the higher electric and magnetic effects, the behaviour under extreme temperature and pressure conditions and the simultaneous interplay of several effects, have until now-if at all-only been studied on very few crystal types. Apart from working on data of long known substances, the prospective material scientist can expect highly interesting work over the next few decades with regard to the search for new crystal types with extreme and novel properties. The book *Kristallphysik* (Crystal Physics) is intended to provide the ground work for the understanding of the distinctiveness of crystalline substances, to bring closer the phenomenological aspects under the influence of symmetry and also to highlight practical considerations for the observation and measurement of the properties. Knowledge of simple physical definitions and laws is presumed as well as certain crystallographic fundamentals, as found, for example, in the books *Kristallgeometrie* (Crystal Geometry) and *Kristallstrukturbestimmung* (Crystal Structure Determination). The enormous amount of material in the realm of crystal physics can, of course, only be covered here in an exemplary way by making certain choices. Fields in which the crystal-specific anisotropy effects remain in the background, such as, for example, the semiconductors and superconductors, are not considered in this book. A sufficient amount of literature already exists for these topics. Also the issue of inhomogeneous crystalline preparations and the inhomogeneous external effects could not be discussed here. Boundary properties as well as the influence of defects connected with growth mechanisms will be first discussed in the volume *Kristallwachstum* (Crystal Growth). The approaches to the structural interpretation of crystal properties based on lattice theory were only touched on in this book. The necessary space for this subject is provided in the volume *Kristallchemie* (Crystal Chemistry) as well as thermodynamic and crystal-chemical aspects of stability. A chapter on methods of preparation is presented at the beginning, which is intended to introduce the experimenter to practical work with crystals. We clearly focus on the problem of orientation with the introduction of a fixed "crystal-physical" reference system in the crystal. For years a well-established teaching method of separating the physical quantities into inducing and induced quantities has been taken over. The connection between these allows a clear definition of the notion of "property." The properties are classified according to the categories "tensorial" and "nontensorial, " whereby such properties which can be directly calculated from tensorial properties, such as, for example, light or sound velocity, can be classified as "derived tensorial" properties. A large amount of space is devoted to the introduction of tensor calculus as far as it is required for the treatment of crystal-physical problems. Important properties of tensors are made accessible to measurement with the intuitive concepts of "longitudinal effect" and "transverse effect." The treatment of group theoretical methods is mainly directed towards a few typical applications, in order to demonstrate the attractiveness and the efficiency of this wonderful tool and thus to arouse interest for further studies. The reader is strongly recommended to work through the exercises. The annex presents tables of proven standard values for a number of properties of selected crystal types. References to tables and further literature are intended to broaden and consolidate the fields treated in this book as well as helping in locating available data. My special thanks go to Dr. P. Preu for his careful and critical reading of the complete text and his untiring help in the production of the figures. A. Möws through her exemplary service on the typewriter was of great support in completion of the manuscript. Finally, I would also like to express my thanks to the people of Chemie Verlag, especially Dr. G. Giesler, for their understanding and pleasant cooperation.

Cologne, summer 1983 *S. Haussühl*

Preface to the English Edition

In the first edition of *Kristallphysik* it was assumed that the reader possessed basic knowledge of crystallography and was familiar with the mathematical tools as well as with simple optical and X-ray methods. The books *Kristallgeometrie* (Crystal Geometry) and *Kristallstrukturbestimmung* (Crystal Structure Determination), both of which have as yet only been published in German, provided the required introduction. The terms and symbols used in these texts have been adopted in Crystal Physics. In order to present to the reader of the English translation the necessary background, a chapter on the basics of crystallography has been prefixed to the former text. The detailed proofs found in Kristallgeometrie (Crystal Geometry) and Kristallstrukturbestimmung (Crystal Structure Determination) were not repeated. Of course, other books on crystallography are available which provide an introduction to the subject matter. Incidentally, may I refer to the preface of the first edition. The present text emerges from a revised and many times amended new formulation. Some proofs where I have given the reader a little help have been made more accessible by additional references. Furthermore, I have included some short sections on new developments, such as, for example, the resonant ultrasound spectroscopy (RUS) method as well as some sections on the interpretation of physical properties. This last measure seemed to make sense because I decided not to bring to print the volumes *Kristallchemie* (Crystal Chemistry) and *Kristallzüchtung* (Crystal Growth) announced in the first edition, although their preparations were at an advanced stage. An important aspect for this decision was that in the meantime several comprehensive and attractive expositions of both subjects appeared and there was therefore no reason, alone from the scope of the work, to publish an equivalent exposition in the form of a book. In addition, the requirement to actualize and evaluate anew the rapid increase in crystallographic data in ever shorter time intervals played a decisive role in my decision. The same applies to the experimental and theoretical areas of crystal growth. Hence, the long-term benefit of an all too condensed representation of these subjects is questionable. In contrast, it is hoped that the fundamentals treated in the three books published so far will provide a sufficient basis for crystallographic training for a long time to come.
I thank Dr. Jürgen Schreuer, Frankfurt, for his many stimulating suggestions with respect to the new formulation of the text. In particular, he compiled the electronic text for which I owe him my deepest gratitude. Finally, I wish to thank Vera Palmer of Wiley-VCH for her cooperation in the publishing of this book.

Siegfried Haussühl

1
Fundamentals

1.1
Ideal Crystals, Real Crystals

Up until a few years ago, crystals were still classified according to their morphological properties, in a similar manner to objects in biology. One often comes across the definition of a crystal as a homogenous space with directionally dependent properties (anisotropy). This is no longer satisfactory because distinctly noncrystalline materials such as glass and plastic may also possess anisotropic properties. Thus a useful definition arises out of the concept of an *ideal crystal* (Fig. 1.1):

An ideal crystal is understood as a space containing a rigid lattice arrangement of uniform atomic cells.

A definition of the lattice concept will be given later. Crystals existing in nature, the *real crystals*, which we will now generally refer to as crystals, very closely approach ideal crystals. They show, however, certain deviations from the rigid lattice arrangement and from the uniform atomic cell structure. The following types of imperfections, i.e., deviations from ideal crystals, may be mentioned:

Imperfections in the uniform structure of the cells. These are lattice vacancies, irregular occupation of lattice sites, errors in chemical composition, deviations from homogeneity by mixed isotopes of certain types

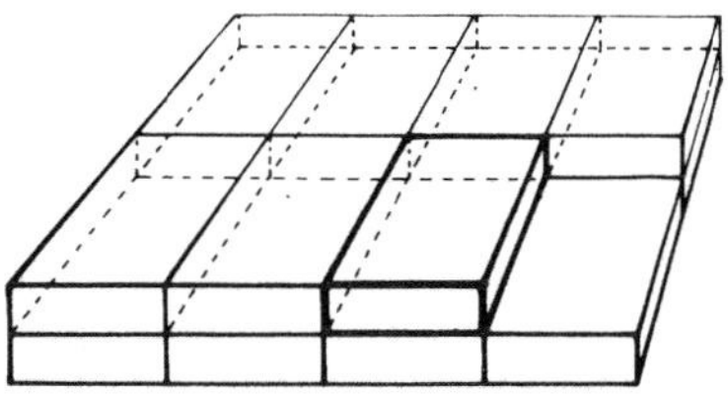

Figure 1.1 Lattice-like periodic arrangement of unit cells.

Physical Properties of Crystals. Siegfried Haussühl.

ISBN: 978-3-527-40543-5

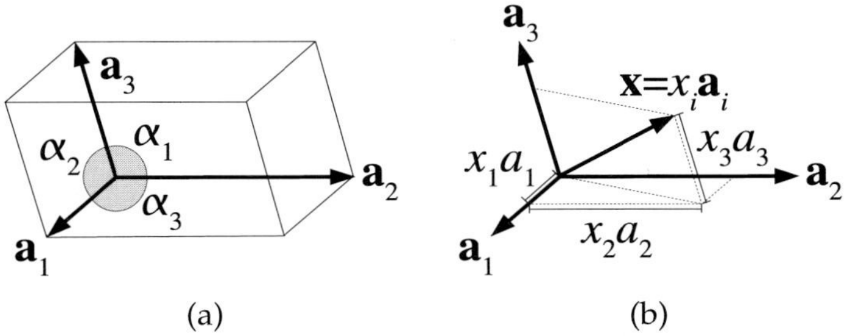

Figure 1.2 (a) Parallelepiped for the definition of a crystallographic reference system and (b) decomposition of a vector into components with respect to the reference system.

of atoms, different excitation states of the building particles (atoms), not only with respect to bonding but also with respect to the position of other building particles (misorientation of building particles).

Imperfections in the lattice structure. These refer to displacement, tilting and twisting of cells, nonperiodic repetition of cells, inhomogenous distribution of mechanical deformations through thermal stress, sound waves, and external influences such as electric and magnetic fields. The simple fact that crystals have finite dimensions results in a departure from the ideal crystal concept because the edge cells experience a different environment than the inner ones.

At this point we mention that materials exist possessing a structure not corresponding to a rigid lattice-type arrangement of cells. Among these are the so-called quasicrystals and substances in which the periodic repetition of cells is impressed with a second noncommensurable periodicity.

To characterize a crystal we need to make some statements concerning structural defects.

One must keep in mind that not only the growth process but also the complete morphological and physical appearance of the crystal is crucially determined by the structure of the lattice, i.e., the form of the cells as well as the spatial arrangement of its constituents.

A unit cell in the sense used here is a parallelepiped, a space enclosed by three pairs of parallel surfaces (Fig. 1.2). The edges originating from one of the corner points determine, through their mutual positions and length, a *crystallographic reference system*. The edges define the *basis vectors* $\boldsymbol{a}_1$, $\boldsymbol{a}_2$, and $\boldsymbol{a}_3$. The angle between the edges are $\alpha_1 = \angle(\boldsymbol{a}_2, \boldsymbol{a}_3)$, $\alpha_2 = \angle(\boldsymbol{a}_1, \boldsymbol{a}_3)$, $\alpha_3 = \angle(\boldsymbol{a}_1, \boldsymbol{a}_2)$. The six quantities $\{a_1, a_2, a_3, \alpha_1, \alpha_2, \alpha_3\}$ form the *metric* of the relevant cell and thus the metric of the appropriate crystallographic reference system which is of special significance for the description and calculation of morphological properties. The position of the atoms in the cell, which characterizes the structure of the corresponding crystal species, is also described in the crystallographic reference system. Directly comprehensible and useful for many questions is the representation of the cell structure by specifying the position of the center of gravity of the atoms in question using so-called parameter vectors. A more detailed description is given by the electron density distribution $\rho(\boldsymbol{x})$ in the cell determined by the methods of crystal structure analysis. The end of the vector $\boldsymbol{x}$ runs through all points within the cell.

In an ideal crystal, the infinite space is filled by an unlimited regular repetition of atomistically identical cells in a gap-free arrangement. Vector methods are used to describe such lattices (see below).

1.2 The First Basic Law of Crystallography (Angular Constancy)

The surface of a freely grown crystal is mainly composed of a small number of practically flat surface elements, which, in the following, we will occasionally refer to as faces. These surface elements are characterized by their normals which are oriented perpendicular to the surface elements. The faces are more precisely described by the following:

1. mutual position (orientation),
2. size,
3. form,
4. micromorphological properties (such as cracks, steps, typical microhills, and microcavities).

The orientation of a certain surface element is given through the angles which its normal makes with the normals of the other surface elements. One finds that arbitrary angles do not occur in crystals. In contrast, the first basic law of crystallography applies:

Freely grown crystals belonging to the same ideal crystal, possess a characteristic set of normal angles (law of angular constancy).

The members belonging to the same ideal crystal form a crystal species. The orientation of the surface elements is thus charcteristic, not, however, the size ratios of the surface elements.

The law of angular constancy can be interpreted from thermodynamic conditions during crystal growth. Crystals in equilibrium with their mother phase or, during growth only slightly apart from equilibrium, can only develop surface elements F_i possessing a relatively minimal specific surface energy σ_i. σ_i is the energy required to produce the ith surface element from 1 cm^2 of the boundary surface in the respective mother phase. Only then does the free energy of the complete system (crystal and mother phase) take on a minimum. The condition for this is

$$\sum F_i \sigma_i = \text{Minimum(Gibbs' condition)},$$

where the numerical value for F_i refers to the size of the ith surface element. From this condition one can deduce Wulff's theorem, which says that the central distances R_i of the ith surface (measured from the origin of growth) are proportional to the surface energy σ_i. According to Gibbs' condition, those surfaces possessing the smallest, specific surface energy are the most stable and largest developed. From simple model calculations, one finds that the less prominent the surface energy becomes, the more densely the respective surface is occupied by building particles effecting strong mutual attraction. The ranking of faces is thus determined by the occupation density. In a lattice, very few surfaces of large occupation density exist exhibiting prominent orientations. This is in accord with the empirical law of angular constancy.

A crude morphological description follows from the concepts *tracht* and *habit*. Tracht is understood as the totality of the existent surface elements and habit as the coarse external appearance of a crystal (e.g., hair shaped, pin shaped, stem shaped, prismatic, columned, leafed, tabular, isometric, etc.).

1.3
Graphical Methods, Stereographic Projection

For the practical handling of morphological findings, it is useful to project the details, without loss of information, onto a plane. Imagine surface normals originating from the center of a sphere intersecting the surface of the sphere. The points of intersection P_i represent an image of the mutual orientation of the surface elements. The surface dimensions are uniquely determined by the central distances R_i of the ith surface from the center of the sphere (Fig. 1.3). One now projects the points of intersection on the sphere on to a flat piece of

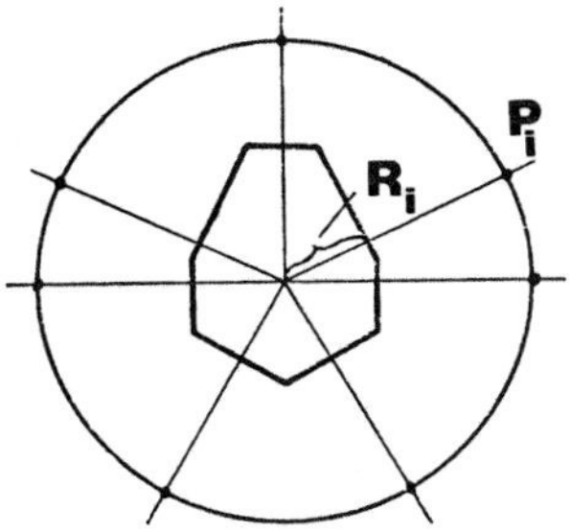

Figure 1.3 Normals and central distances.

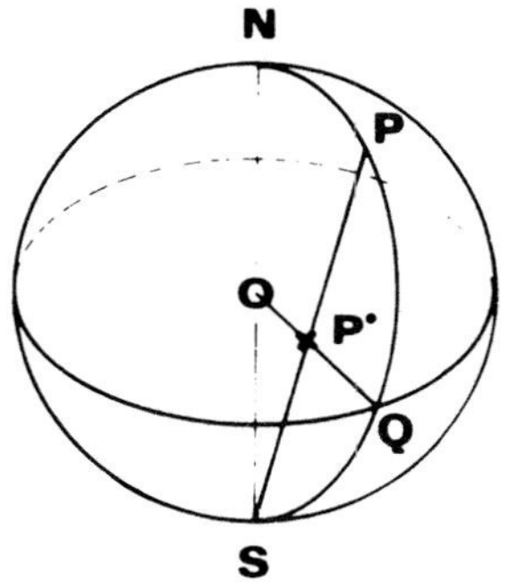

Figure 1.4 Stereographic projection $P^{\cdot}$ of a point P.

paper, the plane of projection. Thus each point on the sphere is assigned a point on the plane of projection. In the study of crystallography, the following projections are favored:

1. the stereographic projection,
2. the gnomonic projection,
3. the orthogonal projection (parallel projection).

Here, we will only discuss the stereographic projection which turns out to be a useful tool in experimental work with crystals. On a sphere of radius R an arbitrary diameter is selected with intersection points N (north pole) and S (south pole). The plane normal to this diameter at the center of the sphere is called the equatorial plane. It is the projection plane and normally the drawing plane. The projection point $P^{\cdot}$ belonging to the point P is the intersection point of the line PS through the equatorial plane (Fig. 1.4).

The relation between P and $P^{\cdot}$ is described with the aid of a coordinate system. Consider three vectors $\boldsymbol{a}_1$,$\boldsymbol{a}_2$, and $\boldsymbol{a}_3$ originating from a fixed point, the origin of the coordinate system. These we have already met as the edges of the elementary cell. The three vectors shall not lie in a plane (not coplanar, Fig. 1.2b). The lengths of $\boldsymbol{a}_i$ $(i = 1, 2, 3)$ and their mutual positions, fixed by the angles α_i, are otherwise arbitrary. One reaches the point P with coordinates (x_1, x_2, x_3) by starting at the origin O and going in the direction $\boldsymbol{a}_1$ a distance $x_1\boldsymbol{a}_1$, then in the direction $\boldsymbol{a}_2$ a distance $x_2\boldsymbol{a}_2$, and finally in the direction $\boldsymbol{a}_3$ by the distance $x_3\boldsymbol{a}_3$. The same end point P is reached by taking any other order of paths.

Each point on the sphere is now fixed by its coordinates (x_1, x_2, x_3). The same applies to the point $P^{\cdot}$ with coordinates $(x_1^{\cdot}, x_2^{\cdot}, x_3^{\cdot})$. For many crystallographic applications it is convenient to introduce a prominent coordinate system, the *Cartesian coordinate system*. Here, the primitive vectors have a length of one unit in the respective system of measure and are perpendicular to each other ($\alpha_i = 90°$). We denote these vectors by $\boldsymbol{e}_1$, $\boldsymbol{e}_2$, $\boldsymbol{e}_3$. The origin is placed in the center of the sphere and $\boldsymbol{e}_3$ points in the direction ON. The vectors $\boldsymbol{e}_1$ and $\boldsymbol{e}_2$ accordingly lie in the equatorial plane. It follows always that $x_3^{\cdot} = 0$. If P is a point on the sphere, then its coordinates obey the spherical equation $x_1^2 + x_2^2 + x_3^2 = R^2$. For $R = 1$ one obtains the following expressions from the relationships in Fig. 1.5:

$$x_1^{\cdot} = \frac{x_1}{1 + x_3}, \quad x_2^{\cdot} = \frac{x_2}{1 + x_3}.$$

These transform to

$$x_1 = \frac{2x_1^{\cdot}}{1 + x_1^{\cdot 2} + x_2^{\cdot 2}}, \quad x_2 = \frac{2x_2^{\cdot}}{1 + x_1^{\cdot 2} + x_2^{\cdot 2}}, \quad x_3 = \frac{1 - x_1^{\cdot 2} - x_2^{\cdot 2}}{1 + x_1^{\cdot 2} + x_2^{\cdot 2}}.$$

In polar coordinates, we define a point by its geographical longitude η and latitude $(90° - \xi)$. Therefore, from Fig. 1.5 we have

$$x_3 = \cos\xi, \quad r = OP^{\cdot} = \sin\xi, \quad x_1 = r\cos\eta = \sin\xi\cos\eta,$$
$$x_2 = r\sin\eta = \sin\xi\sin\eta.$$

Thus

$$x_1^{\cdot} = \frac{\sin\xi\cos\eta}{1 + cos\xi}, \quad x_2^{\cdot} = \frac{\sin\xi\sin\eta}{1 + cos\xi},$$

and

$$\tan\eta = \frac{x_2}{x_1} = \frac{x_2^{\cdot}}{x_1^{\cdot}} \quad \text{and} \quad \cos\xi = x_3 = \frac{1 - x_1^{\cdot} - x_2^{\cdot}}{1 + x_1^{\cdot} + x_2^{\cdot}}.$$

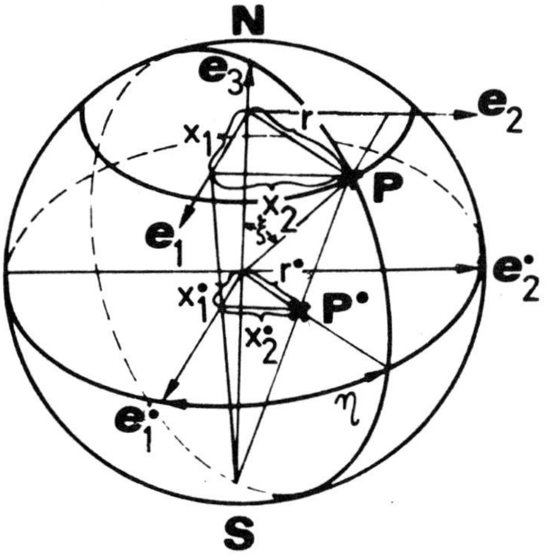

Figure 1.5 Stereographic projection.

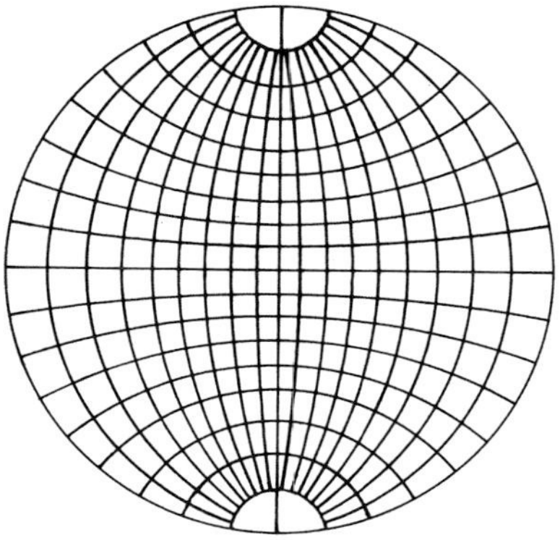

Figure 1.6 Wulff's net.

The stereographic projection is distinguished by two properties, namely the projections are circle true and angle true. All circles on the surface of the sphere project as circles in the plane of projection and the angle of intersection of two curves on the sphere is preserved in the plane of projection. This can be proved with the transformation equations above. In practice, one uses a Wulff net in the equatorial plane, which is a projection of one half of the terrestrial globe with lines of longitude and latitude (Fig. 1.6). Nearly all practical problems of the geometry of face normals can be solved to high precision using a compass and ruler. Frequently, however, it suffices only to work with the Wulff net. The first basic task requires drawing the projection point $P^{\cdot} = (x_1^{\cdot}, x_2^{\cdot})$ of the point $P = (x_1, x_2, x_3)$ (Fig. 1.7). Here, the circle on the sphere passing through the points P, N, and S plays a special role (great circle PSN). It appears rotated about e_3 with respect to the circle passing through the end point of e_1 and through N and S by an angle η, known from $\tan\eta = x_2/x_1$. The projection of this great circle, on which $P^{\cdot}$ also lies, is a line in the projection plane going through the center of the equatorial circle and point Q, the

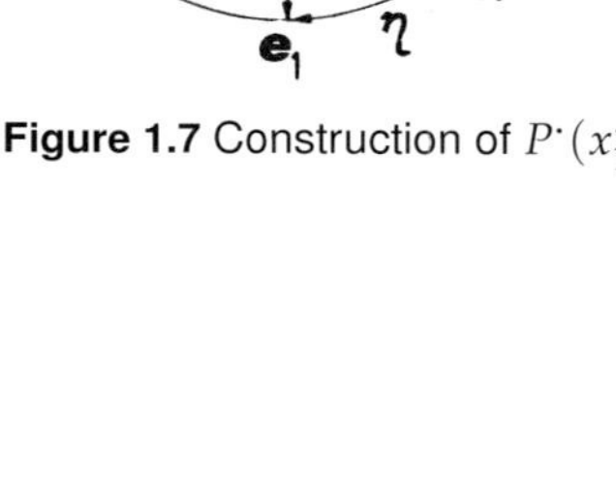

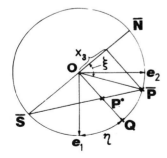

Figure 1.7 Construction of $P^{\cdot}(x_1^{\cdot}, x_2^{\cdot})$ from $P(x_1, x_2 x_3)$.

intersection point of the great circle with the equatorial circle. The point Q remains invariant in the stereographic projection. It has an angular distance of η from the end point of $\boldsymbol{e}_1$. If one now tilts the great circle PSN about the axis OQ into the equatorial plane by 90° one can then construct $P^{\cdot}$ directly as the intersection point of the line OQ with the line $\bar{P}\bar{S}$. $\bar{P}$ and $\bar{S}$ are the points P and S after tilting. One proceeds as follows to obtain a complete stereographic projection of an object possing several faces: the normal of the first face F_1 is projected parallel to $\boldsymbol{e}_1$, so that its projection at the end point of $\boldsymbol{e}_1$ lies on the equatorial circle. The normal of F_2 is also projected onto the equatorial circle at an angular distance of the measured angle between the normals of F_1 and F_2. For each further face F_3, etc. the angles which their normals make with two other normals, whose projections already exist, might be known. Denote the angles between the normals of F_i and F_j by ψ_{ij}. The intersection point P_3 of face F_3 then lies on the small circles having an angular distance ψ_{13} from P_1 and an angular distance ψ_{23} from P_2. Their projections can be easily constructed. One of the two intersection points of these projections is then the sought after the projection point $P_3^{\cdot}$. The reader is referred to standard books on crystal geometry to solve additional problems, especially the determination of angles between surface normals whose stereographic projections already exist.

1.4 The Second Basic Law of Crystallography (Law of Rational Indices)

Consider three arbitrary faces F_1, F_2, F_3 of a freely grown crystal with their associated normals $\boldsymbol{h}_1$, $\boldsymbol{h}_2$, $\boldsymbol{h}_3$. The normals shall not lie in a plane (nontautozonal). Two faces respectively form an intercept edge $\boldsymbol{a}_i$ (Fig. 1.8). The three edge directions define a crystallographic reference system.

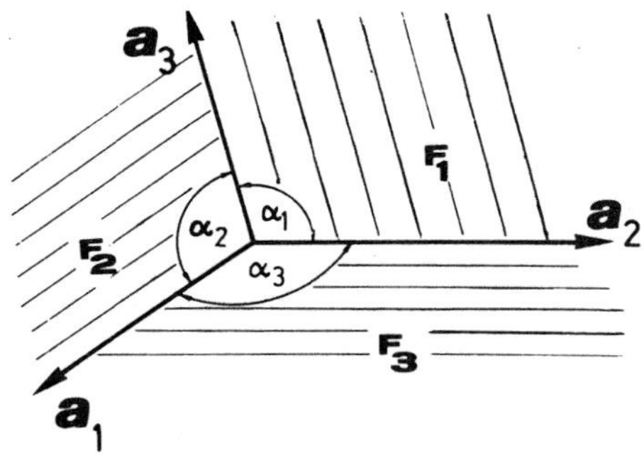

Figure 1.8 Fixing a crystallographic reference system from three non-tautozonal faces.

The system is

$$\boldsymbol{a}_1 \parallel \text{edge}(F_2, F_3),$$
$$\boldsymbol{a}_2 \parallel \text{edge}(F_3, F_1),$$
$$\boldsymbol{a}_3 \parallel \text{edge}(F_1, F_2),$$

in other words $\boldsymbol{a}_i \parallel \text{edge}(F_j, F_k)$. The indices i, j, k run through any triplets of the cyclic sequence 123123123....

$\boldsymbol{a}_i$ are perpendicular to the normals $\boldsymbol{h}_j$ and $\boldsymbol{h}_k$ since they belong to both surfaces F_j and F_k. On the other hand, $\boldsymbol{a}_j$ and $\boldsymbol{a}_k$ span the surfaces F_i with their normals $\boldsymbol{h}_i$. The system of $\boldsymbol{a}_i$ follows from the system of $\boldsymbol{h}_i$ and, conversely, the system of $\boldsymbol{h}_i$ from that of $\boldsymbol{a}_i$ by the operation of setting one of these vectors perpendicular to two vectors of the other respective system. Systems which reproduce after two operations are called *reciprocal systems*. The edges $\boldsymbol{a}_i$ thus form a system reciprocal to the system of $\boldsymbol{h}_i$ and vice versa.

The crystallographic reference system is first fixed by the three angles α_i = angle between $\boldsymbol{a}_j$ and $\boldsymbol{a}_k$. Furthermore, we require the lengths $\mid \boldsymbol{a}_i \mid = a_i$ for a complete description of the system. This then corresponds to our definition of the metric which we introduced previously. We will return to the determination of the lengths and length ratios later. Moreover, the angles α_i can be easily read from a stereographic projection of the three faces F_i. In the same manner, the projections of the intercept points of the edges $\boldsymbol{a}_i$ and thus their orientation can be easily determined.

We consider now an arbitrary face with the normal $\boldsymbol{h}$ in the crystallographic basic system of vectors $\boldsymbol{a}_i$ (Fig. 1.9). The angles between $\boldsymbol{h}$ and $\boldsymbol{a}_i$ are denoted by θ_i. We then have

$$\cos\theta_1 : \cos\theta_2 : \cos\theta_3 = \frac{1}{OA_1} : \frac{1}{OA_2} : \frac{1}{OA_3} = \frac{1}{m_1 a_1} : \frac{1}{m_2 a_2} : \frac{1}{m_3 a_3},$$

where we use the Weiss zone law to set $OA_i = m_i a_i$. The second basic law of crystallography (law of rational indices) now applies.

Two faces of a freely grown crystal with normals $\boldsymbol{h}^I$ and $\boldsymbol{h}^{II}$, which enclose angles θ_i^I and θ_i^{II} with the crystallographic basic vectors $\boldsymbol{a}_i$, can be expressed as the ratios of cosine values to the ratios of integers

$$\frac{\cos\theta_1^I}{\cos\theta_1^{II}} : \frac{\cos\theta_2^I}{\cos\theta_2^{II}} : \frac{\cos\theta_3^I}{\cos\theta_3^{II}} = \frac{m_1^{II}}{m_1^I} : \frac{m_2^{II}}{m_2^I} : \frac{m_3^{II}}{m_3^I}.$$

m_i/m_j are thus rational numbers. The law of rational indices heightens the law of angular constancy to such an extent that, for each crystal species, the characteristic angles between the face normals are subject to an inner rule of conformity. This is a morphological manifestation of the lattice structure of crystals. A comprehensive confirmation of the law of rational indices on numerous natural and synthetic crystals was given by René Juste Hauy (1781).

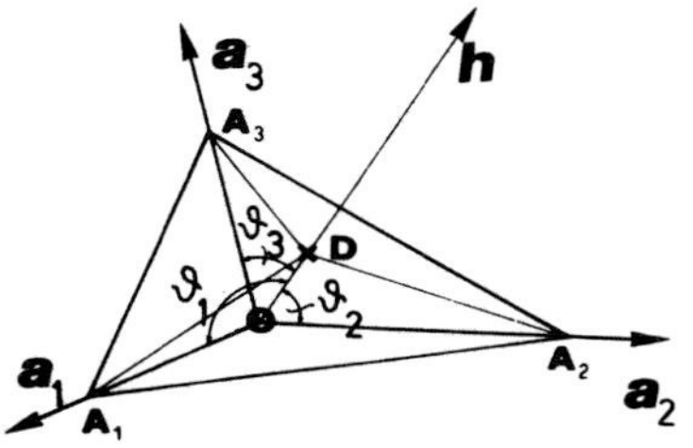

Figure 1.9 Axial intercepts and angles of a face having the normal $\boldsymbol{h}$.

It was found advantageous to introduce the *Miller Indices* (1839) $h_i = t/m_i$ instead of the Weiss indices m_i which fully characterize the position of a face. t is an arbitrary factor. The face in question is then symbolized by $\boldsymbol{h} = (h_1h_2h_3)$. The so-called axes' ratio $a_1 : a_2 : a_3$ now allows one to specify, by an arbitrary choice of indices, a further face F_4 defined by $\boldsymbol{h}^{(4)} = (h_1^{(4)}h_2^{(4)}h_3^{(4)})$. For each face then

$$\cos\theta_1 : \cos\theta_2 : \cos\theta_3 = \frac{h_1}{a_1} : \frac{h_2}{a_2} : \frac{h_3}{a_3}.$$

If the angles of the fourth face are known, one obtains the axes ratio

$$a_1 : a_2 : a_3 = \frac{h_1^{(4)}}{\cos\theta_1^{(4)}} : \frac{h_2^{(4)}}{\cos\theta_2^{(4)}} : \frac{h_3^{(4)}}{\cos\theta_3^{(4)}}.$$

Moreover, the faces F_1, F_2, and F_3 are specified by the Miller indices (100), (010), and (001), respectively.

Now the path is open to label further faces. One measures the angles θ_i and obtains

$$h_1 : h_2 : h_3 = a_1\cos\theta_1 : a_2\cos\theta_2 : a_3\cos\theta_3.$$

As long as morphological questions are in the foreground, one is allowed to multiply through with any number t, so that for h_i the smallest integers, with no common factor, are obtained satisfying the ratio.

1.5 Vectors

1.5.1 Vector Addition

Vectors play an important and elegant role in crystallography. They ease the mathematical treatment of geometric and crystallographic questions. We de-

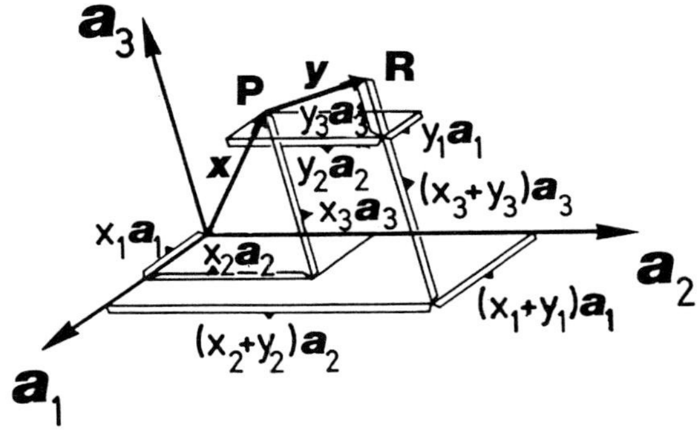

Figure 1.10 Addition of two vectors x and y.

fine a vector by the specifications used earlier for the construction of a point P with coordinates (x_1, x_2, x_3) and a second point Q with coordinates (y_1, y_2, y_3). Now consider the point R with the coordinates $(x_1 + y_1, x_2 + y_2, x_3 + y_3)$ (Fig. 1.10). We reach R after making the construction (x_1, x_2, x_3) and finally attaching the distances $y_1\boldsymbol{a}_1$, $y_2\boldsymbol{a}_2$ and $y_3\boldsymbol{a}_3$ directly to P. One can describe this construction of R as the addition of distances OP and OQ. We now assign to the distance OP the vector $\boldsymbol{x}$, to the distance OQ the vector $\boldsymbol{y}$, and to the distance OR the vector $\boldsymbol{z}$. We then have $\boldsymbol{x} + \boldsymbol{y} = \boldsymbol{z}$. The coordinates are given by $x_i + y_i = z_i$. Quantities which can be added in this manner are called vectors. The order of attaching the vectors is irrelevant.

A vector is specified by its direction and length. Usually it is graphically represented by an arrow over the symbol. Here we write vectors in boldface italic letters. The length of the vector $\boldsymbol{x}$ is called the magnitude of $\boldsymbol{x}$, denoted by the symbol $x =\mid \boldsymbol{x} \mid$. Vectors can be multiplied with arbitrary numbers as is obvious from their component representation. Each component is multiplied with the corresponding factor. A vector of length one is called a unit vector. We obtain a unit vector $\boldsymbol{e}_x$ in the direction $\boldsymbol{x}$ by multiplication with $1/x$ according to $\boldsymbol{e}_x = \boldsymbol{x}/x$.

From the above definition we now formulate the following laws of addition:

1. commutative law: $\boldsymbol{z} = \boldsymbol{x} + \boldsymbol{y} = \boldsymbol{y} + \boldsymbol{x}$ (Fig. 1.11),
2. associative law: $\boldsymbol{x} + (\boldsymbol{y} + \boldsymbol{z}) = (\boldsymbol{x} + \boldsymbol{y}) + \boldsymbol{z}$,
3. distributive law: $q(\boldsymbol{x} + \boldsymbol{y}) = q\boldsymbol{x} + q\boldsymbol{y}$.

The validity of these three laws shall be checked in all further discussions on vector combinations.

Since $-\boldsymbol{x}$ can be taken as a vector antiparallel to $\boldsymbol{x}$ with the same length $(-\boldsymbol{x} + \boldsymbol{x} = 0)$, we have the rule for vector subtraction $\boldsymbol{z} = \boldsymbol{x} - \boldsymbol{y}$ (Fig. 1.11).

Examples for the application of vector addition are as follows.

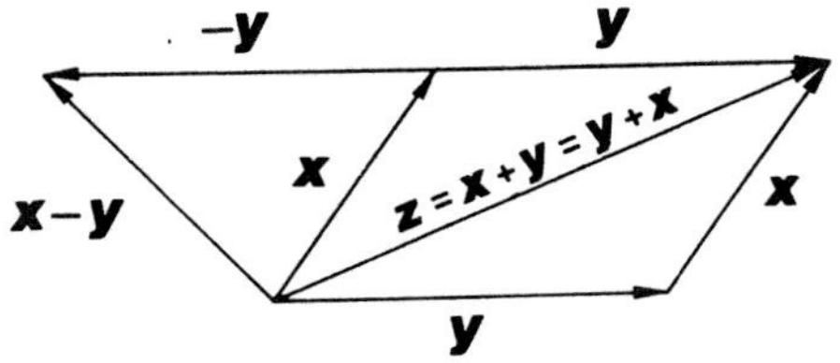

Figure 1.11 Commutative law of vector addition; vector substraction.

1. *Representation of a point lattice* by $\boldsymbol{r} = r_1\boldsymbol{a}_1 + r_2\boldsymbol{a}_2 + r_3\boldsymbol{a}_3$, where r_i run through the integer numbers. The end point of $\boldsymbol{r}$ then sweeps through all lattice points. We use the symbol $[|r_1r_2r_3|]$ for a lattice point and for a lattice row, also represented by $\boldsymbol{r}$, the symbol $[r_1r_2r_3]$. As before, we denote any point with coordinates x_i by (x_1, x_2, x_3).

2. *Decomposition of a vector into components* according to a given reference system. One places through the end points of $\boldsymbol{x}$ planes running parallel to the planes spanned by the vectors $\boldsymbol{a}_j$ and $\boldsymbol{a}_k$. These planes truncate, on the coordinate axes, the intercepts $x_i\boldsymbol{a}_i$ thus giving the coordinates (x_1, x_2, x_3). This decomposition is unique. We thus construct the parallelepiped with edges parallel to the vectors $\boldsymbol{a}_i$ and with space diagonals $\boldsymbol{x}$ (Fig. 1.12).

3. *The equation of a line* through the end points of the two vectors $\boldsymbol{x}_0$ and $\boldsymbol{x}_1$ is given by $\boldsymbol{x} = \boldsymbol{x}_0 + \lambda(\boldsymbol{x}_1 - \boldsymbol{x}_0)$. λ is a free parameter.

4. *The equation of a plane* through the end points of $\boldsymbol{x}_0$, $\boldsymbol{x}_1$, and $\boldsymbol{x}_2$ is given by $\boldsymbol{x} = \boldsymbol{x}_0 + \lambda(\boldsymbol{x}_1 - \boldsymbol{x}_0) + \mu(\boldsymbol{x}_2 - \boldsymbol{x}_0)$. λ and μ are free parameters. In component representation, these three equations correspond to the equation of a plane in the form $u_0 + u_1x_1 + u_2x_2 + u_3x_3 = 0$, which one obtains after eliminating λ and μ (Fig. 1.13). If the components of the three vectors

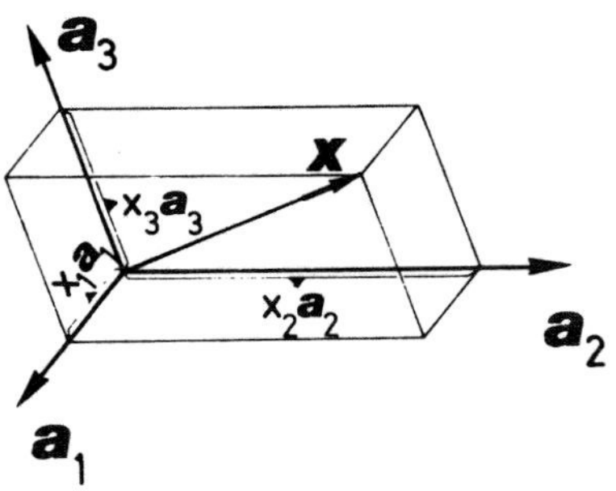

Figure 1.12 Decomposition of a vector into components of a given reference system.

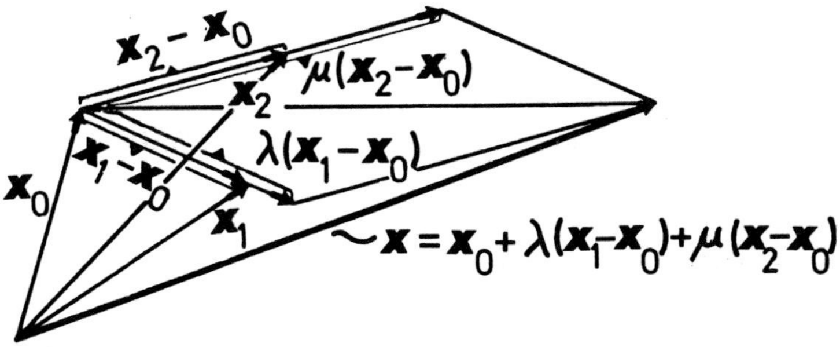

Figure 1.13 Equation of a plane through three points.

have integer values, i.e., we are dealing with a lattice plane, then u_i take on integer values.

1.5.2
Scalar Product

Linear vector functions hold a special place with regard to the different possibilities of vector combinations. They are, like all other combinations of vectors, invariant with respect to the coordinate system in which they are viewed. Linear vector functions are proportional to the lengths of the vectors involved. The simplest and especially useful vector function is represented by the scalar product (Fig. 1.14):

The scalar product $x \cdot y = |x||y|\cos(x, y)$ is equal to the projection of a vector on another vector, multiplied by the length of the other vector.

For simplification we use the symbol (x, y) for the angle between x and y. The commutative law $x \cdot y = y \cdot x$ is satisfied as well as the associative and distributive laws, the latter in the form $x \cdot (y + z) = x \cdot y + x \cdot z$.

The scalar product can now be determined with the aid of the distributive law when the respective vectors in component representation exist in a basic system of known metric. We have $x = \sum x_i a_i = x_i a_i$ (one sums over i, here

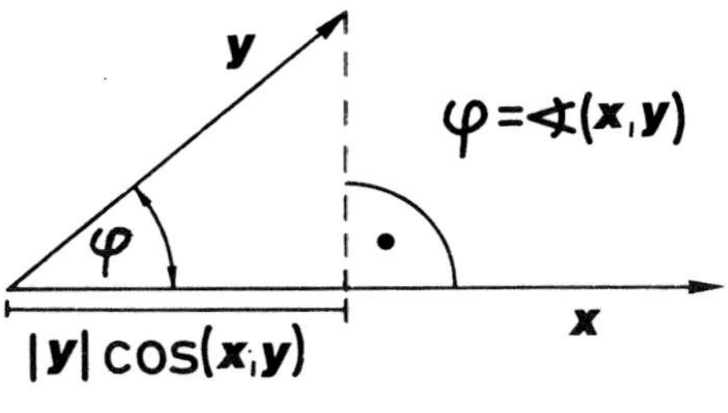

Figure 1.14 Definition of the scalar product of two vectors.

from $i = 1$ to 3; Einstein summation convention!) and $\boldsymbol{y} = y_j\boldsymbol{a}_j$. Then $\boldsymbol{x} \cdot \boldsymbol{y} = (x_i\boldsymbol{a}_i) \cdot (y_j\boldsymbol{a}_j) = x_iy_j(\boldsymbol{a}_i \cdot \boldsymbol{a}_j)$. The products $\boldsymbol{a}_i \cdot \boldsymbol{a}_j$ are, as assumed, known ($\boldsymbol{a}_i \cdot \boldsymbol{a}_j = a_i^2$ for $i = j$ and $\boldsymbol{a}_i \cdot \boldsymbol{a}_j = a_ia_j \cos\alpha_k$ for $i \neq j, k \neq i, j$).

Examples for the application of scalar products are as follows.

1. *Calculating the length of a vector.* We have

$$\boldsymbol{x} \cdot \boldsymbol{x} = \boldsymbol{x}^2 = |x|^2 = x_ix_j(\boldsymbol{a}_i \cdot \boldsymbol{a}_j).$$

2. *Calculating the angle between two vectors $\boldsymbol{x}$ and $\boldsymbol{y}$.* From the definition of the scalar product it follows that $\cos(\boldsymbol{x}, \boldsymbol{y}) = (\boldsymbol{x} \cdot \boldsymbol{y})/(|\boldsymbol{x}||\boldsymbol{y}|)$.

3. *Determining whether two vectors are mutually perpendicular.* The condition for two vectors of nonzero lengths is $\boldsymbol{x} \cdot \boldsymbol{y} = 0$.

4. *Equation of a plane* perpendicular to the vector $\boldsymbol{h}$ and passing through the end point of $\boldsymbol{x}_0$: $(\boldsymbol{x} - \boldsymbol{x}_0) \cdot \boldsymbol{h} = 0$.

5. *Decomposing a vector $\boldsymbol{x}$ into components of a coordinate system.* Assume that the angles δ_i (angles between $\boldsymbol{x}$ and $\boldsymbol{a}_i$) are known. Then one also knows the scalar products $\boldsymbol{x} \cdot \boldsymbol{a}_i = |\boldsymbol{x}|a_i \cos\delta_i$. This gives the following system of equations:

$$\begin{aligned} \boldsymbol{x} &= x_i\boldsymbol{a}_i \\ \boldsymbol{x} \cdot \boldsymbol{a}_j &= x_i\boldsymbol{a}_i \cdot \boldsymbol{a}_j \quad \text{for} \quad j = 1, 2, 3. \end{aligned}$$

   The system for the sought after components x_i always has a solution when $\boldsymbol{a}_i$ span a coordinate system.

1.5.3
Vector Product

Two nonparallel vectors $\boldsymbol{x}$ and $\boldsymbol{y}$ fix a third direction, namely that of the normals on the plane spanned by $\boldsymbol{x}$ and $\boldsymbol{y}$. The vector product of $\boldsymbol{x}$ and $\boldsymbol{y}$ generates a vector in the direction of these normals.

The vector product of $\boldsymbol{x}$ and $\boldsymbol{y}$, spoken "x cross y" and written as $\boldsymbol{x} \times \boldsymbol{y}$, is the vector perpendicular to $\boldsymbol{x}$ and $\boldsymbol{y}$ with a length equal to the area of the parallelogram spanned by $\boldsymbol{x}$ and $\boldsymbol{y}$, thus $|\boldsymbol{x} \times \boldsymbol{y}| = |\boldsymbol{x}||\boldsymbol{y}| \sin(\boldsymbol{x}, \boldsymbol{y})$. The three vectors $\boldsymbol{x}$, $\boldsymbol{y}$, and $\boldsymbol{x} \times \boldsymbol{y}$ form a right-handed system (Fig. 1.15). The vector $\boldsymbol{x} \times \boldsymbol{y}$ lies perpendicular to the plane containing $\boldsymbol{x}$ and $\boldsymbol{y}$ and in such a direction that a right-handed screw driven in the direction of $\boldsymbol{x} \times \boldsymbol{y}$ would carry $\boldsymbol{x}$ into $\boldsymbol{y}$ through a clockwise rotation around the smaller angle between $\boldsymbol{x}$ and $\boldsymbol{y}$.

The vector product is not commutative. In contrast, we have $\boldsymbol{x} \times \boldsymbol{y} = -\boldsymbol{y} \times \boldsymbol{x}$. From the definition, one immediately recognizes the validity of the associative law. It is more difficult to prove the distributive law $\boldsymbol{x} \times (\boldsymbol{y} + \boldsymbol{z}) =$

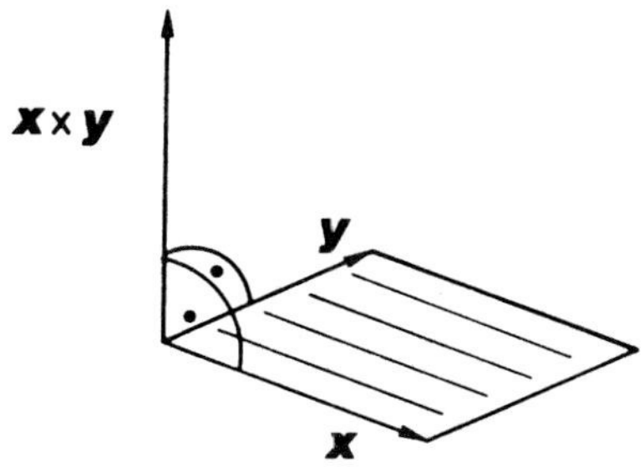

Figure 1.15 Vector product.

$x \times y + x \times z$. We refer the reader to standard textbooks like *Kristallgeometrie* for a demonstration.

For two vectors specified in the crystallographic reference system, i.e., $x = x_i a_i$ and $y = y_i a_i$, we find

$$x \times y = (x_i a_i) \times (y_j a_j) = (x_2 y_3 - x_3 y_2) a_2 \times a_3 + (x_3 y_1 - x_1 y_3) a_3 \times a_1 + (x_1 y_2 - x_2 y_1) a_1 \times a_2. \quad (1.1)$$

Thus the vector products of the basis vectors appear. These we have met before. They are the normals on the three basic faces F_i. It is found useful to introduce these vectors as the base vectors of a new reference system, the so-called reciprocal system (see Section 1.3). For this purpose we need to normalize the length of the new vectors so that the reciprocal of the reciprocal system is in agreement with the basic system. This is accomplished with the following definition of the basic vectors a_i^* of the *reciprocal system*

$$a_i^* = \frac{1}{V(a_1, a_2, a_3)} a_j \times a_k,$$

where i, j, k should observe a cyclic sequence of 1, 2, 3, 1, 2, 3,.... $V(a_1, a_2, a_3)$ is the volume of the parallelepiped spanned by the basic vectors. a_i^* is spoken as "a-i-star."

Correspondingly, for the basic vectors we have $a_i = a_j^* \times a_k^* / V(a_1^*, a_2^*, a_3^*)$. The proof that $(a_i^*)^* = a_i$ is given below.

To calculate V we use the so-called *scalar triple product* of three vectors:

$$\begin{aligned} V &= \text{base surface times the height of the parallelepiped} \\ &= (y \times z) \cdot x = |y|\,|z|\,|\sin(y, z)|\, e \cdot x. \end{aligned}$$

Here e is the unit vector of $y \times z$. If one considers another basic face, then the same result is found, i.e.,

$$(y \times z) \cdot x = (x \times y) \cdot z = (z \times x) \cdot y = x \cdot (y \times z) \quad \text{and so on.}$$

The order of the factors may be cyclically interchanged as well as the operations of the scalar and the vectorial products. A change in the cyclic order results in a change of sign of the product. For $\boldsymbol{x} \cdot (\boldsymbol{y} \times \boldsymbol{z})$ we use the notation $[\boldsymbol{x}, \boldsymbol{y}, \boldsymbol{z}]$. Thus

$$V(\boldsymbol{a}_1, \boldsymbol{a}_2, \boldsymbol{a}_3) = \boldsymbol{a}_1 \cdot (\boldsymbol{a}_2 \times \boldsymbol{a}_3) = [\boldsymbol{a}_1, \boldsymbol{a}_2, \boldsymbol{a}_3].$$

The vector product can be calculated formally using the rules for the calculation of determinants. A third-order determinant $D(u_{ij})$ with the nine quantities u_{ij} has the solution

$$\begin{aligned} D(u_{ij}) &= \begin{vmatrix} u_{11} & u_{12} & u_{13} \\ u_{21} & u_{22} & u_{23} \\ u_{31} & u_{32} & u_{33} \end{vmatrix} \\ &= u_{11}(u_{22}u_{33} - u_{23}u_{32}) - u_{12}(u_{21}u_{33} - u_{23}u_{31}) \\ &\qquad + u_{13}(u_{21}u_{32} - u_{22}u_{31}). \end{aligned}$$

Now using the vectors $\boldsymbol{x} = x_i \boldsymbol{a}_i$ and $\boldsymbol{y} = y_i \boldsymbol{a}_i$ we construct the corresponding scheme and obtain

$$\boldsymbol{x} \times \boldsymbol{y} = V(\boldsymbol{a}_1, \boldsymbol{a}_2, \boldsymbol{a}_3) \begin{vmatrix} \boldsymbol{a}_1^* & \boldsymbol{a}_2^* & \boldsymbol{a}_3^* \\ x_1 & x_2 & y_3 \\ y_1 & y_2 & y_3 \end{vmatrix}.$$

V can be directly calculated from the scalar products of the basic vectors with the aid of Grams determinant. The solution is

$$V^2(\boldsymbol{a}_1, \boldsymbol{a}_2, \boldsymbol{a}_3) = \begin{vmatrix} \boldsymbol{a}_1 \cdot \boldsymbol{a}_1 & \boldsymbol{a}_1 \cdot \boldsymbol{a}_2 & \boldsymbol{a}_1 \cdot \boldsymbol{a}_3 \\ \boldsymbol{a}_2 \cdot \boldsymbol{a}_1 & \boldsymbol{a}_2 \cdot \boldsymbol{a}_2 & \boldsymbol{a}_2 \cdot \boldsymbol{a}_3 \\ \boldsymbol{a}_3 \cdot \boldsymbol{a}_1 & \boldsymbol{a}_3 \cdot \boldsymbol{a}_2 & \boldsymbol{a}_3 \cdot \boldsymbol{a}_3 \end{vmatrix} = a_1^2 a_2^2 a_3^2 \begin{vmatrix} 1 & \cos\alpha_3 & \cos\alpha_2 \\ \cos\alpha_3 & 1 & \cos\alpha_1 \\ \cos\alpha_2 & \cos\alpha_1 & 1 \end{vmatrix}.$$

The vector product has three important applications:

1. Parallel vectors $\boldsymbol{x}$ and $\boldsymbol{y}$ form a vanishing vector product $\boldsymbol{x} \times \boldsymbol{y} = 0$.
2. The normals $\boldsymbol{h}$ of the plane spanned by the vectors $\boldsymbol{x}$ and $\boldsymbol{y}$ are parallel to $\boldsymbol{x} \times \boldsymbol{y}$.
3. The intercept edge $\boldsymbol{u}$ of two planes with the normals $\boldsymbol{h}$ and $\boldsymbol{g}$ is parallel to $\boldsymbol{h} \times \boldsymbol{g}$.

The fundamental importance of the reciprocal system for crystallographic work is made clear by the following statement:

A normal $\boldsymbol{h}$ with the Miller indices $(h_1 h_2 h_3)$ has the component representation $\boldsymbol{h} = h_1 \boldsymbol{a}_1^* + h_2 \boldsymbol{a}_2^* + h_3 \boldsymbol{a}_3^*$.

As proof, we form the scalar product of this equation with $\boldsymbol{a}_i$ and obtain $\boldsymbol{h} \cdot \boldsymbol{a}_i = h_i$, where h_i are rational numbers. From the definition of the scalar product it follows that $\boldsymbol{h} \cdot \boldsymbol{a}_i = |\boldsymbol{h}||\boldsymbol{a}_i| \cos\theta_i$ and thus $\cos\theta_1 : \cos\theta_2 : \cos\theta_3 = h_1/a_1 : h_2/a_2 : h_3/a_3$, i.e., the corresponding face obeys the law of rational indices and h_i correspond to the reciprocal axial intercepts.

Now we consider the length of $\boldsymbol{h}$. The length is related to the distance $OD = d_h$ of the plane from the origin (Fig. 1.9). We have $\cos\theta_i = OD/OA_i = d_h/(m_i a_i) = d_h h_i/a_i$ with $m_i = 1/h_i$. One does not sum over i! On the other hand, from $\boldsymbol{h} \cdot \boldsymbol{a}_i = h_i = |\boldsymbol{h}| a_i \cos\theta_i$ we get the value $\cos\theta_i = h_i/|\boldsymbol{h}| a_i$. Thus the lattice plane distance is $OD = d_h = 1/|h|$. It may be calculated from the so-called quadratic form $(1/d_h)^2 = |\boldsymbol{h}|^2 = (h_i \boldsymbol{a}_i^*) \cdot (h_j \boldsymbol{a}_j^*)$. Here we encounter other triple products which we will now turn to.

1.5.4
Vector Triple Product

The scalar triple product of three vectors $[\boldsymbol{x}, \boldsymbol{y}, \boldsymbol{z}]$ was our first acquaintance with triple products. A further expression is the vector product of a vector with a vector product given by the following theorem, which is called *Entwicklungssatz*:

$$\boldsymbol{x} \times (\boldsymbol{y} \times \boldsymbol{z}) = (\boldsymbol{x} \cdot \boldsymbol{z})\boldsymbol{y} - (\boldsymbol{x} \cdot \boldsymbol{y})\boldsymbol{z}.$$

Applications of the commutability of scalar and vector multiplication are as follows.

1. Scalar product of two vector products

$$(\boldsymbol{u} \times \boldsymbol{v}) \cdot (\boldsymbol{x} \times \boldsymbol{y}) = \boldsymbol{u} \cdot \{\boldsymbol{v} \times (\boldsymbol{x} \times \boldsymbol{y})\} = (\boldsymbol{u} \cdot \boldsymbol{x})(\boldsymbol{v} \cdot \boldsymbol{y}) - (\boldsymbol{u} \cdot \boldsymbol{y})(\boldsymbol{v} \cdot \boldsymbol{x}),$$

2. Vector product of two vector products

$$(\boldsymbol{u} \times \boldsymbol{v}) \times (\boldsymbol{x} \times \boldsymbol{y}) = \{(\boldsymbol{u} \times \boldsymbol{v}) \cdot \boldsymbol{y}\}\boldsymbol{x} - \{(\boldsymbol{u} \times \boldsymbol{v}) \cdot \boldsymbol{x}\}\boldsymbol{y} = [\boldsymbol{u}, \boldsymbol{v}, \boldsymbol{y}]\boldsymbol{x} - [\boldsymbol{u}, \boldsymbol{v}, \boldsymbol{x}]\boldsymbol{y}.$$

With the aid of these identities it is easy to prove that $V(\boldsymbol{a}_1^*, \boldsymbol{a}_2^*, \boldsymbol{a}_3^*) = 1/V(\boldsymbol{a}_1, \boldsymbol{a}_2, \boldsymbol{a}_3)$. For the metric of the reciprocal system we have

$$a_i^* = a_j a_k \sin\alpha_i / V(\boldsymbol{a}_1, \boldsymbol{a}_2, \boldsymbol{a}_3)$$

and

$$a_1^* : a_2^* : a_3^* = \sin\alpha_1/a_1 : sin\alpha_2/a_2 : sin\alpha_3/a_3$$

as well as

$$\cos\alpha_k^* = \frac{\cos\alpha_i \cos\alpha_j - \cos\alpha_k}{\sin\alpha_i \sin\alpha_j},$$

with $i \neq j \neq k \neq i$.

1.6 Transformations

Often it is practical to turn to another reference system that, e.g., is more adapted to the symmetry of the respective crystal or is easier to handle. Let us designate the basic vectors of the old system with $\boldsymbol{a}_i$ and those of the new system with $\boldsymbol{A}_i$. Correspondingly, we write all quantities in the new system with capital letters.

We are now confronted with the following questions:

1. How do we get to the new basic vectors from the old ones, i.e., what form do the functions $\boldsymbol{A}_i(\boldsymbol{a}_j)$ have?
2. What do the old basic vectors look like in the new system, i.e., what form has the inverse transformation $\boldsymbol{a}_i(\boldsymbol{A}_j)$?
3. How do position vectors transform in the basic system $\boldsymbol{x} = x_i\boldsymbol{a}_i = \boldsymbol{X} = X_i\boldsymbol{A}_i$ and what form do the functions $X_i(\boldsymbol{x}_j)$ have?
4. How does one get the inverse transformation $x_i(\boldsymbol{X}_j)$?
5. How do position vectors transform in the reciprocal system $\boldsymbol{h} = h_i\boldsymbol{a}_i^* = H_i\boldsymbol{A}_i^*$ and what form do the functions $H_i(\boldsymbol{h}_j)$ have?
6. What form does the inverse transformation $h_i(\boldsymbol{H}_j)$ have?

To (1) imagine that the basic vectors of the new system are decomposed into components of the old system; thus $\boldsymbol{A}_i = u_{ij}\boldsymbol{a}_j$. Decomposition is possible with the aid of the scalar products $\boldsymbol{a}_i \cdot \boldsymbol{A}_j$. For that purpose, the length of the new basic vectors and the angle between $\boldsymbol{a}_i$ and $\boldsymbol{A}_j$ must be known. We collect the resulting u_{ij} in the transformation matrix U; thus

$$U = (u_{ij}) = \begin{pmatrix} u_{11} & u_{12} & u_{13} \\ u_{21} & u_{22} & u_{23} \\ u_{31} & u_{32} & u_{33} \end{pmatrix}.$$

To (2) the inverse transformation is given by $\boldsymbol{a}_i = U_{ij}\boldsymbol{A}_j = U_{ij}u_{jk}\boldsymbol{a}_k$. This means $U_{ij}u_{jk} = 1$ for $i = k$ and $= 0$ for $i \neq k$. A similar expression is known from the expansion of a determinant $D(u_{ij}) = u_{jk}U'_{ij}$ with $U'_{ij} = (-1)^{i+j}A_{ji}$ for $i = k$ and $U'_{ij} = 0$ for $i \neq k$. Here, $D(u_{ij})$ is the determinant of the transformation matrix and A_{ji} is the subdeterminant (adjunct) after eliminating the jth row and ith column. Thus $U_{ij} = (-1)^{i+j}A_{ji}/D(u_{ij})$. We call $(U_{ij}) = U^{-1} = (u_{ij})^{-1}$ the inverse matrix of (u_{ij}).

To (3) in the basic system we have $\boldsymbol{x} = x_i\boldsymbol{a}_i$ and with $\boldsymbol{a}_i = U_{ij}\boldsymbol{A}_j$ we find $\boldsymbol{x} = x_iU_{ij}\boldsymbol{A}_j = X_j\boldsymbol{A}_j$, i.e., $X_i = U_{ji}x_j$ (after interchanging the indices). The components of the position vector $\boldsymbol{x}$ are transformed with the transposed inverse matrix $(U_{ij})^T = (U_{ji})$.

To (4) we have $\boldsymbol{x} = X_i \boldsymbol{A}_i = X_i u_{ij} \boldsymbol{a}_j = x_j \boldsymbol{a}_j$ and thus $x_i = u_{ji} X_j$. The transposed transformation matrix is used for the inverse transformation.

To (5) the position vector in the reciprocal system is $\boldsymbol{h} = h_j \boldsymbol{a}_j^* = H_i \boldsymbol{A}_i^*$. Scalar multiplication with $\boldsymbol{A}_i$ gives $\boldsymbol{A}_i \cdot \boldsymbol{h} = H_i = u_{ij} \boldsymbol{a}_j \cdot (h_k \boldsymbol{a}_k^*) = u_{ij} h_j$. Because $\boldsymbol{a}_j \cdot \boldsymbol{a}_k^* = 0$ for $j \neq k$ and $= 1$ for $j = k$ it follows that $H_i = u_{ij} h_j$, i.e., the Miller indices transform like the basic vectors. This result deserves special attention.

To (6) $\boldsymbol{h} = H_j \boldsymbol{A}_j^* = h_i \boldsymbol{a}_i^*$. Scalar multiplication with $\boldsymbol{a}_i$ gives $\boldsymbol{a}_i \cdot \boldsymbol{h} = h_i = U_{ij} \boldsymbol{A}_j \cdot (H_k \boldsymbol{A}_k^*) = U_{ij} H_j$ and thus $h_i = U_{ij} H_j$. The inverse transformation occurs naturally as with the corresponding inverse transformation of the basic vectors with the inverse matrix.

1.7 Symmetry Properties

1.7.1 Symmetry Operations

Symmetry properties are best suited for the systematic classification of crystals. Furthermore, the symmetry determines the directional dependence (anisotropy) of the physical properties in a decisive way. Many properties such as, e.g., the piezoelectric effect, the pyroelectric effect, and certain nonlinear optical effects, including the generation of optical harmonics, can only occur in the absence of certain symmetry properties.

We meet the concept of symmetry in diverse fields. The basic notion stems from geometry. Symmetry in the narrow sense is present when we recognize uniform objects in space, which can be transferred by a movement into each other (coincidence) or which behave like image and mirror image. Morphological features of plants and animals (flowers, starfishes, most animals) are examples of the latter. The concept of symmetry may be carried over to nongeometric objects. Accordingly, symmetry in a figurative sense means the repetition of uniform or similar things. This can occur in time and space as, e.g., in music. Also the repetition of a ratio, as in the case of a geometric series, the father–son relationship in a line of ancestors, or the generation of a number sequence from a recursion formula and the relationship of the members between themselves, belong to this concept.

Although it may be fascinating to search for and contemplate such symmetries, we must turn to a narrower concept of symmetry when considering crystallography. We are interested in symmetry as a repetition of similar or uniform objects in space and distinguish between two types of manifesta-

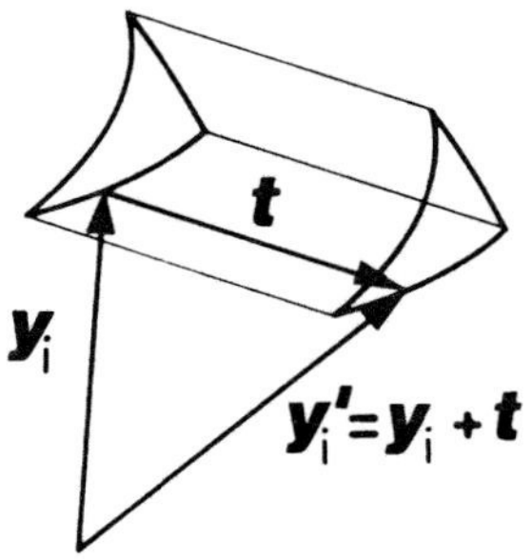

Figure 1.16 Translation.

tions, which, however, exhibit an internal association, namely the geometric symmetry in the narrow sense and the physical symmetry in space. The first case is concerned with the relationship between distances of points and angles between lines that repeat themselves. The second case refers to physical properties of bodies that repeat themselves in different directions. This symmetry arises in part from the structural symmetry of the crystals and in part from the intrinsic symmetry of the physical phenomena. We will come to these questions later. First, we will concern ourselves with geometric symmetry.

Two or more geometric figures or bodies shall be called geometrically uniform (or equivalent) when they differ only with respect to their position. Moreover, figures arising from reflection and centrosymmetry, such as, e.g., right and left hand or a right and left system of the same metric, shall be allowed to be equivalent. Each point specified by the end point of a vector y_i of the first figure shall be assigned a vector y_i' of the second or a further figure such that $|y_i - y_j| = |y_i' - y_j'|$ and $\angle(y_i - y_j, y_k - y_l) = \angle(y_i' - y_j', y_k' - y_l')'$ $(i, j, k, l$ specify four arbitrary points). The respective figures then exhibit correspondingly equal lengths and angles.

The geometric symmetries are now distinguished by the fact that one can describe the association of the equivalent figures with a few basic symmetry operations. Only those operations are permitted that allow an arbitrary repetition. In this sense, an arrangement of equivalent figures in an arbitrary position does not possess symmetry. There are three types of basic symmetry operations

1. *Translation:* We displace each point y_i (considered as the end point of a vector) of a given geometric form by a fixed vector $\boldsymbol{t}$, the translation vector, and come to a second figure with the points $y_i' = y_i + \boldsymbol{t}$ (Fig. 1.16). The required repetition leads to an infinite chain of equivalent figures. The symmetry operation is defined by the vector $\boldsymbol{t}$.

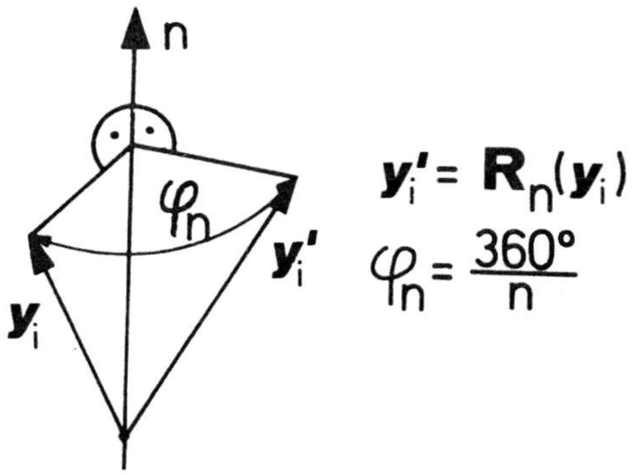

Figure 1.17 Rotation about an axis.

2. *Rotation about an axis:* A rotation through an angle φ about a given axis carries the points y_i of a given geometric figure or body over to the points y_i' of a symmetry-equivalent figure, where the corresponding points have the same distance from the axis of rotation and lie in a plane normal to the axis of rotation (Fig. 1.17). In this type of operation, the points coincide as with translation. Characteristic for the rotation is the position of the axis and the angle of rotation φ. We call $n = 2\pi/\varphi$, where φ is measured in radians, the *multiplicity* of the given axis. The axis of rotation has the symbol n. We write for the operation of rotation $y_i' = R_n(y_i)$. An axis of rotation is known as polar when the direction and reverse direction of the axis of rotation are not symmetry equivalent.

3. *Rotoinversion:* In this operation there exists an inseparable coupling between a rotation as in (2) and a so-called *inversion*. The operation of inversion moves a point y, through a point (inversion center) identical to the origin of coordinates, to get the point $y' = -y$ (Fig. 1.18). The order of both operations is unimportant. We specify the rotoinversion operation by the symbol $\bar{n}$ (read "n bar"). Thus $y_i' = -R_n(y_i) = R_n(-y_i) = R_{\bar{n}}(y_i)$. Occasionally we will introduce a rotation–reflection axis instead of a rotation–inversion axis, i.e., a coupling of rotation and mirror symmetry, normal to the plane of the given axis of rotation. Both operations

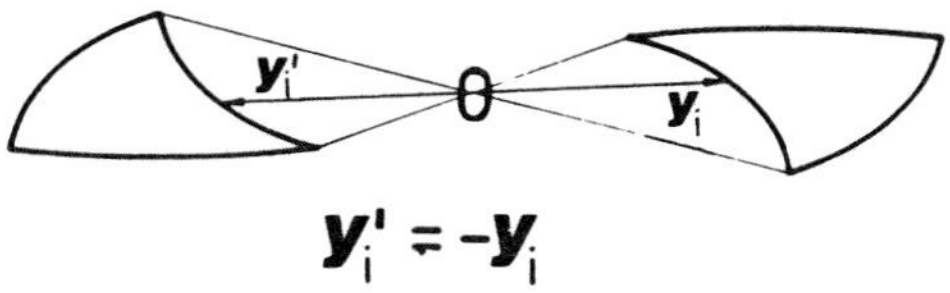

Figure 1.18 Inversion.

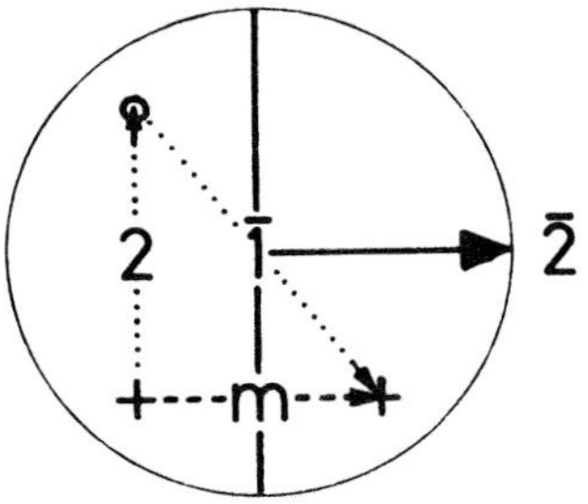

Figure 1.19 Identity of $\bar{2}$ and m (mirror plane).

lead to the same results; however, the multiplicity may be different for the given rotations.

Important special cases of rotoinversion are the inversion $\bar{1}$, in other words, the mirror image about a point, and the rotoinversion $\bar{2}$. The latter is found to be identical to the mirror image about a plane normal to the $\bar{2}$-axis (mirror plane or symmetry plane; Fig. 1.19). The expressions *inversion center* or *center of symmetry* are also used for the inversion. The preferred notation of the mirror image about a plane is m (mirror) instead of $\bar{2}$.

How do these operations express themselves in the components of the vectors y and y'? This will first be demonstrated for the case of a Cartesian reference system. The axis of rotation is parallel to e_3. The rotation carries the basic system $\{e_i\}$ over to a symmetry-equivalent system $\{e_i'\}$ (Fig. 1.20).

$$\begin{aligned} e_1' &= \cos\varphi\, e_1 + \sin\varphi\, e_2 \\ e_2' &= -\sin\varphi\, e_1 + \cos\varphi\, e_2 \\ e_3' &= e_3. \end{aligned}$$

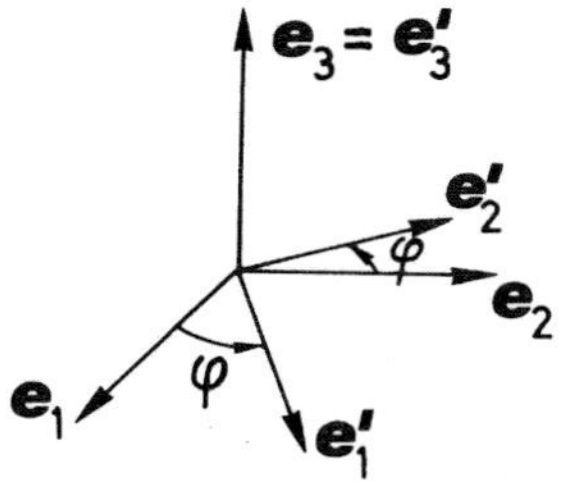

Figure 1.20 Rotation about an axis e_n of a Cartesian reference system.

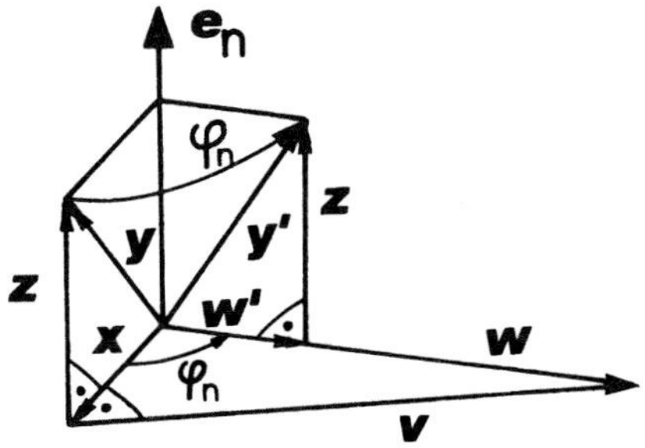

Figure 1.21 Vector relations for a rotation about an arbitrary axis e_n.

Thus the transformation matrix is

$$(u_{ij}) = \begin{pmatrix} \cos\varphi & \sin\varphi & 0 \\ -\sin\varphi & \cos\varphi & 0 \\ 0 & 0 & 1 \end{pmatrix}.$$

What do the coordinates of a point, generated by the rotation, look like in the old system? As we saw in Section 1.6, the inverse transformation is described by the transposed matrix:

$$\begin{aligned} y_1' &= \cos\varphi\, y_1 - \sin\varphi\, y_2 \\ y_2' &= \sin\varphi\, y_1 + \cos\varphi\, y_2 \\ y_3' &= y_3. \end{aligned}$$

We symbolize this by writing

$$\boldsymbol{R}_{n\|e_3} = \begin{pmatrix} \cos\varphi & -\sin\varphi & 0 \\ \sin\varphi & \cos\varphi & 0 \\ 0 & 0 & 1 \end{pmatrix} = (v_{ij}) \quad \text{and} \quad y_i' = v_{ij} y_j.$$

The general case of an arbitrary position of the n-fold axis of rotation $\boldsymbol{e}_n$ may be understood with the aid of vector calculus (Fig. 1.21). Let φ_n be the angle of rotation. We agree upon the clockwise sense as the positive direction of rotation when looking in the direction $+\boldsymbol{e}_n$. One finds

$$\boldsymbol{y}' = [(\boldsymbol{y}\cdot\boldsymbol{e}_n)\boldsymbol{e}_n](1-\cos\varphi_n)] + \cos\varphi_n\,\boldsymbol{y} + (\boldsymbol{e}_n \times \boldsymbol{y})\sin\varphi_n.$$

The individual steps are $\boldsymbol{y}' = \boldsymbol{w}' + \boldsymbol{z}$; $\boldsymbol{z} = \boldsymbol{y} - \boldsymbol{x} = (\boldsymbol{y}\cdot\boldsymbol{e}_n)\boldsymbol{e}_n$; $\boldsymbol{w}' = (\boldsymbol{w}/|\boldsymbol{w}|)|\boldsymbol{x}| = \boldsymbol{w}\cos\varphi_n$; $\boldsymbol{w} = \boldsymbol{x} + \boldsymbol{v}$; $\boldsymbol{v} = (\boldsymbol{e}_n \times \boldsymbol{x})\tan\varphi_n = (\boldsymbol{e}_n \times \boldsymbol{y})\tan\varphi_n$. If one decomposes the above equation for $\boldsymbol{y}'$ into components of an arbitrary coordinate system, whereby the unit vectors of the axis of rotation are $\boldsymbol{e}_n = n_i\boldsymbol{a}_i$ and $\boldsymbol{y} = y_i\boldsymbol{a}_i$, one gets the corresponding transformation matrix $\boldsymbol{R}_n = (v_{ij})$.

For the case of a rotoinversion, we have $\boldsymbol{R}_n = (-v_{ij})$, when the origin of the coordinates is taken as the center of symmetry.

To obtain all symmetry-equivalent points, arising from multiple repetitions of the symmetry operations on y, one must use the same $\boldsymbol{R}_n$ on y' according to $y'' = \boldsymbol{R}_n(y') = \boldsymbol{R}_n^2(y)$ and so on. In general, we have $y'^m = \boldsymbol{R}_n^m(y)$. These matrices are obtained through multiple matrix multiplication.

1.7.2 Point Symmetry Groups

We now turn to the question of which of the three types of symmetry operations discussed above are compatible with each other, i.e., what combinations are simultaneously possible. As a first step we consider only such combinations where at least one point of the given space possessing this symmetry property remains unchanged (invariant). We call these combinations of symmetry operations *point symmetry groups*. When dealing with crystals, the expression *crystal classes* is often used as a matter of tradition.

We should point out that a satisfactory treatment of symmetry theory and its applications to problems in crystal physics and also to problems in atomic and molecular physics is possible especially with the help of *group theory*. In what follows, we will give preference to group theoretical symbols (see also Section 8). Important methods of group theory for crystal physics are treated in Sections 8 and 9.

Textbooks on crystallography give a detailed analysis of the compatibility of different symmetry operations (e.g., Kristallgeometrie). Here we will only remark on the essential procedures and present the most significant results.

The whole complex reduces to the following questions:

(a) In which way are n or $\bar{n}$ compatible with $\bar{1}$, 2, and $\bar{2} = \text{m}$?

(b) Under which conditions can n or $\bar{n}$ simultaneously exist with p or $\bar{p}$ when $n, p \geq 3$? p specifies a second rotation axis of p-fold symmetry.

(c) In (b) can $\bar{1}$, 2, and $\bar{2}$ also occur?

(d) How can operations n, $\bar{n}$ and those combinations permitted under (a), (b), and (c) be combined with a translation?

We will defer case (d) because the invariance of all points is lifted by the translation. With respect to question (a), the following seven cases can be decided at once by direct inspection of stereographic projections:

1. n or $\bar{n}$ with $\bar{1}$,

2. n or $\bar{n}$ parallel to 2,

3. n or $\bar{n}$ perpendicular to 2,
4. n or $\bar{n}$ forms an arbitrary angle with 2.
5. n or $\bar{n}$ parallel to $\bar{2}$ (=m),
6. n or $\bar{n}$ perpendicular to $\bar{2}$ (=m),
7. n or $\bar{n}$ forms an arbitrary angle with $\bar{2}$ (=m).

With a single principal axis n or $\bar{n}$, the following 7 permissible combinations result from the 14 possibilities above:

n (only one n-fold axis),

n/m (read "n over m," symmetry plane perpendicular to an n-fold axis),

$n\mathrm{m}$ (symmetry plane contains the n-fold axis),

$n2$ (2-fold axis perpendicular to the n-fold axis),

n/mm (symmetry plane perpendicular to the n-fold axis, a second symmetry plane contains the n-fold axis),

$\bar{n}$ (only one n-fold rotoinversion axis),

$\bar{n}2$ (2-fold axis perpendicular to the n-fold rotoinversion axis).

All other combinations turn out to be coincidences to the seven just mentioned. One finds that apart from the "generating" symmetry operations, other symmetry operations are necessarily obtained which can also be used to generate the given combination. For example, $\bar{n}2 = \bar{n}\mathrm{m}$ or $2\bar{1} = 2/\mathrm{m}$. Normally we use the shorthand symbols with the respective generating symbols. The complete symbols, which comprise all compatible symmetry operations of a certain combination, play an important role in some areas of crystallography (structure determination, group theoretical methods). The *Hermann–Mauguin notation* used here is the international standard. The older notation of Schoenflies is still used by chemists and spectroscopists but will not be discussed in this book.

Before we turn our attention to case (b) let us consider which n-fold rotation axes or n-fold rotoinversion axes can occur in crystals, i.e., in lattices. From experience, one deduces the *third basic law of crystallography*:

In crystals one observes only 1-, 2-, 3-, 4-, and 6-fold symmetry axes.

The proof that no other n-fold symmetry is compatible with the lattice arrangement of uniform cells is as follows: We consider two parallel axes A_1 and A_2 of n- (or $\bar{n}$) fold symmetry which possess the smallest separation of such symmetry axes in the given lattice. We allow the symmetry operations to

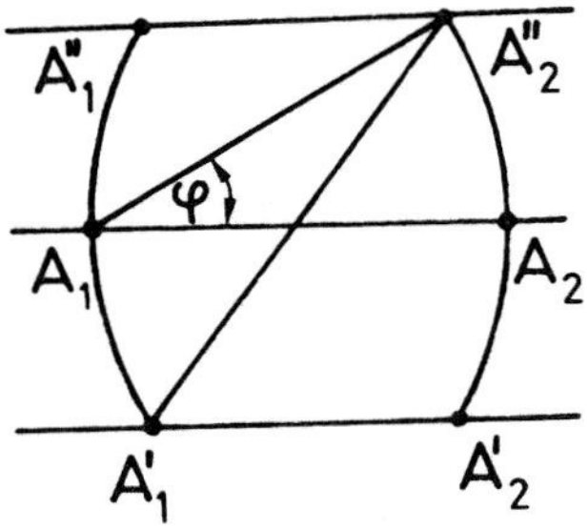

Figure 1.22 Compatibility of n-fold axes in lattices.

Table 1.1 Compatible multiplicities of n-fold axes in lattices.

n	φ	$\|r'/r\|$	$\|r'''/r\|$
1	360°	1	0
2	180°	3	2
3	120°	2	$\sqrt{3}$
4	90°	1	$\sqrt{2}$
5	72°	≈0,38	≈1,18
6	60°	0	1
$n > 6$	<60°	$0 < \|r'/r\| < 1$	$0 < \|r'''/r\| < 1$

work on each other and get further symmetry axes according to $A_1(A_2) \rightarrow A_2'$, $A_2(A_1) \rightarrow A_1''$ and A_2'' and A_1' by a rotation in the opposite sense (Fig. 1.22). The new symmetry axes must either coincide or at least have the same separation as the axes A_1 and A_2. Using the notation in Fig. 1.22 we have

$$\begin{aligned} r &= A_1A_2; \\ r' &= A_1'A_2' = 2r\cos\varphi - r; \\ r'' &= A_1'A_2'' = \sqrt{r^2 + 4r^2\sin^2\varphi} = r\sqrt{5 - 4\cos\varphi} \geq r; \\ r''' &= A_2A_2' = \sqrt{(r - r\cos\varphi)^2 + r^2\sin^2\varphi} = r\sqrt{(2(1-\cos\varphi)} = 2r\sin\varphi/2. \end{aligned}$$

Table 1.1 presents the values $|r'/r|$ and $|r'''/r|$ as a function of the multiplicity. They must be $\geq$1 or 0.

Similar considerations for rotoinversion axes lead to the same end result, namely, that crystals can only have 1-, 2-, 3-, 4-, and 6-fold symmetry axes. Other n-fold symmetries can, however, exist in noncrystalline forms. Even molecules can possess, e.g., 5-fold and higher symmetry axes not permitted in crystals. From the combinations in (a) and under the restrictions just mentioned, only 27 crystallographic point symmetry groups exist. These are listed in the annex.

Now to case (b): The combination of rotation axes n (or rotoinversion axes $\bar{n}$) with rotation axes p (or $\bar{p}$) for the case $n, p \geq 3$ leads to a mutual multiplicity

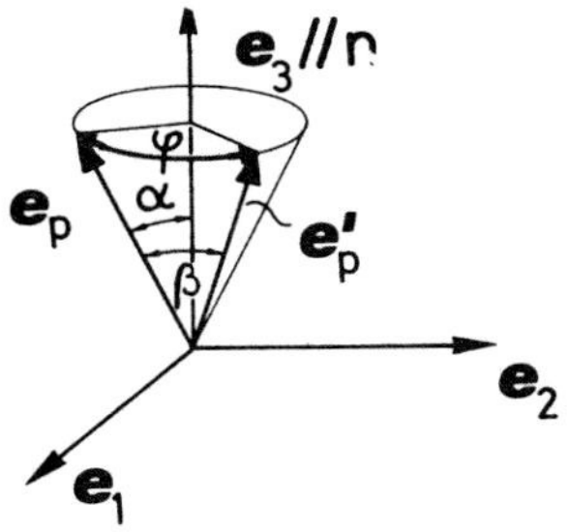

Figure 1.23 Compatibility of n-fold and p-fold axes.

of the axes, namely to at least n different p axes and p different n axes. In a Cartesian coordinate system let one axis n lie parallel to e_3 and a second axis p lie in the plane spanned by e_1 and e_3 perpendicular to e_2 (Fig. 1.23). Let the angle between these axes be α. Applying the operation n on the axis p gives us a second axis p'. Let the unit vectors along these axes be e_n, e_p, and e'_p. With $e_p = \sin\alpha e_1 + \cos\alpha e_3$ and

$$R_{n\|e_3} = \begin{pmatrix} \cos\varphi & -\sin\varphi & 0 \\ \sin\varphi & \cos\varphi & 0 \\ 0 & 0 & 1 \end{pmatrix}$$

one gets $e_{p'} = R_n(e_p) = \sin\alpha\cos\varphi\ e_1 + \sin\alpha\sin\varphi\ e_2 + \cos\alpha\ e_3$. We now calculate the angle β between p and p'. The result is $e_p \cdot e'_p = \cos\beta = \sin^2\alpha\ \cos\varphi + \cos^2\alpha = 1 + \sin^2\alpha\ (\cos\varphi - 1)$. From this equation and with $(1 - \cos u) = 2\sin^2 u/2$ we derive the relationship $\sin\beta/2 = \pm\sin\alpha\sin\varphi/2$, where $\varphi = 2\pi/n$.

We first consider the simple case of the combination of a 3-fold axis with another axis $p \geq 2$, where for $p = 2$ the condition $\alpha = 0$ or $90°$ was already discussed. Thus several symmetry-equivalent 3-fold axes are created, which on a sphere, whose center is the common intercept point, fix an equal-sided spherical triangle, whose center also specifies the intercept point of a 3-fold axis. In this spherical triangle $\alpha = \beta$ (Fig. 1.24). For the case $n \geq 4$ let α be the smallest angular distance between two of the symmetry equivalent axes n. Then the angular distance α' between two axes resulting from the application of one on the other axis, respectively, must either vanish, i.e., both axes must coincide, or we have $\alpha' \geq \alpha$. However, the largest possible angular distance is $90°$. As one can easily see from a stereographic projection (Fig. 1.25), the only possibility for $n \geq 4$ rotation axes is that both axes coincide since $\alpha' < \alpha$ in each case.

This means that in case (b) the intercept points of the symmetry-equivalent n-fold axes ($n \geq 3$) always form an equal-sided spherical triangle ($\alpha = \beta$). From the relationship derived above, we have for $\alpha = \beta$: $\cos\alpha/2 =$

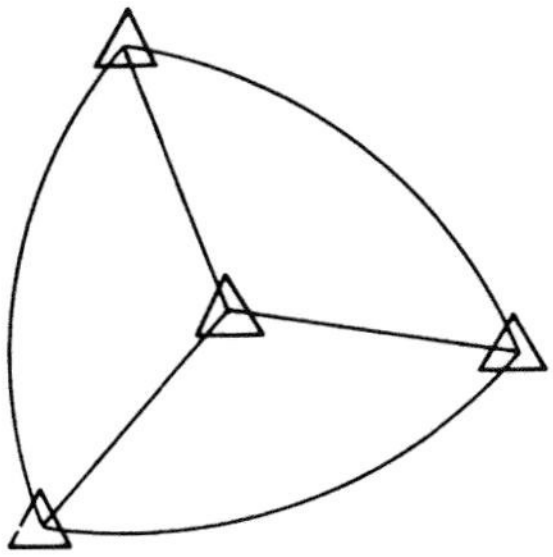

Figure 1.24 Combination of two 3-fold axes.

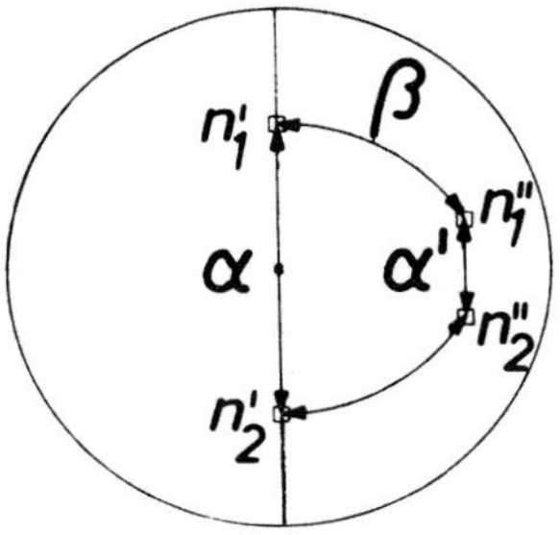

Figure 1.25 Combination of two n-fold axes with $n \geq 4$.

$\pm 1/(2 \sin \varphi/2)$. Table 1.2 lists the possible angles α as a function of the n-fold symmetry.

A combination of several symmetry-equivalent rotation axes with $n \geq 3$ is only allowed for $n = 1, 2, 3, 4, 5, 6$. The angles appearing are prescribed.

To case (c): The discussion of the combinations of n-fold axes ($n \geq 3$) requires a complement, since with the n-fold axes only the case $\alpha = \beta$ was settled. In the center of the equilateral spherical triangle, formed by the intercept points of the 3-fold axes on the sphere, there exists a further 3-fold axis, which with the other axes specifying the spherical triangle includes the angle α' with $\sin \alpha' = 2\sqrt{2}/3$, $\cos \alpha' = 1/3$. This is recognized by applying the formula

Table 1.2 Angles between possible n-fold axes. Concerning 2-fold axes see the results obtained in case a).

n	$\varphi = 2\pi/n$	$\cos \alpha/2$	$\alpha = \beta$
2	180°	$\pm 1/2$	120°; 240°
3	120°	$\pm 1/\sqrt{3}$	109,47°; 250,53°
4	90°	$\pm 1/\sqrt{2}$	90°; 270°
5	72°	$\pm 1/(2 \sin 36°)$	63,43°; 296,57°
6	60°	± 1	0°; 360°
$n > 6$	<60°	$\lvert \cos \alpha/2 > 1 \rvert$	–

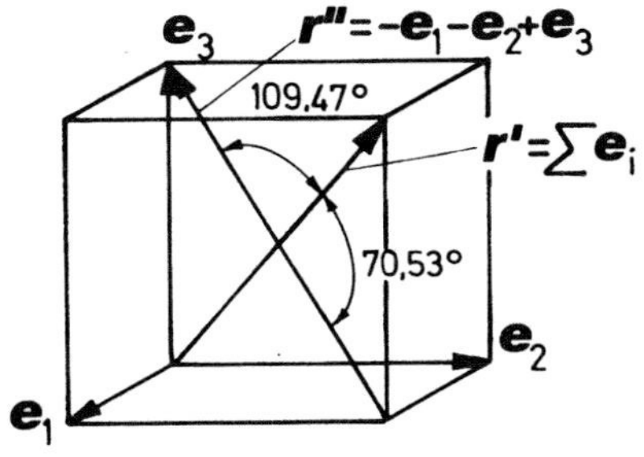

Figure 1.26 Symmetry of a cube.

derived above for the rotation about an n-fold axis according to Fig. 1.24. The result is $\alpha' \approx 70.53°$. The question is now whether even with this small angular distance, 3-fold axes exist whose intercept points on a sphere also form an equilateral spherical triangle. With the same formula just used, one gets in this case, for a further 3-fold intercept axis in the center of the triangle, the angle $\alpha'' \approx 41.81°$ from $\sin\alpha'' = 2/3$. This is in fact the smallest angle of two 3-fold axes occurring in the icosahedral groups through the combination of several 3-fold axes.

Just as with the case of the 3-fold axes, the equilateral spherical triangles of the intercept points of the symmetry-equivalent n-fold axes n_1', n_2', and n_1'' also possess a 3-fold axis in their centers (Fig. 1.25). Thus in all combinations of n-fold axes ($n > 3$), 3-fold axes are always present, which mutually include the angles just discussed. For further discussions of the combination possibilities of n-fold axes ($n > 3$) with other symmetry operations it is useful to consider the symmetry properties of a cube. Let $\boldsymbol{e}_i$ be the edge vectors of a unit cube (identical to the Cartesian basis vectors); then the directions of the space diagonals of the cube may be represented by $\boldsymbol{r} = \pm\boldsymbol{e}_1 \pm \boldsymbol{e}_2 \pm \boldsymbol{e}_3$ (Fig. 1.26). One obtains for the angle α between three different space diagonals $\cos\alpha = \pm 1/3$. The values $-1/3$ and $+1/3$ give for α approximately 109.47° and the complementary angle of 180°. In Table 1.2 we had $\cos\alpha/2 = \pm 1/\sqrt{3}$; therefore $\cos\alpha = -1/3$ (with $\cos u = -1 + 2\cos^2 u/2$). Thus the space diagonals of the cube, themselves 3-fold rotation axes, intersect at angles identical to those for 3-fold axes given in Table 1.2.

The system of four space diagonals of the cube represents the simplest point symmetry group (abbreviated as PSG in what follows) of the combination of two polar 3-fold axes ("polar" means direction and inverse direction are not equivalent). As one can easily show with the aid of the transformation formulae or a stereographic projection, this arrangement also contains three 2-fold axes, which run parallel to the cube edges, that is, at half the angle of the larger angle between two 3-fold axes. This PSG is given the Hermann–Mauguin symbol 23 (Fig. 1.27).

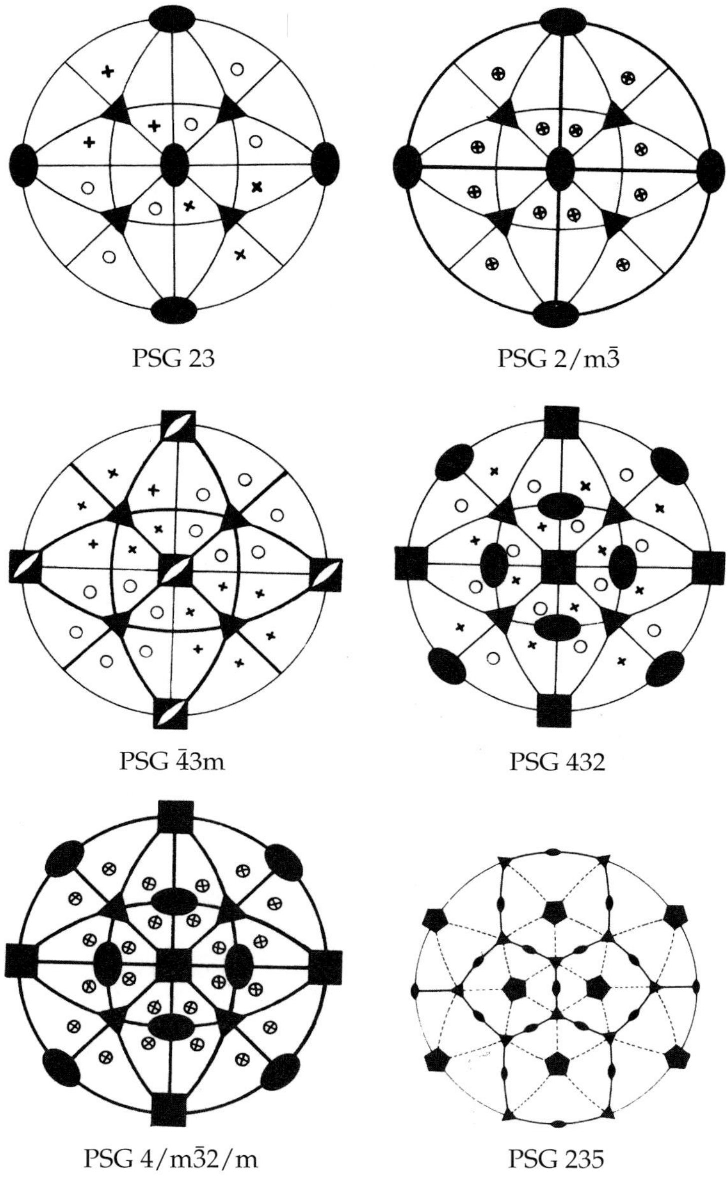

Figure 1.27 Stereographic projection of the symmetry operations in cubic point symmetry groups and in icosahedral PSG 235. The symmetry operations as well as the intercept points of the assembly of symmetry-equivalent normals are drawn. The intercept points on the northern hemisphere are indicated by a cross, those on the southern hemisphere with an empty circle.

The combination of 23 with symmetry planes perpendicular to the 2-fold axes or with $\bar{1}$ leads to PSG $2/m\bar{3}$ (Fig. 1.27). With symmetry planes, each containing two 3-fold axes, we get PSG $\bar{4}3m$, where the 2-fold axes of 23 turn into $\bar{4}$-axes (Fig. 1.27). In $2/m\bar{3}$, short symbol m3, the 3-fold axes are nonpolar; in $\bar{4}3m$, short symbol $\bar{4}3$, they are polar just as in 23. If we introduce 4-fold axes instead of 2-fold axes we get PSG 432, short symbol 43 (Fig. 1.27). Here, additional 2-fold axes are generated at half of the smaller angles between two 3-fold axes. The 3-fold axes are nonpolar. Moreover, the 4-fold axes form an angle of 90°, as demanded in Table 1.2. Finally, symmetry planes perpendicular to the 4-fold axes can also be combined. This leads to PSG $4/m\bar{3}2/m$, short symbol $4/m3$ or m3m, the highest symmetry group in crystals (Fig. 1.27). We also obtain $4/m3$ with the inclusion of $\bar{1}$ to 43. The five point symmetry groups just discussed comprise the *cubic crystal system*.

For completion, let us discuss the noncrystallographic PSG of the combination of 5-fold axes. The basic framework here is also the arrangement of four space diagonals of the cube. In each field of the cube two further 3-fold axes are constructed so that the intercept points of neighboring axes mark spherical triangles on the sphere with the angular distance $\alpha' \approx 70.52°$ ($\sin\alpha' = 2\sqrt{2}/3$) just discussed. This results in a smallest angular distance of $\alpha'' \approx 41.81°$ ($\sin\alpha'' = 2/3$) between the inserted axes and the axes along the space diagonals. Thus arrangements of five 3-fold axes are formed whose intercept points on the sphere give the corners of regular pentagons (Fig. 1.27). The angle between two 5-fold axes may be easily calculated from the known angular distance of the 3-fold axes with the aid of the rotation formula. It is in agreement with the result of Table 1.2, namely $\cos\alpha/2 = \pm 1/(2\sin 36°)$. This symmetry group has the symbol 235. It exhibits six 5-fold, ten 3-fold, and fifteen 2-fold axes. Introducing symmetry planes perpendicular to the 2-fold axes results in PSG $2/m35$. Both these so-called icosahedral groups play an important role in the structure of viruses, in certain molecular structures, such as, e.g., in the B_{12}-structures of boron and in certain quasicrystals.

There exist a total of *32 different crystallographic PSGs*. These are divided into *seven crystal systems* depending on the existence of a certain minimum symmetry (Table 1.3). These systems are associated with the seven distinguishable symmetry classes of the crystallographic reference systems. These systems are specified by prominent directions, the so-called *viewing directions*, along which possibly existing symmetry axes or normals on symmetry planes are running. It turns out that each system *has at most three different viewing directions*.

Table 1.3 The seven crystal systems.

System	Minimal symmetry	Conditions for lattice parameters of symmetry-adapted reference system (viewing directions)	PSG
Triclinic	1	a_i, α_i not fixed (1. arbitrary, 2. arbitrary, 3. arbitrary)	$1, \bar{1}$
Monoclinic	2 or m = $\bar{2}$	$\alpha_1 = \alpha_3 = 90°$ (1. $\boldsymbol{a}_2 \parallel 2$ or $\bar{2}$, 2. arbitrary, 3. arbitrary)	2, m, 2/m
Orthorhombic	22 or mm = $\bar{2}\bar{2}$	$\alpha_i = 90°$ (1. $\boldsymbol{a}_1 \parallel 2$ or $\bar{2}$, 2. $\boldsymbol{a}_2 \parallel 2$ or $\bar{2}$, 3. $\boldsymbol{a}_3 \parallel 2$ or $\bar{2}$)	22, mm, 2/mm
Trigonal (rhombohedral)	3 or $\bar{3}$	$a_i = a, \alpha_i = \alpha$ (1. $\boldsymbol{a}_1 + \boldsymbol{a}_2 + \boldsymbol{a}_3 \parallel 3$ or $\bar{3}$, 2. $\boldsymbol{a}_1 - \boldsymbol{a}_2 \perp 3$, and $\boldsymbol{a}_3$, 3. $2\boldsymbol{a}_3 - \boldsymbol{a}_1 - \boldsymbol{a}_2 \perp 3$ and $\boldsymbol{a}_1 - \boldsymbol{a}_2$)	3, 3m, 32, $\bar{3}$, $\bar{3}$m = $\bar{3}$2
Tetragonal	4 or $\bar{4}$	$a_1 = a_2, \alpha_i = 90°$ (1. $\boldsymbol{a}_3 \parallel 4$ or $\bar{4}$, 2. $\boldsymbol{a}_1$, 3. $\boldsymbol{a}_1 + \boldsymbol{a}_2$)	4, 4/m, 4m, 42, 4/mm, $\bar{4}$, $\bar{4}$m = $\bar{4}$2
Hexagonal	6 or $\bar{6}$	$a_1 = a_2, \alpha_1 = \alpha_2 = 90°, \alpha_3 = 120°$ (1. $\boldsymbol{a}_3 \parallel 6$ or $\bar{6}$, 2. $\boldsymbol{a}_1$, 3. $2\boldsymbol{a}_1 + \boldsymbol{a}_2 \perp \boldsymbol{a}_2$)	6, 6/m, 6m, 62, 6/mm, $\bar{6}$, $\bar{6}$m = $\bar{6}$2
Cubic	23	$a_i = a, \alpha_i = 90°$ (1. $\boldsymbol{a}_1 \parallel$ edge of cube, 2. $\boldsymbol{a}_1 + \boldsymbol{a}_2 + \boldsymbol{a}_3 \parallel 3$, 3. $\boldsymbol{a}_1 + \boldsymbol{a}_2$)	23, $\bar{4}$3, 43, m3, 4/m3 = m3m

1.7.3
Theory of Forms

We will now turn to the discussion of morphological properties. The complete set of symmetry-equivalent faces to a face $(h_1h_2h_3)$ in a point symmetry group is designated as a form with the symbol $\{h_1h_2h_3\}$. The entirety of the symmetry-equivalent vectors to a lattice vector $[u_1u_2u_3]$ is symbolized as $\langle u_1u_2u_3\rangle$; correspondingly, $\langle|u_1u_2u_3|\rangle$ means the entirety of the symmetry-equivalent points to the point $[|u_1u_2u_3|]$.

To calculate the symmetry-equivalent faces, lattice edges, or points we use the transformations already discussed with a transition from the basic system to a symmetry-equivalent system (Section 1.6).

In a symmetry-equivalent system the cotransformed face normals and vectors, respectively, possess the same coordinates as in the basic system. Thus one gets the symmetry-equivalent faces and vectors respectively or points by enquiring about the indices or coordinates of the transformed quantities in the old system. These result from the inverse transformation, thus in the case of the Miller indices, with the inverse transformation matrix U^{-1} and in the case of the vectors or points, with the transposed matrix U^T. We will call the number of symmetry-equivalent objects generated by a symmetry operation the order h of the given operation. Repeated application gives us all symmetry-

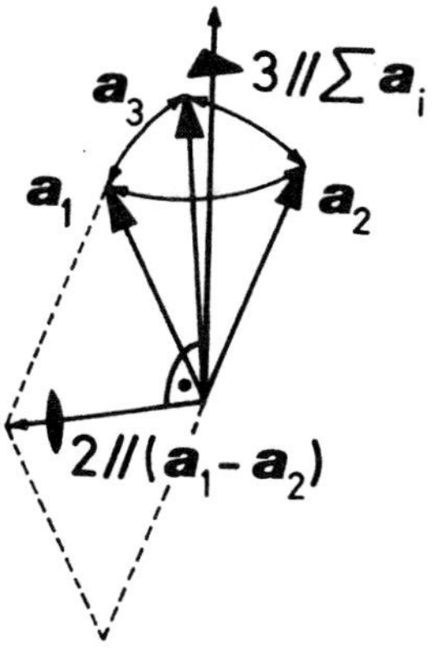

Figure 1.28 Symmetry operations of PSG 3m in a trigonal–hexagonal reference system.

equivalent quantities; thus $\{h_1h_2h_3\} = \{(U^{-1})^m(h_1h_2h_3)\} = \{U^{-m}(h_1h_2h_3)\}$ and correspondingly $\langle u_1u_2u_3\rangle = \langle (U^T)^m[u_1u_2u_3]\rangle$ with $m = 1, 2, \ldots, h$.

If several symmetry-generating operations exist, then the calculation of all symmetry-equivalent quantities requires that the additional symmetry operations be applied to the quantities already generated by the other operations. This is demonstrated by the example $LiNbO_3$, PSG 3m (Fig. 1.28). In a trigonal–hexagonal reference system, with a metric defined by $a_1 = a_2, \alpha_1 = \alpha_2 = 90°, \alpha_3 = 120°$, we have

$$U(3||\boldsymbol{a}_3) = \begin{pmatrix} 0 & 1 & 0 \\ \bar{1} & \bar{1} & 0 \\ 0 & 0 & 1 \end{pmatrix}, \quad U(3)^{-1} = \begin{pmatrix} \bar{1} & \bar{1} & 0 \\ 1 & 0 & 0 \\ 0 & 0 & 1 \end{pmatrix},$$

$$U(3)^T = \begin{pmatrix} 0 & \bar{1} & 0 \\ 1 & \bar{1} & 0 \\ 0 & 0 & 1 \end{pmatrix}, \quad U(\bar{2}||\boldsymbol{a}_1) = \begin{pmatrix} \bar{1} & 0 & 0 \\ 1 & 1 & 0 \\ 0 & 0 & 1 \end{pmatrix} = U(\bar{2}||\boldsymbol{a}_1)^{-1}.$$

With $U(3)^{-1}$ one finds for $\{h_1h_2h_3\}$ the faces $(h_1h_2h_3)$, $(\bar{h}_1 + \bar{h}_2.h_1h_3)$, $(h_2.\bar{h}_1 + \bar{h}_2.h_3)$ and with $U(\bar{2}||\boldsymbol{a}_1)^{-1}$ the additional faces $(\bar{h}_1.h_1 + h_2.h_3)$, $(h_1 + h_2.\bar{h}_2h_3)$, $(\bar{h}_2\bar{h}_1h_3)$. These six faces together form, in the general case, a ditrigonal pyramid (Fig. 1.28). A two-digit or combined Miller index is separated by a dot from the other indices.

If one selects a trigonal–rhombohedral reference system with $a_1 = a_2 = a_3$ and $\alpha_1 = \alpha_2 = \alpha_3 = \alpha$, which is permitted for trigonal crystals, then the symmetry operations of the PSG 3m have the following form:

$$U_{3||(\boldsymbol{a}_1+\boldsymbol{a}_2+\boldsymbol{a}_3)} = \begin{pmatrix} 0 & 1 & 0 \\ 0 & 0 & 1 \\ 1 & 0 & 0 \end{pmatrix} \quad \text{and} \quad U_{\bar{2}||(\boldsymbol{a}_1-\boldsymbol{a}_2} = \begin{pmatrix} 0 & 1 & 0 \\ 1 & 0 & 0 \\ 0 & 0 & 1 \end{pmatrix}.$$

Table 1.4 The meroedries of the seven crystal systems.

System	tri-clinic	mono-clinic	ortho-rhombic	tri-gonal	tetra-gonal	hexa-gonal	cubic
Holoedrie	$\bar{1}$	2/m	2/mm	$\bar{3}$m	4/mm	6/mm	4/m3
Hemimorphie	–	–	2m	3m	4m	6m	–
Paramorphie	–	–	–	$\bar{3}$	4/m	6/m	m3
Enantiomorphie	1	2	22	32	42	62	43
Hemiedrie II	–	m	–	–	$\bar{4}$2	$\bar{6}$2	$\bar{4}$3
Tetartoedrie	–	–	–	3	4	6	23
Tetartoedrie II	–	–	–	–	$\bar{4}$	$\bar{6}$	–

For $\{h_1h_2h_3\}$ one finds, with the inverse operations, the faces $(h_1h_2h_3)$, $(h_2h_3h_1)$, $(h_3h_1h_2)$, $(h_2h_1h_3)$, $(h_3h_2h_1)$ and $(h_1h_3h_2)$. The difference clearly indicates that one must also specify the reference system used when characterizing faces of trigonal crystals.

Let us now consider the different forms in the different PSGs of a system. For this purpose, we will first investigate the relationships between the PSG of highest symmetry and the PSGs of lower symmetry in the same system. These PSGs, the holohedries, are $\bar{1}$, 2/m, 2/mm, $\bar{3}$m, 4/mm, 6/mm, 4/m3. If one removes single minor symmetry elements from the holohedries, one gets the PSGs of lower symmetry of the same system. If 2-fold axes are missing or symmetry planes (only one in each case), the resulting forms are hemihedries, that is PSGs, in which only half the number of surfaces occur as in the holohedries. If one removes two minor symmetry elements, one gets the tetartohedries, PSGs with a quarter of the number of faces as in the holohedries. These are known as merohedral PSG depending on the type of symmetry elements removed or remaining. Each system has a maximum of seven PSGs including the holohedries. These are classified in Table 1.4.

The following holds true: In each holohedry there exists a spherical triangle whose repetition by the generating symmetry elements covers the whole sphere just once. In the merohedries, the symmetry elements cover only one part, namely, half the sphere in the hemihedries and a quarter of the sphere in the tetartohedries. The triangles are referred to as elementary triangles and represented in Fig. 1.29 . The arrangement of these triangles is characteristic for each point symmetry group. Their number corresponds to the order of the point symmetry group. Each face normal is associated with one of the following seven distinguishable positions in the spherical triangle of the holohedries (Fig. 1.30):

1. Corner 1,
2. Corner 2,
3. Corner 3,

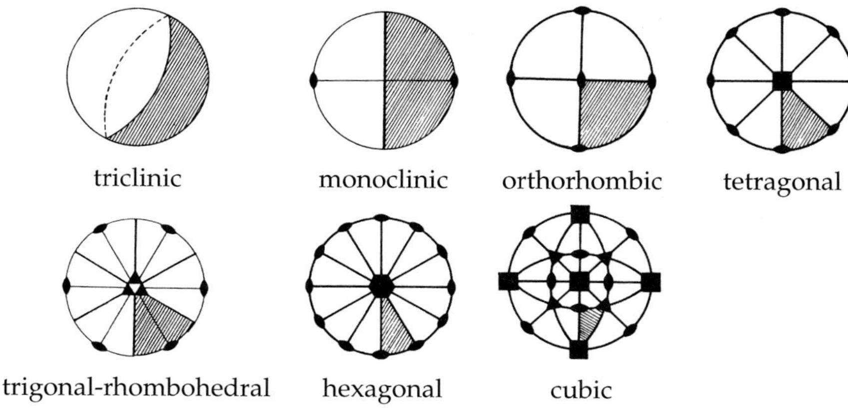

Figure 1.29 Elementary triangles in the seven crystal systems.

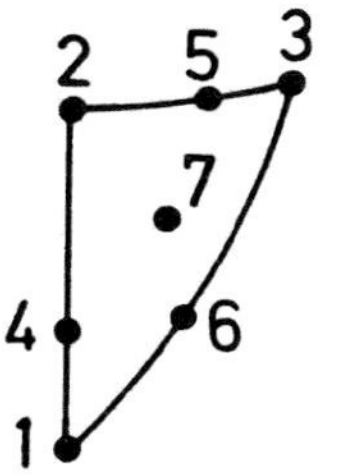

Figure 1.30 The seven positions in an elementary triangle.

4. On the side between 1 and 2,

5. On the side between 2 and 3,

6. On the side between 3 and 1,

7. Inside the triangle.

In each system, except triclinic and monoclinic, positions 1, 2, and 3 are associated with a fixed direction, i.e., the given faces have distinct Miller indices. The side positions 4, 5, and 6 possess one degree of freedom. Only the third position is not bound to any restrictions (two degrees of freedom). Special forms evolve from the first six positions. Position 7 generates general forms, characteristic for the given PSG and also for the distribution of the elementary triangles. Tables 1.5 and 1.6 present the seven forms for the orthorhombic and cubic systems. The forms have the following nomenclature (according to Groth):

Table 1.5 The seven forms of the orthorhombic PSG.

Pos.		22	2m	2/mm
1	{100}	pinacoid	pinacoid	pinacoid
2	{010}	pinacoid	pinacoid	pinacoid
3	{001}	base pinacoid	base pedion	base pinacoid
4	$\{h_1 0 h_3\}$	prisma II. position	doma II. position	prisma II. position
5	$\{h_1 h_2 0\}$	prisma III. position	prisma III. position	prisma III. position
6	$\{0 h_2 h_3\}$	prisma I. position	doma I. position	prisma I. position
7	$\{h_1 h_2 h_3\}$	disphenoid	pyramide	dipyramide

Pedion: Single face, not possessing another symmetry-equivalent face.

Pinacoid: Face with a symmetry-equivalent counter face, generated by $\bar{1}$, 2, or m.

Dome: Pair of faces generated by a mirror plane.

Prism: Tautozonal entirety of symmetry-equivalent faces (all faces intercept in parallel edges, which define the direction of the zone axis).

Pyramid: Entirety of symmetry-equivalent faces, whose normals, with a prominent direction, the pyramid axis, enclose the same angle.

Dipyramid: Double pyramid, generated by a mirror plane perpendicular to the pyramid axis.

Sphenoid: A pair of nonparallel faces generated by a 2-fold axis.

Disphenoid: Two sphenoids evolving separately from a further 2-fold axis.

Scalenohedron: Two pyramids evolving from $\bar{n}2$ with $n \neq 4q - 2$ (n, q integers). Dipyramids are generated for the case $n = 4q - 2$ ($n > 2$).

Streptohedron: Two pyramids, mutually rotated by half the angle of the rotation axes (n odd number). This form is called rhombohedron in den PSGs $\bar{3}$, $\bar{3}$m, and 32.

The special nomenclature of the forms of the cubic system is mentioned in Table 1.6. Parallel projections of these forms are presented in Fig. 1.31. Table 1.7 gives an overview of the 32 crystallographic point symmetry groups.

Table 1.6 The forms of the cubic and icosahedral PSG. The Miller indices are only valid for cubic crystals.

Pos.		23	m3	$\bar{4}3$	43	4/m3	235 and m35
1	$\{100\}$	cube	cube	cube	cube	cube	regular pentagon-dodecahedron
2	$\{110\}$	rhombic dodecahedron	rhombic dodecahedron	rhombic dodecahedron	rhombic dodecahedron	rhombic dodecahedron	regular icosahedron
3	$\{111\}$	tetrahedron	octahedron	tetrahedron	octahedron	octahedron	rhombentria-contahedron
4	$\{h_1 0 h_3\}$	pentagon-dodecahedron	pentagon-dodecahedron	tetrakis-hexahedron	tetrakis-hexahedron	tetrakis-hexahedron	pentakis-dodekahedron
5	$\{h_1 h_2 h_2\}$ $h_1 > h_2$	tris-tetrahedron	deltoidikosi-tetrahedron	tris-tetrahedron	deltoidikosi-tetrahedron	deltoidikosi-tetrahedron	tris-icosahedron
6	$\{h_1 h_2 h_2\}$ $h_1 < h_2$	deltoid-dodecahedron	tris-octahedron	deltoid-dodekahedron	tris-octahedron	tris-octahedron	deltoidhexa-contahedron
7	$\{h_1 h_2 h_3\}$	tetrahedral pentagon-dodecahedron	disdodeca-hedron	hexakis-tetrahedron	pentagon-ikositetra-hedron	hexakis-octahedron	235: pentagonal hexacontahedron m35: hekatonikosahedron

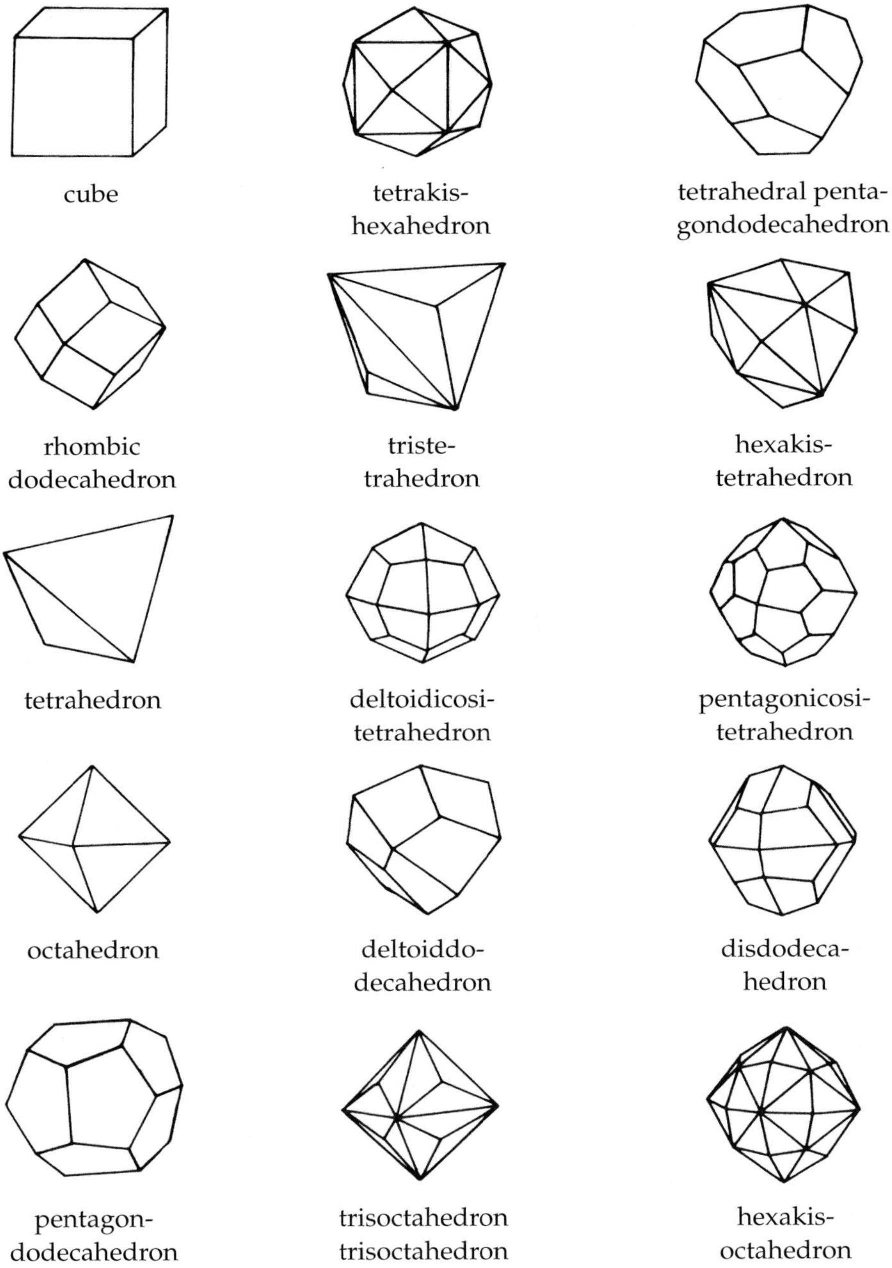

Figure 1.31 The 15 different forms of the cubic system.

Table 1.7 The 32 crystallographic point symmetry groups (crystal classes) with examples (symbols in parentheses give Schoenflies symbol).

Symmetry symbol (shorthand notation)		Groth designation	Examples Minerals	Synthetic crystals
1	(C_1)	triclinic pedial	Wegscheiderite: $Na_2CO_3 \cdot 3NaHCO_3$	$CaS_2O_3 \cdot 6H_2O$
			Parahilgardite: $Ca_2ClB_5O_8(OH)_2$	$LiHC_2O_4 \cdot H_2O$
$\bar{1}$	(C_1)	triclinic pinacoidal	Mikrokline: $KAlSi_3O_8$	$CuSO_4 \cdot 5H_2O$
			Wollastonite: $Ca_3Si_3O_9$	$B(OH)_3$
2	(C_2)	monoclinic	Harmotome: $BaAl_2Si_6O_{16} \cdot 6H_2O$	$Li_2SO_4 \cdot H_2O$
		sphenoidic	Brushite: $CaHPO_4 \cdot 2H_2O$	tartaric acid ($C_4H_6O_6$)
m	(C_s)	monoclinic	Hilgardite: $Ca_2ClB_5O_8(OH)_2$	$K_2S_4O_6$
		domatic	Klinoedrite: $Ca_2Zn_2(OH)_2Si_2O_7 \cdot H_2O$	$LiH_3(SeO_3)_2$
2/m	(C_{2h})	monoclinic	gypsum: $CaSO_4 \cdot 2H_2O$	$KHCO_3$
		prismatic	Diopside: $CaMgSi_2O_6$	SnF_2
22	(D_2)	orthorhombic	Epsomite: $MgSO_4 \cdot 7H_2O$	Seignette salt ($NaKC_4H_4O_6 \cdot 4H_2O$)
		disphenoidal	Descloizite: $Pb(Zn,Cu)(OH)VO_4$	$Sr(HCOO)_2$
mm	(C_{2v})	orthorhombic	Dyskrasite: Ag_3Sb	$LiHCOOH_2O$
		pyramidal	Shortite: $Na_2Ca_2(CO_3)_3$	$MgBaF_4$
2/mm	(D_{2h})	orthorhombic	baryte: $BaSO_4$	K_2SO_4
		dipyramidal	olivine: $(Mg,Fe)_2SiO_4$	$Ca(HCOO)_2$
3	(C_3)	trigonal	Parisite: $CaCe_2F_2(CO_3)_3$	$NaJO_4 \cdot 3H_2O$
		pyramidal	Roentgenite: $Ca_2Ce_3F_3(CO_3)_5$	$LiBO_2 \cdot 8H_2O$
32	(D_3)	trigonal	α-Quartz: SiO_2	benzil
		trapezohedral	cinnabarite: HgS	$Rb_2S_2O_6$
3m	(C_{3v})	ditrigonal	tourmaline	$LiNbO_3$
		pyramidal	Proustite: Ag_3AsS_3	$KBrO_3$

Table 1.7 *(continued)*

Symmetry symbol (shorthand notation)		Groth designation	Examples Minerals	Synthetic crystals
$\bar{3}$	(C_{3i})	trigonal rhombohedral	dolomite: $CaMg(CO_3)_2$	$NiSnCl_6\text{-}_6H_2O$
			Dioptase: $Cu_6Si_6Oi_8\text{-}\ _6H_2O$	Li_2MoO_4
$\bar{3}$m	(D_{3d})	trigonal scalenoedric	calcite: $CaCO_3$	$NaNO_3$
			corundum: $A_{12}O_3$	NH_4SnF_3
4	(C_4)	tetragonal pyramidal	–	CS_2 $BjsO_{28}$
			–	iodosuccinimide $(CH_2CO)_2NJ$
42	(D_4)	tetragonal trapezohedral	Mellite: $A_{12}C_{12}O_{12}\ _{18}H_2O$	$NiSO_4{}'_6H_2O$
			phosgenite: $Pb_2Cl_2CO_3$	TeO_2
4m	(C_{4v})	ditetragonal pyramidal	Diaboleite: $_2PbC_1OH$ Cu $(OH)_2$	$Ba_6Ti_2Nb_8O_3o$
			Fresnoite: $Ba_2TiSi_2O_8$	NbP
4/m	(C_{4h})	tetragonal dipyramidal	Scheelite: $CaWO_4$	$NaJO_4$
			Fergusonite: $YNbO_4$	AgC_1O_3
4/mm	(D_{4h})	ditetragonal dipyramidal	rutile: TiO_2	MnF_2
			zirkon: $ZrSiO_4$	HgJ_2
$\bar{4}$	(S_4)	tetragonal disphenoidal	Cahnite: $Ca_2(AsO_4)(B(OH)_4)$	$LiNH_2$
			–	pentaerythritol $C(CH_2OH)_4$
$\bar{4}$2	(D_{2d})	tetragonal scalenohedral	chalkopyrite: $CuFeS_2$	KH_2PO_4
			Stannine: Cu_2FeSnS_4	urea $CO(NH_2)_2$
6	(C_6)	hexagonal pyramidal	Mixite	$LiJO_3$
			Nepheline: $KNa_3(AlSiO_4)_4$	$A_1(JO_3)_3\text{-}_2HJO_3\text{-}_6H2O$
62	(D_6)	hexagonal trapezohedral	β-Quartz: SiO_2	$TaSi_2$
			Kaliophilite: $KAlSiO_4$	$LaPO_4$
6m	(C_{6v})	dihexagonal pyramidal	Wurtzite: ZnS	ZnO
			Greenockite: CdS	$LiCKV\ _3H_2O$

Table 1.7 *(continued)*

Symmetry symbol (shorthand notation)		Groth designation	Examples: Minerals	Synthetic crystals
6/m	(C_{6h})	hexagonal	Apatite: $Ca_5(PO_4)_3F$	$PrBr_3$
		dipyramidal	Jeremejewite: $AIBO_3$	$Ce_2(SO_4)_3{'}_9H_2O$
6/mm	(D_{6h})	bihexagonal	beryl: $Al_2Be_3Si_6Oi_8$	Mg
		dipyramidal	Covelline: CuS	graphite
$\bar{6}$	(C_{3h})	trigonal	–	Li_2O_2
		dipyramidal	–	$NaLuF_4$
$\bar{6}2$	(D_{3h})	ditrigonal	Benitoite: $BaTiSi_3O_9$	K_2ThF_6
		dipyramidal	Bastnäsite: $CeFCO_3$	$LiNaCO_3$
23	(T)	cubic tetrahedral	Ullmannite: NiSbS	NaC_1O_3
		pentagonal dodecahedral	Langbeinite: $K_2Mg_2(SO_4)_3$	Na_3SbS_4-$_9H_2$ O
m3	(T_h)	cubic	pyrite: FeS_2	alum
		bisdodecahedral	Cobaltin: CoAsS	$Pb(NO_3)_2$
$\bar{4}3$	(T_d)	cubic	zinc blende: ZnS	urotropine
		hextetrahedral	Eulytine: Bi_4SiO_4	GaAs
43	(O)	cubic pentagonal	Petzite: Ag_3AuTe_2	NH_4CI (?)
		icositetrahedral		
4/m3	(O_h)	cubic hexakis-	spinel: $MgAl_2O_4$	NaCl
		oktahedral	garnet	CaF_2

1.7.4
Morphological Symmetry, Determining the Point Symmetry Group

Sometimes the morphological symmetry of the freely grown crystals is lower than the associated point symmetry group. This case is called hypomorphy. It can occur when the mother phase, from which the crystals grow, possesses asymmetric molecules (e.g., from an aqueous solution of potassium sulfate with orthorhombic disphenoids, which simulate the PSG 22; but the true symmetry is mmm). If the crystals show a morphologically higher symmetry than the corresponding point symmetry group, then we have a hypermorphy, such as, e.g., with the α-alums. The hypermorphy is much more prevalent than the hypomorphy.

The PSG of a crystal can be determined as follows:

1. *Morphological diagnosis* with the aid of general forms or certain combinations of special forms. If the result is not unequivocal, one can try to obtain general forms by means of spherical growth experiments. This is done by preparing probes with spherically shaped regions from the crystal to be analyzed, and placing them in a slightly supersaturated solution of the given substance for further growth. One gets small plane surface elements (spherical caps) with normals belonging to faces with relatively minimal surface energy. The PSG can often, but not always, be derived from the distribution of these spherical caps.

2. *Investigating the surface symmetry* of different forms (growth assessories, etch figures, epitaxial growth figures, impact and pressure figures, directional dependence of the mechanical, chemical or physical erosion).

3. *Investigating the physical properties*. The suitable measures are discussed in Section 12.3.

1.7.5
Symmetry of Space Lattices (Space Groups)

1.7.5.1 Bravais Types

We return now to point (d) of Section 1.7.2. The discussion concerned the combination of rotation axes and rotoinversion axes with translations. Since the operation of translation can be repeated an arbitrary number of times, bodies with translational symmetry always possess unlimited extension in the rigorous sense. If only one translation vector t_1 exists, then we are dealing with a one-dimensional lattice (lattice chain), whose symmetry-equivalent objects may take three-dimensional forms. Two different translations t_1 and t_2 result in a two-dimensional lattice (net). The general three-dimensional translation lattice exhibits three noncoplanar translations t_1, t_2, and t_3, which, repeated any number of times, always reproduce the lattices. For the description of the symmetry relationships and the consideration of the properties of the lattice, and thus that of the ideal crystal, it suffices to investigate a single cell of the lattice, the elementary cell.

We represent the entirety of the lattice with the end points of the vectors $\boldsymbol{r} = r_1\boldsymbol{a}_1 + r_2\boldsymbol{a}_2 + r_3\boldsymbol{a}_3 = r_i\boldsymbol{a}_i$, where r_1, r_2, r_3 are arbitrary integers and $\boldsymbol{a}_1$, $\boldsymbol{a}_2$, $\boldsymbol{a}_3$ form a basic system of three nonplanar vectors. An elementary cell of this lattice is the parallelepiped spanned by $\boldsymbol{a}_1$, $\boldsymbol{a}_2$, $\boldsymbol{a}_3$. This translation is called simple primitive because one cell contains just one lattice point. How do these translation lattices differ with respect to their symmetry properties? As we have already seen in our discussion of crystal systems, there exist, due to symmetry properties, only seven distinguishable metric types, and thus only seven distinguishable systems of primitive cells. These seven systems are called *primitive Bravais types* after the French mathematician Auguste Bravais. Each possesses the holohedral symmetry of the given crystal system (Fig. 1.32).

The question now remains to be answered, whether for primitive lattices with specific metric values, further distinguishable types exist. One can also pose this question in another light. Do certain axes ratios $a_1{:}a_2{:}a_3$ and angles α_i of the primitive Bravais types exist such that in the given lattice one finds larger cells with higher symmetry than the primitive cells? These larger cells, however, can only possess the holohedral symmetry of the crystal system. They contain more than one lattice point per cell and are thus designated as *multiple-primitive Bravais types*. How many different types of these exist? We imagine that a multiple-primitive cell is constructed from a simple-primitive cell by the addition of further lattice points. This new lattice must also be a translation lattice. Let $\boldsymbol{r} = r_i\boldsymbol{a}_i$ be a first primitive lattice. If one joins at the end point of $\boldsymbol{p}$, itself not a lattice vector, a further lattice of the same type, then one constructs a lattice consisting of points $\boldsymbol{r}$ and $\boldsymbol{r}' = \boldsymbol{r} + \boldsymbol{p}$. A repetition of the translation $\boldsymbol{p}$ must again lead to a lattice point, if the constructed lattice is to represent a translation lattice. Thus $\boldsymbol{r}' + \boldsymbol{p} = \boldsymbol{r} + 2\boldsymbol{p} = \boldsymbol{r}''$ and therefore $2\boldsymbol{p} = \boldsymbol{r}'' - \boldsymbol{r} = \boldsymbol{r}'''$ is a lattice vector of the first translation lattice, i.e., $\boldsymbol{p}$ must be equal to half a lattice vector of the first lattice. One now selects $\boldsymbol{p}$ so that it lies within the unit cell of the first lattice. This is not a restriction, since each other cell can be considered as an elementary cell. For $\boldsymbol{p}$, we then have the following possibilities:

p = half basic vector, thus $\boldsymbol{p} = \boldsymbol{a}_i/2$ with $i = 1, 2, 3$.

p = half diagonal vector of a basic face of the unit cell, thus $\boldsymbol{p} = (\pm\boldsymbol{a}_i \pm \boldsymbol{a}_j)/2$ with $i \neq j$.

p = half space-diagonal vector of the elementary cell, thus $\boldsymbol{p} = (\pm\boldsymbol{a}_1 \pm \boldsymbol{a}_2 \pm \boldsymbol{a}_3)/2$.

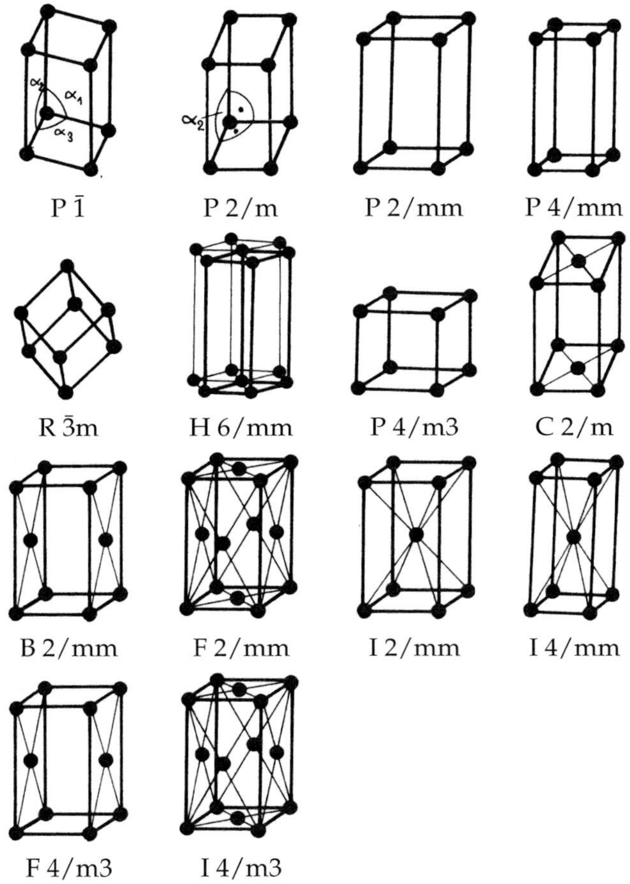

Figure 1.32 The 14 Bravais types, 7 primitive, and 7 multiple primitive.

This operation of construction is called *centering*. One can distinguish a total of seven multiple-primitive lattices, also presented in Fig. 1.32.

The following symbols are used to specify Bravais types:

P: primitive lattice

A, B, C: one-sided face-centered lattice, depending on the orientation of the face parallel to (100), (010), or (001),

R: instead of P in a lattice with trigonal–rhombohedral unit cell,

H: instead of P in a hexagonal lattice with trigonal–hexagonal unit cell,

F: face-centered lattice,

I: body-centered lattice.

These symbols are provided with the Hermann–Mauguin symbols for the holohedral symmetry of the given cell (Fig. 1.32).

1.7.5.2 **Screw Axes and Glide Mirror Planes**

Consider an arbitrary translation lattice with the properties of an *ideal crystal*. In primitive translation lattices the space is empty except for the lattice points. We now imagine that the cells of a general translation lattice are occupied with an arrangement (motif) which is described by further discrete points inside the primitive cells. These are represented by the parameter vectors $\boldsymbol{p}_i$ associated with a property A_i. For example, occupation by the center of gravity of certain types of atoms or by a continuous function $A(\boldsymbol{x})$ is dependent on the position vector such as, e.g., a time-averaged electron density. Which symmetry properties can now appear in such general lattices? Apart from the symmetry operations already discussed for the point symmetry groups, one must now consider the possible combinations resulting from translations with rotations and rotoinversions.

First we consider the combination of rotation axes n with a translation $\boldsymbol{t}$ applied to the end point of an arbitrary vector $\boldsymbol{y}$ in a general translation lattice. The transformation is thus $\boldsymbol{y}' = R_n(\boldsymbol{y}) + \boldsymbol{t}$. We call this a *screw operation* (Fig. 1.33a). It turns out that only the component $\boldsymbol{t}_{\parallel n}$ (parallel to the screw axis) brings anything new. After an n-fold screw operation, any point along the screw axis is displaced by $n\boldsymbol{t}$. Arbitrary repetition creates a chain of equidistant points. The distance between points must be equal to the length a of a lattice vector in the direction of the screw axis or a multiple thereof; thus $n\boldsymbol{t} = p\boldsymbol{a}$, where $p = 0, 1, 2, 3, \ldots, n-1$. We obtain the condition $\boldsymbol{t} = p\boldsymbol{a}/n$. We introduce the symbol n_p for such a screw operation (screw axis). A total of 15 rotation and screw axes exist: 2, 2_1, 3, 3_1, 3_2, 4, 4_1, 4_2, 4_3, 6, 6_1, 6_2, 6_3, 6_4,

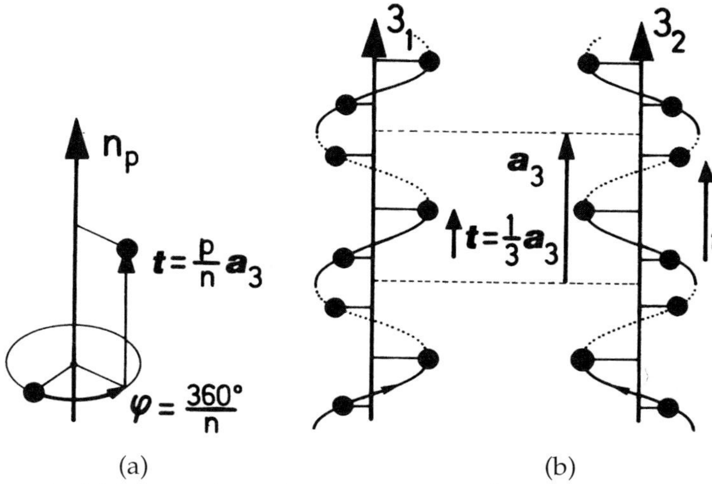

Figure 1.33 (a) Screw axis n_p with screw component $\parallel a_3$; (b) screw operation 3_1 and 3_2 (right and left screw).

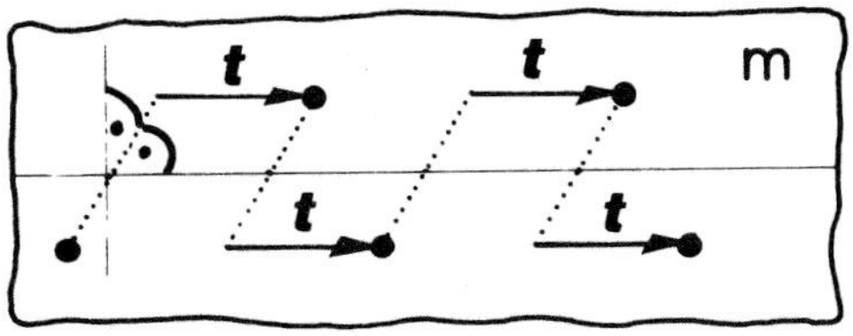

Figure 1.34 Glide plane.

65. The screw axes n_p and $n_{(n-p)}$ are distinguished only by the sense of screw rotation (right and left screws). Examples are presented in Fig. 1.33b.

The combination of translation and rotoinversion only leads, in the case of $\bar{2}$ = m (mirror plane), to new operations with the component $t_{\|m}$ (perpendicular to $\bar{2}$, i.e., lying in the mirror plane). A repetition of the operation $y' = R_m(y)$ results in a zigzag chain of points with a separation of $2\boldsymbol{t}$. These chains must have the period a of a lattice vector, i.e., $\boldsymbol{t} = \boldsymbol{a}/2$. These operations are called *glide planes* (Fig. 1.34). Here, it also suffices to restrict oneself to lattice vectors lying within the elementary cell.

1.7.5.3 **The 230 Space Groups**

We now ask which distinguishable combinations of all the operations discussed so far exist in general translation lattices (with arbitrary motif of the elementary cells). One must also examine the compatibility of the different operations in a similar manner as with the derivation of the point symmetry groups. This work was completed independently by Fedorow (1890) and Schoenflies (1891). The result was 230 distinguishable combinations, the space groups. The symbols used today follow the Hermann–Mauguin symbols for the point symmetry groups. The first symbol represents the Bravais type of the lattice, hence P, R, H, A, B, C, F, I. These are followed by the symbols for the symmetry operations, ordered according to the three prominent viewing directions. Here also, it suffices in most cases only to give the symbol for the generating operation. To avoid alternative interpretations one should correctly note the absence of symmetry operations in a certain viewing direction with the symbol 1. Detailed explanations on the arrangement of the symmetry elements in the space groups and the selection of the origin as well as an overview on equivalent settings and further properties are found in the *International Tables for X-ray Crystallography*.

The important question of how one derives from a space group the appropriate point symmetry group, which describes the macroscopic symmetry of a finite space lattice, is quite simple to answer. In the macroscopic world, screw axes appear as ordinary rotation axes and glide mirror planes as ordinary symmetry planes. Rotoinversion axes are also directly expressed in the symmetry of the point groups. In the macroscopic world an eventual centering of the unit cells seems to remain hidden. However, the Bravais-type as well as the screw axes and glide mirror planes often manifest themselves quite clearly in morphological and other macroscopic properties.

For the determination of the space group of a crystal, knowledge of the point symmetry group, the morphological metric and morphological details, especially the existence or absence of an inversion center, is an advantage. However, modern methods of structure analysis, especially with the aid of X-ray and neutron diffraction techniques, supported by computers and simulations, practically always allow us to disclose the symmetry of the space groups.

1.8
Supplements to Crystal Geometry

Many aspects of crystal geometry could not be taken up in the previous section. It may, however, be useful to point out some special features for the material scientist working in the field of crystal physics.

- *Bravais indices.* According to a suggestion from Bravais one can introduce in the trigonal–hexagonal coordinate system a fourth basic vector $\boldsymbol{a}_1' = -\boldsymbol{a}_1 - \boldsymbol{a}_2$ apart from the three base vectors $\boldsymbol{a}_1$, $\boldsymbol{a}_2$, and $\boldsymbol{a}_3$, produced by a rotation of $\boldsymbol{a}_2$ about $\boldsymbol{a}_3$ by an angle of 120°. The direction of $\boldsymbol{a}_1'$ is symmetry equivalent to $\boldsymbol{a}_1$ and $\boldsymbol{a}_2$. One can thus obtain, e.g., the symbols of symmetry-equivalent faces by the permutation of the first three Miller indices. Since the Miller indices transform as the basic vectors, we have $h_1' = -h_1 - h_2$. From the normal triple indices $(h_1 h_2 h_3)$ we get the Bravais indices $(h_1 h_2 h_1' h_3)$. In order to avoid confusion, one should always specify the reference system with triple indices used for trigonal or hexagonal crystals.

- *Twin formation.* The regular growth of several individuals of a crystal species orientated with respect to one another according to distinct rules is called twinning. Twin formation can be recognized using optic polarization microscopy and X-ray methods, and often macroscopically by re-entrant angles. A careful analysis of the mutual orientations of the parts leads to the corresponding twinning rule that the metric of the first part connects with the metric of the second part. A detailed presentation of twin and domain formation is given in Volume D, *International Tables for Crystallography* (2003).

- *Plane groups, line groups.* If one limits the translation symmetry to two vectors or one vector, one gets 17 or 2 different combination possibilities, respectively, the so-called plane groups and line groups. They play a special role in the projection of space groups onto a plane.

- *Black–white groups, color groups.* The purely geometrical notion of symmetry may be broadened by connecting the geometric symmetry operations with the change of a certain property. The simplest type occurs when one generates a geometric symmetry-equivalent arrangement and simultaneously reverses the sign of a property. Thus one generates from $+1$ the sign -1 and vice versa. Instead of $+1$ and -1, other arbitrary two-valued properties, such as, e.g., black–white, on–off, up–down (spin orientation), can occur. Further broadening of the notion of symmetry leads to the color groups, in which the geometric symmetry operation is connected with a change in color or other multivalent properties, which repeats itself with the order of symmetry of the given operation. A clear representation of the color groups is made possible, e.g., when one attaches a fourth coordinate q to the geometric operation which delivers the value q^{m+1} with the m-fold application of the operation, where, e.g., in the case of the n-fold rotation axis $q^{n+1} = q$ and thus we must have $q^n = 1$. Here, the possible color values are specified by the unit roots $e^{2\pi im/n}$ $(1 \leq m \leq n)$. The associated transformation matrix is then

$$R_n^F = \begin{pmatrix} v_{11} & v_{12} & v_{13} & 0 \\ v_{21} & v_{22} & v_{23} & 0 \\ v_{31} & v_{23} & v_{33} & 0 \\ 0 & 0 & 0 & e^{i\alpha} \end{pmatrix}$$

with $e^{i\alpha} = \cos\alpha + i\sin\alpha$, $i = \sqrt{-1}$ and $\alpha = 2\pi/n$.

There exist a total of 58 real black–white symmetry groups apart from the ordinary 32 PSGs, as well 32 further groups, the so-called gray point symmetry groups, created from the ordinary PSGs by the simultaneous inclusion of a negative property to a positive property. Thus we have a total of 122 distinguishable PSGs. The space groups have 1651 distinguishable cases, the Heesch–Shubnikov groups.

1.9 The Determination of Orientation with Diffraction Methods

The orientation of a macroscopic crystal, with a well-developed natural morphology, can very often be found with the help of the angles measured between neighboring faces or face normals. Additional information is obtained by studying thin slices under the polarization microscope. Finally, other properties, such as the propagation velocity of elastic waves in a certain direction can contribute to the determination of orientation when the given elastic properties of the crystal are already known. Further details will be discussed in the following sections. More precise statements and with fundamentally higher certainty are obtained with the aid of X-ray methods. These, however, must be directly applicable to large crystals, or one must try to produce very small crystals of dimensions less than 1 mm as used in X-ray structure analysis, with a fixed relation to the object under study. This measure, however, requires special care. Thus, we will only outline the two most important methods for large crystals. These are the *Laue method* and the *Bragg method*.

In the Laue method, discussed at length in many textbooks, a beam of non-monochromatic X-rays (essentially the spectrum of the Bremsstrahlung) is incident on a probe and one observes the scattered reflections in different directions, each of which can be assigned to a certain assembly of lattice planes. This occurs with the aid of photographic films or sensitive detector systems, which also allow automatic evaluation. The respective gnomonic or stereographic projections of the normals of the lattice planes derived from the analysis allow us to determine the orientation of the crystallographic reference system of the object. The preferred method is to conduct the investigation in the transmission technique because, in general, it delivers more and sharper diffraction reflections than in the reflexion technique.

In the Bragg method one observes the diffraction in reflection with monochromatic radiation. The crystal is rotated in small steps about different axes, at a fixed direction of the primary X-ray beam, until the first reflection occurs with sufficient intensity. From the Bragg condition $2d(\boldsymbol{h})\sin\theta = \lambda$, where λ is the wavelength of the radiation and 2θ is the diffraction angle (angle between the incident beam and the reflected beam), one gets, because of $d(\boldsymbol{h}) = 1/|\boldsymbol{h}|$, not only the orientation of the given lattice plane, but also the associated lattice plane spacing. If one succeeds in obtaining further reflections after directed rotation or tilting, then the orientation of the crystallographic reference system can be fixed. If the crystal possesses well-developed faces, one can measure the d-value at these without further manipulation, thus considerably simplifying the work. Nowadays, automatic equipment is available for this task.

2
Sample Preparation

The success of an experimental investigation often depends crucially on whether suitable objects are available. Therefore, special attention must be paid to the production and preparation of the specimens. In the next two sections we will discuss established methods for the preparation and orientation of crystals. Prospective material scientists should make themselves familiar with these methods.

2.1
Crystal Preparation

Specimens for measurement purposes as well as discrete devices for various applications are mainly required in the form of thin plates, thin rods, or rectangular parallelepipeds. The production of these and also more complicated forms out of a crystal blank takes place through the following processes: cleaving, sawing, drilling, turning on a lathe, grinding, and polishing.

(a) *Cleaving* when a cleavage exists with a suitable orientation of the cleavage planes. Many crystals, such as, e.g., alkali halides of NaCl type, CaF_2, diamond, mica, MoS_2, and especially crystals with layered structures, possess a cleavage or a direction of preferred crack tendency. Cleavage planes are always crystallographically prominent faces parallel to net planes of large occupation density. The cleaving process must be practised with great care. Soft crystals can be cleaved with a razor blade, harder crystals with a stable blade, whereby the blade is held parallel to the cleavage plane. When cleaving soft crystals such as guanidinium iodide or gypsum, the blade is slowly pressed through the specimen, whereas cleaving harder crystals, such as, e.g., LiF or CaF_2, requires a short and fast impact. The cleavage can be easily guided in the desired direction by making a fine notch along the cleavage plane. In any case, a massive base plate is of advantage.

(b) *Sawing with a thread saw.* When working with a thread saw, the mechanical stress on the crystal is practically negligible as opposed to cleaving. The thread saw should be used for the stress-free preparation of specimens, as well as with all specimens possessing good solubility or where chemical dissolution is possible. The thread, moistened with a suitable solution, is

Physical Properties of Crystals. Siegfried Haussühl.

ISBN: 978-3-527-40543-5

moved automatically back and forth by commercial equipment and the crystal clamped in a moveable holder is pressed with an adjustable force against the thread. Cooling occurring during endothermal dissolution processes must be suppressed as far as possible when working with delicate crystals. This is done, for example, by using an almost saturated solution and by stripping off drops of liquid on the thread immediately in front of the crystal. This is especially important with highly volatile solvents such as ethyl alcohol or acetone. Difficult to dissolve crystals, such as calcite ($CaCO_3$) or quartz (SiO_2), can be cut using diluted hydrochloric acid or hydrofluoric acid. Suitable sawing materials are twisted natural fibers. Synthetic fibers can be used with neutral or alkaline solutions. Metal wires, e.g., made from tungsten (∅ 0.1–0.2 mm) are still well proven and work with a water solution, especially with crystals that are hardly dissoluble. In the case of very hard crystals, such as silicates or metals, one uses a solution consisting of a suspension of corundum or diamond powder (grain size ca. 10 μm) in a highly viscous liquid such as paraffin oil or dextrin boiled in water.

(c) *Sawing with a blade saw* (normally diamond coated). The usual blade saw with outside peripheral cut and also such saws which cut with the inner periphery of a circular ring to improve the stability of the blade are only suitable for materials of low mechanical and thermal sensitivity. In any case, one must work with minimal sawing pressure and ample cooling liquid.

(d) *Spark erosion techniques.* Spark discharge methods can be used on some materials, such as alloys, to preferentially divide and separate individuals. This method also plays an important role in the smoothing of surfaces and can even achieve the quality of polishing.

(e) *Laser ablation.* Some substances allow a targeted ablation of material by laser bombardment; mainly used on very small objects in microtechnology applications.

(f) *Drilling and turning.* As in wood and metal working, crystals with different mechanical properties can also be machined with a drill or lathe. In general, sharp and fine diamond cutting tools should be used. Furthermore, when working on the lathe, one has the possibility of using grinding paste and emery paper. Hollow drills have also been found useful in the machining of cylindrical specimens of various thicknesses. Just like drilling in glass, these must operate with a plentiful supply of abrasives and cooling fluid. The drill core is then the desired specimen. Other tools such as fret saws and files are only used in special cases.

(g) *Surface grinding.* This requires flat grinding plates, which one can produce with high precision preferentially from glass or brass. Other grinding forms, e.g., with concave or convex surfaces can be produced in a similar manner. A suitable grinding plate should be provided for each grain size of the abrasive and if possible a separate working surface so that coarse grains do not mix with the fine grains in the grinding process. Thus a high degree of cleanliness is necessary. Flat grinding plates with diamond coatings, where the danger of carrying over coarse grains is minimal, are also commercially available. For the preparation of single specimens it is still appropriate to work by hand. Only in series production it is worthwhile to work simultaneously with several objects on the grinding machine. When working by hand, the expenditure of fastening the piece in a grinding holder can be spared as long as the pieces are large enough. Small crystals and exacting work demand the use of a grinding holder, a brass ring, where the underside is machined flat. The specimen is imbedded in the ring with the aid of a suitable cement (beeswax, synthetic resins which harden or melt at low temperatures depending on the thermal stability of the crystal), so that the underside of the ring, which limits the grinding process, fixes the desired flat surface. These holders can also be fastened in automatic grinding machines.

The grain size of the abrasive must be relatively homogenous to prevent deep mechanical damage to the surface. For this reason, grinding with coarse grain should be avoided as far as possible. In any case, one should use ample amount of grinding fluid. Water, propanol, and ethyl glycol have been found to be especially useful. The grinding process must be continuously controlled. For this purpose one uses precision angles made of steel, goniometers, straightedges, micrometers, and thickness gauges for the quantitative determination of thickness. These instruments can measure, or keep to specifications, angles to an accuracy of a few arcseconds and lengths to an accuracy of $0.5\,\mu$m. Employing optical interferometer devices a still better precision can be achieved.

Polishing is often required as the last step in the preparation. The methods used are mainly those proven in optics. The process can be considered as a refined form of grinding when we work with polishing plates made of plastic or pitch (for optical purposes). The only polishing fluids one can use are those which do not attack the polishing plate and have little aggressive effect on the specimen. Ethyl alcohol or higher alcohols are often used for water-soluble crystals. The recommended polishing agent is a fine wash of suspended chromium oxide, iron oxide, cerium oxide, aluminum oxide, or diamond paste. The polishing process is also carried out in several steps, from coarse- to fine-grained agents. Medium-hard crystals such as fluorite require a grain size of about $0.5\,\mu$m. The special measures needed to achieve an optic polish of the highest quality, e.g., with a flatness of 1/20 of the wavelength in the visible spectrum, must be learned in special training courses. They, however, are only more refined methods of those discussed above.

In the technical application of crystals, the method of preparation can be considerably simplified when one succeeds in getting the desired form already during the manufacturing process.

2.2 Orientation

The term 'orientation' is used in the *passive* sense for the description of the geometrical position of a crystal referred to a crystallographic reference system, and in the *active* sense for the alignment of certain faces to the crystallographic reference system by cutting or grinding. We can, for example, search for the orientation of the basic vectors of a prominent crystallographic reference system by studying morphological details or by means of X-ray analysis (Laue method, Bragg method). The investigation of physical properties, especially with optical and piezoelectric methods, which we will come to later, can be of great help in this regard. At present, we will assume that the orientation of the crystal, in the passive sense, is known and the task is now to prepare an oriented specimen, e.g., a parallelepiped, with specified faces. We proceed as follows: first, instead of the crystallographic reference system defined by the three vectors $\boldsymbol{a}_i$, which is normally associated with one of the seven prominent crystallographic systems due to the given minimum symmetry, we introduce a new reference system, the crystal-physical reference system. This is a Cartesian system with basic vectors $\boldsymbol{e}_i$ attached to the vectors $\boldsymbol{a}_i$ according to the following convention:

$$\boldsymbol{e}_2 \parallel \boldsymbol{a}_2^*, \quad \boldsymbol{e}_3 \parallel \boldsymbol{a}_3, \quad \boldsymbol{e}_1 \parallel \boldsymbol{e}_2 \times \boldsymbol{e}_3$$

($\boldsymbol{a}_2^*$ is normal to the surface (010) and is the second basic vector of the reciprocal system!)

We use, practically without exception, this Cartesian reference system for the description of physical properties as opposed to morphological properties and diffraction phenomena. If another orientation of the reference system is selected, this must be specially noted. It is often convenient to specify the faces in the crystal-physical reference system by their Miller indices. To distinguish between the crystallographic Miller indices, we use a symbol with a dash, e.g. $(110)'$ means a face whose normal bisects the angle between $\boldsymbol{e}_1$ and $\boldsymbol{e}_2$.

Now to get from an arbitrary crystal, say a melt boule, where often no morphological details are recognizable, to a defined reference base for further orientation work, it is useful to introduce another reference system which we shall call the laboratory system. This system with the Cartesian vectors $\boldsymbol{e}_i^{\mathrm{L}}$ is randomly assigned to the object, however, matched as far as possible to the crystal-physical system. This is done by grinding two arbitrary surfaces $\mathrm{F}_1^{\mathrm{L}}$ and $\mathrm{F}_2^{\mathrm{L}}$ on the object, preferentially perpendicular to each other. Let $\boldsymbol{e}_1^{\mathrm{L}}$ run parallel to the normal of the surface $\mathrm{F}_1^{\mathrm{L}}$ and $\boldsymbol{e}_3^{\mathrm{L}}$ parallel to the intercepting edge of both faces. $\boldsymbol{e}_2^{\mathrm{L}}$ is thus perpendicular to both vectors in such a way that just like in the $\boldsymbol{e}_i$ system, a right-handed system is created. The situation is sketched in Fig. 2.1.

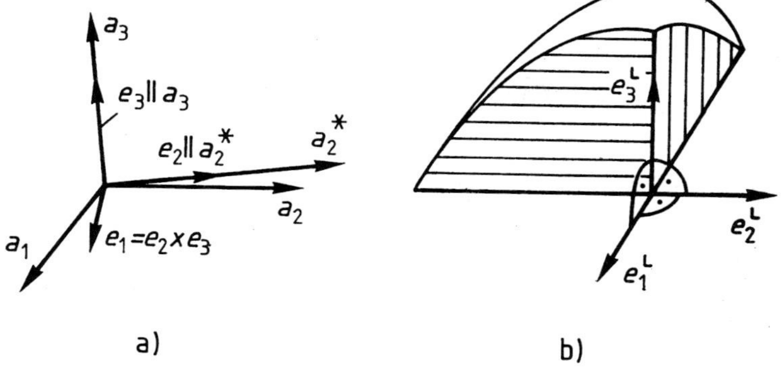

Figure 2.1 (a) Orientation of the Cartesian crystal–physical reference system in the crystallographic basic system. (b) The laboratory reference system.

To prepare the oriented specimen it is now practical to find the relationship between both Cartesian systems in order to carry out quickly and clearly all the required calculations. Thus one must set up the system $e_i = u_{ij}e_j^{\mathrm{L}}$ (where $i = 1,2,3$) with the coefficients u_{ij}. This occurs with the help of the Laue method or the Bragg method.

The task is often to represent the unit vector of a face normal $\boldsymbol{h} = h_i\boldsymbol{a}_i^*$ or the unit vector of a lattice line $\boldsymbol{u} = u_i\boldsymbol{a}_i$ in the crystal-physical system. For this purpose we calculate the unit vectors $\boldsymbol{h}/|\boldsymbol{h}| = u_{1i}\boldsymbol{e}_i = \boldsymbol{e}_1'$ and $\boldsymbol{u}/|\boldsymbol{u}| = v_{1i}\boldsymbol{e}_i = \boldsymbol{e}_1'$, respectively. From the definition of the crystal-physical system $\boldsymbol{e}_1 = \boldsymbol{e}_2 \times \boldsymbol{e}_3$, $\boldsymbol{e}_2 = \boldsymbol{a}_2^*/|\boldsymbol{a}_2*|$ and $\boldsymbol{e}_3 = \boldsymbol{a}_3/|\boldsymbol{a}_3|$ and with the help of the known scalar products $\boldsymbol{h}\cdot\boldsymbol{e}_i$ and $\boldsymbol{u}\cdot\boldsymbol{e}_i$, we have

$$\begin{aligned}
u_{11} &= \frac{1}{Va_2^*|\boldsymbol{h}|}(h_1a_3 - h_3a_1\cos\alpha_2),\\
u_{12} &= \frac{1}{|\boldsymbol{h}|}(h_1a_1^*\cos\alpha_3^* + h_2a_2^* + h_3a_3^*\cos\alpha_1^*),\\
u_{13} &= \frac{h_3}{a_3|\boldsymbol{h}|},\\
v_{11} &= \frac{1}{V^*a_3|\boldsymbol{u}|}(u_1a_2^* - u_2a_1^*\cos\alpha_3^*),\\
v_{12} &= \frac{u_2}{a_2^*|\boldsymbol{u}|},\\
v_{13} &= \frac{1}{|\boldsymbol{u}|}(u_1a_1\cos\alpha_2 + u_2a_2\cos\alpha_1 + u_3a_3).
\end{aligned}$$

a_i, α_i, and V and a_i^*, α_i^*, and V^*, respectively, are the values of the metric of the basic system and the reciprocal system. These relationships also allow, without further requisites, additional measurements, referred to the selected crystallographical system, in the direction of the normals of natural crystal faces or along natural crystal edges.

All the required angle data for a special cut can now be expressed in the laboratory system and controlled by simple angle measurements during the grinding process. The practical procedure is explained below with the help of three typical examples of melt boules of cubic crystals, which often show no morphological details with respect to the determination of orientation.

(a) *Melt boule of an alkali halides of NaCl type.* These crystals possess an excellent cleavage of the cubic faces $\{100\}$. We therefore try to discover, by carefully cleaving at the edge of the boule, a cleavage face recognizable by a flat, mirror-smooth fracture. We then search for a second surface perpendicular to this face. The normal to these faces defines the crystal-physical reference system. The laboratory system, in this case, is for reasons of practicality, also fixed by the cleavage faces ($\boldsymbol{e}_i = \boldsymbol{e}_i^{\mathrm{L}}$).

(b) *Melt boule of CsBr or CsI.* These crystals exhibit no well-developed cleavage. If one inserts the melt boule in a saturated aqueous solution of CsBr or CsI and cools the system down to several degrees, one observes after a few hours mirror-smooth surface elements of the rhombododecahedron $\{110\}$. The cubic basic vectors bisect the angles between the normals of such surface elements, which form angles of $90°$. Thus the orientation of the crystal-physical system is known.

(c) *Melt boule of AgCl or AgBr.* These crystals do not possess a well-developed cleavage at room temperature nor sufficient solubility in water. In a saturated ammoniacal solution of AgCl or AgBr, they form, after a short time of evaporation, mirror-smooth surface elements of the cube faces $\{100\}$. As in (a) these allow a direct determination of crystal orientation.

The accuracy of the determination of crystal orientation can be easily checked using X-ray techniques such as the Laue method or the Bragg method.

We now imagine that the fixed crystal-physical reference system is established in space in the crystal. We thus speak of an initial reference system. During measuring processes or in technical applications, the crystal can experience certain changes which often have an effect on the quantities used to fix the orientation. Normally, we assume that the influence of such processes is so small that the orientation of the reference system, as well as the properties of the crystal, measured in this system, is only affected to a nonmeasurable extent. If this is not true, then one requires a careful analysis of the special situation, as in the case of nonlinear elasticity (see Section 4.6.3), where primary mechanical deformations are superimposed by secondary deformations.

3 Definitions

3.1 Properties

The physical phenomena of matter are made apparent in experiments and their quantitative formulation in measurements. The results of such measurements can often lead to direct statements concerning the so-called properties of the given substance. We now want to look more closely at the term property. For this purpose, we imagine those actions, embracing the group of independent or *inducing quantities*, which we can arbitrarily perform on a probe. Let these be specified by the symbol $A_{j'j''j'''\ldots}$, or in short notation A_j, where the indices j', j'', and so on are introduced to more precisely characterize the quantities. Such independent quantities are, for example, volume, temperature, temperature gradients, hydrostatic pressure, pressure gradients, general mechanical stress states, velocity, rotational velocity, electric, and magnetic field strengths.

The inducing quantities give rise to effects which we measure with the aid of dependent or *induced quantities* $B_{i'i''i'''\ldots}$, in short notation B_i. Examples for B_i are: caloric heat content, mechanical deformation, heat flow density, mass flow density, electric current density, electric polarization, and magnetization. The relationship between inducing and induced quantities is described by

$$B_i = f_{i;jkl\ldots}(A_j, A_k, A_l, \ldots).$$

Consequently, the function $f_{i;jkl\ldots}$ specifies those properties of the body which under the action of the quantities A_j, A_k, A_l, and so on produce the quantities B_i. This concept for the definition of properties has been found to be sufficient for most macroscopic phenomena as we shall see in the following.

There also exists a further group of properties, which are derived as functions $g(f_1, f_2, \ldots)$ of certain f_i (here, e.g., f_1, f_2 ...) such as light velocity and sound velocity. We shall call These properties *derived properties*. Sometimes the question has to be settled as to whether the role of the inducing and induced quantities, which may be assigned to each other in pairs, e.g., mechan-

Physical Properties of Crystals. Siegfried Haussühl.

ISBN: 978-3-527-40543-5

ical stress and mechanical deformation or electric field strength and current density, is interchangeable. The formal description often allows such inversions; however, the physical realization of the inversion is often not easy to achieve, as one knows from the example of mechanical deformation and mechanical stress. We will return to this matter in detail in the discussion of actual properties.

As a simple example for the concept of inducing and induced quantities let us consider *specific heat* and *electric conductivity*. The specific heat C_p (per g, at constant pressure) connects an arbitrary temperature rise ΔT, as an inducing quantity, to the consequent appearance of an increase in the caloric capacity ΔQ_p (per g, at constant pressure) according to

$$\Delta Q_p = C_p \Delta T.$$

Here, f is a linear function of the independent variable ΔT. This association is approximately valid in a small temperature interval. An exchange of inducing and induced quantities is not only formally but also physically feasible.

The electrical conductivity s establishes the connection between electric charge current density $\boldsymbol{I}$ (unit: charge per s and mm^2) and the electric field strength $\boldsymbol{E}$ as the inducing quantity according to

$$\boldsymbol{I} = s\boldsymbol{E}.$$

This form of Ohm's law is approximately correct for isotropic, i.e., directionally independent, conductivity. With crystals of lower symmetry, as e.g., arsenic, antimony or bismuth (PSG $\bar{3}$m) or with $LiIO_3$ (PSG 6) one observes in the direction of the threefold or sixfold axis other values of electric conductivity as in the direction perpendicular to these. Related to the crystal-physical system, the conductivity along the three- or sixfold axis is described by $I_3 = s_{33}E_3$ and perpendicular to these by $I_1 = s_{11}E_1$. We thus find an anisotropy, represented by the introduction of the mutually independent components of the electric field E_i and of the current density vector I_i ($i = 1, 2, 3$). We will discuss in detail the question of how the anisotropy looks like in general, i.e., which relationship exists between $\boldsymbol{I}$ and $\boldsymbol{E}$ in an arbitrary direction.

The functions f and therefore the properties are divided into two groups:

I. Tensor Properties. A multilinear relationship exists between B_i and the A_j of the type

$$B_i = f^0_{i;j} A_j + f^0_{i;jk} A_j A_k + f^0_{i;jkl} A_j A_k A_l + \cdots$$

(One sums according to the summation convention.) The $f^0_{i;jkl\ldots}$ are constant coefficients and all nonaffected inducing quantities are held constant. This representation corresponds to a Taylor series in B_i according to

$$B_i = \sum_\nu \frac{1}{\nu!} \frac{\partial^\nu B_i}{\partial A_j \partial A_k \cdots} A_j A_k \cdots,$$

with the characteristic that no constant term is present. The expansion occurs at the zero value of all A_j.

The coefficients $f^0_{i;jkl\ldots}$ with fixed indices $jkl\cdots$ represent a certain property. In the case of a single index j we are dealing with a property of the first order, with index pairs jk we are dealing with properties of the second order, and so on. Many mechanical, electrical, optical, and thermal properties, also those of a very complicated nature, belong to the tensor properties.

II. Nontensor Properties. Here, the relationship between inducing and induced quantities is more complicated. Nontensor properties are, e.g., growth properties, rate of dissolution (etching behavior), boundary surface properties, plasticity, abrasive hardness, scratch hardness, and properties connected with energetic activation thresholds, e.g., in emission and absorption processes.

3.2 Reference Surfaces and Reference Curves

Many properties, especially the tensor properties, possess an anisotropy, which can be represented by a surface in space.

In the case of a simple directional dependence, the reference surface furnishes an overview of the complete anisotropy. We obtain the reference surface as the entirety of the end points of radius vectors $\boldsymbol{r}$, spreading from a fixed point, with lengths equal to the value of the property for the given direction. As an example we mention the *rate of dissolution* which describes the etching behavior on a given face of a crystal in a distinct solvent (see Fig. 3.1; the measurement method is described in Section 6.2.

In the same manner as with the rate of dissolution, one can also represent, e.g., abrasive strength, indentation hardness (a measure for the plasticity), and the tensile strength of thin cylinders with the aid of a simple reference surface. A special reference surface is generated by a freely growing crystal through its outer boundary surface, which represents the mean velocity of crystallization in any direction.

If there exists, for each measurement direction, several values of the given property, then one gets a multishell reference surface. For example, one needs for the representation of the velocity of light in crystals a double-shelled reference surface because for each propagation direction two different values of the velocity of light exist. In the case of sound velocity, a triple-shell surface is required because for each propagation direction three different values of sound velocity are possible. We will handle these and other examples in more detail later.

For some properties, in which several directions simultaneously come into play as, for example, with sawing velocity and scratch hardness, a compli-

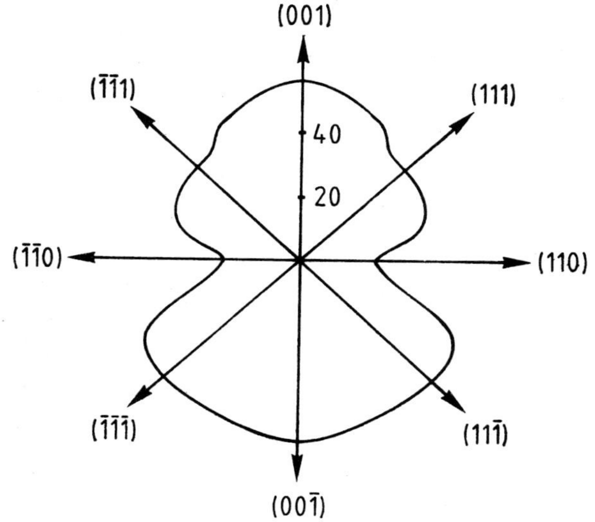

Figure 3.1 Anisotropy of dissolution velocity of $\mathrm{Sr(HCOO)_2 \cdot 2H_2O}$, PSG 22, in water at 293 K. The radial distances give the loss in weight in mg per 100 ml $\mathrm{H_2O}$ in directions within the zone $[1\bar{1}0]$. A distinct minimum is observed on the principal growth faces $\{110\}$. The anisotropy reflects the symmetry of the twofold axis along $[001]$. The lack of a center of inversion is immediately recognizable.

cated representation in the form of reference curves must be chosen, because the measurement results depend not only on the orientation of the crystal face but also on the cutting direction of the saw or the direction of scratching (Section 6.3).

3.3 Neumann's Principle

The consequences of crystal symmetry on the physical properties, in which the influence of boundaries can be neglected, are governed by a fundamental postulate of crystal physics, known as Neumann's principle (F. Neumann, 1798–1895):

"The space symmetry of the physical properties of a crystal cannot be less than the structural symmetry of the crystal."

As a justification, it suffices to note that the given property is a consequence of the atomic arrangement of the lattice particles and their bonding, i.e., the electronic states. Other reasons for the occurrence of the properties do not exist, as long as we look at the empty space, in which the material is embedded, as an intrinsically isotropic medium. This assumption is essentially fulfilled for our purposes. In all macroscopic properties, the point group symmetry replaces the position of the structural symmetry described by the space group symmetry. This is due to the fact that translation symmetry is macroscopically not immediately recognizable and the special operations of screw axes and glide mirror planes, in the macroscopic sense, act as ordinary symmetry axes or mirror planes. From Neumann's principle we learn immediately that the reference surfaces must at least possess the point symmetry group of the given crystal as long as boundary-independent properties are considered. The mathematical formulation for this fact is as follows: the reference surfaces are invariant under the transformations of the point symmetry group.

Thus each continuous reference surface is to be approximated to an arbitrary degree by symmetry-fitted fundamental polynomials of the given symmetry group (Exercise 2). In order to make use of this relationship, it is important, as far as possible, to define the properties so that they are boundary independent. Furthermore, we always assume that the observed properties fulfill this requirement if no special indications are given for another situation. When nothing else is said, we also assume that the properties of the crystal, and therefore also their intrinsic symmetry are not measurably changed by the action of the inducing quantities. Thus for the description of the phenomena we will also view the crystal-physical reference system as fixed and invariable.

3.4 Theorem on Extreme Values

In directions of symmetry axes with $n \geq 3$ the properties take on relative extreme values, i.e., in all sufficiently adjacent directions we find either larger or smaller values of the given properties, except for the isotropic case which does not concern us here. In directions of twofold axes, the properties take on either extreme values or saddle point values, i.e., the reference surface possesses at such points along the principal curvature lines perpendicular to one another relative extreme values that can also be of a different character (maximum or minimum). In all directions within a symmetry plane, each property takes on a relative extreme value when passing through the principal curvature line perpendicular to the symmetry plane. There is no general rule for extreme values in other directions or the position of absolute extreme values.

As proof we imagine that the reference surface is differentiable to sufficient order. Then in each point there exist two defined principal curvatures, a maximum and a minimum. We now replace the reference surface by an elliptic or hyperbolic paraboloid tangential to the surface in the given point and with principal curvatures corresponding to those of the surface (*Dupin indicatrix*).

In the coordinate system, whose basic vectors e'_1 and e'_2 are given by the tangents along the directions of principal curvature (e'_3 is perpendicular to the tangential plane), the paraboloid is given by

$$ax_1'^2 + bx_2'^2 + cx_3' = 0.$$

In the case of n-fold axes with $n \geq 3$, the paraboloid has the form of a rotation paraboloid ($a = b$). All points of the reference surface in a certain neighborhood of the intercept point of an n-fold axis lie approximately on the paraboloid, that again, in the point of contact, is to be approximated by a sphere of a radius corresponding to the principal curvature. This radius is either smaller or larger than the length of the radius vector of the reference surface, i.e., the radius vector leads either to a maximum or a minimum. In the intercept point of a twofold axis, the axis e'_3 of the Dupin indicatrix runs parallel to the twofold axis. This is compatible with the existence of an extreme value or a saddle point. For directions within a symmetry plane, the vectors e'_1 and e'_2 of the Dupin indicatrix lie parallel and perpendicular, respectively, to the symmetry plane. Thus it follows, as stated, that when passing through the principal curvature line perpendicular to the symmetry plane there must appear a relative extreme value for all directions in the symmetry plane. Furthermore, in the directions within a symmetry plane there exist an even quantity of relative extreme values with the same number of maxima and minima. There exists at least one maximum and one minimum.

3.5 Tensors

As previously mentioned, many important physical properties can be described by *tensors*. In the next few sections we will get to know some properties of tensors, which will be extremely useful for further work. The beginner should familiarize himself with this mathematical tool as soon as possible. Often the endeavor to evoke a picture of the nature of tensors is futile and leads to a certain aversion. For this reason, such attempts should not be stimulated at the beginning. Rather it is recommended, at first only to pay attention to the definitions that follow and the rules of calculation resulting from these. We will then derive those quantities which provide us a useful conception of tensors. As a first step we will introduce tensors on the basis of their transformation properties. For this purpose we will recapitulate the behavior of basic vectors and coordinates of the position vector when changing a reference system. The introduction of a new reference system with the basic vectors $\boldsymbol{a}'_i$ takes place advantageously with the help of the transformation matrix (u_{ij}) which generates the new basic vectors from the old according to $\boldsymbol{a}'_i = u_{ij}\boldsymbol{a}_j$. The reverse transformation as well as the transformation of the position vec-

Table 3.1 Formulae for the transformation into a new reference system. The coordinates of the position vector in the reciprocal system are designated by x_i^* for the sake of uniformity. In crystallography normally the symbols h_i are used. (U_{ij}), the inverse matrix of (u_{ij}), is obtained by $U_{ij} = (-1)^{i+j} A_{ji}/|u_{ij}|$, where A_{ji} is the adjunct determinant resulting from the transformation matrix after dropping the jth row and ith column. $|u_{ij}|$ is the determinant of the transformation matrix.

$\boldsymbol{a}_i' = u_{ij}\boldsymbol{a}_j$	$\boldsymbol{a}_i = U_{ij}\boldsymbol{a}_j'$	$x_i' = U_{ji}x_j$	$x_i = u_{ji}x_j'$
$\boldsymbol{a}_i^{*\prime} = U_{ji}\boldsymbol{a}_j^*$	$\boldsymbol{a}_i^* = u_{ji}\boldsymbol{a}_j^{*\prime}$	$x_i^{*\prime} = u_{ij}x_j^*$	$x_i^* = U_{ij}x_j^{*\prime}$

tors in the old and new systems and also in the associated reciprocal systems is compiled in Table 3.1 (refer to Section 1.6 for the derivation of the formulae).

Quantities that transform as the basic vectors of the initial system with the matrix (u_{ij}) are called covariant, and those that transform as the basic vectors of the associated reciprocal system with the matrix (U_{ji}) are called contravariant. The coordinates of the position vector in the basic system are the contravariant coordinates, and the coordinates of the same position vector in the reciprocal system are the covariant coordinates. A simplification of the situation occurs when Cartesian reference systems are used. We consider the transformation $\boldsymbol{e}_i' = u_{ij}\boldsymbol{e}_j$ with the reverse $\boldsymbol{e}_i = U_{ij}\boldsymbol{e}_j'$. Then we have

$$\boldsymbol{e}_i' \cdot \boldsymbol{e}_j = u_{ij} = \boldsymbol{e}_j \cdot \boldsymbol{e}_i' = U_{ji}.$$

Thus in Cartesian systems, covariant and contravariant transformations are not distinguishable. This is a good reason to prefer Cartesian reference systems in practical crystal physics. Transformation matrices with the above properties are called unitary ($\boldsymbol{U} = (\boldsymbol{U}^{-1})^T$).

We now consider the ensembles of quantities $t_{ijk...s}$, where each individual attribute, labeled by the index positions $i, j, k \ldots s$, runs through a certain range of values. An example for such quantities is index cards of a file system, as used, for example in animal husbandry. The first index is provided, let us say, for the date of birth, the second for gender, the third for the breed, and so on. Under all these varieties there exists a marked group, whose members, also called elements, exhibit a special internal relationship. We are dealing with tensors. The individual quantities $t_{ijk...s}$ are called *tensor components*. We specify the ensemble of components, the tensors, by $\{t_{ijk...s}\}$. If the tensor components possess m index positions, then we are dealing with a *tensor of the mth rank*. The range of values of all indices covers the numbers $1, 2, 3, \ldots, n$, where n is the dimension of the associated space. The tensors are thus assigned to an n-dimensional space, which, for example, is spanned by the Cartesian vectors $\boldsymbol{e}_i$ ($i = 1, 2, \ldots, n$). The tensors distinguish themselves from the other quantities by their transformation behavior when changing the reference system. The definition of tensors in this way may, at first, seem artificial. We will see, however, that just this property of tensors is of fundamental importance for all further work, especially for direct applications. The definition reads as follows: Ensembles, whose components $t_{ijk\ldots s}$ transform like the corresponding coordinate products during a change of the reference system, are tensors. With a covariant transformation of all index positions, the components $t_{ijk\ldots s}$ are assigned to the product $x_i^* x_j^* x_k^* \cdots x_s^*$, and with a contravariant transformation they are assigned to the product $x_i x_j x_k \cdots x_s$. In principle, a certain transformation behavior (covariant or contravariant) is allowed for each individual index position. When changing the reference system $t_{ijk\ldots s}$ converts in

$$t'_{ijk\ldots s} = u_{ii^*} u_{jj^*} u_{kk^*} \cdots u_{ss^*} t_{i^*j^*k^*\ldots s^*}$$

with a covariant transformation and to

$$t'_{ijk\ldots s} = U_{i^*i} U_{j^*j} U_{k^*k} \cdots U_{s^*s} t_{i^*j^*k^*\ldots s^*}$$

with a contravariant transformation.

One sums over all indices $i^*, j^*, k^*, \ldots, s^*$ from 1 to n. This results directly from

$$x'_i x'_j x'_k \cdots x'_s = U_{i^*i} U_{j^*j} U_{k^*k} \cdots U_{s^*s} x_{i^*} x_{j^*} x_{k^*} \cdots x_{s^*}$$

and the corresponding expression for the covariant coordinate product. The indices primed with a star only serve to distinguish the indices and are not to be confused with those used to specify the quantities of the reciprocal system. In mixed variant transformations, the component of the corresponding transformation matrix is to be inserted for each index position.

The inverse transformation is carried out in an analogous manner. We have

$$t_{ij\ldots s} = U_{ii^*} U_{jj^*} \cdots U_{ss^*} t'_{i^*j^*\ldots s^*}$$

for the covariant transformation and

$$t_{ij\ldots s} = u_{i^*i} u_{j^*j} \cdots u_{s^*s} t'_{i^*j^*\ldots s^*}$$

for the contravariant transformation.

If we work in Cartesian reference systems, which we will do in the following almost without exception, no difference between both transformation methods exists. Moreover, the introduction of general coordinates, such as cylindrical or polar coordinates, brings advantages only in exceptional cases.

In practical work we are always confronted with the question of whether a tensorial connection exists between inducing and induced quantities. For example, are we dealing with tensor components with quantities s_{ij} obeying a linear relationship $I_i = s_{ij} E_j$ between the components of the current density vector and the electric field? Situations like these can be elegantly handled with the help of the following theorem.

3.6 Theorem on Tensor Operations

If in an arbitrary reference system the mth rank tensor $p_{ij\ldots s}$ is connected to the nth rank tensor $q_{\alpha\beta\ldots\sigma}$ according to

$$p_{ij\ldots s} = r_{ij\ldots s;\alpha\beta\ldots\sigma} q_{\alpha\beta\ldots\sigma}$$

by the quantities $r_{ij\ldots s;\alpha\beta\ldots\sigma}$, which carry a total of $(m+n)$ index positions, then the ensemble $\{r_{ij\ldots s;\alpha\beta\ldots\sigma}\}$ also represents a tensor. The proof is as follows: We imagine having performed a transformation in a Cartesian basic system with the basic vectors $\boldsymbol{e}_i' = u_{ii^*}\boldsymbol{e}_{i^*}$. The relation given above then reads

$$p'_{ij\ldots s} = r'_{ij\ldots s;\alpha\beta\ldots\sigma} q'_{\alpha\beta\ldots\sigma}.$$

We have

$$p'_{ij\ldots s} = u_{ii^*}u_{jj^*}\cdots u_{ss^*}p_{i^*j^*\ldots s^*}.$$

For $p_{i^*j^*\ldots s^*}$ we insert the above expression and obtain after exchanging indices

$$p'_{ij\ldots s} = u_{ii^*}u_{jj^*}\cdots u_{ss^*}r_{i^*j^*\ldots s^*;\alpha^*\beta^*\ldots\sigma^*}q_{\alpha^*\beta^*\ldots\sigma^*}.$$

Now we substitute the components $q_{\alpha^*\beta^*\ldots\sigma^*}$ with the help of the inverse transformation by

$$u_{\alpha\alpha^*}u_{\beta\beta^*}\cdots u_{\sigma\sigma^*}q'_{\alpha\beta\ldots\sigma},$$

and get

$$p'_{ij\ldots s} = u_{ii^*}u_{jj^*}\cdots u_{ss^*}u_{\alpha\alpha^*}u_{\beta\beta^*}\cdots u_{\sigma\sigma^*}r_{i^*j^*\ldots s^*;\alpha^*\beta^*\ldots\sigma^*}q'_{\alpha\beta\ldots\sigma}.$$

Comparing this with the expression given above in an arbitrary Cartesian reference system yields

$$r'_{ij\ldots\sigma} = u_{ii^*}u_{jj^*}\cdots u_{\sigma\sigma^*}r_{i^*j^*\ldots\sigma^*}.$$

This relation is nothing else but the transformation rule for tensors. Thus the $r_{ij\ldots\sigma}$ are in fact components of a tensor.

Let us now consider some examples of such tensor connections which will illustrate the special usefulness of the theorem just discussed.

We will proceed with increasing rank and distinguish between the group of inducing and induced quantities and the group of tensor properties.

As a general example, consider the Taylor expansion of a scalar function

$$F(\boldsymbol{x}) = \sum_{(ij\ldots s),n} \frac{1}{n!}\frac{\partial^n F}{\partial x_i \partial x_j \cdots \partial x_s} x_i x_j \cdots x_s$$

about the point $x = 0$. As is well known, one sums over all permutations of the indices for each n. If we go over to a new reference system $F(x) = F(x')$ is preserved. On the left-hand side we have a zero-rank tensor, a quantity that does not change with a transformation. The individual differential quotients for each fixed n are tensor components because they are connected with the coordinate products $x_i x_j \cdots x_s$ which themselves represent tensor components.

It is obvious that the coordinate products of a vector or arbitrary vectors are also tensor components, likewise the differential quotients of tensors with respect to the components of other tensors, as e.g., the differential quotients with respect to time, temperature, pressure or with respect to the components of the position vector, and so on. Furthermore, the products of components of two or more tensors also represent tensor components just as the differential quotients of components of a vector or of different vectors or tensors, respectively. A further possibility of creating tensors from existing tensors is the operation of tensor contraction. Let there be two tensors $\{a_{ij\ldots rs}\}$ and $\{b_{\alpha\beta\ldots\rho\sigma}\}$ of arbitrary rank. The connection of these tensors according to

$$c_{ij\ldots r;\alpha\beta\ldots\rho} = a_{ij\ldots rs} b_{\alpha\beta\ldots\rho s},$$

which contains a summation over a common index position (or in the general case, several common index positions: multiple contraction), leads also to a tensor, whose rank for each summation position is two less than the sum of the ranks of both initial tensors. The proof for a simple contraction is based upon the transformation behavior. We write on the right-hand side of the above equation the inverse transformations

$$a_{ij\ldots rs} = u_{i^*i} u_{j^*j} \cdots u_{r^*r} u_{s^*s} a'_{i^*j^*\ldots r^*s^*}$$

and

$$b_{\alpha\beta\ldots\rho s} = u_{\alpha^*\alpha} u_{\beta^*\beta} \cdots u_{\rho^*\rho} u_{\sigma^*s} b'_{\alpha^*\beta^*\ldots\rho^*\sigma^*}$$

and get

$$c_{ij\ldots r\alpha\beta\ldots\rho} = a'_{i^*j^*\ldots r^*s^*} b'_{\alpha^*\beta^*\ldots\rho^*\sigma^*} u_{i^*i} u_{j^*j} \cdots u_{r^*r} u_{\alpha^*\alpha} u_{\beta^*\beta} \cdots u_{\rho^*\rho} (u_{s^*s} u_{\sigma^*s}).$$

The expression in the brackets represents a summation over s. When transforming from Cartesian systems, it takes on the value 0 for $s^* \neq \sigma^*$ and the value 1 for $s^* = \sigma^*$. Thus

$$c_{ij\ldots r\alpha\beta\ldots\rho} = a'_{i^*j^*\ldots r^*s^*} b'_{\alpha^*\beta^*\ldots\rho^*s^*} u_{i^*i} u_{j^*j} \ldots u_{r^*r} u_{\alpha^*\alpha} u_{\beta^*\beta} \ldots u_{\rho^*\rho}.$$

This is the formula for the inverse transformation of the tensor components $c'_{i^*j^*\ldots\rho^*}$; thus the contraction process results in a new tensor. The proof is carried out in a corresponding manner for the case of several contractions with arbitrarily chosen index positions. If one works in a non-Cartesian system, the index positions, over which one sums, must have different transformation behavior (covariant or contravariant) in both tensors $\{a\}$ and $\{b\}$. The contraction operation plays an important role in the generation of invariants which we will consider later.

We should note that the theorem on tensor operations represents nothing else but multiple contraction operations. The formulation of the problem, however, is different to the contraction case just discussed.

We now come to concrete examples which will be discussed in detail in later sections. For the sake of brevity, we will specify inducing or induced quantities with I and properties with II.

Zero-Rank Tensors (Scalars)

I: Temperature, pressure, volume, electrical potential, as well as differences and differential quotients of these quantities,

II: Specific weight ρ, specific heat C_p and other energy densities, all mean values of anisotropic properties in space, e.g., mean speed of light, mean abrasive strength, mean electrical conductivity; scalar invariants; all properties of isotropic substances.

First-Rank Tensors (Vectors)

I: Position vector $\boldsymbol{x}$, differential quotient of vectors with respect to scalars (e.g., velocity, acceleration), angular vector $\boldsymbol{u}$, impulse, angular momentum $\boldsymbol{L}$, gradients of scalar fields (temperature, pressure), electric field strength $\boldsymbol{E}$, magnetic field strength $\boldsymbol{H}$, current density vectors (heat, charge, mass), electric moment, magnetic moment. (Explanation: the gradient of a scalar function F is the vector grad $F(x) = \frac{\partial F}{\partial x_i}\boldsymbol{e}_i$).

II: Pyroelectric effect $\{\pi_i\}$: Change in electric polarization $\boldsymbol{P}$ induced by a change in temperature ΔT according to $\Delta P_i = \pi_i \Delta T$. Since the left side represents a tensor of rank 1 and ΔT is a scalar, the quantities π_i are components of a tensor of rank 1. A crystal possessing such a property is assigned a fixed vector. Similarly, there exists a pyromagnetic effect and the piezoelectric effect by a change in the hydrostatic pressure Δp according to $\Delta P_i = q_i \Delta p$.

Second-Rank Tensors

I: Products of components of a vector or components of two vectors $\{x_i x_j\}$ or $\{x_i y_j\}$; differential quotients of components of a vector $\{\frac{\partial E_i}{\partial x_j}\}$ (vector gradient) and differential quotients of types $\{\frac{\partial^2 F}{\partial x_i \partial x_j}\}$, $\{\frac{\partial^2 F}{\partial x_i \partial y_j}\}$, $\{\frac{\partial F}{\partial t_{ij}}\}$; deformation tensor $\{\varepsilon_{ij}\}$ and mechanical stress tensor $\{\sigma_{ij}\}$; quadrupole moment; differential quotients of second-rank tensors with respect to a scalar.

II: *Thermal Conductivity* $\{\lambda_{ij}\}$: $Q_i = -\lambda_{ij}(\text{grad}\, T)_j$; $\boldsymbol{Q}$ is the heat current density vector. In this case, as with some of the following examples, the combination of two vectors is by means of a second-rank tensor $\{\lambda_{ij}\}$.

Electrical Conductivity $\{s_{ij}\}$: $I_i = s_{ij}E_j = -s_{ij}(\operatorname{grad} U)_j$; $\boldsymbol{I}$ is the charge density vector, U the electric potential.

Mass Permeability $\{q_{ij}\}$: $S_i = -q_{ij}(\operatorname{grad} p)_j$; $\boldsymbol{S}$ is the mass density vector.

Dielectricity $\{\epsilon_{ij}\}$: $D_i = \epsilon_{ij}E_j$; $\boldsymbol{D}$ is the vector of the electric displacement, $\boldsymbol{E}$ the vector of the electric field strength.

Magnetic Permeability $\{\mu_{ij}\}$: $B_i = \mu_{ij}H_j$; $\boldsymbol{B}$ is the vector of magnetic induction, $\boldsymbol{H}$ the vector of magnetic field strength.

Thermal Expansion $\{\alpha_{ij}\}$: $\varepsilon_{ij} = \alpha_{ij}\Delta T$; the mechanical deformations ε_{ij} are proportional to the scalar ΔT. Thus $\{\alpha_{ij}\}$ is also a second-rank tensor, since $\{\varepsilon_{ij}\}$ is a tensor as will be shown later.

Volta Striction $\{\beta_{ij}\}$: $\varepsilon_{ij} = \beta_{ij}\Delta U$; this is analogous to thermal expansion. ΔU is the electrical potential difference.

Linear Compressibility at Hydrostatic Pressure $\{K_{ij}\}$: $\varepsilon_{ij} = K_{ij}\Delta p$; this is also analogous to thermal expansion. Δp is the hydrostatic pressure difference.

Moment of Inertia $\{\Theta_{ij}\}$: $L_i = \Theta_{ij}u_j$; $\boldsymbol{u}$ is the angular vector, $\boldsymbol{L}$ the angular momentum vector.

First-Order Displacement Vector $\{v_{ij}\}$: $\xi_i = v_{ij}x_j$; $\boldsymbol{\xi}$ is the displacement vector describing the displacement of the end point of $\boldsymbol{x}$.

Third-Rank Tensors

I: Coordinate products $\{x_i x_j x_k\}$ or $\{x_i y_j z_k\}$ and the differential quotients of the types

$$\left\{\frac{\partial^3 F}{\partial x_i \partial x_j \partial x_k}\right\}, \left\{\frac{\partial^2 E_i}{\partial x_j \partial x_k}\right\}, \left\{\frac{\partial E_i}{\partial t_{jk}}\right\}, \text{ and } \left\{\frac{\partial E_i E_j}{\partial x_k}\right\}$$

and so on.

II. *Piezoelectric Tensor* $\{d_{ijk}\}$: $P_i = d_{ijk}\sigma_{jk}$; $\boldsymbol{P}$ is the vector of the electric polarization, $\{\sigma_{jk}\}$ the mechanical stress tensor. Here we see the combination of a second-rank tensor with a first-rank tensor.

Inverse Piezoelectric Tensor $\{\hat{d}_{ijk}\}$: $\varepsilon_{ij} = \hat{d}_{ijk}E_k$. The deformation $\{\varepsilon_{ij}\}$ proportional to the electric field strength is called first-order electrostriction. It is directly related to the piezoelectric effect.

First-Order Electrooptic Tensor $\{r_{ijk}\}$: $\Delta a_{ij} = r_{ijk}E_k$; $\{\Delta a_{ij}\}$ describes the change of the dielectric behavior in the optical region and thus $\{r_{ijk}\}$ the variation of the velocity of light as a function of the components of the electric field.

Nonohmic Conductivity $\{s_{ijk}\}$: $I_i = s_{ij}E_j + s_{ijk}E_jE_k$; the first term on the right describes the ohmic conductivity, the tensor $\{s_{ijk}\}$ the deviation from Ohm's law in quadratic dependence of the components of the electric field strength.

Hall Tensor $\{R_{ijk}\}$: $E_i = R_{ijk}I_jH_k$; the electric field strength generated by a current in the presence of a magnetic field is represented by a third-rank tensor.

Nonlinear Dielectric Tensor $\{\epsilon_{ijk}\}$: $D_i = \epsilon_{ij}E_j + \epsilon_{ijk}E_jE_k$; $\{\epsilon_{ijk}\}$ describes the deviation from linear behavior.

Fourth-Rank Tensors

I: Product of the coordinates of vectors (also mixed) and their differential quotients as well as the products of tensor components and corresponding differential quotients analogous to third-rank tensors e.g.,

$$\{x_ix_jx_kx_l\},\ \left\{\frac{\partial^4 F}{\partial x_i\partial x_j\partial x_k\partial x_l}\right\},\ \left\{\frac{\partial^2 F}{\partial t_{ij}\partial t_{kl}}\right\},\ \left\{\frac{\partial^3 E_i}{\partial x_j\partial x_k\partial x_l}\right\}.$$

II. *Elasticity Tensor* $\{c_{ijkl}\}$: $\sigma_{ij} = c_{ijkl}\varepsilon_{kl}$; there exists a linear combination of the components of the mechanical deformation tensor with the components of the mechanical stress tensor. The reversal is $\varepsilon_{ij} = s_{ijkl}\sigma_{kl}$. In static experiments, the second expression is mostly used and in dynamic experiments the first expression is favored.

Nonlinear Fourth-Rank Dielectric Tensors $\{\epsilon_{ijkl}\}$ and $\{f_{ijkl}\}$:

$$D_i = \epsilon_{ijkl}E_jE_kE_l \quad \text{or} \quad D_i = f_{ijkl}E_j\frac{\partial E_k}{\partial x_l}.$$

The first tensor describes, to a third-order approximation, the dielectric behavior in homogenous electric fields. The second tensor describes the dependence of products of the field strength and its variation (vector gradient).

Piezooptic and Photoelastic Tensors $\{p_{ijkl}\}$ and $\{q_{ijkl}\}$: $\Delta a_{ij} = q_{ijkl}\sigma_{kl}$ or $\Delta a_{ij} = p_{ijkl}\varepsilon_{kl}$; the change of the optical polarization constants Δa_{ij} under the influence of mechanical stresses or deformations is described to a first-order approximation by a fourth-rank tensor.

Second-Order Electrostriction $\{\hat{d}_{ijkl}\}$: $\varepsilon_{ij} = \hat{d}_{ijkl}E_kE_l$; this is the second-order approximation of electrostriction mentioned above.

Piezoelectric Effect by Electric Prepolarization $\{D_{ijkl}\}$: $P_i = D_{ijkl}E_j\sigma_{kl}$; electric polarization is generated by mechanical stress in the presence of an electric field.

Change in Magnetic Resistance $\{R_{ijkl}\}$: $E_i = R_{ijkl}I_jH_kH_l$; this is a property similar to the Hall effect, however, to second order of the components of the magnetic field strength.

Higher Rank Tensors

I: Products of tensor components and their differential quotients, especially those of mixed quantities.

II: *Second-Order Piezoelectric Effect* $\{d_{ijklm}\}$: $P_i = d_{ijklm}\sigma_{jk}\sigma_{lm}$; the second-order approximation of the piezoelectric effect is described by a fifth-rank tensor and the following approximations by tensors of odd rank.

Nonlinear Elasticity Tensor (Deviation from Hooke's law) $\{c_{ijklmn}\}$: for higher approximations, the relationship between deformation and stress is given by

$$\sigma_{ij} = c_{ijkl}\varepsilon_{kl} + c_{ijklmn}\varepsilon_{kl}\varepsilon_{mn} + c_{ijklmnop}\varepsilon_{kl}\varepsilon_{mn}\varepsilon_{op} + \cdots$$

Especially the easily observed nonlinear acoustic effects are reproduced with the help of sixth- and eighth-rank tensors.

Most higher tensor properties occur as higher approximations of simpler properties. Their tensorial representations follow analogous to the examples presented above.

3.7
Pseudo Tensors (Axial Tensors)

Our definition of tensors requires a supplement. Some quantities exist which only transform like normal tensor components when the coordinate system keeps its chirality (right- or left-handed system). We will explain this by considering the vector products $z = x \times y$ of two position vectors x and y in the basic system. In the definition of the vector products, the handedness of the system spanned by three vectors x, y, and $x \times y$ is involved. Thus a change in the handedness of the reference system in a transformation must be taken into consideration. When applying a rotation–inversion operation $R_{\bar{n}}$, which corresponds to the product of a normal rotation and an inversion, the right-handed reference system goes over into a left-handed system and vice versa. In the case of the inversion $R_{\bar{1}}$, described by the transformation matrix

$$R_{\bar{1}} = (u_{ij}) = \begin{pmatrix} \bar{1} & 0 & 0 \\ 0 & \bar{1} & 0 \\ 0 & 0 & \bar{1} \end{pmatrix},$$

we have $a_i' = -a_i$, $x_i' = -x_i$, and $y_i' = -y_i$. According to the prescription for calculating the vector product

$$z = x \times y = V \begin{vmatrix} a_1^* & a_2^* & a_3^* \\ x_1 & x_2 & x_3 \\ y_1 & y_2 & y_3 \end{vmatrix} = z_i a_i^*$$

we also form, with the vectors $x' = -x_i a_i$ and $y' = -y_i a_i$, which results from the inversion of x and y, the vector product and get $x' \times y' = z$.

Thus $z_i' = z_i$ as opposed to $x_i' = -x_i$. According to this convention, the vector product produces a change in sign by an inversion compared to the transformation of a normal vector. The same applies to the case of an arbitrary rotation–inversion, as one can easily confirm. A change in sign does not occur with pure rotations. This special feature of the transformation of a vector, represented as the vector product of two normal vectors, is taken into consideration by multiplying the transformation formulae with the determinant $|u_{ij}|$ of the transformation matrix, which takes on the value $+1$ with a pure rotation, and the value -1 with a rotation–inversion.

Because of this difference to a normal vector, one designates the vector product of normal vectors as a *pseudo vector* or a first-rank *pseudo tensor*. Sometimes the designation "*axial vector*" is used contrary to a normal tensor, which is then termed "polar." This designation stems from the rotation vector $\boldsymbol{u}$, whose length specifies the rotation velocity or rotation angle and whose direction specifies the rotation axis, whereby the rotation is defined as clockwise, seen along the rotation vector.

Even with pseudo vectors of higher rank, which are always connected with first-rank tensors, a change in sign occurs with a change of handedness of the reference system, when the transformation formulae used so far are applied. The correct transformed quantities are also obtained when, in addition, the factor $|u_{ij}|$ is applied. For the transformation of pseudo tensors we then have

$$t'_{ij\ldots s} = |u_{ij}| u_{ii^*} u_{jj^*} \cdots u_{ss^*} t_{i^*j^*\ldots s^*}.$$

With the aid of a second-rank asymmetric tensor, where $t_{ij} = -t_{ji}$, we can represent the vector product $\boldsymbol{z} = \boldsymbol{x} \times \boldsymbol{y}$ by $z_i = t_{ij} y_j$. We assume the second-rank tensor written as a matrix:

$$(t_{ij}) = \begin{pmatrix} 0 & -x_3 & x_2 \\ x_3 & 0 & -x_1 \\ -x_2 & x_1 & 0 \end{pmatrix}.$$

When we perform an inversion in this representation of the vector product we get $z'_i = t'_{ij} y'_j = z_i$, where $t'_{ij} = -t_{ij}$ and thus the sign is different from $t'_{ij} = u_{ii^*} u_{jj^*} t_{i^*j^*}$. As a consequence, we can consider the vector product of two normal vectors (first-rank tensors) as a tensorial combination of a vector with a second-rank asymmetric pseudo tensor. The axial character of a tensor is propagated through the tensor operation. The following relationships derivable directly from the transformation behavior hold:

$$(p) = (p)(p), \quad (a) = (p)(a), \quad (a) = (a)(p), \quad (p) = (a)(a)$$

as well as further products derived from the above relationships. Here p and a are symbols for a polar (normal) and an axial tensor, respectively.

The most important group of pseudo tensors is associated with magnetic quantities. The vector of the magnetic field strength $\boldsymbol{H}$ and the vector of magnetic induction $\boldsymbol{B}$ are pseudo vectors.

One recognizes the axial character of $\boldsymbol{H}$ and $\boldsymbol{B}$ from Maxwell equations, the fundamental equations of electrodynamics. The equations are

$$\frac{1}{c}\frac{\partial \boldsymbol{D}}{\partial t} + I = \operatorname{rot} \boldsymbol{H},$$

$$\frac{1}{c}\frac{\partial \boldsymbol{B}}{\partial t} = -\operatorname{rot} \boldsymbol{E}.$$

where t is the time, $\boldsymbol{D}$, $\boldsymbol{I}$, and $\boldsymbol{E}$ are the vectors of electric displacement, current density, and electric field, and c the velocity of light; the symbol rot (read "rotation") represents the differential operator $\boldsymbol{\nabla}\times$, where $\boldsymbol{\nabla}$ signifies the differentiation vector $\boldsymbol{\nabla} = \frac{\partial}{\partial x_i}\boldsymbol{e}_i$, which we became acquainted with in the construction of gradients. In a Cartesian coordinate system, we have, for example,

$$\operatorname{rot}\boldsymbol{E} = \boldsymbol{\nabla} \times \boldsymbol{E} = \begin{vmatrix} \boldsymbol{e}_1 & \boldsymbol{e}_2 & \boldsymbol{e}_3 \\ \frac{\partial}{\partial x_1} & \frac{\partial}{\partial x_2} & \frac{\partial}{\partial x_3} \\ E_1 & E_2 & E_3 \end{vmatrix}$$

a pseudo vector (because of the vector product); on the other hand rot $\boldsymbol{H}$ is not a pseudo vector, because the axiality of $\boldsymbol{H}$ and rot compensate each other according to $(a)(a) = (p)$.

Further examples of pseudo tensors are the optical activity in the representation of the gyration tensor $\{\gamma_{ij}\}$, the scalar triple product of three vectors and the Levi-Cività symbol

$$e_{ijk} = \boldsymbol{e}_i \cdot (\boldsymbol{e}_j \times \boldsymbol{e}_k) \quad \text{with} \quad i, j, k = 1, 2, 3.$$

3.8
Symmetry Properties of Tensors

This section deals with the question of how far mathematical or physical arguments as well as symmetry properties of the crystals lead to relationships among the tensor components. In particular, we have to examine how these relationships reduce the number of independent tensor components.

3.8.1
Mathematical and Physical Arguments: Inherent Symmetry

Tensors often exhibit internal relationships among the index positions independent of the medium and its symmetry. For example, consider the so-called "symmetric" second-rank tensor whose components obey $t_{ij} = t_{ji}$ (six independent components). Each second-rank tensor can be decomposed into a symmetric part $t_{(ij)}$ and an antisymmetric part $t_{[ij]}$, according to

$$t_{ij} = \frac{t_{ij} + t_{ji}}{2} + \frac{t_{ij} - t_{ji}}{2} = t_{(ij)} + t_{[ji]}.$$

$t_{[ij]}$ has only three independent components.

Symmetric and antisymmetric tensors keep their symmetry character even with a change in the reference system, as can be easily checked. We previously pointed out the connection between the vector product and an antisymmetric second-rank tensor. In higher rank tensors, symmetric and antisymmetric parts can be separated with respect to certain pairs of positions, as, for example, with a third-rank tensor

$$d_{ijk} = \frac{d_{ijk} + d_{ikj}}{2} + \frac{d_{ijk} - d_{ikj}}{2} = d_{i(jk)} + d_{i[jk]}.$$

A completely symmetric n-rank tensor exists when

$$t_{ij\ldots s} = t_{(ij\cdots s)},$$

where $(ij\ldots s)$ is an arbitrary permutation of the arrangement $ij\ldots s$. Such a tensor is termed total symmetric. An example is the tensor of the products of the coordinates of a vector $\boldsymbol{x}$ with the components

$$t_{ij\ldots s} = t_{(ij\ldots s)} = x_i x_j \cdots x_s = x_j x_i \cdots x_s = \cdots$$

and so on.

One can also define higher rank antisymmetric tensors in a similar manner. If, for example,

$$t_{ijk} = t_{[ijk]} = \frac{1}{3!}(t_{ijk} + t_{jki} + t_{kij} - t_{ikj} - t_{jik} - t_{kji}),$$

we are dealing with a fully antisymmetric tensor. As an example, we mention the Levi-Cività symbol $\{e_{ijk}\}$. This property is also conserved in each new reference system.

Apart from the permutation of indices within certain pairs of index positions, permutations of pairs or certain groups of pairs can also occur. As an example, we mention the elasticity tensor $\{c_{ijkl}\}$ with the property $c_{ijkl} = c_{(ij)(kl)} = c_{(kl)(ij)}$, i.e., the elasticity tensor is symmetric in the pairs of the 1 and 2 positions as well as the 3 and 4 positions, whereby the pairs are mutually permutable, too. An explanation will be given later.

Some quantities connected with the transport processes of charges, heat quantities or masses, for example, current densities and related properties such as magnetic fields, change their sense of direction when the time scale is reversed (time reversal). Accordingly, we distinguish between two groups:

- time invariant properties with $f(-t) = f(t)$, (t time) and
- non time invariant properties with $f(-t) = -f(t)$.

The generalization of these properties is such that one assigns, e.g., to each index position of a tensor a certain behavior with respect to time reversal. We will return to this point in Section 5.3.

3.8.2
Symmetry of the Medium

According to *Neumann's principle*, the property tensors must at least possess the symmetry group of the given crystal. This means that the tensor components must be invariant with respect to all symmetry operations of the given point symmetry group. In crystals, the only macroscopic symmetry operations that come into consideration are rotations $\boldsymbol{R}_n$ and rotation–inversions $\boldsymbol{R}_{\bar{n}}$.

For these special symmetry operations, the following relations must be fulfilled if they appear in the given point symmetry group:

$$t'_{ijk\ldots s} = u_{ii^*}u_{jj^*}\cdots u_{ss^*}t_{i^*j^*\ldots s^*} = t_{ij\ldots s}.$$

If h symmetry operations exist and if the maximum number of independent components is Z ($= n^m$ with an m-rank tensor in an n-dimensional space), then one gets a total of hZ such linear equations.

If h is the order of the symmetry group, the system of equations also contains the identity for each tensor component once, resulting in a total of $(h-1)Z$ nontrivial equations. In principle, it suffices only to apply the generators of the symmetry group. However, it is often useful to apply further symmetry operations such as, e.g., powers $\boldsymbol{R}_n^2$, $\boldsymbol{R}_n^3$, and so on, in order to simplify solving the system. If $h \geq 3$, the equations exhibit a strong linear dependence, often resulting in an enormous reduction in the number of independent components.

As a first example we consider the operation of an inversion center $\bar{1}$, whose transformation matrix is given by

$$\boldsymbol{R}_{\bar{1}} = \begin{pmatrix} \bar{1} & 0 & 0 \\ 0 & \bar{1} & 0 \\ 0 & 0 & \bar{1} \end{pmatrix}.$$

As a result we have

$$t_{ij\ldots s} = (-1)^m t_{ij\ldots s}$$

because $u_{ii} = -1$ and $u_{ij} = 0$ for all $i \neq j$. This means that all odd-rank tensors completely vanish, when an inversion center exists. In particular, there exist no pyroelectric and piezoelectric effects. Furthermore, first-order electro-optical effects and the first-order nonlinear optical effects, as well as the deviation from first-order ohmic conductivity do not occur.

The situation is different with odd-rank pseudo tensors. They exist in the presence of $\bar{1}$ and the even-rank pseudo tensors vanish.

The inversion center has no influence on even-rank tensors. We can therefore say that these tensors are centro-symmetric, independent of the symmetry of the medium.

As a second example we consider the operation of a twofold axis parallel to e_1 and a symmetry plane perpendicular to e_1, represented as a rotation–inversion $\bar{2}$ parallel to e_1. The associated transformation matrices are

$$R_{2\|e_1} = \begin{pmatrix} 1 & 0 & 0 \\ 0 & \bar{1} & 0 \\ 0 & 0 & \bar{1} \end{pmatrix} \quad \text{and} \quad R_{\bar{2}\|e_1} = \begin{pmatrix} \bar{1} & 0 & 0 \\ 0 & 1 & 0 \\ 0 & 0 & 1 \end{pmatrix}.$$

The transformation of the tensor components gives

$$t_{ij\ldots s} = (-1)^p t_{ij\ldots s} \text{ and } t_{ij\ldots s} = (-1)^q t_{ij\ldots s}, \text{ respectively,}$$

where p stands for the number of the indices 2 and 3 and q for the number of the indices 1.

For p and q even we get the identity, i.e., these tensor components are not affected by the symmetry properties. The tensor components vanish for odd p and q. With even-rank tensors, if p is even or odd we also have q even or odd respectively. This means that even-rank tensors exhibit, in the case of a twofold axis or a mirror plane, the combined symmetry of a twofold axis and a symmetry plane perpendicular to this axis, in other words 2/m symmetry. With odd-rank tensors, the situation is different. If p is even, then q must be odd and vice versa because $p + q = m = 2r + 1$ (r integer). The components which vanish in the case of a twofold axis, exist in the case of a symmetry plane and vice versa. Thus the odd-rank tensors behave complementary with respect to the operation of a twofold axis or a symmetry plane. We will now consider, how large is the number of nonvanishing tensor components in both cases.

We first inquire for the number $Z_m(p)$ of components of an m-rank tensor possessing p times the index 1. For $p = 1$ we can choose the index 1 at m different positions. The other positions, namely $(m-1)$, can take on the index 2 or 3, thus having two degrees of freedom. This gives a total of $Z_1 = m \cdot 2^{(m-1)}$ tensor components with an index 1. We proceed in the same manner for $p > 1$. There exists m possibilities for the choice of the first index 1, for the second only $(m-1)$, for the third $(m-2)$, and so on. The number of these possibilities must, however, be divided by the number of nondistinguishable arrangements created, namely $p!$. The other index positions again have two degrees of freedom, so that a total of

$$Z_m(p) = \frac{m(m-1)(m-2)\cdots(m-p+1)}{p!}2^{m-p} = \frac{m!}{p!(m-p)!}2^{m-p}$$

components are concerned.

For the quotient we introduce the symbol $\binom{m}{p}$ used in combinatorial analysis. Thus the total number of tensor components possessing an odd number of indices 1 is

$$Z'_m = \sum_{p'} \binom{m}{p'} 2^{(m-p')},$$

where p' runs through all odd numbers in the interval $1 \leq p' \leq m$.

Accordingly, one gets for the number of components where the index 1 appears an even number of times

$$Z''_m = \sum_{p''} \binom{m}{p''} 2^{(m-p'')} \text{ with } 0 \leq p'' \leq m; \ p'' \text{ even.}$$

Since the index 1 occurs either an even or odd number of times in a component we have

$$Z'_m + Z''_m = Z = 3^m.$$

From this equation one finds for $m = 0, 1, 2, \ldots$:

$$Z'_m = (3^m - 1)/2 \text{ and hence also } Z''_m = (3^m + 1)/2.$$

If one investigates the operation of a twofold axis parallel to $\boldsymbol{e}_1$ or a symmetry plane perpendicular to $\boldsymbol{e}_1$ on a tensor in an n-dimensional space, where the transformation matrix is given by $u_{11} = 1$, $u_{ii} = -1$ for $i \neq 1$ and $u_{ij} = 0$ or $u_{11} = -1$, $u_{ii} = 1$ and $u_{ij} = 0$ $(i \neq j)$, respectively, one gets

$$Z'_m = \sum_{1 \leq p' \leq m} \binom{m}{p'} (n-1)^{(m-p')} = (n^m - (n-2)^m)/2$$

for the number with odd m and

$$Z''_m = (n^m + (n-2)^m)/2$$

for the number with even m. Our equations above are thus special cases for $n = 3$. The general validity of the equations for arbitrary m and n is easy to show with the help of a proof by induction when one substitutes

$$\binom{m+1}{p} = \binom{m}{p} + \binom{m}{p-1}$$

in the summation and checks the validity of the equation for $m = 0$ and 1 beforehand. We can now immediately specify the number of nonvanishing and independent tensor components Z_2 and $Z_{\bar{2}}$ when a twofold axis or symmetry plane is present. We have

$$Z_2 = Z_{\bar{2}} = (n^m + (n-2)^m)/2 \text{ for even } m,$$

however

$$Z_2 = (n^m - (n-2)^m)/2$$

and

$$Z_{\bar{2}} = (n^m + (n-2)^m)/2 \text{ for odd } m.$$

Now we include a second twofold axis parallel to e_2 or a second symmetry plane perpendicular to e_2. We then get for the point symmetry group 22 or $\bar{2}\bar{2}$=mm the number of nonvanishing independent components of tensors in a three-dimensional space

$$Z_{22} = Z_{\bar{2}\bar{2}} = (3^m + 3)/4 \text{ for even,}$$

however,

$$Z_{22} = (3^m - 3)/4 \text{ and } Z_{\bar{2}\bar{2}} = (3^m + 1)/4 \text{ for odd } m.$$

As proof we use the result for the number of components where one distinct index occurs an odd number of times, $Z' = (Z-1)/2$. If two indices should occur an odd number of times—here, the indices 1 and 2—then instead of Z one writes Z' and gets

$$(Z')' = (Z'-1)/2 = ((3^m - 1)/2 - 1)/2 = (3^m - 3)/4.$$

This is identical to the number of nonvanishing and independent components Z_{22} of an odd-rank tensor.

When two indices i and j are only allowed to occur an odd number of times, the third index must also occur an odd number of times. This means that Z_{22} and $Z_{\bar{2}\bar{2}}$ complement each other to $Z = n^m$ for odd m. For even m, all indices must occur an even number of times. Thus one must eliminate from the collection Z_2 those with an odd number of index 2, namely $Z' = (Z_2 - 1)/2$. This is accordingly $Z_{22} = Z_{\bar{2}\bar{2}} = Z_2 - Z' = (3^m + 3)/4$. Table 3.2 presents an overview of the number and type of independent tensor components for tensors up to rank 4.

For symmetry operations of three-, four- or sixfold rotation axes or rotation–inversion axes, the general relations are more complicated. We will treat these cases later in concrete examples. In all symmetry groups containing 2, $\bar{2}$, 22, or $\bar{2}\bar{2}$ as subgroups, the symmetry reduction is naturally only to be applied to tensor components existing in the respective subgroups. In Section 8.3 we will become acquainted with a group theoretical method to calculate the number and type of independent tensor components for arbitrary symmetry groups.

Table 3.2 Independent and non-vanishing tensor components for the cases of a twofold axis (2) parallel to e_1, a mirror plane ($\bar{2}$) perpendicular to e_1, two twofold axes (22) parallel to e_1 and e_2 and two mirror planes ($\bar{2}\bar{2}$ = mm) perpendicular to e_1 and e_2. The number of independent tensor components is given in parentheses.

Conditions for existing $t_{ij\ldots s}$:
a) 2 or $\bar{2}$ parallel e_k, m even: index k even times.
b) 2 parallel e_k, m odd: index k odd times.
c) $\bar{2}$ parallel e_k, m odd: index k odd times.
d) 22 or $\bar{2}\bar{2}$ parallel e_k and e_l, m even: indices k and l even times.
e) 22 parallel e_i, m odd: all indices even times.
f) $\bar{2}\bar{2}$ parallel e_k and e_l, m odd: indices k and l even times.

	$2 \parallel e_1$	$\bar{2} \parallel e_1$	$22 \parallel e_1$ and e_2	$\bar{2}\bar{2} \parallel e_1$ and e_2
Z_m, m even	$(3^m+1)/2$	$(3^m+1)/2$	$(3^m+3)/4$	$(3^m+3)/4$
Z_m, m odd	$(3^m-1)/2$	$(3^m+1)/2$	$(3^m-3)/4$	$(3^m+1)/4$
$m=0$	t (1)	t (1)	t (1)	t (1)
$m=1$	t_1 (1)	t_2, t_3 (2)	(0)	t_3 (1)
$m=2$	$t_{11}, t_{22}, t_{33}, t_{23}, t_{32}$ (5)	$t_{11}, t_{22}, t_{33}, t_{23}, t_{32}$ (5)	t_{11}, t_{22}, t_{33} (3)	t_{11}, t_{22}, t_{33} (3)
$m=3$	$t_{123}, t_{132}, t_{231}, t_{213}, t_{312}, t_{321}, t_{122}, t_{212}, t_{221}, t_{133}, t_{313}, t_{331}, t_{111}$ (13)	$t_{112}, t_{121}, t_{211}, t_{113}, t_{131}, t_{311}, t_{223}, t_{232}, t_{322}$ $t_{332}, t_{323}, t_{233}, t_{222}, t_{333}$ (14)	$t_{123}, t_{132}, t_{231}, t_{213}, t_{312}, t_{321}$, (6)	$t_{113}, t_{131}, t_{311}, t_{223}, t_{232}, t_{322}, t_{333}$ (7)
$m=4$	$t_{1111}, t_{1122}, t_{1133}, t_{2211}, t_{3311}, t_{1123}, t_{1132}, t_{2311}, t_{3211}, t_{1213}, t_{1321}, t_{1231}, t_{1321}, t_{2131}, t_{3121}, t_{2311}, t_{3211}, t_{1212}, t_{1221}, t_{2112}, t_{2121}, t_{1313}, t_{1331}, t_{3113}, t_{3131}, t_{2222}, t_{2233}, t_{3322}, t_{2223}, t_{2232}, t_{2322}, t_{3222}, t_{2323}, t_{2332}, t_{3223}, t_{3232}, t_{3333}, t_{3332}, t_{3323}, t_{3233}, t_{2333}$ (41)	$t_{1111}, t_{1122}, t_{1133}, t_{2211}, t_{3311}, t_{1123}, t_{1132}, t_{2311}, t_{3211}, t_{1213}, t_{1321}, t_{1231}, t_{1321}, t_{2131}, t_{3121}, t_{2311}, t_{3211}, t_{1212}, t_{1221}, t_{2112}, t_{2121}, t_{1313}, t_{1331}, t_{3113}, t_{3131}, t_{2222}, t_{2233}, t_{3322}, t_{2223}, t_{2232}, t_{2322}, t_{3222}, t_{2323}, t_{2332}, t_{3223}, t_{3232}, t_{3333}, t_{3332}, t_{3323}, t_{3233}, t_{2333}$ (41)	$t_{1111}, t_{1122}, t_{2211}, t_{1212}, t_{1221}, t_{2121}, t_{2112}, t_{1133}, t_{3311}, t_{1313}, t_{1331}, t_{3113}, t_{3131}, t_{2222}, t_{2233}, t_{3322}, t_{2323}, t_{2332}, t_{3223}, t_{3232}, t_{333}$ (21)	$t_{1111}, t_{1122}, t_{2211}, t_{1212}, t_{1221}, t_{2121}, t_{2112}, t_{1133}, t_{3311}, t_{1313}, t_{1331}, t_{3113}, t_{3131}, t_{2222}, t_{2233}, t_{3322}, t_{2323}, t_{2332}, t_{3223}, t_{3232}, t_{333}$ (21)

$\bar{2}$ parallel e_1 signifies a mirror plane perpendicular to e_1.

3.9
Derived Tensors and Tensor Invariants

Through the operation of multiplication and tensor contraction with itself as well as with nonspecific tensors, such as the tensors of the Kronecker symbol (spherical tensor), the Levi-Cività tensor, the position vector, and their combinations, one obtains, according to the rules just discussed, new tensors that are often more accessible to an interpretation than the given initial tensor.

The group of tensor powers comes into play, with respect to the generating quantities, in higher order effects, e.g., the tensor $\{E_iE_j\}$ of the components of the electric field strength. For an m-rank tensor, the *quadric tensor* is

$$t^2_{ij\ldots s,i^*j^*\ldots s^*} = t_{ij\ldots s}t_{i^*j^*\ldots s^*}$$

a $2m$-rank tensor. In an analogous manner, one can form the *zth power* of a tensor

$$t_{ij\ldots s,i^*j^*\ldots s^*,i^{**}j^{**}\ldots s^{**}\ldots} = t_{ij\ldots s}t_{i^*j^*\ldots s^*}t_{i^{**}j^{**}\ldots s^{**}\ldots}.$$

In self-contraction, one deals with the generation of tensors of the type

$$t_{ij\ldots ri^*j^*\ldots r^*} = t_{ij\ldots rs}t_{i^*j^*\ldots r^*s} \quad \text{contraction over the } m\text{th position (index } s).$$

Multiple contractions can also occur, whereby the indices, over which are to be summed, can take on different positions. Of importance is the complete contraction over all positions

$$Q = t_{ij\ldots s}t_{ij\ldots s},$$

a scalar invariant, independent of the reference system. In the case of a vector $\boldsymbol{x}$, Q corresponds to the square of the magnitude (length) of the vector $Q = x_ix_i = x_1^2 + x_2^2 + x_3^2$. Accordingly, Q can in general be designated as the square of the magnitude of a tensor. With respect to the contraction with other tensors, special emphasis is paid to the total contraction with tensors of the corresponding products of the components of the position vector:

$$F = t_{ij\ldots s}x_ix_j\cdots x_s.$$

This expression represents for a fixed F a surface of mth order, the so-called *tensor surface*. If one inserts, instead of $t_{ij\ldots s}$, the components of the associated total symmetric tensor, one gets the same tensor surface. This means that the tensor surface reproduces the complete tensor properties only in the case of a total symmetric tensor. The tensor surface allows one to represent one of the most important tensor properties, the so-called *longitudinal effect*, in a quite instructive manner as we shall see in the next section.

Further interesting contractions can be generated with the help of the tensors of the Kronecker symbol $\delta_{ij} = 1$ for $i = j$ and $= 0$ for $i \neq j$ or the Levi-Cività symbol $e_{ijk} = \boldsymbol{e}_i \cdot (\boldsymbol{e}_j \times \boldsymbol{e}_k)$. As an example, we mention the scalar invariants for even-rank tensors

$$I = t_{ijkl\ldots rs}\delta_{ij}\delta_{kl}\cdots\delta_{rs}.$$

The operation $t_{ij\ldots s}\delta_{im} = t_{mj\ldots s}$ exchanges an index. This can be carried out any number of times. Important contraction types for fourth-rank tensors are

$$A_{jl} = t_{ijkl}\delta_{ik},\quad B_{kl} = t_{ijkl}\delta_{ij},\quad C_{ik} = t_{ijkl}\delta_{jl},\quad \text{and}\quad D_{ij} = t_{ijkl}\delta_{kl};$$

here we are dealing with second-rank tensor invariants arising from fourth-rank tensors.

As an example of the application of the Levi-Cività symbol, we consider the vector $\boldsymbol{q}$ (pseudo vector), assigned to a second-rank tensor, possessing the components $q_k = t_{ij}e_{ijk}$ as well as the vector product of the vectors $\boldsymbol{x}$ and $\boldsymbol{y}$, namely $\boldsymbol{z} = \boldsymbol{x} \times \boldsymbol{y}$, that can also be represented by $z_k = x_i y_i e_{ijk}$ in a Cartesian reference system. The triple vector product of $\boldsymbol{x}$, $\boldsymbol{y}$, and $\boldsymbol{z}$ gives $\boldsymbol{x} \cdot (\boldsymbol{y} \times \boldsymbol{z}) = x_i y_i z_k e_{ijk}$, an invariant representation in any Cartesian reference system.

Furthermore, we should mention invariants derived from fourth-rank tensors

$$S_{mn} = t_{ijkl}e_{ijm}e_{kln} \quad \text{and} \quad T_{mn} = t_{ijkl}e_{ikm}e_{jln}$$

as well as similar operations, which in part, are of practical importance.

Other examples will be treated in our discussion of concrete properties. Here we want to emphasize the special importance of scalar invariants because they indicate a directionally independent value, which can normally be interpreted as a spatial mean value of a certain tensor property.

At this point we must forgo a systematic discussion of invariants, although just now a number of very fascinating problems emerge, e.g., the question of the number of independent variables of a tensor, the decomposition and construction of a tensor with a basis of invariants. Naturally, we can safely answer the first question by saying that the number of independent invariants cannot be larger than the number of independent tensor components. Useful aids in the discussion of such relationships are available from group theory, which we shall return to in a later section.

3.10
Longitudinal and Transverse Effects

If two vectors $\boldsymbol{A}$ and $\boldsymbol{B}$ are connected via a second-rank tensor according to $B_i = t_{ij}A_j$, then both vectors run parallel only in distinct directions. For arbitrary directions of $\boldsymbol{A}$ we can analyze the decomposition of $\boldsymbol{B}$ into components parallel and perpendicular to $\boldsymbol{A}$. We get

$$\boldsymbol{B}_{\|A} = B_i A_i \boldsymbol{A}/A^2 = t_{ij}A_i A_j \boldsymbol{A}/A^2.$$

We specify the unit vector in the direction of $\boldsymbol{A}$ by $\boldsymbol{e}'_1 = \boldsymbol{A}/A = u_{1i}\boldsymbol{e}_i$. We then get

$$\boldsymbol{B}_{\|A} = u_{1i}u_{1j}t_{ij}\boldsymbol{A}.$$

We call the quantity $u_{1i}u_{1j}t_{ij} = t'_{11}$ the longitudinal effect of the given tensor property. It describes, e.g., in the case of electrical conductivity, the conductivity in the direction of the electric field strength. We now emphasize that this longitudinal effect is normally easily amenable to measurements. By transformation, it can be immediately calculated for any direction. Correspondingly, there exists a transverse effect for a component of $\boldsymbol{B}$ perpendicular to $\boldsymbol{A}$ in the direction $\boldsymbol{e}'_2$, which can be arbitrarily chosen to be perpendicular to $\boldsymbol{e}'_1$. We get the quantity t'_{21}, in which the vector $\boldsymbol{A}$ as the generating quantity parallel to $\boldsymbol{e}'_1$ and the part $\boldsymbol{B}$ parallel to $\boldsymbol{e}'_2$ are connected to each other. We call $t'_{21} = u_{2i}u_{1j}t_{ij}$ the transverse effect in the direction $\boldsymbol{e}'_2$ by excitation in the direction $\boldsymbol{e}'_1$.

The general definition of longitudinal and transverse effects looks like this: We consider the longitudinal component $A'_{111\ldots1}$ of a generating quantity in the direction $\boldsymbol{e}'_1$ and observe the associated longitudinal component of the generated quantity $B'_{111\ldots1}$ or the component $B'_{222\ldots2}$ in the direction $\boldsymbol{e}'_1$ or $\boldsymbol{e}'_2$, respectively.

In the operation $B'_{111\ldots1} = t'_{111\ldots1}A'_{111\ldots1}$, $t'_{111\ldots1} = u_{1i}u_{1j}\cdots u_{1s}t_{ij\ldots s}$ is the longitudinal component for the direction $\boldsymbol{e}'_1 = u_{1i}\boldsymbol{e}_i$, and in $B'_{222\ldots2} = t'_{222\ldots2111\ldots1}A'_{111\ldots1}$, $t'_{222\ldots2111\ldots1}$ is the transverse component for the direction $\boldsymbol{e}'_2 = u_{2i}\boldsymbol{e}_i$ by excitation in the direction $\boldsymbol{e}'_1 = u_{1i}\boldsymbol{e}_i$. Although there exists only one longitudinal effect for a direction $\boldsymbol{e}'_1$, one can calculate transverse effects in directions perpendicular to $\boldsymbol{e}'_1$.

An interesting relation exists between the tensor surface and the longitudinal effect. We consider a direction along the position vector $\boldsymbol{x} = x_i\boldsymbol{e}_i = |\boldsymbol{x}|\boldsymbol{e}'_1$. Here, $\boldsymbol{e}'_1 = u_{1i}\boldsymbol{e}_i$ with $u_{1i} = x_i|\boldsymbol{x}|^{-1}$. Thus for the longitudinal component of an m-rank tensor we get

$$t'_{111\ldots1} = u_{1i}u_{1j}\ldots u_{1s}t_{ij\ldots s} = x_ix_j\ldots x_st_{ij\ldots s}|\boldsymbol{x}|^{-m}.$$

We substitute the numerator of the quotient by $F = x_ix_j\cdots x_st_{ij\ldots s}$, the scalar invariant of the tensor surface, and get $t'_{111\ldots1} = F|\boldsymbol{x}|^{-m}$. This means that the longitudinal component in the direction $\boldsymbol{x}$ is equal to F divided by the mth power of the distance of the end point of $\boldsymbol{x}$ on the tensor surface from the origin of the reference system. Thus the tensor surface gains an intuitive physical meaning.

4
Special Tensors

In this chapter we will discuss the more important tensors and tensor properties. We begin with zero-rank tensors and work our way through tensors of higher rank. The sequence is essentially chosen according to physical aspects.

4.1
Zero-Rank Tensors

We investigate two types of zero-rank tensors, namely, the intrinsic isotropic properties, such as, e.g., the specific heat at constant hydrostatic pressure, the specific weight and the chemical composition, on the one hand and the spatial mean values of anisotropic properties as well as all other scalar invariants on the other hand. We also look at all differential quotients of these properties with respect to scalar quantities, such as hydrostatic pressure, temperature, and electric potential.

In order to achieve a certain degree of completeness in the discussion of the more important experimental methods, let us consider the comparatively simple measurement methods for the specific weight [1] and the specific heat.

One obtains a precision value of the *specific weight* with a sufficiently accurate measurement of the crystallographic metric given by

$$\rho = \frac{Z \cdot \text{molar weight}}{V \cdot L},$$

where Z is the number of stoichiometric units in the unit cell, V the volume of the elementary cell and L Avogadro's number 6.0222×10^{23} molecules per mole. V can normally be determined to a fraction of one-tenth of a percent from high-precision measurements of the lattice constants and thus also the specific weight. Very small crystals with dimensions of about 0.1 mm are suitable for these measurements.

1) In practice, the difference between the specific weight and density does not play a role when the weights used for measurement are calibrated in mass units (g).

Physical Properties of Crystals. Siegfried Haussühl.

ISBN: 978-3-527-40543-5

Today, such precision measurements are possible within about an hour with the aid of an automatic diffractometer as long as the crystals are homogenous and their chemical composition is reliably known. If problems occur, or if a higher precision is required, one must employ the methods of direct density determination. Here we will discuss the buoyancy method that, especially with large crystals with a weight of at least 10 g, allows the highest accuracy with a relative error of less than 10^{-5}. If possible, the specimen should be large. It should not possess inclusions and its surface should be smooth and free of impurities. The specimen is first weighed in air and then in a homogenous, practically nonsolvent liquid. It is important to ensure that the surface is completely wetted (no gas bubbles). Let both weights be G_A and G_L. The crystal is again weighed in air to check that no loss has occurred due to dissolution. The actual weight G is determined from G_A after the addition of the buoyancy in air, or the buoyancy in the liquid, respectively:

$$G = G_A + \rho_A V_B = G_L + \rho_L V_B.$$

Thus

$$\rho = \frac{G}{V_B} = \frac{G_A}{G_A - G_L}(\rho_L - \rho_A) + \rho_A.$$

V_B is the volume of the body, ρ_L and ρ_A are the specific weights of the liquid and the air, respectively (ρ_A at 293 K and normal pressure is about 1.21 g per liter). In practice, it has been found expedient to use a calibrated standard liquid that remains unchanged over many years, e.g., xylene or petrol. Crystal samples of known density, such as quartz or LiF, are suitable for the measurement of ρ_L. For many materials water does not come into consideration, despite its well-known density, because of its good solvent power and the necessity to degas it by boiling before each measurement. Therefore, it is only suitable in special cases. If a pure liquid, with sufficiently low solubility, cannot be found, one can, as a last resort, use a saturated solution of the given crystal.

Measurements with the suspension method, where a liquid composed of two components is mixed such that a specimen just floats in the mixture, as well as pycnometer measurements for powder specimens usually yield inaccurate values, so that these methods are not normally taken into consideration.

Excellent commercial equipment is available for the measurement of the *specific heat* under constant pressure. For precision measurements, as required, e.g., in the determination of the transition enthalpy, an exact and highly sensitive temperature measurement is decisive. Thermistors (semiconductors) where the resistance exhibits an extremely high temperature dependence are well proven for this task. These elements must be calibrated by the experimentalist. This is done appropriately with the help of calibrated quartz thermometers, where the temperature dependence of the resonant frequencies of

α-quartz plates is used for the temperature measurement. Temperature differences of up to 10^{-4} K can be measured, fulfilling essentially all requirements for a highly precise measurement of the specific heat, provided the crystals are of sufficient size and quality, with weights of at least 0.05 g.

Of the few pseudo scalars playing a role in practical work, we mention the optical activity of isotropic media which will be discussed in more detail in Section 4.3.6.7.

4.2 First-Rank Tensors

4.2.1 Symmetry Reduction

As we have already seen, odd-rank polar tensors vanish in centro-symmetric point symmetry groups. Also in all acentric point symmetry groups, containing the subgroup 22 (22, 42, $\bar{4}2$, 62, 23, 43, $\bar{4}3$), the existence of a first-rank polar tensor is impossible, as is evident from Table 3.2. We also know the effect of 2, $\bar{2}$, and $\bar{2}\bar{2}$ from Table 3.2. Consequently, we have only to check the three- and four-fold rotation axes and rotation–inversion axes and their combinations with 2, $\bar{2}$ and $\bar{2}\bar{2}$. We place these rotation axes parallel to e_3. With

$$\boldsymbol{R}_{\pm 3\|e_3} = \begin{pmatrix} -1/2 & \pm\sqrt{3}/2 & 0 \\ \mp\sqrt{3}/2 & -1/2 & 0 \\ 0 & 0 & 1 \end{pmatrix}$$

one finds $t_1' = t_1 = -1/2\,t_1 \pm \sqrt{3}/2\,t_2$, thus $t_1 = t_2 = 0$. $+$ applies to a clockwise (right) rotation, and $-$ to a counter-clockwise (left) rotation. $t_3' = t_3$ remains as the only independent component. In PSG 32 and $3/\mathrm{m} \equiv \bar{6}$, the two-fold axis and the symmetry plane, respectively, also destroy the existence of t_3. All t_i vanish for $\bar{3}$ (Laue-class!).

With

$$\boldsymbol{R}_4 = \begin{pmatrix} 0 & 1 & 0 \\ \bar{1} & 0 & 0 \\ 0 & 0 & 1 \end{pmatrix}$$

one gets $t_1' = t_1 = t_2$, $t_2' = t_2 = -t_1$, and thus $t_1 = t_2 = 0$, which is already a consequence of the two-fold axis parallel to 4. On the other hand $t_3' = t_3$ exists.

In $\bar{4}$ with

$$R_{\bar{4}} = \begin{pmatrix} 0 & \bar{1} & 0 \\ 1 & 0 & 0 \\ 0 & 0 & \bar{1} \end{pmatrix}$$

we have $t_3 = 0$.

With first-rank pseudo tensors, the inversion center provides no reduction. $\bar{2}$ parallel to e_i (symmetry plane perpendicular to e_i) only allows the existence of t_i. The same component is also preserved by a rotation about e_i. Thus first-rank pseudo tensors have the same form in all symmetry groups of type n/m.

Finally, let us discuss the effect of cylindrical symmetry. Here, the rotation through an arbitrary angle φ about a fixed axis, running, e.g., parallel to e_3, gives rise to a symmetry-equivalent situation. The transformation matrix is

$$R_{\infty} = \begin{pmatrix} \cos\varphi & \sin\varphi & 0 \\ -\sin\varphi & \cos\varphi & 0 \\ 0 & 0 & 1 \end{pmatrix}.$$

Thus $t'_1 = t_1 = \cos\varphi\, t_1 + \sin\varphi\, t_2$ for each value of φ, in other words $t_1 = t_2 = 0$. However, the component t_3 does not vanish.

The result of these considerations is noted in Table 4.1. It should be stressed that a polar vector, except in triclinic and monoclinic system, always takes on a distinct orientation, namely that of the given principal axis. The ten PSGs with nonvanishing first-rank polar tensors are called *pyroelectric groups* (1, 2, m, mm, 3, 3m, 4, 4m, 6, 6m).

4.2.2
Pyroelectric and Related Effects

As a concrete example we discuss the *pyroelectric effect*. If one heats certain crystals, e.g., tourmaline, in which the effect was first observed, opposite electric charges are generated at certain opposed face elements ΔF. Within a certain temperature interval, one observes, to a first approximation, a linear relationship between the change in charge density and the temperature difference. We describe the electric polarization accompanying a change in charge ΔQ by the vector

$$\Delta P = K \frac{\Delta Q}{\Delta F} f;$$

f is the unit vector of the face normal on ΔF, K is the constant of the given system of measurement (in the MKS system: $K = 1$, in the CGS system: $K = 4\pi$). We compare this quantity to the electric moment $M = KQx$, where x

Table 4.1 Non-vanishing components of first-rank tensors in the 32 point symmetry groups ($t = t_i e_i$).

PSG	Polar tensor	Pseudo tensor	PSG	Polar tensor	Pseudo tensor
1	t_1, t_2, t_3	t_1, t_2, t_3	$6 \parallel e_3$	t_3	t_3
$\bar{1}$	–	t_1, t_2, t_3	62	–	–
$2 \parallel e_3$ or e_2	t_3 or t_2	t_3 or t_2	6m	t_3	–
$m \equiv \bar{2} \parallel e_3$ or e_2	t_1, t_2 or t_1, t_3	t_3 or t_2	6/m	–	t_3
$2/m \parallel e_3$ or e_2	–	t_3 or t_2	$\bar{6}$	–	t_3
22	–	–	$\bar{6}$m	–	–
$\bar{2}\bar{2} \equiv$ mm2, $2 \parallel e_3$	t_3	–	6/mm	–	–
2/mm	–	–	23	–	–
$3 \parallel e_3$	t_3	t_3	m3	–	–
32	–	–	43	–	–
3m	t_3	–	$\bar{4}$3m	–	–
$\bar{3}$	–	t_3	4/m3	–	–
$\bar{3}$m	–	–	$\infty \parallel e_3$	t_3	t_3
$4 \parallel e_3$	t_3	t_3	∞2	–	–
42	–	–	∞m	t_3	–
4m	t_3	–	∞/m	–	t_3
4/m	–	t_3	$\bar{\infty}$	–	t_3
$\bar{4}$	–	t_3	$\bar{\infty}$m	–	–
$\bar{4}$m	–	–	∞/mm	–	–
4/mm	–	–			

specifies the distance vector of the center of charge on the opposite faces. A rectangular parallelepiped probe with edges x_i parallel to the Cartesian basic vectors e_i then has an electric moment with the components

$$\Delta M_i = K \Delta Q_i x_i,$$

where the charges on the opposed faces are $+\Delta Q_i/2$ and $-\Delta Q_i/2$. The electric moment per volume element is then

$$\frac{\Delta M_i}{\Delta V} = K \frac{\Delta Q_i x_i}{x_i x_j x_k} = K \frac{\Delta Q_i}{\Delta F_i} = \Delta P_i \text{ with } \Delta F_i = x_j x_k \ (i \neq j, k).$$

A change in the electric moment per volume element caused by a change in temperature should be recognized as a change in the electric polarization $\Delta \boldsymbol{P}$.

Thus we describe the pyroelectric properties of a crystal by the relation

$$\Delta P_i = \pi_i \Delta T + \rho_i (\Delta T)^2 + \sigma_i (\Delta T)^3 + \cdots,$$

where the tensors $\{\pi_i\}$, $\{\rho_i\}$, $\{\sigma_i\}$ $\cdots$ give the pyroelectric effect to first-, second-, third- and further-order approximations.

This definition also provides a simple measurement procedure, an aspect, which should be carefully noted with the introduction of any new property. For the practical realization of the measurement of the pyroelectric effect, it is appropriate to use a plane-parallel thin plate with a rather large diameter-to-thickness ratio. The plate is to be metallized on both sides, where the boundary parts remain free for the purpose of insulation of both electrodes. One now measures the change in charge density after a change in temperature, e.g., with the aid of a commercial charge amplifier or via the electric voltage produced at the electrodes. Note, however, that the dielectric constant of the plate in the direction of the normals must be considered (see Section 4.3.3).

Hence, one directly obtains the longitudinal component of the pyroelectric effect in the direction $\boldsymbol{e}_i'$ of the plate normals. It is

$$\pi_1' = u_{1i}\pi_i.$$

The reference face $F = \pi_i x_i$, a plane perpendicular to the vector $\boldsymbol{\pi} = \pi_i \boldsymbol{e}_i$, represents the longitudinal effect through

$$\pi_1' = F/|\boldsymbol{x}|;$$

on the other hand F is equal to the scalar product of $\boldsymbol{\pi}$ and $\boldsymbol{x}$, thus $\pi_i' = |\boldsymbol{\pi}|\cos\zeta$, where ζ is the angle between the normals on the reference face (parallel to $\boldsymbol{\pi}$) and $\boldsymbol{x}$. From this, one recognizes the decrease of the pyroelectric effect with increasing deviation from the direction of the maximum effect. Any other form of the test object, e.g., a partially metallized sphere, would result in a complicated relationship between measured quantity and tensor property.

The determination of the pyroelectric effect is possible in all PSGs, except 1 and m, by means of a single measurement along the respective principal axis in which $\boldsymbol{\pi}$ also lies. In PSG 1 we carry out the measurement of the longitudinal effect along three noncoplanar directions $\boldsymbol{e}_j' = u_{ji}\boldsymbol{e}_i$ and get the following system of equations for the determination of the three components π_i:

$$\begin{aligned}\pi_1' &= u_{1i}\pi_i,\\ \pi_2' &= u_{2i}\pi_i,\\ \pi_3' &= u_{3i}\pi_i\end{aligned}$$

(here, the directions $\boldsymbol{e}_i'$ need not be chosen to be orthogonal). The solution is:

$$\boldsymbol{\pi} = \frac{1}{D}\begin{vmatrix}\pi_1' & u_{12} & u_{13}\\ \pi_2' & u_{22} & u_{23}\\ \pi_3' & u_{32} & u_{33}\end{vmatrix}\boldsymbol{e}_1 - \frac{1}{D}\begin{vmatrix}\pi_1' & u_{11} & u_{13}\\ \pi_2' & u_{21} & u_{23}\\ \pi_3' & u_{31} & u_{33}\end{vmatrix}\boldsymbol{e}_2 + \frac{1}{D}\begin{vmatrix}\pi_1' & u_{11} & u_{12}\\ \pi_2' & u_{21} & u_{22}\\ \pi_3' & u_{31} & u_{32}\end{vmatrix}\boldsymbol{e}_3.$$

D is the determinant of the matrix (u_{ij}). In PSG m, two measurements in directions within the symmetry plane will be sufficient.

The pyroelectric effect consists of two parts: the primary or true pyroelectric effect resulting from a change in the arrangement of dipole moments and the secondary effect, resulting from a change in charge density due to thermal expansion (see Section 4.3.11). All crystals with large pyroelectric effects, such as triglycine sulfate $(NH_2CH_2COOH)_3H_2SO_4$, and its analogs, where H_2SO_4 is substituted by H_2SeO_4 or H_2BeF_4, and in addition $LiNbO_3$ and $Pb_5Ge_3O_{11}$ also show ferroelectric properties (see Section 4.3.4). Pyroelectric constants of some important crystal species are listed in Table 12.5 (annex).

In the last few years pyroelectric crystals have found special interest in the construction of highly sensitive radiation detectors in all spectral regions, especially for high energy radiation. With the aid of very small crystal plates it is possible to count individual photons (even in the visible spectral range) and from the magnitude of the pyroelectric effect to measure the energy of the photons (energy discrimination). A further application is possible in connection with the conversion of thermal energy into electrical energy when the charges appearing due to a temperature change are collected and stored. For example, the natural day–night temperature variations, which can be artificially modulated by suitable control processes, offer a simple and cheap inducing source. The efficiency, i.e., that fraction of the required heat input per heating cycle, that can be converted into electrical energy, is at present still under 1%. If one could succeed in finding crystals with substantially larger effects, a broad commercial utilization of the pyroelectric effect to generate electrical energy would be conceivable, provided production and processing costs are sufficiently low.

Further examples of first-rank tensors are the piezoelectric effect under hydrostatic pressure Δp as well as both analogous magnetic effects:

$$B_i = \beta_i \Delta T \quad \text{pyromagnetic effect and}$$
$$B_i = \gamma_i \Delta p \quad \text{piezomagnetic effect.}$$

$\{\beta_i\}$ and $\{\gamma_i\}$ are pseudo tensors, of which at present no reliable results are known.

4.3 Second-Rank Tensors

4.3.1 Symmetry Reduction

First we want to gain a general view of the form of the polar tensors and the pseudo tensors in all point symmetry groups. We choose the Cartesian reference system according to the convention introduced in Section 2.2. With the polar tensors of even rank it is sufficient to discuss only the 11 Laue groups due to their centrosymmetric behavior. Pseudo tensors of even rank vanish in the Laue groups. In the enantiomorphic PSGs they take on the same form as

Table 4.2 Non-vanishing components of second-rank tensors in standard setting.

PSG	Polar tensor	Pseudo tensor
1	$t_{ij},\ i,j=1,2,3$	$t_{ij},\ i,j=1,2,3$
$\bar{1}$	$t_{ij},\ i,j=1,2,3$	–
$2 \parallel e_2$	$t_{11},t_{22},t_{33},t_{13},t_{31}$	$t_{11},t_{22},t_{33},t_{13},t_{31}$
$\bar{2} \parallel e_2$	$t_{11},t_{22},t_{33},t_{13},t_{31}$	$t_{12},t_{21},t_{23},t_{32}$
$2/\mathrm{m} \parallel e_2$	$t_{11},t_{22},t_{33},t_{13},t_{31}$	–
22	t_{11},t_{22},t_{33}	t_{11},t_{22},t_{33}
mm2	t_{11},t_{22},t_{33}	t_{12},t_{21}
2/mm	t_{11},t_{22},t_{33}	–
$3 \parallel e_3$	$t_{11}=t_{22},t_{33},t_{12}=-t_{21}$	$t_{11}=t_{22},t_{33},t_{12}=-t_{21}$
32	$t_{11}=t_{22},t_{33}$	$t_{11}=t_{22},t_{33}$
3m	$t_{11}=t_{22},t_{33}$	$t_{12}=-t_{21}$
$\bar{3}$	$t_{11}=t_{22},t_{33},t_{12}=-t_{21}$	–
$\bar{3}$m	$t_{11}=t_{22},t_{33}$	–
$4 \parallel e_3$	$t_{11}=t_{22},t_{33},t_{12}=-t_{21}$	$t_{11}=t_{22},t_{33},t_{12}=-t_{21}$
42	$t_{11}=t_{22},t_{33}$	$t_{11}=t_{22},t_{33}$
4m	$t_{11}=t_{22},t_{33}$	$t_{12}=-t_{21}$
4/m	$t_{11}=t_{22},t_{33},t_{12}=-t_{21}$	–
$\bar{4}$	$t_{11}=t_{22},t_{33},t_{12}=-t_{21}$	$t_{11}=-t_{22},t_{12}=t_{21}$
$\bar{4}\mathrm{m}, 2 \parallel e_1$	$t_{11}=t_{22},t_{33}$	$t_{11}=-t_{22}$
4/mm	$t_{11}=t_{22},t_{33}$	–
$6 \parallel e_3, \infty \parallel e_3$	$t_{11}=t_{22},t_{33},t_{12}=-t_{21}$	$t_{11}=t_{22},t_{33},t_{12}=-t_{21}$
$62, \infty 2$	$t_{11}=t_{22},t_{33}$	$t_{11}=t_{22},t_{33}$
$6\mathrm{m}, \infty\mathrm{m}$	$t_{11}=t_{22},t_{33}$	$t_{12}=-t_{21}$
$6/\mathrm{m}, \infty/\mathrm{m}$	$t_{11}=t_{22},t_{33},t_{12}=-t_{21}$	–
$\bar{6}, \overline{\infty}$	$t_{11}=t_{22},t_{33},t_{12}=-t_{21}$	–
$\bar{6}\mathrm{m}, \overline{\infty}\mathrm{m}$	$t_{11}=t_{22},t_{33}$	–
$6/\mathrm{mm}, \infty/\mathrm{mm}$	$t_{11}=t_{22},t_{33}$	–
23	$t_{11}=t_{22}=t_{33}$	$t_{11}=t_{22}=t_{33}$
m3	$t_{11}=t_{22}=t_{33}$	–
43	$t_{11}=t_{22}=t_{33}$	$t_{11}=t_{22}=t_{33}$
$\bar{4}3$	$t_{11}=t_{22}=t_{33}$	–
4/m3	$t_{11}=t_{22}=t_{33}$	–
isotropic without $\bar{1}$	$t_{11}=t_{22}=t_{33}$	$t_{11}=t_{22}=t_{33}$
isotropic with $\bar{1}$	$t_{11}=t_{22}=t_{33}$	–

the polar tensors. Thus we have to study only the 11 enantiomorph PSG's (1, 2, 22, 3, 32, 4, 42, 6, 62, 23, 43) for all even-rank tensors and, in addition, the remaining ten non-Laue groups (m, 2m = mm, 3m, 4m, $\bar{4}$, $\bar{4}$m, 6m, $\bar{6}$, $\bar{6}$m, $\bar{4}$3) for even-rank pseudo tensors.

We now show the symmetry reduction for the most important PSG's. The reader can handle all other cases with ease. A complete survey of all second-rank tensors is given in Table 4.2. For better clarity, it is often useful to write the components of a second-rank tensor in matrix notation, thus

$$\{t_{ij}\} = \begin{pmatrix} t_{11} & t_{12} & t_{13} \\ t_{21} & t_{22} & t_{23} \\ t_{31} & t_{32} & t_{33} \end{pmatrix}.$$

Monoclinic System (PSG 2, m = $\bar{2}$, 2/m)
The effect of two-fold rotation axes and rotation–inversion axes parallel to $\boldsymbol{e}_i$ for polar tensors of even rank was previously discussed in Section 3.8.2. The result is: the indices i may only occur with even numbers.

We choose the standard setting $\boldsymbol{e}_2 \parallel \boldsymbol{a}_2$ parallel to the two-fold rotation axis or rotation–inversion axis, respectively (normal on the mirror plane). Thus for polar tensors there exists only the following five components containing the index 2 even numbered: t_{11}, t_{22}, t_{33}, t_{13} and t_{31}. With a symmetric tensor we have $t_{13} = t_{31}$ (four independent components). For a pseudo-tensor $\{t_{ij}\}$ in PSG 2 there exists no difference to a polar tensor. This is not the case in PSG m. Using the transformation condition for pseudo tensors we have

$$t'_{ii} = t_{ii} = -t_{ii}, \quad t'_{12} = t_{12}, \quad t'_{21} = t_{21}, \quad t'_{13} = t_{13} = -t_{13},$$
$$t'_{31} = t_{31} = -t_{31} \quad \text{as well as} \quad t'_{23} = t_{23} \quad \text{and} \quad t'_{32} = t_{32}.$$

Thus all tensor components vanish in PSG m except t_{12}, t_{21}, t_{23}, and t_{32}. Polar tensors and pseudo tensors show complementary forms in PSG m.

Orthorhombic System (PSG 22, mm, 2/mm)
In the standard setting $\boldsymbol{e}_i \parallel \boldsymbol{a}_i$ only such components exist in PSG 22 where the indices 1 and 2 (and hence 3) occur an even number of times, namely t_{11}, t_{22}, and t_{33}.

In PSG mm one gets for the pseudo tensors in the standard setting ($\boldsymbol{e}_1$ and $\boldsymbol{e}_2$ perpendicular to the symmetry planes) only the two nonvanishing components t_{12} and t_{21}.

Trigonal System (PSG 3, 32, 3m, $\bar{3}$, $\bar{3}$m)
The three-fold rotation axis or rotation–inversion axis may run parallel to $\boldsymbol{e}_3$. The respective symmetry operations are

$$\boldsymbol{R}_{\pm 3 \parallel e_3} = \begin{pmatrix} -1/2 & \pm\sqrt{3}/2 & 0 \\ \mp\sqrt{3}/2 & -1/2 & 0 \\ 0 & 0 & 1 \end{pmatrix}$$

and

$$\boldsymbol{R}_{\pm \bar{3} \parallel e_3} = \begin{pmatrix} 1/2 & \mp\sqrt{3}/2 & 0 \\ \pm\sqrt{3}/2 & 1/2 & 0 \\ 0 & 0 & \bar{1} \end{pmatrix}.$$

The signs + and − designate a clockwise or anti-clockwise rotation, respectively.

The conditions for polar tensors in the case of a three-fold rotation axis are

$$t'_{11} = t_{11} = 1/4\,t_{11} + 3/4\,t_{22} \mp \sqrt{3}/4(t_{12} + t_{21})$$
$$t'_{22} = t_{22} = 3/4\,t_{11} + 1/4\,t_{22} \pm \sqrt{3}/4(t_{12} + t_{21})$$
$$t'_{23} = t_{23} = -1/2\,t_{23} \mp \sqrt{3}/2\,t_{13}$$
$$t'_{13} = t_{13} = -1/2\,t_{13} \pm \sqrt{3}/2\,t_{23}$$

(and analogous equations for t'_{32} and t'_{31}),

$$t'_{33} = t_{33}.$$

The result must be independent of the direction of rotation. Thus the terms with alternating signs vanish. For the PSG's 3 and $\bar{3}$ it follows that $(t_{12} + t_{21}) = t_{23} = t_{13} = t_{32} = t_{31} = 0$. Consequently, the following tensor components exist: $t_{11} = t_{22}$, t_{33}, and $t_{12} = -t_{21}$.

The PSG 32 with the two-fold axis parallel to $\boldsymbol{e}_1$ (standard setting) also requires $t_{12} = t_{21} = 0$ (index 1 must occur even numbered!). If $\boldsymbol{e}_1$ is not placed along the two-fold axis, but anywhere perpendicular to the three-fold axis, one obtains the same result.

Pseudo tensors of PSG 3m must simultaneously satisfy the conditions for PSG 3 and PSG m. Hence only the components t_{12} and t_{21} remain with $t_{12} = -t_{21}$.

Tetragonal System (PSG 4, 4/m, 42, 4m, $\bar{4}$, $\bar{4}$m, 4/mm)

The symmetry operations of the four-fold rotation axis or rotation–inversion axis parallel to $\boldsymbol{e}_3$ are

$$\boldsymbol{R}_{\pm 4\|e_3} = \begin{pmatrix} 0 & \pm 1 & 0 \\ \mp 1 & 0 & 0 \\ 0 & 0 & 1 \end{pmatrix} \quad \text{and} \quad \boldsymbol{R}_{\pm \bar{4}\|e_3} = \begin{pmatrix} 0 & \mp 1 & 0 \\ \pm 1 & 0 & 0 \\ 0 & 0 & \bar{1} \end{pmatrix}.$$

Since the operations 4 and $\bar{4}$ also contain a two-fold axis parallel to $\boldsymbol{e}_3$, it suffices to consider only the tensor components containing the index 3 even numbered, namely t_{11}, t_{22}, t_{33}, t_{12}, and t_{21}. For polar tensors of PSG 4 the transformation gives

$$t'_{11} = t_{11} = t_{22}, \; t'_{22} = t_{22} = t_{11}, \; t'_{33} = t_{33} \quad \text{and} \quad t'_{12} = t_{12} = -t_{21}.$$

Accordingly, only three independent components exist: $t_{11} = t_{22}$, t_{33}, and $t_{12} = -t_{21}$. 2 or $\bar{2} \equiv$ m parallel to $\boldsymbol{e}_1$ causes t_{12} to vanish.

Pseudo tensors of PSG 4m fulfill the conditions for PSG 4 as well as for PSG mm. Hence only $t_{12} = -t_{21}$ remain as nonvanishing components. Pseudo tensors of PSG $\bar{4}$ are subject to the conditions $t_{11} = -t_{22}$, $t_{33} = 0$, $t_{12} = t_{21}$,

$t_{23} = t_{13} = t_{32} = t_{31} = 0$. The PSG $\bar{4}$m ($\equiv \bar{4}$2m) contains the subgroup 22, thus of those components existing in PSG $\bar{4}$ only $t_{11} = -t_{22} \neq 0$ remain. If we chose the base vectors $\boldsymbol{e}'_1$ and $\boldsymbol{e}'_2$ perpendicular to the symmetry planes, only the components t'_{12} and t'_{21} with $t'_{12} = t'_{21}$ would exist.

Hexagonal System (PSG 6, 6/m, 62, 6m, $\bar{6}$, $\bar{6}$m, 6/mm)

Since a six-fold axis is equivalent to two- and three-fold axes with the same orientation, the symmetry reduction can be considerably simplified. In PSG 6, the conditions for the subgroups 2 and 3 in the case of polar tensors result in the following independent components: $t_{11} = t_{22}$, t_{33}, $t_{12} = -t_{21}$. In PSG 62, t_{12} vanishes with $\boldsymbol{e}_1$ parallel to the two-fold axis (index 1 odd-numbered occurence).

For pseudo tensors of PSG 6m, the symmetry plane perpendicular to $\boldsymbol{e}_1$ demands: $t_{ii} = 0$, $t_{12} = -t_{21}$. Pseudo tensors in PSG $\bar{6} \equiv 3/\text{m}$ and $\bar{6}$m do not exist, because the subgroup m (normals of the symmetry plane parallel to $\boldsymbol{e}_3$) only allows components to exist containing the index 3 once. The three-fold axis does not allow such components, as we have seen.

Cubic System (PSG 23, m3, $\bar{4}$3, 43, m3m)

The three-fold axis common to all cubic PSG's is represented by

$$\boldsymbol{R}_{3\|[111]} = \begin{pmatrix} 0 & 1 & 0 \\ 0 & 0 & 1 \\ 1 & 0 & 0 \end{pmatrix} .$$

Since all cubic PSG's contain the subgroup 22, only the components t_{ii} can appear. These are transformed into each other by the three-fold axis so that only one independent component $t_{11} = t_{22} = t_{33}$ is possible. With polar tensors, addition of further symmetry operations does not result in new conditions. The pseudo tensors vanish totally in PSG $\bar{4}$3, because the operation $\bar{4}$ with the condition $t_{11} = -t_{22}$ contradicts the relation $t_{11} = t_{22} = t_{33}$.

Furthermore, all cubic crystals behave like isotropic substances with respect to second-rank tensor properties. In particular, a pure longitudinal effect exists in all directions.

The case of the cylindrical symmetry groups is homologous to the situation in the corresponding PSG's of the hexagonal system (see Exercise 4).

4.3.2
Tensor Quadric, Poinsots Construction, Longitudinal Effects, Principal Axes' Transformation

The quadric $t_{ij}x_ix_j = F$ is fixed by the six independent quantities t_{11}, t_{22}, t_{33}, $t_{12} + t_{21}$, $t_{13} + t_{31}$, $t_{23} + t_{32}$. They completely represent the symmetrical part of a second-rank tensor. As we shall see, almost all important second-

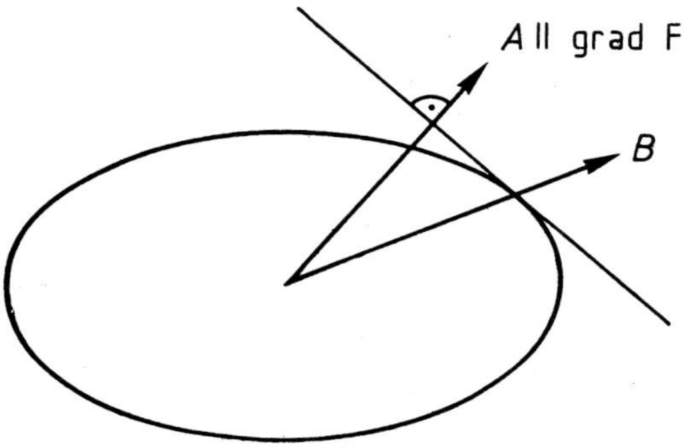

Figure 4.1 Poinsot construction. The tangent plane to the ellipsoid stands perpendicular to the plane of projection.

rank tensors are symmetric. The representation quadric provides an instructive geometric visualization of the tensor in a way which cannot be achieved with higher rank tensors. In the following, we assume a certain knowledge concerning the form of these second-order surfaces (ellipsoids, hyperboloids, paraboloids, spheres, cylinders, and their degeneracies). Let the tensor $\{t_{ij}\}$ operate on the vectors $\boldsymbol{A}$ and $\boldsymbol{B}$ according to $A_i = t_{ij}B_j$. With symmetric tensors, one can then recognize the position of the vector $\boldsymbol{A}$ with known $\boldsymbol{B}$ from the quadric with the help of the *Poinsot construction*. Let the vector $\boldsymbol{B}$, starting at the origin of the coordinate system, intercept the quadric at point P (Fig. 4.1). We imagine the tangent plane drawn at this point on the quadric. The normal on this tangent plane then runs parallel to the wanted vector $\boldsymbol{A}$. The proof for this relationship is as follows: We first consider the differential $\Delta F = (dF/dx_i)\,\Delta x_i$, the first-order approximation of the Taylor expansion of F with respect to the coordinates x_i. We can write ΔF as a scalar product:

$$\Delta F = \operatorname{grad} F \cdot \Delta \boldsymbol{x} = (\operatorname{grad} F)_i \Delta x_i,$$

where grad F (read "Gradient F") denotes the vector $(dF/dx_i)\,\boldsymbol{e}_i$. If the end point of the vector $\boldsymbol{x} + \Delta\boldsymbol{x}$ also lies in the quadric, then $\Delta\boldsymbol{x}$ is parallel to the tangential plane, thus $\Delta F = 0$ and hence the scalar product $\operatorname{grad} F \cdot \Delta\boldsymbol{x} = 0$. Therefore grad F is perpendicular to the quadric and thus runs parallel to the normal on the tangent plane. From $F = t_{ij}x_ix_j$ one gets $dF/dx_i = 2t_{ij}x_j$. Because $\boldsymbol{x} = c\boldsymbol{B}$ (c is a constant) we have $(\operatorname{grad} F)_i = 2ct_{ij}B_j = 2cA_i$, thus $\operatorname{grad} F = 2c\boldsymbol{A}$.

A second important property of the quadric is to read for every direction the corresponding longitudinal effect. As shown in Section 3.10, for any arbitrary direction $\boldsymbol{e}'_i$, the corresponding longitudinal effect is $t'_{ii} = F|\boldsymbol{x}|^{-2}$. The vector $\boldsymbol{x}$ runs from the origin, in the direction $\boldsymbol{e}'_i$, to point P on the quadric.

The complete measurement of symmetric second-rank tensors or the symmetric part of second-rank tensors is possible with the help of longitudinal effects. These possess the advantage that they can be easily measured in practice with sufficient accuracy. Let us demonstrate the procedure on the general example of a symmetric tensor in the triclinic system. The tensor is represented in the crystal-physical reference system by the basic vectors $\boldsymbol{e}_i$. The measurement of the longitudinal effect along $\boldsymbol{e}_i$ leads directly to the components t_{ii} (t_{11}, t_{22}, t_{33}). Further measurements of longitudinal effects along the bisectors of any two basic vectors, that is, in the direction $\boldsymbol{e}'_{i\pm} = \sqrt{2}/2(\boldsymbol{e}_j \pm \boldsymbol{e}_k)$ with $i \neq j,k$ and i, j, k cyclic, deliver the longitudinal effects

$$t'_{ii\pm} = \tfrac{1}{2}(t_{jj} + t_{kk}) \pm t_{jk}$$

(application of the transformation formula). The unknown components t_{jk} are found from

$$t'_{ii+} - t'_{ii-} = 2t_{jk}.$$

The sums $t'_{ii+} + t'_{ii-} = t_{jj} + t_{kk}$ allow a control of the first series of measurements. Hence, all six independent components can be determined. Since the three principal components t_{11}, t_{22}, and t_{33} are already known from the measurements along $\boldsymbol{e}_i$, only a few measurements are needed in the direction of the bisectors to obtain t_{jk}. In another version, only the longitudinal effects along the six different bisectors are used. Of course, one can also choose longitudinal effects in other directions. The evaluation is then less transparent. We will come to concrete examples later. The measurement of the asymmetric part of a second-rank tensor is not possible without the inclusion of transverse effects.

A transformation of the reference system can bring a second-order surface into a special position in which all geometric parameters are directly evident. This is the *principal axes' transformation*. In such a principal axes' reference system, the longitudinal effects along the basic vectors of the reference system take on extreme values. An equivalent statement is that in the direction of the basic vectors pure longitudinal effects appear, i.e., no transverse effects occur when the inducing quantities act along these basic vectors. The condition for a pure longitudinal effect in the tensor relation $A_i = t_{ij}B_j$ is $A_i = \lambda B_i$, where λ is the corresponding longitudinal effect (eigenvalue). Hence one obtains three equations $A_i = \lambda B_i = t_{ij}B_j$ for $i = 1,2,3$ or $(t_{ij} - \lambda\delta_{ij})B_j = 0$, where δ_{ij} is the Kronecker symbol.

For the calculation of the extreme values of the longitudinal effect, which are connected with the square of the radius vector $\boldsymbol{x}$ of the quadric $F = t_{ij}x_ix_j$, we construct, according to the rules of variational calculus, an auxiliary function $H = F - \lambda\boldsymbol{x}^2$, which also takes on an extreme value, as long as F remains constant. That is, the end point of $\boldsymbol{x}$ lies on the quadric and in addition, $\boldsymbol{x}^2$

itself takes on an extreme value. λ is still an open factor, not dependent on $\boldsymbol{x}$, a so-called Lagrangian multiplier. The condition for an extreme value is then $\partial H/\partial x_i = 0$, thus

$$\frac{\partial H}{\partial x_i} = \frac{\partial F}{\partial x_i} - \lambda 2 x_i = 2t_{ij}x_j - 2\lambda x_i = 0 \quad \text{for} \quad i = 1,2,3,$$

where we here assume $\{t_{ij}\}$ to be a symmetric tensor. We write this system of three linear equations in the form $(t_{ij} - \lambda\delta_{ij})x_j = 0$. These conditions are identical with those of a pure longitudinal effect. The system with $\boldsymbol{x} \neq 0$ only has a solution when the associated determinant vanishes, that is

$$\begin{vmatrix} t_{11} - \lambda & t_{12} & t_{13} \\ t_{12} & t_{22} - \lambda & t_{23} \\ t_{13} & t_{23} & t_{33} - \lambda \end{vmatrix} = 0.$$

We obtain a third-order equation in λ, the so-called characteristic equation. It is an invariant, as one recognizes from the transformation behavior of determinants. In particular, the coefficients of the characteristic equation are also scalar invariants of the tensor, i.e., quantities that assume the same value in any arbitrary Cartesian reference system. The characteristic equation possesses three solutions specified by λ_1, λ_2, and λ_3. Written out, the characteristic equation is

$$\begin{aligned} -\lambda^3 + \lambda^2(t_{11} + t_{22} + t_{33}) - \lambda(-t_{23}^2 + t_{22}t_{33} - t_{13}^2 + t_{11}t_{33} - t_{12}^2 + t_{11}t_{22}) \\ + |t_{ij}| = 0, \end{aligned}$$

where $|t_{ij}|$ stands for the determinant of the matrix of the tensor components. From the alternative writing of the characteristic equation in the form $(\lambda_1 - \lambda)(\lambda_2 - \lambda)(\lambda_3 - \lambda) = 0$ we see that the three invariants have the following form:

$$\begin{aligned} I_1 &= t_{11} + t_{22} + t_{33} &&= \lambda_1 + \lambda_2 + \lambda_3, \\ I_2 &= t_{11}t_{22} + t_{22}t_{33} + t_{11}t_{33} - t_{12}^2 - t_{13}^2 - t_{23}^2 &&= \lambda_1\lambda_2 + \lambda_2\lambda_3 + \lambda_1\lambda_3, \\ I_3 &= |t_{ij}| &&= \lambda_1\lambda_2\lambda_3. \end{aligned}$$

The solution of the characteristic equation can be quickly found with the aid of an iteration method using the prescription $A_i^{(n+1)} = t_{ij}A_j^{(n)}$ which produces the new vector $\boldsymbol{A}^{(n+1)}$ of the next iteration step from the vector $\boldsymbol{A}^{(n)}$. For increasing n, $\boldsymbol{A}^{(n+1)}$ more closely approaches a direction with an extreme radius vector of the quadric, as one can read from the Poinsot construction. With this method, one obtains the direction for the maximum longitudinal effect. In practice, this is today naturally carried out with the help of a personal computer, which gives the solutions in a few seconds. For each eigenvalue λ

one obtains the direction of the associated basic vector (eigenvector) from the system of equations according to

$$x_1 : x_2 : x_3 =$$
$$[(t_{22} - \lambda)(t_{33} - \lambda) - t_{23}^2] : [(t_{23}t_{13} - t_{12}(t_{33} - \lambda)] : [t_{12}t_{23} - t_{13}(t_{22} - \lambda)].$$

This follows from the representation of our system of equations in the form of scalar products of the vector $\boldsymbol{x}$ with the vectors $\boldsymbol{C}_i$, having the components $C_{ij} = t_{ij} - \lambda\delta_{ij}$. We have $\boldsymbol{C}_i \cdot \boldsymbol{x} = 0$ for $i = 1, 2, 3$, i.e., $\boldsymbol{x}$ is perpendicular to the vectors $\boldsymbol{C}_i$ and thus runs parallel to one of the vector products of two of these vectors $\boldsymbol{C}_i$. The above proportionality results from $\boldsymbol{x} \parallel \boldsymbol{C}_2 \times \boldsymbol{C}_3$. The directions of the extreme values are mutually perpendicular. This is recognized as follows: Let λ' and λ'' be two arbitrary eigenvalues of the characteristic equation and $\boldsymbol{x}'$ and $\boldsymbol{x}''$ the associated eigenvectors. We then have the equations

$$t_{ij}x'_j - \lambda' x'_i = 0, \quad t_{ij}x''_j - \lambda'' x''_i = 0.$$

We multiply the ith equation of the first system with x''_i, the second with x'_i, sum the three equations and get

$$t_{ij}x'_jx''_i - \lambda' x'_i x''_i = 0 \quad \text{and} \quad t_{ij}x''_j x'_i - \lambda'' x''_i x'_i = 0.$$

We form the difference of these expressions and note that

$$t_{ij}x'_jx''_i = t_{ij}x''_jx'_i \qquad (\text{because } t_{ij} = t_{ji});$$

thus

$$(\lambda' - \lambda'')x'_ix''_i = 0.$$

If $\lambda' \neq \lambda''$, it follows that

$$x'_ix''_i = \boldsymbol{x}' \cdot \boldsymbol{x}'' = 0;$$

therefore the eigenvectors $\boldsymbol{x}'$ and $\boldsymbol{x}''$ are mutually perpendicular. If we place the basic vectors of our Cartesian reference system parallel to these eigenvectors, the transverse components t'_{ij} $(i \neq j)$ vanish and the relationship between the vectors $\boldsymbol{A}$ and $\boldsymbol{B}$ is $A_i = t'_{ii}B_i$ with $t'_{ii} = \lambda_i$. This means that the physical efficacy of a symmetric second-rank tensor is fully described by the three eigenvalues (also called principal values) when, in addition, the position of the principal axes (eigenvectors) of the tensor in the crystal-physical system is known.

The symmetry properties of crystals also determine of course, according to Neumann's principle, the form and position of the quadric in the crystal. For example, the quadric of all second-rank tensors with a three-, four-

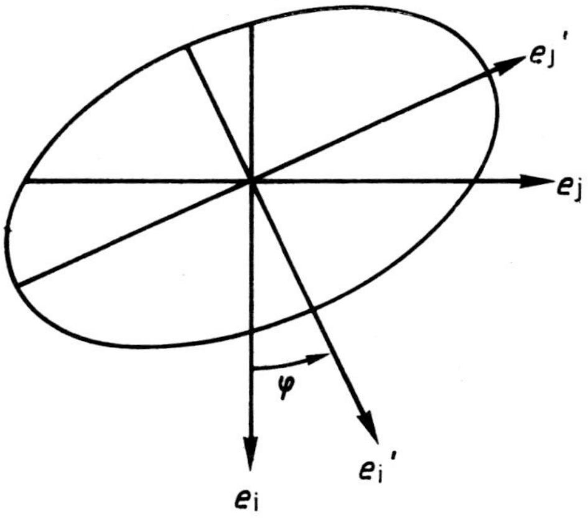

Figure 4.2 Principal axes transformation within a plane perpendicular to e_k.

or six-fold rotation- or rotation–inversion axis possesses rotational symmetry with just this symmetry axis as the rotation axis. In the orthorhombic system, the principal axes of the quadric run parallel to the basic directions of the orthorhombic system (parallel to the two-fold rotation- or rotation–inversion axes respectively). Furthermore, the arrangement of the crystals, as based on the specifications in Table 4.2, corresponds in all systems, except the triclic and monoclinic, to the principal axes' system of second-rank tensors.

Of special practical importance is the *plane principal axis transformation*. We consider a cut through the quadric perpendicular to the basic vector $\boldsymbol{e}_k$. An example of such a cut is represented by the ellipse in Fig. 4.2. The other two basic vectors $\boldsymbol{e}_i'$ and $\boldsymbol{e}_j'$ shall be placed parallel to the principal axes of the given sectional plane by a rotation of an angle φ (positive in the clockwise direction looking along $\boldsymbol{e}_k$!) about $\boldsymbol{e}_k$. The transformation formulae are

$$\boldsymbol{e}_i' = \cos\varphi\,\boldsymbol{e}_i + \sin\varphi\,\boldsymbol{e}_j, \quad \boldsymbol{e}_j' = -\sin\varphi\,\boldsymbol{e}_i + \cos\varphi\,\boldsymbol{e}_j, \quad \boldsymbol{e}_k' = \boldsymbol{e}_k.$$

The principal axis position is reached when the tensor components t_{ij}' $(i \neq j)$ vanish. For a symmetric tensor we get by tensor transformation

$$\begin{aligned} t_{ii}' &= \cos^2\varphi\,t_{ii} + \sin^2\varphi\,t_{jj} + 2\sin\varphi\cos\varphi\,t_{ij} \\ t_{jj}' &= \sin^2\varphi\,t_{ii} + \cos^2\varphi\,t_{jj} - 2\sin\varphi\cos\varphi\,t_{ij} \\ t_{ij}' &= -\sin\varphi\cos\varphi\,t_{ii} + \sin\varphi\cos\varphi\,t_{jj} + (\cos^2\varphi - \sin^2\varphi)t_{ij} \end{aligned}$$

(with $t_{ij} = t_{ji}$).

With $2\sin\varphi\cos\varphi = \sin 2\varphi$ and $\cos^2\varphi - \sin^2\varphi = \cos 2\varphi$, we have from the condition $t'_{ij} = 0$ the angle of rotation

$$\tan 2\varphi = \frac{2t_{ij}}{t_{ii} - t_{jj}}.$$

From this and with the help of the relations

$$\cos^2\varphi = \frac{1+\cos 2\varphi}{2} \quad \text{and} \quad \sin^2\varphi = \frac{1-\cos 2\varphi}{2}$$

one finds

$$t'_{ii} = \frac{t_{ii} + t_{jj}}{2} + \frac{t_{ij}}{\sin 2\varphi},$$
$$t'_{jj} = \frac{t_{ii} + t_{jj}}{2} - \frac{t_{ij}}{\sin 2\varphi}.$$

It is also interesting that with each arbitrary rotation about the axis $\boldsymbol{e}_k$ the sum of the principal components remains constant: $t'_{ii} + t'_{jj} = t_{ii} + t_{jj}$. In the monoclinic system, plane and general principal axes' transformations are identical, if $\boldsymbol{e}_k$ is laid parallel to 2 or $\bar{2}$, respectively. Examples will follow in the following sections.

4.3.3
Dielectric Properties

An electric field induces in matter a charge displacement described by the electric polarization $\boldsymbol{P}$ which we introduced previously. The electric field is represented by the vector of the electric field strength. This points into the direction of the force $\boldsymbol{K}$ experienced by a positive electric point charge q in the electric field according to $\boldsymbol{K} = q\boldsymbol{E}$. The relationship between the electric charge density and the electric field strength in a plate capacitor leads us to the vector of the electric displacement. One plate of the capacitor with the surface F and charge Q possesses a charge density of Q/F (ignoring boundary effects). The vector $\boldsymbol{D} = K\frac{Q}{F}\boldsymbol{e}$ perpendicular to the plate, where $\boldsymbol{e}$ specifies the surface normal of the plate, is in this case the vector of electric displacement. K is a constant dependent on the system of units (MKSA system: $K = 1$; cgs system: $K = 4\pi$). The field strength $\boldsymbol{E}$ takes on the same direction. Inside the capacitor the field strength is practically constant, so that the potential difference of a unit charge on one plate in relation to the charge on the opposite plate is

$$U = \int_{x_1}^{x_2} \boldsymbol{E} \cdot d\boldsymbol{x} = |\boldsymbol{E}|d,$$

where d is the distance between the plates. The electric voltage U at the capacitor determines the field strength. The charge Q and thus the magnitude of the electric displacement is proportional to the applied voltage, as one can demonstrate experimentally as well as by a simple consideration of electrostatics. Accordingly, in vacuum $\boldsymbol{D}$ and $\boldsymbol{E}$ are parallel, hence $\boldsymbol{D} = \epsilon_0 \boldsymbol{E}$. The quantity ϵ_0 is called the absolute dielectric constant. Its numerical value in the MKSA system is

$$\epsilon_0 = \frac{1}{4\pi \cdot c^2} 10^7 = 8.854 \times 10^{-12} \mathrm{AsV^{-1}m^{-1}}$$

(c is the velocity of light in vacuum $2.998 \times 10^8 \mathrm{ms^{-1}}$); in the cgs system

$$\epsilon_0 = 1 \quad \text{(dimensionless)}.$$

If we now place an isotropic medium between the plates, $\boldsymbol{D}$ and $\boldsymbol{E}$ remain parallel. The total electric displacement is determined by the contribution of the polarization of the medium and we have the relationship $\boldsymbol{D} = \epsilon_0 \boldsymbol{E} + \boldsymbol{P}$. Since, to a first approximation, $\boldsymbol{P}$ may be considered as proportional to the electric field, we can write $\boldsymbol{D} = \epsilon \boldsymbol{E}$ with $\boldsymbol{P} = (\epsilon - \epsilon_0)\boldsymbol{E}$. We call ϵ the absolute dielectric constant of the given isotropic medium. It is useful to refer ϵ to the value of the vacuum constant ϵ_0. $\epsilon_{\mathrm{rel}} = \epsilon/\epsilon_0$ is the relative dielectric constant (DC). Sometimes it is advantageous to use the relation $\boldsymbol{P} = \chi \epsilon_0 \boldsymbol{E}$. χ is called the electric susceptibility. With the above relation we have $\chi = \epsilon_{\mathrm{rel}} - 1$.

In anisotropic media, instead of $\boldsymbol{D} = \epsilon \boldsymbol{E}$ we must introduce the general vector function $D_i = \epsilon_{ij} E_j$. $\{\epsilon_{ij}\}$ is the *dielectric tensor.* $\{\epsilon_{ij}\}$ is symmetric, as one can recognize from the energy density W_{el} of the electric field in a plate capacitor.

A change in the electric displacement ΔD_i leads to a reversible change of the energy density

$$\Delta W_{\mathrm{el}} = E_i \Delta D_i = \epsilon_{ij} E_i \Delta E_j;$$

hence

$$\frac{\partial W_{\mathrm{el}}}{\partial E_j} = \epsilon_{ij} E_i.$$

The total energy density is

$$W_{\mathrm{el}} = \int E_i dD_i = \int E_i \epsilon_{ij} dE_j = \frac{1}{2} \epsilon_{ij} E_i E_j.$$

Because

$$\frac{\partial^2 W_{\mathrm{el}}}{\partial E_j \partial E_i} = \frac{\partial^2 W_{\mathrm{el}}}{\partial E_i \partial E_j}$$

we have $\epsilon_{ij} = \epsilon_{ji}$.

The measurement of the dielectric properties takes place preferably with the help of the longitudinal effect of thin crystal plates which fill the inner space of a plate capacitor. The capacitance of a simple plate capacitor is described to a good approximation by $C = \epsilon_{\mathrm{rel}} C_0$. C_0 is the capacitance of the evacuated capacitor. It is expedient to measure the capacitance as a frequency determining term of a high frequency oscillator via a bridge circuit. In order to keep the highly interfering depolarization phenomena at the boundary regions of the specimens small, the diameter-to-thickness ratio of the plates should be at least 10.

The plate capacitor is realized by metallization of the large surfaces of a thin plane-parallel crystal plate (coated with a conducting silver paste or vacuum evaporated by silver or gold). In this case $C = \epsilon_{\mathrm{rel}} \epsilon_0 F/d$, where d and F refer to the thickness and area of the plate, respectively. If the weight G of the plate and the density ρ are known, one obtains very accurate values for $F = V/d = G/\rho d$ (V being the volume of the plate).

For many purposes, the *immersion method* (Fig. 4.3a) achieves sufficient accuracy. The specimen, in the form of a thin plate, is placed in a cylindrical measurement cell. The isolated main electrode should be nearly fully covered by the specimen. The remaining space in the cell is filled, bubble free,

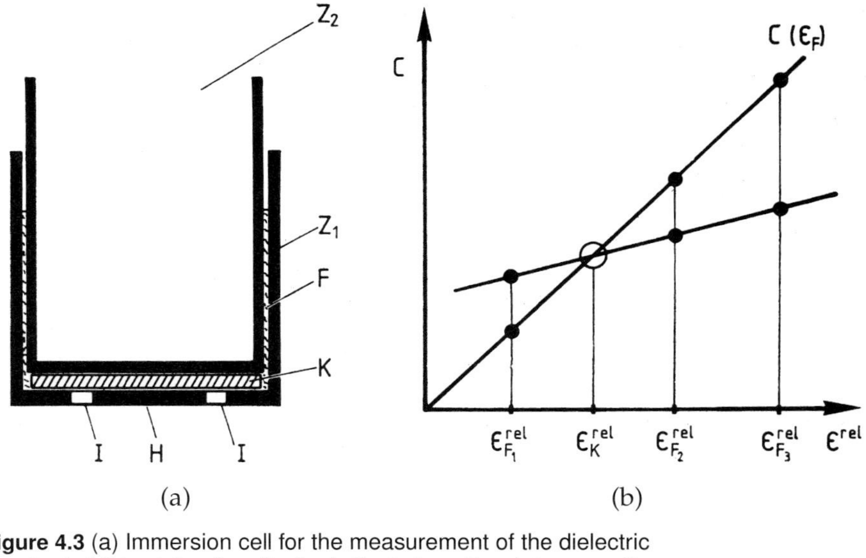

Figure 4.3 (a) Immersion cell for the measurement of the dielectric constant of thin plates (longitudinal effect along the plate normal). H main electrode, I isolation, K crystal plate, F immersion liquid, Z_1 external cylinder, Z_2 internal cylinder. Z_1 and Z_2 represent the base electrode. (b) Graphic evaluation of the data obtained by the immersion method. C is the measured capacity of the arrangement in arbitrary units.

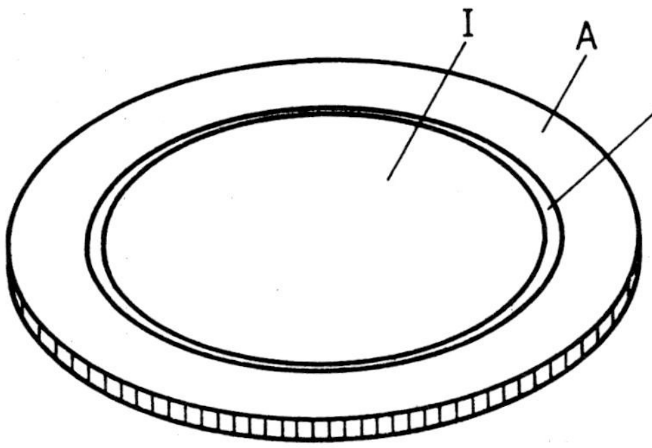

Figure 4.4 Arrangement of the electrodes for the measurement of the longitudinal dielectric constant of thin plates. I main electrode, A outside or guard electrode, and S free gap.

with a liquid of known dielectric constant. Using a bridge circuit, one now measures the capacitance of this capacitor, filled with crystal and liquid and only with liquid, respectively. The knowledge of the absolute capacitance is not necessary; rather it suffices to read a reproducible setting of the variable plate capacitor in the bridge circuit. This procedure is carried out with several liquids. The liquids are selected such that the effective dielectric constant of the specimen lies between those of the liquids. One now plots the measured readings of the compensation capacitor against the known dielectric constants of the liquids and gets to a good approximation a straight line (Fig. 4.3b). In all measurements, the cell is to be adjusted to the thickness of the specimen. Now the measured values for the specimen are plotted as a function of the dielectric constants of the liquids. The line connecting the points may show a slight curvature. The abscissa of the intersection point of both curves yields the wanted longitudinal effect.

If the specimens are homogenous, an accuracy of up to 1 per mille can be achieved with this method without further precautions when the values for the liquids are sufficiently well known. In this case, the specimens need not be prepared with conducting electrodes. Within limits, reasonable values can also be obtained using plates of ill-defined geometry with pieces broken off and unequal thickness. This applies, in particular, to cubic crystals with their isotropic behavior.

If one requires higher accuracy, as, e.g., for the measurement of the pressure- or temperature dependence of the dielectric properties, a method using a guard electrode is recommended in which nearly all interferences due to boundary effects are suppressed (see Fig. 4.4). An arrangement for highly precise measurements was described by Andeen *et al.* (1971). This so-called *substitution method* with a three-electrode arrangement works on a principle similar to the immersion method. The capacitance C is measured between the main electrode and the base electrode, whereby, in magnitude and phase, the same alternating voltage present at the main electrode is applied to the guard electrode, separated by a small gap from the main electrode. As a consequence, a largely parallel and homogenous electric field forms in the region of the crystal above the main electrode. Altogether, one requires the following four measurements: 1. Filling with air:

$$C_1 = \epsilon_0 \epsilon_{\mathrm{A}} F / d_0,$$

2. filling with crystal plate and air:

$$1/C_2 = (d_0 - d_1)/(\epsilon_0 \epsilon_A F) + d_1/(\epsilon_0 \epsilon'_{\mathrm{rel},11} F),$$

3. filling with liquid:

$$C_3 = \epsilon_0 \epsilon_{\mathrm{Fl}} F / d_0,$$

4. filling with crystal and liquid:

$$1/C_4 = (d_0 - d_1)/(\epsilon_0 \epsilon_{\mathrm{Fl}} F) + d_1/(\epsilon_0 \epsilon'_{\mathrm{rel},11} F),$$

where ϵ_0 is the absolute dielectric constant of the vacuum, ϵ_{A}, ϵ_{Fl}, $\epsilon'_{\mathrm{rel},11}$ are the relative dielectric constants of air, the liquid, and the crystal along the plate normal $\boldsymbol{e}'_1$. d_0 is the separation between main and base electrode, d_1 is the thickness of the plane-parallel crystal plate, and F is the effective area of the main electrode. The capacitances in cases 2 and 4 result from the formula of capacitances in serial connection ($1/C = 1/C_a + 1/C_b$). The four relations allow the elimination of d_0, d_1, ϵ_{Fl}, and F.

The wanted dielectric constant of the plate (longitudinal effect) is then

$$\epsilon'_{\mathrm{rel},11} = \frac{1 - C_3/C_1 + C_3/C_2 - C_3/C_4}{C_1/C_2 - C_3/C_4} \epsilon_{\mathrm{A}}.$$

For ϵ_{A} one can write, to a sufficient approximation, the value 1.000 53 for 293 K and atmospheric pressure.

Up to now we assumed that the dielectric tensor possesses real components. Let us give a short comment on the general case of a complex dielectric constant. It only plays a role in time-dependent processes. Let the components ϵ_{ij} be represented as $\epsilon_{ij} = \epsilon'_{ij} + i\epsilon''_{ij}$, where ϵ'_{ij} and ϵ''_{ij} are real quantities.

We imagine a periodic electric field $\boldsymbol{E} = \boldsymbol{E}_0 e^{2\pi i \nu t}$ producing an electric displacement $\boldsymbol{D} = \boldsymbol{D}_0 e^{2\pi i \nu t - i\delta}$ in a crystal. ν is the frequency and δ the phase difference of both waves. For simplicity, we consider the situation of a pure longitudinal effect with $\boldsymbol{D} = \epsilon \boldsymbol{E}$, where $\epsilon = \epsilon' + i\epsilon''$. We then have

$$\frac{D_0}{E_0} e^{-i\delta} = (\epsilon' + i\epsilon'')$$

and hence

$$\epsilon'' = \frac{-D_0}{E_0} \sin\delta \quad \text{and} \quad \tan\delta = \frac{-\epsilon''}{\epsilon'}$$

(because $e^{-i\delta} = \cos\delta - i\sin\delta$).

We now calculate the dielectric loss per second, the dissipated energy

$$L = \frac{1}{T_0}\int_0^{T_0} \boldsymbol{I}\cdot\boldsymbol{E}dt.$$

T_0 is the period of oscillation.

Here we require the relation between the current density vector $\boldsymbol{I}$ and the electric field strength as expressed in Ohm's law (see Section 4.3.7).

We have $I_i = (s'_{ij} + is''_{ij})E_j$, where the conductivity may also be described by a complex tensor quantity. In the case of a pure longitudinal effect , one has $\boldsymbol{I} = (s' + is'')\boldsymbol{E}$. Furthermore, the current density is proportional to the time derivative of the electric displacement, thus

$$\boldsymbol{I} = K^{-1}\frac{d\boldsymbol{D}}{dt} = K^{-1}(\epsilon' + i\epsilon'')\frac{d\boldsymbol{E}}{dt}.$$

As a result we have $s' = -K^{-1}\cdot 2\pi\nu\epsilon''$ and $s'' = K^{-1}\cdot 2\pi\nu\epsilon'$. K is the constant of the chosen system of units.

Only the real part is relevant for the calculation of the dissipated energy. Hence we obtain

$$L = \frac{1}{T_0}\int_0^{T_0} s' E_0^2 \cos^2 2\pi\nu t dt = \frac{1}{2}E_0^2 s' = K^{-1}\pi\nu D_0 E_0 \sin\delta.$$

Thus, a fraction of the energy of the electric field, proportional to $\sin\delta$, is continuously converted into ohmic heat. The measurement of $\sin\delta$ follows directly from the comparison of the phase position of $\boldsymbol{E}$ and the displacement current or by a comparison of the reactive power and effective power.

As an example, we discuss the measurement of the dielectric tensor on triclinic lithium hydrogenoxalate-hydrate ($LiHC_2O_4 \cdot H_2O$), PSG 1. Table 4.3 presents the measured longitudinal effects for altogether eight different directions.

The crystal-physical reference system is connected to the crystallographic reference system used by Thomas (1972) according to the convention described in Section 2.2.

The measurements were made on circular plates with diameters of about 20 mm and thicknesses of about 1.5 mm. The immersion liquids used were m-xylene (ϵ_{rel} =2.37), chlorobenzene (ϵ_{rel} =5.71), and ethyl dichloride (ϵ_{rel} = 10.65). A plate of NaCl with ϵ_{rel} =5.87 was used for control. The data are valid for 20°C.

Table 4.3 Dielectric measurements on triclinic $LiHC_2O_4 \cdot H_2O$ at 293 K. Values of the longitudinal effect $\epsilon'_{rel,11}$ along the direction $e'_1 = u_{1i}e_i$ (plate normal) for a frequency of 10 MHz.

Nr.	u_{11}	u_{12}	u_{13}	$\epsilon'_{rel,11}$
1	1	0	0	6.77
2	0	1	0	5.43
3	0	0	1	4.91
4	0	$\sqrt{2}/2$	$\sqrt{2}/2$	4.70
5	0	$\sqrt{2}/2$	$-\sqrt{2}/2$	5.33
6	−0.6996	0.7145	0	6.58
7*)	−0.4591	0.1662	0.873	3.89
8	−0.2767	0.8035	0.5263	4.71

*) Normal on the cleavage face $(\bar{1}01)$

Measurements 1, 2, and 3 directly yield $\epsilon_{rel,11}$, $\epsilon_{rel,22}$, and $\epsilon_{rel,33}$. From 4 and 5 and using the relation mentioned in Section 4.3.2 one gets $\epsilon_{rel,23} = -0.41$; similarly, from 6 $\epsilon_{rel,12} = -0.54$. Applying the general formula for the longitudinal effect of measurement in direction 7 $\epsilon_{rel,13} = +1.73$ is obtained. The longitudinal effect in direction 8, calculated from these values of the dielectric constants, is in good agreement with the experimental value.

Through a principal axes' transformation one finds the principal values

$$\epsilon'_{rel,11} = 8.10, \quad \epsilon'_{rel,22} = 5.26\,; \text{ and } \epsilon'_{rel,33} = 3.75.$$

The associated eigenvectors, referred to the crystal-physical reference system are

$$\begin{aligned} e'_1 &= 0{,}822e_1 - 0{,}246e_2 + 0{,}514e_3, \\ e'_2 &= -0{,}232e_1 - 0{,}968e_2 - 0{,}094e_3, \\ e'_3 &= 0{,}520e_1 - 0{,}044e_2 - 0{,}853e_3. \end{aligned}$$

We thus find a distinct anisotropy with an absolute maximum in a direction close to the cleavage surface and a minimum slightly perpendicular to the cleavage plane $(\bar{1}01)$ (see Table 4.3). We will return to these anisotropy effects when discussing other properties.

A simple model has proved successful for the interpretation of the dielectric properties of isotropic media. It merely takes into account the polarizability of quasi-spherically shaped lattice particles as invariant quantities in the local electric field. The local electric field is composed of the external field and the field produced by the polarized dielectric. The local electric field is approximately given by $\boldsymbol{E}_{loc}$

$$\boldsymbol{E}_{loc} = \boldsymbol{E} + \frac{1}{3\epsilon_0}\boldsymbol{P}.$$

Let an isotropic lattice particle contribute the amount $\boldsymbol{p}_j = \alpha_j \boldsymbol{E}_{loc}$ to the total polarization, under the assumption that the local field has about the same strength for each atomistic lattice particle within the medium; α_j is the polarizability of the jth particle. The total polarization $\boldsymbol{P}$ is then

$$\boldsymbol{P} = \boldsymbol{E}_{\text{loc}} \sum_j N_j \alpha_j ,$$

where N_j represents the number of the jth particle per volume unit. Taking into account the relation $\boldsymbol{P} = (\epsilon - \epsilon_0)\boldsymbol{E}$, one finds

$$\frac{\epsilon_{\text{rel}} - 1}{\epsilon_{\text{rel}} + 2} = \frac{1}{3\epsilon_0} \sum_j N_j \alpha_j .$$

If one is dealing with a substance composed of equal particles, as, e.g., with molecular crystals, one obtains the *Clausius–Mosotti formula*

$$\frac{\epsilon_{\text{rel}} - 1}{\epsilon_{\text{rel}} + 2} = \frac{L\rho}{3\epsilon_0 M} \alpha ,$$

where $L = 6.022 \times 10^{23} \text{mol}^{-1}$ denotes Avogadro's number, M the molar weight, ρ the density, and α the polarizability of the molecule. Let N be the number of molecules per volume element, then $\rho L = NM$. The quantity

$$\frac{\epsilon_{\text{rel}} - 1}{\epsilon_{\text{rel}} + 2} \frac{M}{\rho} = \frac{L}{3\epsilon_0} \alpha$$

is termed the molar polarization.

It enables us to experimentally determine the polarizability of simple compounds from the dielectric constant. If one knows the polarizability of the lattice particles, then with the help of the general formula, one can also estimate the mean dielectric constant of these materials. An important aspect here is the law of addition of polarizability, which, however, does not take into account the mutual interaction of the lattice particles nor the anisotropy of the polarizability.

4.3.4 Ferroelectricity

The temperature dependence of the polarizability and thus also of the dielectric constants of many isotropic substances may be described, according to Debye's approach, approximately by $\alpha = \alpha_0 + \bar{\boldsymbol{p}}^2/3kT$. $\bar{\boldsymbol{p}}$ is a mean value of the dipole moment, k is Boltzmann's constant. If $\bar{\boldsymbol{p}} \neq 0$ and $\partial\alpha_0/\partial T$ is sufficiently small, then the polarizability and hence the dielectric constants decrease weakly with increasing temperature, as one recognizes from $\partial\epsilon_{\text{rel}}/\partial\alpha$ according to the Clausius–Mosotti relation. On the other hand, in crystals with vanishing $\bar{\boldsymbol{p}}$ one often observes, over a wide temperature range, a slight, approximately uniform increase of ϵ_{rel} with increasing temperature.

A totally different behavior of the temperature dependence is observed in certain crystal species, such as potassium calcium tartrate-dihydrate (Rochelle salt), triglycine sulfate, $BaTiO_3$, or KH_2PO_4. An example is shown in Fig. 4.5. When approaching a certain transition temperature, extremely large dielectric constants (DC) are observed, in some cases several powers of ten higher than in normal substances, where DC values rarely surmount the value of 50 (see Tables 12.6, 12.7). In addition, one also finds a rapid decrease of the DC with increasing frequency. In many cases, the temperature dependence of the susceptibility can be approximately represented by a *Curie–Weiss law* of the type

$$\chi = \frac{C_0}{T - T_0},$$

where C_0 is the Curie–Weiss constant and T_0 the Curie temperature.

One just about always sees a hysteresis, i.e., the electric polarization follows the electric field with a certain delay up to the transition point of saturation. Often one also sees a remenance, i.e., the polarization remains unchanged by a change in the direction of the electric field. From the analogy to ferromagnetic properties, this anomalous dielectric behavior was given the name *ferroelectricity*. In the meantime, several hundred ferroelectric substances are known, which even between themselves exhibit very different dielectric properties. Some materials, such as $BaTiO_3$, Rochelle salt, and $NH_4H_2PO_4$ (ADP) as well as their structural derivatives have found wide technical application not only because of their dielectric properties but also because of their piezoelectric properties (see Section 4.4.1).

The atomistic interpretation of ferroelectricity presumes that the crystals possess permanent dielectric dipoles belonging to individual particle or particle complexes. An external electric field can order these dipoles in a parallel direction. The temperature change works against this order. Below a certain temperature T_0, most ferroelectrics self-order these dipoles in parallel directions within domains. Thus these domains possess a primary electric moment and belong to one of the ten pyroelectric point symmetry groups (see Section 4.2.2). The different domains can be considered as electric twins. Above the temperature T_0, the domains vanish and the distribution of the dipoles is static (paraelectric phase). When the pyroelectric easy-axis direction vanishes, the crystal takes on a higher symmetry group which must be a supergroup of the symmetry group of the ferroelectric phase. The total moment per volume element due to the parallel ordering of the dipoles is called spontaneous polarization. This normally occurs along a distinct crystallographic direction.

If the dipoles order themselves in a regular sequence of alternating sign, one talks about *antiferroelectricity* analogous to the situation with antiferromagnetic effects.

4.3.5
Magnetic Permeability

Although the origin of magnetism is quite different certain magnetic phenomena can be described analogously to electrostatic ones. The magnetic quantity

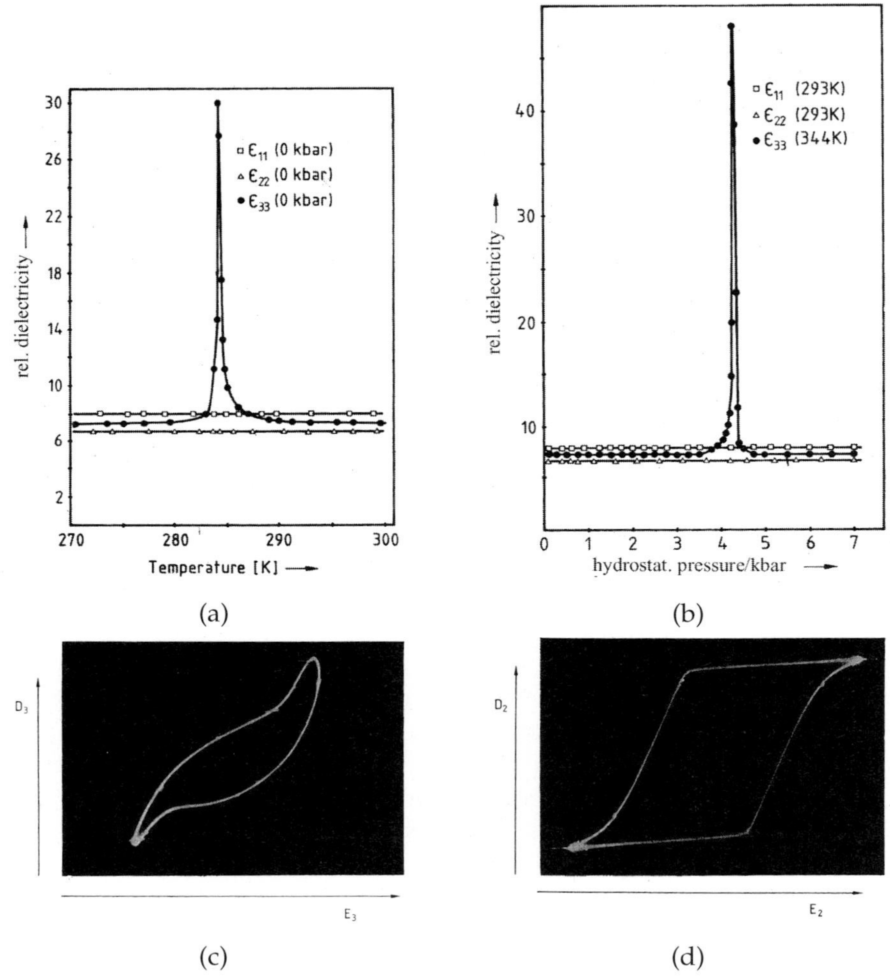

Figure 4.5 Dependence of the dielectric constants on temperature (a) and hydrostatic pressure (b) in $Li_2Ge_7O_{15}$ (paraelectric phase: PSG mmm, ferroelectric phase: PSG mm2; Preu & Haussühl, 1982). ϵ_{33} exhibits a steep anomalous increase at the phase transition. (c) and (d) ferroelectric hysteresis of polarization and external electric field in $C(NH_2)_3Al(SO_4)_2 \cdot 6H_2O$ (GASH) and triglycine sulfate (TGS) in the ferroelectric phase (PSG 3m and 2, respectively; $\boldsymbol{E} \parallel 3$ and $\parallel 2$, respectively). The oscillograms have been recorded at 293 K employing the arrangement suggested by Schubring et al. (1964). GASH shows the rather rare case of an asymmetric hysteresis loop.

corresponding to the electric field strength is the magnetic field strength $\boldsymbol{H}$ pointing into the direction of the force $\boldsymbol{F}$ which the magnetic field exerts on an imaginary magnetic point-like north pole of strength p:

$$\boldsymbol{F} = p\boldsymbol{H}.$$

The field strength at an arbitrary position around a bar magnet points in the direction of the field lines leaving the north pole and ending in the south pole. The quantity p corresponds to the electric charge as in the definition of the electric field strength. Since no isolated point-like magnetic poles exist, the effect of the magnetic field is demonstrated simply by considering the force exerted by a magnetic field on a small bar magnet of magnetic moment $\mathcal{M} = p\boldsymbol{l}$. $\boldsymbol{l}$ is a vector pointing from the imaginary point-like south pole of the magnet to the point-like north pole. The north pole is pulled in the direction of the magnetic field while the south pole experiences a corresponding repulsion. The resulting torque $\boldsymbol{M}$, perpendicular to $\boldsymbol{H}$ and to the magnetic moment $\mathcal{M}$ has the magnitude $|\boldsymbol{H}||\boldsymbol{l}|| \sin \varphi|$, where φ is the angle between $\boldsymbol{l}$ and $\boldsymbol{H}$. Thus $\boldsymbol{M} = \mathcal{M} \times \boldsymbol{H}$ (Fig. 4.6).

If we now place a body in a magnetic field, the magnetic moments of the atoms or molecules also experience a torque which may lead to an alignment of the magnetic moments. The total magnetic moment of the body per volume element is called the magnetization $\boldsymbol{C}$. This corresponds to the vector of the electric polarization. In weak magnetic interactions one expects a linear relationship between the magnetic field $\boldsymbol{H}$ and the magnetization, thus $C_i = \chi_{ij}\mu_0 H_j$. $\{\chi_{ij}\}$ is the *tensor of magnetic susceptibility* and μ_0 is the permeability of the vacuum. By analogy to the electric case, we introduce a further quantity, the magnetic induction $\boldsymbol{B}$, corresponding to the vector of the electric displacement. The definition is

$$B_i = \mu_{ij} H_j = \chi_{ij}\mu_0 H_j + \mu_0 H_i = (\chi_{ij} + \delta_{ij})\mu_0 H_j.$$

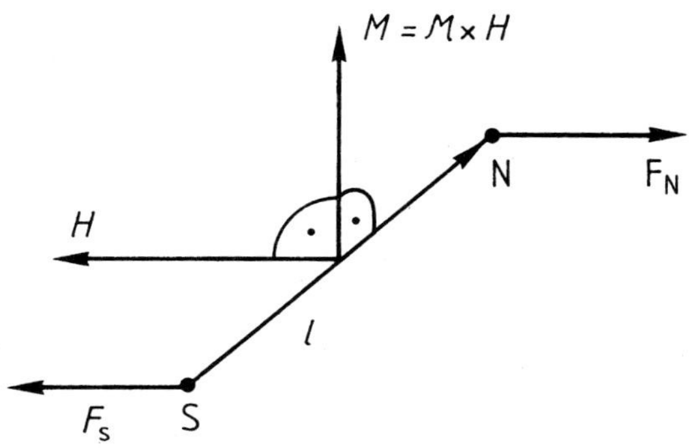

Figure 4.6 Body torque M, which is exerted on a bar magnet $\mathcal{M} = p\boldsymbol{l}$ in a magnetic field $\boldsymbol{H}$.

$\{\mu_{ij}\}$ is the *tensor of the magnetic permeability*.

The tensors $\{\chi_{ij}\}$ and $\{\mu_{ij}\}$ are of polar character because $\boldsymbol{H}$ as well as $\boldsymbol{C}$ and hence $\boldsymbol{B}$ are pseudo tensors (axial vectors). The latter is a consequence of the 2. Maxwell equation

$$\operatorname{rot} \boldsymbol{H} = \boldsymbol{I} + \frac{1}{c}\frac{\partial \boldsymbol{D}}{\partial t};$$

$\boldsymbol{I}$ specifies the electric current density (Ampère's law). Since $\boldsymbol{I}$ and $\boldsymbol{D}$ are polar vectors, rot $\boldsymbol{H}$ must also be polar (see Section 3.7).

Essentially, one distinguishes three groups of magnetic behavior.

1. *Diamagnetism:* Primarily, the given substances possess no magnetic moment. In a magnetic field, a weak magnetization is induced opposing the field. The magnetization is only slightly temperature dependent. Diamagnetic substances are repelled by a magnetic field, the χ-values are negative (of the order 10^{-6}). Typical representatives are most salts of the A subgroup of the periodic table with inorganic anions. Bismuth possesses a particularly large effect.

2. *Paramagnetism:* The magnetization induced by the field is aligned parallel to the field. The magnetization is more strongly temperature dependent than in the case of diamagnetism. Atoms or molecules in the medium possess a magnetic moment. Paramagnetic substances are attracted towards the magnetic field. The χ-values are positive (of the order 10^{-5}). Typical representatives are the transition elements, in particular those of the iron group and the rare earths as well as their salts.

3. *Ferromagnetism:* The magnetic moments possess, within certain temperature ranges, opposite orientations, e.g., parallel or antiparallel arrangements (ferro, antiferromagnetism) or a combination of both (ferrimagnetism). Due to the cooperative alignment of the magnetic moments, the magnetization is much stronger than in the case of paramagnetism. If all moments are aligned parallel, no further increase in magnetization is possible (saturation). From this we recognize that the linear relationship $B_i = \mu_{ij}H_j$ is not sufficient for ferromagnetic substances. Rather, the components μ_{ij} are quite complicated functions of $\boldsymbol{H}$.

In addition, hysteresis effects also occur, i.e., the magnetization lags behind the external magnetic field. The persistence of the magnetization after switching off the external magnetic field is called remanence. Hysteresis is a characteristic phenomenon of ferromagnetism, analogous to the situation of ferroelectricity. In some iron alloys the χ-values can rise up to 10^6. Typical ferromagnetics are iron, nickel, cobalt, and rare earth metals as well as the alloys of these elements. Furthermore, ferromagnetic properties appear in certain compounds of otherwise nonferromagnetic transition elements such as CrTe or MnP.

The measurement of magnetic properties is based on the magnetization generated in the magnetic field. For ferromagnetic substances with large values of susceptibility simple arrangements are sufficient. For example, one can measure the change of magnetic induction inside a cylindrical coil after inserting the probe in a way that is similar to measuring the change in capacitance of a capacitor. The resulting torque experienced by a magnetometer needle is a direct measure of the change in $\boldsymbol{B}$. Calibration with known probes then allows the determination of the susceptibility. The measurement of the inductance of a coil allows a simple and reliable determination of the frequency dependence of the susceptibility.

With diamagnetic and paramagnetic substances, the sensitivity of this method is too low. In order to detect the expected anisotropy in noncubic crystals it is necessary to generate, as far as possible, purely longitudinal effects of magnetization, i.e., the magnetic field and the magnetization should be aligned parallel through the whole probe. We now consider the simplest case of a dia- or paramagnetic crystal without primary magnetic moment. A small probe in a magnetic field experiences a magnetic moment of $\mathcal{M} = V\chi_{\text{eff}}\mu_0\boldsymbol{H} = p\boldsymbol{l}$. V is the volume of the probe and χ_{eff} is the longitudinal effect of the susceptibility in the direction of the magnetic field. In the case of an inhomogenous magnetic field we imagine $\boldsymbol{H}$ represents a mean value over the volume of the probe. Let the magnetic field possess the strength $\boldsymbol{H}_1$ at the south pole and the strength $\boldsymbol{H}_2$ at the north pole, with $\boldsymbol{H}_2 \parallel \boldsymbol{H}_1 \parallel \boldsymbol{l}$. The resulting force on the probe is then

$$\boldsymbol{F} = p(\boldsymbol{H}_2 - \boldsymbol{H}_1) = p|\boldsymbol{l}|\frac{(\boldsymbol{H}_2 - \boldsymbol{H}_1)}{|\boldsymbol{l}|} \approx V\chi_{\text{eff}}\mu_0|\boldsymbol{H}|\frac{d\boldsymbol{H}}{dx},$$

where $(\boldsymbol{H}_2 - \boldsymbol{H}_1)/|\boldsymbol{l}|$ is replaced by the differential quotient $d\boldsymbol{H}/dx$ for the transition to infinitesimal $\boldsymbol{l}$. x is the coordinate along the vector of the magnetization. For practical measurements it is desirable to achieve, as far as possible, a constant value of $d\boldsymbol{H}/dx$ over the whole probe. A particularly simple arrangement is again the cylindrical coil, in which one suspends a cylindrical probe symmetric to the coil axis. A large part of the probe remains outside the coil in order to make the region of the inhomogenous field as effective as possible (Fig. 4.7).

If one works with a cylindrical probe of length L and cross-section Q and sets up the magnetic field so that $d\boldsymbol{H}/dx$ is constant over each cross-section perpendicular to the cylindrical axis, then one can calculate the total force on the probe according to

$$|\boldsymbol{F}| = \int_V \chi_{\text{eff}}\mu_0|\boldsymbol{H}|\left|\frac{d\boldsymbol{H}}{dx}\right| dV = \frac{Q}{2}\chi_{\text{eff}}\mu_0|\boldsymbol{H}^2 - \boldsymbol{H}_0^2|,$$

where dV is replaced by $Q\,dx$. $\boldsymbol{H}$ and $\boldsymbol{H}_0$ are the field strengths at the top and bottom ends of the probe respectively, and can be measured with calibrated probes. The force $\boldsymbol{F}$ due to the vertical field gradients $d\boldsymbol{H}/dx$ of the cylindrical

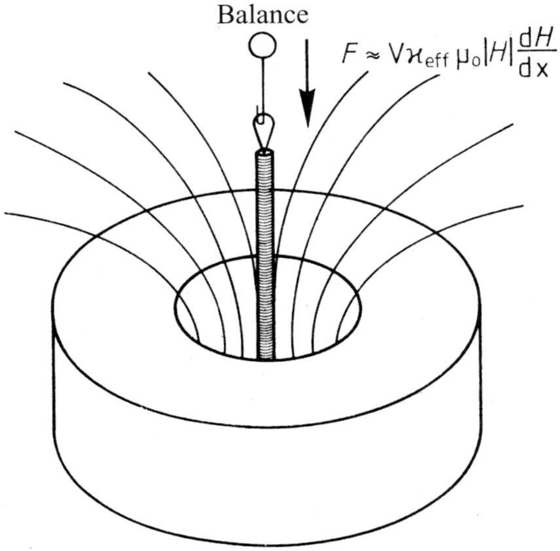

Figure 4.7 Measurement of the magnetization of a cylindrical probe in an inhomogeneous magnetic field.

coil is measured by weighing, in the case of horizontal field gradients with the aid of a torsion balance.

4.3.6
Optical Properties: Basic Laws of Crystal Optics

The optical properties of crystals are essentially determined by their dielectric behavior. In view of the fundamental importance of these properties for practical crystal optics, e.g., in polarization microscopy as well as for many crystal-physical effects, in particular higher order optical effects (e.g., electro-optics, piezo-optics, nonlinear optics) and spectroscopic phenomena, we must carefully discuss some of the basic laws of crystal optics in this section. We will restrict ourselves mainly to the phenomena of wave propagation, which is coupled with the vibrations of the electric displacement vector $\boldsymbol{D}$. These are easier to handle theoretically than energy propagation.

Let us first consider the case of a nonabsorbing (electrically nonconducting) and centrosymmetric medium. The material properties are given by the dielectric tensor and the tensor of the magnetic permeability, hence by the relations $D_i = \epsilon_{ij}E_j$ and $B_i = \mu_{ij}H_j$, whereby all quantities are to be understood as frequency-dependent functions. Other material properties which interact with electromagnetic waves, as represented by light rays, are excluded for the present. In a further simplification we assume that the tensor $\{\mu_{ij}\}$ at the high frequencies of the optical range is quasi-isotropic and for all crystals, described to sufficient accuracy by the vacuum value μ_0 according to $B_i = \mu_0 H_i$. This is also in agreement with experimental observations.

The electrical processes of light propagation obey the fundamental equations of electrodynamics (*Maxwell's equations*, which in the cgs system are (with $\mu_0 = 1$):

$$\operatorname{rot} \boldsymbol{E} = -\frac{1}{c}\frac{\partial \boldsymbol{H}}{\partial t} \quad \text{(induction law)}$$

$$\operatorname{rot} \boldsymbol{H} = \frac{1}{c}\frac{\partial \boldsymbol{D}}{\partial t} \quad \text{(Ampères law of current density and magnetic field)}$$

$$\operatorname{div} \boldsymbol{D} = 0 \quad \text{(no free charge density; otherwise } \operatorname{div} \boldsymbol{D} = 4\pi\rho', \text{ where } \rho' \text{ specifies the electric charge density)}$$

$$\operatorname{div} \boldsymbol{H} = 0 \quad \text{(no magnetic charges exist).}$$

c is the velocity of light, and t the time. The reader is referred to a standard physics textbook for the derivation of Maxwell equations. The most important experimental situation is the propagation of a plane wave, which to a good approximation occurs when the dimensions of the homogenous medium perpendicular to the direction of propagation are far greater than the wavelength of the given radiation. For the magnetic field we take a plane wave of the form

$$\boldsymbol{H} = \boldsymbol{H}_0 e^{2\pi i(\boldsymbol{k}\cdot\boldsymbol{x} - \nu t)}.$$

$\boldsymbol{k}$ is the propagation vector, perpendicular to the wave front with length $|\boldsymbol{k}| = 1/\lambda$. λ is the wavelength, ν the frequency. With $\operatorname{rot} \boldsymbol{v} = \nabla \times \boldsymbol{v}$ and the differential operator

$$\nabla = e_j \frac{\partial}{\partial x_j} \quad \text{(summed over j!)}$$

we obtain from Ampères law

$$\operatorname{rot} \boldsymbol{H} = 2\pi i \boldsymbol{k} \times \boldsymbol{H} = \frac{1}{c}\frac{\partial \boldsymbol{D}}{\partial t}.$$

Hence, it follows that $\boldsymbol{H}$ and $\boldsymbol{D}$ possess the same time and space dependence, thus

$$\boldsymbol{D} = \boldsymbol{D}_0 e^{2\pi i(\boldsymbol{k}\cdot\boldsymbol{x} - \nu t)} \quad \text{and hence only} \quad \frac{\partial \boldsymbol{D}}{\partial t} = -2\pi i \nu \boldsymbol{D}.$$

Because $D_i = \epsilon_{ij} E_j$ and $B_i = \mu_{ij} H_j$ we also have

$$\boldsymbol{E} = \boldsymbol{E}_0 e^{2\pi i(\boldsymbol{k}\cdot\boldsymbol{x} - \nu t)} \quad \text{and} \quad \boldsymbol{B} = \boldsymbol{B}_0 e^{2\pi i(\boldsymbol{k}\cdot\boldsymbol{x} - \nu t)}.$$

We recognize further that $\boldsymbol{D}$ is perpendicular to $\boldsymbol{k}$ and $\boldsymbol{H}$ and the vectors $\boldsymbol{k}$, $\boldsymbol{D}$, $\boldsymbol{H}$ form a right-handed system. From

$$\operatorname{div} \boldsymbol{H} = \frac{\partial H_j}{\partial x_j} = 2\pi i k_j H_j = 2\pi i \boldsymbol{k} \cdot \boldsymbol{H} = 0$$

follows that $\boldsymbol{k}$ and $\boldsymbol{H}$ are orthogonal. This means the waves of $\boldsymbol{D}$ and $\boldsymbol{H}$ are always of pure transversal type.

To which conditions are now $\boldsymbol{D}$ and $\boldsymbol{E}$ subject? Differentiation with respect to time gives, from Ampères law,

$$\frac{\partial^2 \boldsymbol{D}}{\partial t^2} = c \operatorname{rot} \frac{\partial \boldsymbol{H}}{\partial t}$$

and after inserting $\partial \boldsymbol{H}/\partial t$, we get from the induction law

$$\frac{\partial^2 \boldsymbol{D}}{\partial t^2} = -c^2 \operatorname{rot}\operatorname{rot} \boldsymbol{E}.$$

Using the expansion rule, we find

$$\begin{aligned}\operatorname{rot}\operatorname{rot} \boldsymbol{E} = \nabla \times (\nabla \times \boldsymbol{E}) &= \nabla(\nabla \cdot \boldsymbol{E}) - \nabla^2 \boldsymbol{E} \\ = \operatorname{grad} 2\pi i \boldsymbol{k} \cdot \boldsymbol{E} - \sum_j \frac{\partial^2 \boldsymbol{E}}{\partial x_j^2} &= -4\pi^2 (\boldsymbol{k} \cdot \boldsymbol{E})\boldsymbol{k} + 4\pi^2 \boldsymbol{k}^2 \boldsymbol{E}.\end{aligned}$$

Thus we have

$$\frac{\partial^2 \boldsymbol{D}}{\partial t^2} = -4\pi^2 \nu^2 \boldsymbol{D} = c^2 4\pi^2 \{(\boldsymbol{k} \cdot \boldsymbol{E})\boldsymbol{k} - \boldsymbol{k}^2 \boldsymbol{E}\},$$

hence

$$\frac{\nu^2}{c^2} \boldsymbol{D} = \boldsymbol{k}^2 \boldsymbol{E} - (\boldsymbol{k} \cdot \boldsymbol{E})\boldsymbol{k}.$$

After introducing the unit vector in the propagation direction $\boldsymbol{g} = \boldsymbol{k}/|\boldsymbol{k}|$, we obtain the wave equation

$$\frac{v^2}{c^2} \boldsymbol{D} = \boldsymbol{E} - (\boldsymbol{g} \cdot \boldsymbol{E})\boldsymbol{g},$$

where $\nu\lambda = v$ is the propagation velocity of the wave.

If one eliminates $\boldsymbol{D}$ by $D_i = \epsilon_{ij} E_j$, one obtains a system of three linear equations for E_i. The system only has solutions for $\boldsymbol{E} \neq 0$ when its determinant vanishes. We then obtain, for any arbitrary propagation direction $\boldsymbol{g}$, an equation for $v^2/c^2 = 1/n^2$. The ratio n of the velocity of light c in vacuum and propagation velocity v in the medium is known as the *refractive power* or *refractive index*. The equation referred to is therefore also called the *index equation*.

Assume now the axes of our Cartesian reference system have been chosen parallel to the principal axes of the dielectric tensor, where we have $D_j = \epsilon_{jj} E_j$ and $\epsilon_{ij} = 0$ for $i \neq j$. We can then replace E_j by D_j/ϵ_{jj} in the above equation and for each component write

$$D_j \left\{ \frac{v^2}{c^2} - \frac{1}{\epsilon_{jj}} \right\} = -(\boldsymbol{g} \cdot \boldsymbol{E}) g_j.$$

In the directions of the principal axes we have $\boldsymbol{D} \parallel \boldsymbol{E}$, thus because $\boldsymbol{g} \cdot \boldsymbol{D} = 0$ we also have $\boldsymbol{g} \cdot \boldsymbol{E} = 0$; in addition we have $1/\epsilon_{jj} = v^2/c^2$, in case $D_j \neq 0$. We then obtain for the accompanying propagation velocity v_j

$$v_j^2 = \frac{c^2}{\epsilon_{jj}}.$$

The quantities $c/v_j = n_j$ are the *principal refractive indices*. For these $n_j = \sqrt{\epsilon_{jj}}$ is true in the system of units selected here. Especially for the vacuum $n = 1 = \sqrt{\epsilon_0}$ and thus $n_j = \sqrt{\epsilon_{\mathrm{rel},jj}}$ (*Maxwell's relation* for $\mu = 1$).

Consider now the general case where $\boldsymbol{E}$ does not run parallel to $\boldsymbol{D}$. From the above equation, we have for an arbitrary $\boldsymbol{g}$

$$D_1 : D_2 : D_3 = \frac{g_1}{v_1^2 - v^2} : \frac{g_2}{v_2^2 - v^2} : \frac{g_3}{v_3^2 - v^2},$$

where v is the velocity of the wave propagating in direction $\boldsymbol{g}$.

For the determination of v^2 as a function of $\boldsymbol{g}$ we form $\boldsymbol{D} \cdot \boldsymbol{g} = 0$ and obtain

$$\frac{g_1^2}{v_1^2 - v^2} + \frac{g_2^2}{v_2^2 - v^2} + \frac{g_3^2}{v_3^2 - v^2} = 0,$$

resulting in a second-order equation in v^2

$$\sum_{i=1,2,3} g_i^2 (v_j^2 - v^2)(v_k^2 - v^2) = 0 \qquad (j = i+1,\, k = i+2 \bmod 3).$$

For an arbitrary direction $\boldsymbol{g}$, this equation yields two intrinsic values v'^2 and v''^2, which in general, are different. A total of four solutions exist $v = \pm v'$ and $v = \pm v''$. The propagation velocities in the direction $\boldsymbol{g}$ and in the opposite direction $-\boldsymbol{g}$ are equal in magnitude. How are now the accompanying vectors $\boldsymbol{D}'$ and $\boldsymbol{D}''$ oriented to v' and v''? They are mutually perpendicular as seen by forming the scalar product $\boldsymbol{D}' \cdot \boldsymbol{D}''$. We have

$$\begin{aligned} \boldsymbol{D}' \cdot \boldsymbol{D}'' &= \sum_j Q'Q'' \frac{g_j^2}{(v_j^2 - v'^2)(v_j^2 - v''^2)} \\ &= \frac{Q'Q''}{(v'^2 - v''^2} \sum_j \left[\frac{g_j^2}{(v_j^2 - v'^2)} - \frac{g_j^2}{(v_j^2 - v''^2)} \right] = 0, \end{aligned}$$

provided $v'^2 \neq v''^2$.

Q' and Q'' are factors determining the length of $\boldsymbol{D}'$ and $\boldsymbol{D}''$; the terms in the square brackets vanish individually. The case $v'^2 = v''^2$ leaves the position of $\boldsymbol{D}'$ and $\boldsymbol{D}''$ open (degeneracy!). Furthermore, $\boldsymbol{D}'$ and $\boldsymbol{D}''$ lie parallel to the half axes of the sectional ellipse cut out of the tensor quadric $F = a_{ij}x_i x_j$ (ellipsoid) by the central plane perpendicular to the propagation direction $\boldsymbol{g}$, characterized by $\boldsymbol{g}\cdot\boldsymbol{x} = 0$ and containing the center of the ellipsoid.

The components a_{ij} establish the connection of $\boldsymbol{E}$ and $\boldsymbol{D}$ according to $E_j = a_{jk}D_k$. They are called *polarization constants*. As proof we calculate the directions in which the radius vector of the sectional ellipse takes on an extreme value; these directions coincide with the directions of the major and minor half axes of the ellipse. Similar to the principal axes' transformation, the conditions for extreme values here are: $\boldsymbol{x}^2 =$ extremum, auxiliary condition $\boldsymbol{g}\cdot\boldsymbol{x} = 0$ (ellipse lies in the plane perpendicular to $\boldsymbol{g}$) and $a_{ij}x_i x_j = F$ (ellipse belongs to the tensor quadric). Again we assume that our coordinate system is the principal axes' system of the dielectric tensor. We then have $a_{jj} = 1/\epsilon_{jj}$, $a_{jk} = 0$ for $j \neq k$. The tensor quadric is then $\sum_j x_j^2/\epsilon_{jj} = F$ and with $n_j^2 = \epsilon_{jj}$ and $F = 1$

$$\sum_j \frac{x_j^2}{n_j^2} = 1.$$

In this form it is called the *indicatrix*. We now introduce an auxiliary function H, which also takes on an extreme value:

$$H = \sum_j x_j^2 - \lambda_1 \boldsymbol{g}\cdot\boldsymbol{x} - \lambda_2\left(\sum_j x_j^2/n_j^2 - 1\right).$$

The conditions $\partial H/\partial x_j = 0$ give for $j = 1, 2, 3$

$$2x_j - \lambda_1 g_j - 2\lambda_2 x_j/n_j^2 = 0.$$

We multiply the jth equation with x_j, sum over $j = 1, 2, 3$, and obtain

$$\sum_j 2x_j^2(1 - \lambda_2/n_j^2) - \lambda_1 \sum_j g_j x_j = 0.$$

The second term vanishes because $\boldsymbol{g}\cdot\boldsymbol{x} = 0$. Therefore

$$\sum_j x_j^2 = \lambda_2 \sum_j x_j^2/n_j^2, \quad \text{hence} \quad \lambda_2 = \sum_j x_j^2 = \boldsymbol{x}^2.$$

We recognize the meaning of $\boldsymbol{x}^2$ as follows: by scalar multiplication of the wave equation with $\boldsymbol{D}$ one gets

$$\boldsymbol{D}^2/n^2 = \boldsymbol{E}\cdot\boldsymbol{D} - (\boldsymbol{g}\cdot\boldsymbol{E})(\boldsymbol{g}\cdot\boldsymbol{D}).$$

The last term vanishes. Furthermore,

$$\boldsymbol{D}^2/n^2 = \boldsymbol{E}\cdot\boldsymbol{D} = \sum_j a_{jj} D_j^2,$$

and thus

$$1/n^2 = \sum_j a_{jj} D_j^2 / \boldsymbol{D}^2 = a'_{11},$$

where a'_{11} represents the longitudinal effect along $\boldsymbol{e}'_1 = \boldsymbol{D}/|\boldsymbol{D}|$. On the other hand this is obtained from the radius vector of the tensor quadric according to $a'_{11} = F/\boldsymbol{x}^2$. With $F = 1$ we have $\boldsymbol{x}^2 = n^2$, i.e., the lengths of the major and minor half axes of the sectional ellipse are equal to the refractive indices of both possible waves propagating along $\boldsymbol{g}$. From the above system of equations, one obtains for the coordinates of the half axes

$$2x_j = \lambda_1 g_j \Bigg/ \left(1 - \frac{n^2}{n_j^2}\right)$$

and thus

$$x_1 : x_2 : x_3 = \frac{g_1}{1 - n^2/n_1^2} : \frac{g_2}{1 - n^2/n_2^2} : \frac{g_3}{1 - n^2/n_3^2}.$$

This ratio is identical to the relation derived above for $D_1 : D_2 : D_3$. This is recognized when one substitutes the refractive indices by the corresponding velocities. Hence the directions of vibrations $\boldsymbol{D}'$ and $\boldsymbol{D}''$ run parallel to the half axes of the sectional ellipse.

We summarize these results to the *basic law of crystaloptics* for nonconducting centrosymmetric crystals:

For each propagation direction $\boldsymbol{g}$ in a crystal there exist two linear polarized waves, whose $\boldsymbol{D}$-vectors (directions of vibration)) run parallel to the half axes of the associated sectional ellipse of the indicatrix. The associated refractive indices are equal to the lengths of these half axes. If the refractive indices n' and n'' belonging to $\boldsymbol{g}$ are equal, then the sectional ellipse is in the form of a circle and the direction of vibrations within the plane of the circle is not fixed (degenerate).

Thus the propagation of light in a crystal is determined only by the form and position of the indicatrix. We can divide the centrosymmetric crystals into the following optical classes (see also Section 4.3.6.4):

1. *Optically isotropic:* cubic crystals. The indicatrix is a sphere. The position of the $\boldsymbol{D}$-vectors is not fixed.

2. *Optically uniaxial:* trigonal, tetragonal, and hexagonal crystals. The indicatrix is an ellipsoid of revolution; the axis of revolution runs parallel to the three-, four-, or six-fold axis.

3. *Optically biaxial:* orthorhombic, monoclinic, and triclinic crystals. The indicatrix is a triaxial ellipsoid. In orthorhombic crystals the principal axes of the indicatrix lie parallel to the orthorhombic basic vectors. In monoclinic crystals the distinct monoclinic basic vector $\boldsymbol{a}_2$ (parallel to a two-fold axis or

to a normal on the symmetry plane) must run parallel to one principal axis of the indicatrix. In triclinic crystals the indicatrix is not fixed by any symmetry considerations. (The term "biaxial" refers to the existence of two directions called *optic axes*, in which the sectional ellipse takes on the form of a circle; see Section 4.3.6.4).

As we have seen, the vector of the electric displacement $\boldsymbol{D}$, also within the crystal, is perpendicular to the propagation direction $\boldsymbol{k}$ of the electromagnetic wave and to the vector of the magnetic field strength $\boldsymbol{H}$. In crystals, as opposed to isotropic media, the vector of the electric field strength $\boldsymbol{E}$ usually deviates from $\boldsymbol{D}$. However, $\boldsymbol{E}$ is also perpendicular to $\boldsymbol{H}$, as one recognizes from the Maxwell equation $\operatorname{rot} \boldsymbol{E} = -\frac{1}{c}\partial \boldsymbol{H}/\partial t$ after introducing the expression for plane waves. Hence the three vectors $\boldsymbol{D}$, $\boldsymbol{E}$, and $\boldsymbol{k}$ lie in the plane perpendicular to the magnetic field $\boldsymbol{H}$. The position of $\boldsymbol{E}$ with given $\boldsymbol{k}$ and $\boldsymbol{D}$ is obtained by the Poinsot construction (see Section 4.3.2) in the plane perpendicular to $\boldsymbol{H}$. For this purpose we draw the ellipse cut out by the plane perpendicular to $\boldsymbol{H}$ and passing through the center of the tensor quadric (Fig. 4.8). $\boldsymbol{E}$ is then perpendicular to the tangent of this ellipse at the intercept point of the vector $\boldsymbol{D}$ (tensor quadric $F = a_{ij}x_ix_j$, $E_i = a_{ij}D_j$). For our description thus far we have given preference to the $\boldsymbol{D}$-vector because it allows a simpler representation of the crystal-optic properties.

The flow of energy of the electromagnetic wave is described by the Poynting vector $\boldsymbol{S} = (c/4\pi)\boldsymbol{E} \times \boldsymbol{H}$. $\boldsymbol{S}$ shows the direction of the ray $\boldsymbol{s}$ along which the electromagnetic energy (per second and unit area) propagates. In the Poinsot construction we can now draw directly the position of $\boldsymbol{S}$, namely perpendicular to $\boldsymbol{H}$ and $\boldsymbol{E}$ (see Fig. 4.8).

Ray vector $\boldsymbol{s}$ and propagation vector $\boldsymbol{k}$ thus stand in a close geometrical relationship from which one can derive a set of interesting rules which we are unable to pursue here. The phenomenon of double refraction is taken up in Exercise 5 (see Section 4.3.2).

4.3.6.1 Reflection and Refraction

An electromagnetic wave, incident on a boundary surface, causes two phenomena: reflection and refraction. Both phenomena may be considered as sufficiently well known for the case of isotropic media. *Law of reflection:* Angle of incidence α_{I} = Angle of reflection α_{Ir},

$$\text{\textit{Snellius' law of refraction:}} \quad \frac{\sin \alpha_{\mathrm{I}}}{\sin \alpha_{\mathrm{II}}} = \frac{v_{\mathrm{I}}}{v_{\mathrm{II}}} = \frac{n_{\mathrm{II}}}{n_{\mathrm{I}}},$$

where v_{I} and v_{II} are the wave velocities in media I and II (Fig. 4.9). In order to also calculate the intensities of the reflected and refracted waves we draw upon the boundary conditions for the quantities $\boldsymbol{H}$, $\boldsymbol{D}$, and $\boldsymbol{E}$. We imagine a plane boundary surface of a crystal II being struck by a wave with the wave

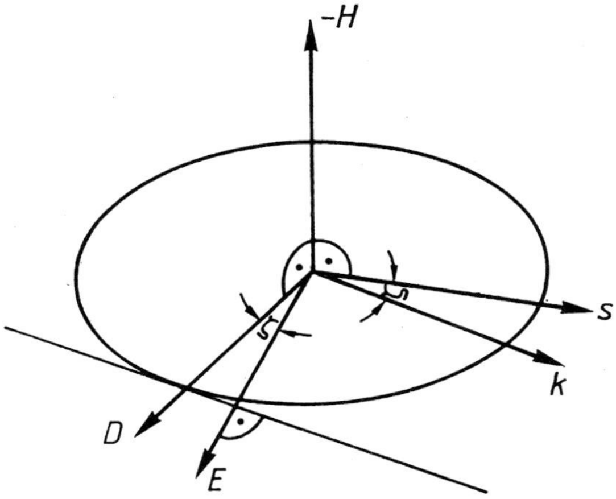

Figure 4.8 Construction of ray vector s (|| Poynting vector) according to the Poinsot construction from the position of D and k within the plane perpendicular to the magnetic field H.

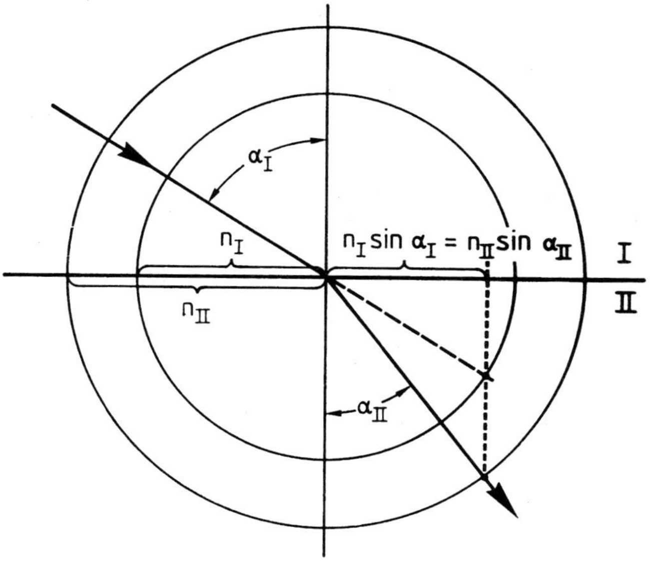

Figure 4.9 Refraction at the boundary surface of two isotropic media I and II.

normal k_I coming from an external isotropic medium I. The plane spanned by the normal on the boundary surface and k_I is called the incidence plane. At first, let the direction of vibration of D_I be arbitrary. In general, the tangential components of E and H, i.e., the projection of E or H on the boundary surface, pass through the boundary surface continuously, in other words, without spontaneous change. The reason for this behavior is that the closed line integral $\oint \boldsymbol{E} \cdot d\boldsymbol{x}$ vanishes since the path of integration outward runs along one border of the boundary surface and returns along the other border. We then have

$$E_{\text{tg}}(\text{I})\Delta x - E_{\text{tg}}(\text{II})\Delta x = 0,$$

hence $E_{\text{tg}}(\text{I}) = E_{\text{tg}}(\text{II})$; Δx is the path of integration along the border. The contributions from crossing the boundary surface twice can be neglected. The same applies to $\boldsymbol{H}$. In contrast to $\boldsymbol{E}$ and $\boldsymbol{H}$, the normal components of the vectors of the electric displacement $\boldsymbol{D}$ and of the magnetic induction $\boldsymbol{B}$ pass continuously through the boundary surface. As proof, we make use of the condition $\operatorname{div} \boldsymbol{D} = 0$ (for nonconducting media) and integrate over a thin plate whose parallel surfaces enclose the boundary surface on both sides. According to Gauss's law, the volume integral is

$$\int_{\text{Plate}} \operatorname{div} \boldsymbol{D} dV = \int_{\text{Surface}} \boldsymbol{D} \cdot d\boldsymbol{f},$$

where dV is a volume element of the plate and $d\boldsymbol{f}$ is the normal on a surface element of area $|d\boldsymbol{f}|$. For the limiting value of an arbitrarily thin plate one obtains

$$\int_{\text{Surface}} \boldsymbol{D} \cdot d\boldsymbol{f} = D_{\text{In}}\Delta f - D_{\text{IIn}}\Delta f;$$

Δf is the surface of one side of the plate. Since $\operatorname{div} \boldsymbol{D} = 0$ we have $D_{\text{In}} = D_{\text{IIn}}$. Analogously for $\boldsymbol{B}$ we have $B_{\text{In}} = B_{\text{IIn}}$. With the help of these conditions the relations at the boundary surface can now be elucidated.

Let the three participating waves, namely the incident wave from medium I, the wave reflected back into medium I and the wave refracted into medium II be specified by their propagation vectors $\boldsymbol{k}_{\text{I}}$, $\boldsymbol{k}_{\text{Ir}}$, and $\boldsymbol{k}_{\text{II}}$ (Fig. 4.10). Let the basic vectors of a Cartesian reference system be selected such that $\boldsymbol{e}_1$ and $\boldsymbol{e}_2$ lie in the boundary surface, whereby $\boldsymbol{e}_1$ is perpendicular to the plane of incidence. $\boldsymbol{e}_3$ points in the direction of the normal on the boundary surface.

In the following, we discuss the important case of an isotropic medium I adjacent to an anisotropic medium II. At first we prove that $\boldsymbol{k}_{\text{I}}$, $\boldsymbol{k}_{\text{Ir}}$, and $\boldsymbol{k}_{\text{II}}$ lie in the plane of incidence. We imagine the electric field strength $\boldsymbol{E}$ is resolved into components perpendicular and parallel to the plane of incidence ($\boldsymbol{E}_{\text{tg}} \parallel \boldsymbol{e}_1$, $\boldsymbol{E}_{\perp} \cdot \boldsymbol{e}_1 = 0$). As we shall see, both components behave differently with respect to reflection and refraction.

In the case $\boldsymbol{E}_{\text{I}} \parallel \boldsymbol{e}_1$ we also have $\boldsymbol{D}_{\text{I}} \parallel \boldsymbol{e}_1$, and furthermore $\boldsymbol{H}_{\text{I}} \cdot \boldsymbol{e}_1 = 0$ and $\boldsymbol{H} \parallel \boldsymbol{B}$. From $\boldsymbol{D}_{\text{I}} \parallel \boldsymbol{D}_{\text{Ir}}$ and $\boldsymbol{D}_{\text{Ir}} \cdot \boldsymbol{k}_{\text{Ir}} = 0$ we note that $\boldsymbol{k}_{\text{Ir}}$ also lies in the plane of incidence. The condition $D_{\text{In}} = D_{\text{IIn}}$ means here that $\boldsymbol{D}_{\text{II}}$ lies in the boundary

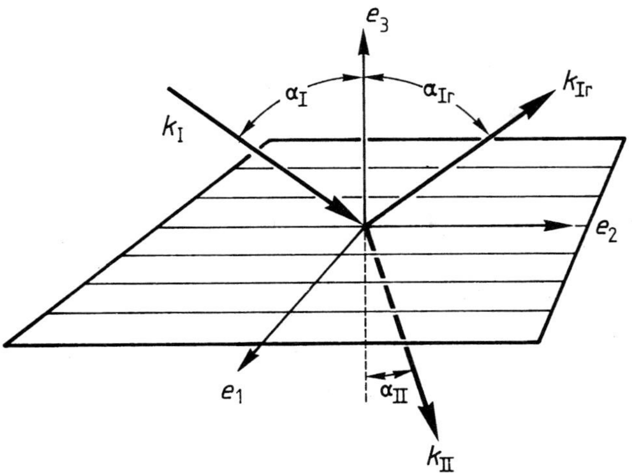

Figure 4.10 Propagation vectors in reflection and refraction on a boundary surface.

surface. $\boldsymbol{H}_{\text{II}}$ remains in the plane of incidence as a consequence of $H_{\text{IItg}} = H_{\text{Itg}}$. Because $\boldsymbol{H}_{\text{II}} \cdot \boldsymbol{D}_{\text{II}} = 0$ we also have $\boldsymbol{D}_{\text{II}}$ perpendicular to the plane of incidence. Hence $\boldsymbol{k}_{\text{II}}$ lies in the plane of incidence ($\boldsymbol{D}_{\text{II}} \cdot \boldsymbol{k}_{\text{II}} = 0!$).

In the case $\boldsymbol{E}_{\text{I}} \cdot \boldsymbol{e}_1 = 0$, $\boldsymbol{H}_{\text{I}}$ and hence $\boldsymbol{B}_{\text{I}}$ are perpendicular to the plane of incidence ($\boldsymbol{H}_{\text{I}} \parallel \boldsymbol{e}_1$). In medium I we have $\boldsymbol{H}_{\text{I}} \parallel \boldsymbol{H}_{\text{Ir}}$. Because $\boldsymbol{H}_{\text{Ir}} \cdot \boldsymbol{k}_{\text{Ir}} = 0$, $\boldsymbol{k}_{\text{Ir}}$ must lie in the plane of incidence. The condition $B_{\text{In}} = B_{\text{IIn}}$ requires that $\boldsymbol{B}_{\text{II}}$ lies in the boundary surface and hence also $\boldsymbol{H}_{\text{II}}$. Conservation of the tangential component of $\boldsymbol{H}$, thus, $H_{\text{I}} = H_{\text{IItg}}$ ($H_{\text{I}} = H_{\text{Itg}}!$), means that we have $\boldsymbol{H}_{\text{II}} \parallel \boldsymbol{H}_{\text{I}}$ and that $\boldsymbol{k}_{\text{II}}$ also lies in the plane of incidence ($\boldsymbol{H}_{\text{II}} \cdot \boldsymbol{k}_{\text{II}} = 0!$).

For the following derivation of the formulae for the amplitude of the reflected and refracted waves we assume that both media are isotropic.

A. E perpendicular to the plane of incidence, H in the plane of incidence

Let the incident wave be given by

$$\boldsymbol{E}_{\text{I}} = \boldsymbol{E}_{\text{I}0} e^{2\pi i (\boldsymbol{k}_{\text{I}} \cdot \boldsymbol{x} - \nu t)},$$

the reflected wave by

$$\boldsymbol{E}_{\text{Ir}} = \boldsymbol{E}_{\text{Ir}0} e^{2\pi i (\boldsymbol{k}_{\text{Ir}} \cdot \boldsymbol{x} - \nu t)},$$

and the refracted wave by

$$\boldsymbol{E}_{\text{II}} = \boldsymbol{E}_{\text{II}0} e^{2\pi i (\boldsymbol{k}_{\text{II}} \cdot \boldsymbol{x} - \nu t)}.$$

The $\boldsymbol{k}$'s are of the form $\boldsymbol{k} = k_2\boldsymbol{e}_2 + k_3\boldsymbol{e}_3$. Because of the continuity of E_{tg} we then have for $x_3 = 0$:

$$E_{\text{I}0}e^{2\pi i k_{\text{I}2}x_2} + E_{\text{Ir}0}e^{2\pi i k_{\text{Ir}2}x_2} = E_{\text{II}0}e^{2\pi i k_{\text{II}2}x_2}.$$

This condition can only be satisfied for arbitrary $\boldsymbol{E}_{\text{I}}$ and $\boldsymbol{k}_{\text{I}}$ when the exponential factors are equal, hence $k_{\text{I}2} = k_{\text{Ir}2} = k_{\text{II}2}$. This means, the tangential component of the propagation vector remains conserved. Because

$$k_{\text{I}2} = |\boldsymbol{k}_{\text{I}}| \sin\alpha_{\text{I}}, \quad k_{\text{Ir}2} = |\boldsymbol{k}_{\text{Ir}}| \sin\alpha_{\text{Ir}}, \quad \text{and} \quad k_{\text{II}2} = |\boldsymbol{k}_{\text{II}}| \sin\alpha_{\text{II}}$$

one obtains for both conditions

$$\alpha_{\text{I}} = \alpha_{\text{II}} (\text{law of reflection}) \quad \text{and}$$
$$\frac{\sin\alpha_{\text{I}}}{\sin\alpha_{\text{II}}} = \frac{|\boldsymbol{k}_{\text{II}}|}{|\boldsymbol{k}_{\text{I}}|} = \frac{n_{\text{II}}}{n_{\text{I}}} (\text{law of refraction}).$$

Furthermore, we have $E_{\text{I}0} + E_{\text{Ir}0} = E_{\text{II}0}$. A second condition is gained from the continuity of the tangential component of $\boldsymbol{H}$. Since $\boldsymbol{k}$, $\boldsymbol{E}$, and $\boldsymbol{H}$ form a right-handed system, one finds for $x_3 = 0$

$$-\cos\alpha_{\text{I}} H_{\text{I}0}e^{2\pi i k_{\text{I}2}x_2} + \cos\alpha_{\text{I}} H_{\text{Ir}0}e^{2\pi i k_{\text{Ir}2}x_2} = -\cos\alpha_{\text{II}} H_{\text{II}0}e^{2\pi i k_{\text{II}}2x_2}.$$

This also applies in the case that medium II is anisotropic and two refracted waves arise, because the exponential factors must be equal in each case. Now in order to connect this condition and the one above, it is necessary to use the relationship between the amplitudes of the electric and magnetic field strengths of an electromagnetic wave. From the induction law

$$\operatorname{rot}\boldsymbol{E} = -\frac{1}{c}\frac{\partial\boldsymbol{H}}{\partial t} \quad \text{we get} \quad \boldsymbol{k}\times\boldsymbol{E} = \frac{\nu}{c}\cdot\boldsymbol{H}.$$

For the absolute values in isotropic media one gets

$$|\boldsymbol{H}| = |\boldsymbol{E}||\boldsymbol{k}|(c/\nu) \quad \text{and with} \quad |\boldsymbol{k}|/\nu = 1/v \quad (\text{because } \lambda\nu = v),$$
$$|\boldsymbol{H}| = \frac{c}{v}|\boldsymbol{E}| = n|\boldsymbol{E}|.$$

Thereby we convert the condition for the tangential component of $\boldsymbol{H}$ into

$$n_{\text{I}}\cos\alpha_{\text{I}}(-E_{\text{I}0} + E_{\text{Ir}0}) = -n_{\text{II}}\cos\alpha_{\text{II}}E_{\text{II}0}.$$

Together with $E_{\text{I}0} + E_{\text{Ir}0} = E_{\text{II}0}$ and with $n_{\text{II}}/n_{\text{I}} = \sin\alpha_{\text{I}}/\sin\alpha_{\text{II}}$, one obtains for $\boldsymbol{E}$ perpendicular to the plane of incidence the *first Fresnel formula*

$$E_{\text{I}0} : E_{\text{II}0} : E_{\text{Ir}0} = \sin(\alpha_{\text{I}} + \alpha_{\text{II}}) : (\sin(\alpha_{\text{I}} + \alpha_{\text{II}}) + \sin(\alpha_{\text{II}} - \alpha_{\text{I}})) : \sin(\alpha_{\text{II}} - \alpha_{\text{I}}).$$

B. E in the plane of incidence, H perpendicular to the plane of incidence.
We can proceed analogously to the previous derivation when we correspondingly exchange $\boldsymbol{E}$ and $\boldsymbol{H}$. Preserving the tangential component of $\boldsymbol{H}$ and $\boldsymbol{E}$, respectively, gives

$$\begin{aligned} H_{\text{Itg}} + H_{\text{Irtg}} &= H_{\text{IItg}} \qquad (\text{here } \boldsymbol{H} = \boldsymbol{H}_{\text{tg}}\,!) \quad \text{and} \\ E_{\text{Itg}} &= \cos\alpha_{\text{I}} E_{\text{I0}} e^{2\pi i k_{\text{I}2} x_2} - \cos\alpha_{\text{I}} E_{\text{Ir0}} e^{2\pi i k_{\text{Ir}2} x_2} = \cos\alpha_{\text{II}} E_{\text{II0}} e^{2\pi i k_{\text{II}2} x_2}. \end{aligned}$$

If we again substitute the amplitudes of the magnetic field strength by those of the electric field strength and use the addition theorem of trigonometric functions we get for $\boldsymbol{E}$ in the plane of incidence the *second Fresnel formula*

$$E_{\text{I0}} : E_{\text{II0}} : E_{\text{Ir0}} = \tan(\alpha_{\text{I}} + \alpha_{\text{II}}) : \left\{ \frac{\tan(\alpha_{\text{I}} + \alpha_{\text{II}})}{\cos(\alpha_{\text{I}} - \alpha_{\text{II}})} + \frac{\tan(\alpha_{\text{II}} - \alpha_{\text{I}})}{\cos(\alpha_{\text{I}} + \alpha_{\text{II}})} \right\} : \tan(\alpha_{\text{I}} - \alpha_{\text{II}}).$$

With the aid of the Fresnel formulae we can now calculate the intensity of the reflected and refracted waves for the case of isotropic media by constructing the respective Poynting vectors and also take into account the change of the ray cross-section. The cross-section q_{II} in medium II is

$$q_{\text{II}} = q_{\text{I}} \cos\alpha_{\text{II}} / \cos\alpha_{\text{I}},$$

where q_{I} is the cross-section of the incident ray. Moreover, the energy conservation of course holds. This is guaranteed by the Fresnel equations, as one can easily check ($|\boldsymbol{S}_{\text{I}}| = |\boldsymbol{S}_{\text{Ir}}| + |\boldsymbol{S}_{\text{II}}|$).

We have for the intensity I_{Ir} of the reflected wave in the case $\boldsymbol{E}$ perpendicular to the plane of incidence

$$\frac{I_{\text{Ir}}}{I_{\text{I}}} = \left(\frac{\sin(\alpha_{\text{I}} - \alpha_{\text{II}})}{\sin(\alpha_{\text{I}} + \alpha_{\text{II}})} \right)^2 = \left(\frac{1 - \dfrac{n_{\text{II}} \cos\alpha_{\text{II}}}{n_{\text{I}} \cos\alpha_{\text{I}}}}{1 + \dfrac{n_{\text{II}} \cos\alpha_{\text{II}}}{n_{\text{I}} \cos\alpha_{\text{I}}}} \right)^2$$

using $\sin\alpha_{\text{I}} / \sin\alpha_{\text{II}} = n_{\text{II}} / n_{\text{I}}$.

The result for perpendicular incidence ($\alpha_{\mathrm{I}} = 0$) is

$$\frac{I_{\mathrm{Ir}}}{I_{\mathrm{I}}} = \left(\frac{1 - n_{\mathrm{II}}/n_{\mathrm{I}}}{1 + n_{\mathrm{II}}/n_{\mathrm{I}}}\right)^2 .$$

For the reflection at a plate of refractive index 1.50 adjacent to vacuum or air one thus obtains only the small fraction of about 4%. On the other hand, one observes for glancing incidence ($\alpha_{\mathrm{I}} = 90°$) independent of $n_{\mathrm{II}}/n_{\mathrm{I}}$ $\frac{I_{\mathrm{Ir}}}{I_{\mathrm{I}}} = 1$, hence the full reflection of the incident wave. From the second Fresnel formula one gets the same result for both these limiting cases. The course of $I_{\mathrm{Ir}}/I_{\mathrm{I}}$ between both limiting values is, however, different for both directions of vibration. In particular, as a consequence of the second formula for $\boldsymbol{E}$ in the plane of incidence, the amplitude of the reflected wave

$$I_{\mathrm{Ir}} = I_{\mathrm{I}} \left(\frac{\tan \alpha_{\mathrm{II}} - \alpha_{\mathrm{I}})}{\tan(\alpha_{\mathrm{II}} + \alpha_{\mathrm{I}})}\right)^2 \quad \text{vanishes,}$$

as ($\alpha_{\mathrm{II}} + \alpha_{\mathrm{I}}$) approaches 90°, that is, the reflected and refracted waves are perpendicular to one another.

If one substitutes in the law of refraction $\sin \alpha_{\mathrm{I}} / \sin \alpha_{\mathrm{II}} = n_{\mathrm{II}}/n_{\mathrm{I}}$ the angle α_{II} by $(90 - \alpha_{\mathrm{I}})$, we get Brewster's law for the vanishing of the reflected wave $\tan \alpha_{\mathrm{I}} = n_{\mathrm{II}}/n_{\mathrm{I}}$. Because of the conservation of energy, the whole intensity of the incident wave goes into the refracted wave. If the incident wave is composed of light with arbitrary directions of vibration then the ray reflected under the Brewester angle contains only the direction of vibration perpendicular to the plane of incidence. This is the simplest method to produce polarized light.

A further phenomena is the case of *total reflection*. If $n_{\mathrm{II}} < n_{\mathrm{I}}$, there exists a limiting angle for α_{I}, above which no refracted wave can occur. The condition is $\alpha_{II} = 90°$ and thus $\sin \alpha'_{\mathrm{I}} = n_{\mathrm{II}}/n_{\mathrm{I}}$. A method to determine refractive indices is based on the measurement of the angle α'_{I} of the total reflection.

We now return to refraction when a light wave is incident upon an anisotropic medium. According to the above, with a known plane of incidence, we merely have to take into account the condition for the preservation of the tangential component of the propagation vector of the incident wave. With the help of the index surface we establish which refractive indices in the plane of incidence are consistent with the condition $k_{\mathrm{Itg}} = k_{\mathrm{IItg}}$. The index surface is a double-shelled centrosymmetric surface, constructed such that the lengths of the radius vectors $\boldsymbol{x}$ are equal to the refractive indices of the given directions of propagation, hence $x_i = n g_i$. In the principal axes' reference system of the indicatrix, one obtains the index surface from the velocity equation derived above for a given direction of propagation $\sum_i g_i^2/(v_i^2 - v^2) = 0$, when we write $v = c/n$, $v_i = c/n_i$, $x_i = n g_i$, and $\boldsymbol{x}^2 = n^2$. The result is the equation for the index surface:

$$\sum_i \frac{x_i^2 n_i^2}{\sum_j x_j^2 - n_i^2} = 0$$

or after multiplication with the common denominator and separation of a factor $\sum_i x_i^2$

$$(x_1^2 n_1^2 + x_2^2 n_2^2 + x_3^2 n_3^2)(x_1^2 + x_2^2 + x_3^2) - x_1^2 n_1^2 (n_2^2 + n_3^2) \\ - x_2^2 n_2^2 (n_3^2 + n_1^2) - x_3^2 n_3^2 (n_1^2 + n_2^2) + n_1^2 n_2^2 n_3^2 = 0.$$

The index surface of an isotropic medium, hence also cubic crystals, is a sphere of radius n. In optical uniaxial crystals (one principal symmetry axis 3,4, or 6), the index surface consists of a sphere $x_1^2 + x_2^2 + x_3^2 = n_1^2 = n_2^2$ and the ellipsoid $x_1^2/n_3^2 + x_2^2/n_3^2 + x_3^2/n_1^2 = 1$, as is seen by setting $n_1 = n_2$. With these crystals of trigonal, tetragonal, and hexagonal symmetry we have accordingly for each direction of propagation a wave with the fixed refractive index n_1. This wave is called the ordinary wave. If $n_1 < n_3$, the crystal is said to be optical positive (optical positive character) and in the other case $n_1 > n_3$ optical negative. We will return to the diagnostic value of this classification and the general definition of the optical character later.

Now to construct the refracted wave, we place the center of the index surface on the intercept line of the boundary surface and the plane of incidence. The curves of intersection of the plane of incidence with the index surface are cut, in medium II, by the line parallel to the axis of incidence at an interval of $n_{\mathrm{I}} \sin \alpha_{\mathrm{I}} = (c/v) k_{\mathrm{Itg}}$ (Fig. 4.11). The lines from the center of the index surface to the intersection points give the propagation direction of the wave refracted in medium II. Therefore, the condition for the conservation of the tangential component of $\boldsymbol{k}_{\mathrm{I}}$ and hence the law of refraction is fulfilled:

$$n_{\mathrm{I}} \sin \alpha_{\mathrm{I}} = (c/v) k_{\mathrm{Itg}} = (c/v) k_{\mathrm{IItg}} = n_{\mathrm{II}} \sin \alpha_{\mathrm{II}}.$$

In anisotropic media, two waves appear in the general case, except in total reflection.

The associated ray vectors are obtained with the aid of the Poinsot construction. For this purpose, the directions of vibrations of both refracted waves must be constructed. This occurs with the help of the indicatrix and the sectional ellipses perpendicular to both propagation vectors $\boldsymbol{k}'_{\mathrm{II}}$ and $\boldsymbol{k}''_{\mathrm{II}}$ analogous to Fig. 4.8. Besides wave double refraction, one also observes ray double refraction. In the case of perpendicular incidence of the wave we have $\alpha_{\mathrm{I}} = \alpha_{\mathrm{II}} = 0$, i.e., the refracted waves keep the direction of the incident wave, although they possess different propagation velocities and different directions of vibration of the $\boldsymbol{D}$-vector. Hence we observe no wave double refraction. If

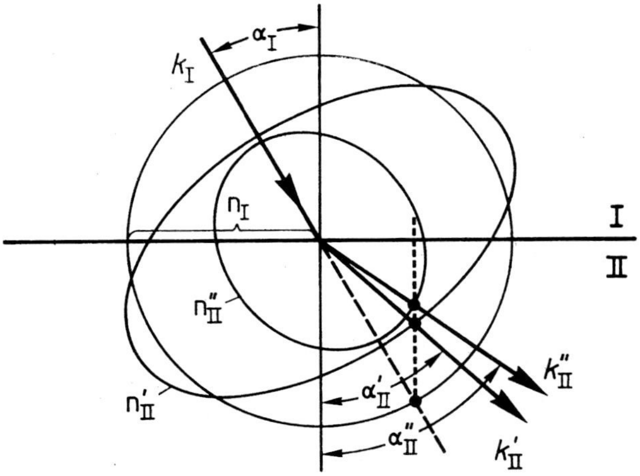

Figure 4.11 Construction of the refraction at the transition from an isotropic medium I into a double-refracting medium II. Intersection of the incident plane with the double-shelled index surface. Conservation of the tangential component of $\boldsymbol{k}_\mathrm{I}$.

the principal axes of the indicatrix lie oblique to the axis of incidence, then with perpendicular incidence, we also have ray double refraction. When constructing the refracted waves at the reverse side of the plate we merely have to take into account the conservation of the tangential component of the $\boldsymbol{k}$-vectors as before.

Fig. 4.12 shows the separation of rays through a plane-parallel plate under perpendicular incidence. This arrangement also allows, in a simple manner, the production of linearly polarized light.

We now calculate the angle of ray double refraction at perpendicular incidence. The position of both $\boldsymbol{D}$-vectors results from the condition

$$D_1 : D_2 : D_3 = g_1 n_1^2/(n^2 - n_1^2) : g_2 n_2^2/(n^2 - n_2^2) : g_3 n_3^2/(n^2 - n_3^2),$$

where for n we set the values n' and n'' from the index equation for the given direction of propagation.

With the notation $E_i = a_{ij} D_j$ one finds the directions of the associated $\boldsymbol{E}$-vectors (a_{ij} are the polarization constants; in the principal axes' system $a_{ii} = 1/n_i^2$ and $a_{ij} = 0$ for $i \neq j$). Both ray vectors $\boldsymbol{s}'$ and $\boldsymbol{s}''$ are given by $\boldsymbol{s}' \parallel \boldsymbol{E}' \times \boldsymbol{H}'$ and $\boldsymbol{s}'' \parallel \boldsymbol{E}'' \times \boldsymbol{H}''$, whereby the vectors $\boldsymbol{H}'$ and $\boldsymbol{H}''$ run parallel to $\boldsymbol{D}''$ and $\boldsymbol{D}'$ respectively. Thus, one finds for the angle of ray double refraction ζ, with the help of the scalar product $\boldsymbol{s}' \cdot \boldsymbol{s}''$:

$$\cos\zeta = \frac{(\boldsymbol{E}' \times \boldsymbol{H}') \cdot (\boldsymbol{E}'' \times \boldsymbol{H}'')}{|\boldsymbol{E}' \times \boldsymbol{H}'||\boldsymbol{E}'' \times \boldsymbol{H}''|} \qquad \text{(see also Exercise 11).}$$

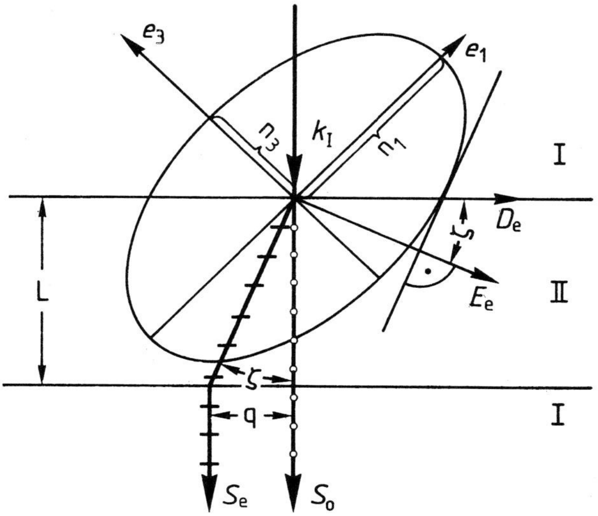

Figure 4.12 Ray double refraction and ray separation on a plane-parallel plate of an optical uniaxial crystal at perpendicular incidence (the indicatrix is a rotation ellipsoid!). The ordinary ray s_o (direction of vibration perpendicular to the plane of incidence) experiences no refraction. The direction of vibration of the extraordinary ray s_e lies within the plane of incidence. s_e itself runs within the crystal parallel to the trace of the tangent of the indicatrix at the piercing point of $\boldsymbol{D}_e$ (Poinsot construction). The ray separation after refraction at the reverse side is $q = L \tan \zeta$ (L thickness of plate).

4.3.6.2 **Determining Refractive Indices**

Besides the possibilities discussed above (e.g., total reflection or reflection coefficients using the Fresnel formulae), the practical determination of refractive indices is mainly made by deflection through a prism. In isotropic media, minimal deflection allows very exact measurements, whereby the light ray runs symmetric to the prism. Furthermore, the adjustment of the minimal deflection is, to a first approximation, noncritical with respect to small deviations from the exact position for minimal deflection (see Exercise 12).

On the other hand, for anisotropic crystals, perpendicular incidence has proven rather successful, because here no wave double refraction takes place, i.e., the wave normals of both possible waves in the crystal run parallel to the normal on the front prism face (Fig. 4.13a). Hence, we know with high reliability the direction of the wave normals in the crystal. Let the prism angle be φ. Because $\sin \alpha_{\mathrm{I}} / \sin \alpha_{\mathrm{II}} \doteq n_{\mathrm{II}} / n_{\mathrm{I}}$ and $\alpha_{\mathrm{II}} = \varphi$ one obtains for the deflection angle $\alpha = \alpha_{\mathrm{I}} - \varphi$ and thus $\sin(\alpha + \varphi) / \sin \varphi = n_{\mathrm{II}} / n_{\mathrm{I}}$. The deflection angle can be directly determined with the help of a goniometer, where a detector system is moved on the periphery of a circle, or by the aid of length measurements (Fig. 4.13b).

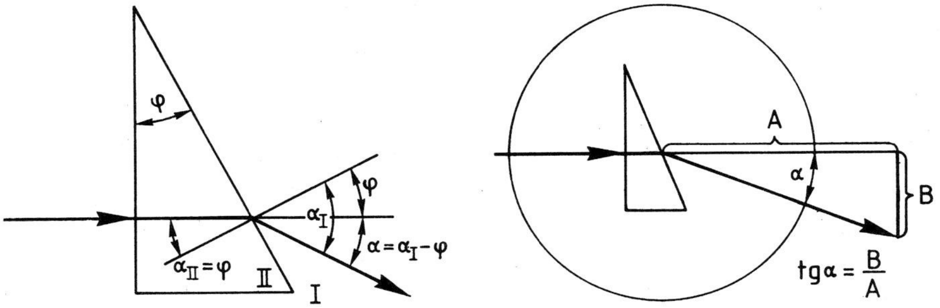

(a) (b)

Figure 4.13 (a) Prism method with perpendicular incidence. (b) Measurement of the deflection angle by a goniometer or by the aid of a measurement of the lengths A and B.

We now discuss the general case of the measurement of the complete set of refractive indices of a triclinic crystal. We start with the crystal-physical reference system and prepare three prisms with the refractive edge $\boldsymbol{e}_j$ ($j = 1, 2, 3$) and a face perpendicular to $\boldsymbol{e}_i$, the respective transmission direction at perpendicular incidence (Fig. 4.14). In a first step, we determine for each transmission direction $\boldsymbol{e}_i$ the directions of vibrations of both possible $\boldsymbol{D}$-vectors, $\boldsymbol{D}'$ and $\boldsymbol{D}''$. This is performed best on a thin, plane-parallel crystal plate by the aid of polarization microscope under crossed polarizers. If the direction of vibration of $\boldsymbol{D}'$ or $\boldsymbol{D}''$ agrees with the transmission direction of the first or second polarizer, the plate appears completely dark after the second polarizer: Extinction position (see Section 4.3.6.3). Let the angle between $\boldsymbol{e}_j$ and the transmission direction of the first polarizer in the extinction position be ψ_i, whereby the fixed position of the respective vectors as shown in Fig. 4.14 is to be maintained. In order to avoid confusion, the indices i, j, k are to be chosen in a cyclic sequence. In a second step we measure the deflection angles α_i' and α_i'' for each of the three transmission directions. From these we obtain the associated refractive indices n_i' und n_i''. The assignment of the extinction angle ψ_i to $\boldsymbol{D}'$ is to be controlled by the aid of a polarizer placed in front of the prism. The polarizer is rotated until the intensity of the respective ray (deflection angle α') vanishes. $\boldsymbol{D}'$ is then perpendicular to the transmission direction of the polarizer. We now have available a total of nine quantities, namely ψ_i, n_i' and n_i'' ($i = 1, 2, 3$), for the determination of the six coefficients a_{ij} of the indicatrix. The sectional ellipse perpendicular to $\boldsymbol{e}_i$ is given in the crystal-physical system by

$$a_{jj}x_j^2 + a_{kk}x_k^2 + 2a_{jk}x_jx_k = 1$$

(do not sum!; $x_i = 0$ in the equation $a_{ij}x_ix_j = 1$!).

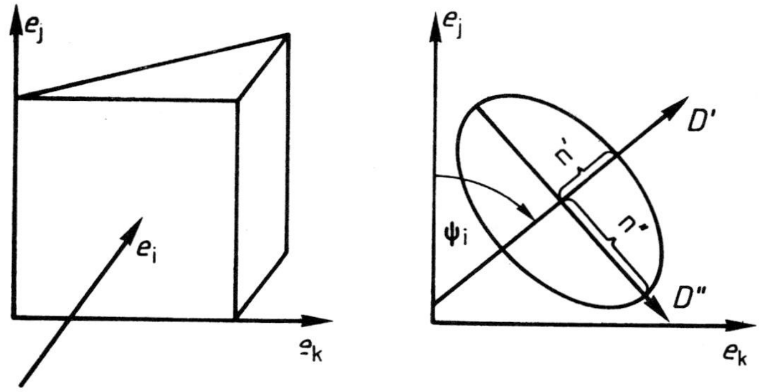

Figure 4.14 Measurement of indices of refraction on triclinic crystals by the prism method with perpendicular incidence. Pay attention to the sense of rotation of the angle ψ_i when applying the inverse transformation.

A plane principal axes' transformation

$$\begin{aligned} \boldsymbol{e}_i' &= \boldsymbol{e}_i \\ \boldsymbol{e}_j' &= \cos\psi_i \boldsymbol{e}_j + \sin\psi_i \boldsymbol{e}_k \\ \boldsymbol{e}_k' &= -\sin\psi_i \boldsymbol{e}_j + \cos\psi_i \boldsymbol{e}_k \end{aligned}$$

transforms the equation of the sectional ellipse into $a_{jj}'x_j'^2 + a_{kk}'x_k'^2 = 1$. We have $a_{jj}' = 1/n_i'^2$ and $a_{kk}' = 1/n_i''^2$.

We find the necessary coefficients with the help of the inverse transformation (see Section 4.3.2)

$$\begin{aligned} a_{jj} &= \cos^2\psi_i a_{jj}' + \sin^2\psi_i a_{kk}'; \\ a_{kk} &= \sin^2\psi_i a_{jj}' + \cos^2\psi_i a_{kk}'; \\ a_{jk} = a_{kj} &= \tfrac{1}{2}(a_{jj}' - a_{kk}')\sin 2\psi_i. \end{aligned}$$

One then obtains the three principal coefficients a_{11}, a_{22}, and a_{33}, each twice, independent of one another and thus we have a good control of the reliability of the measurement.

In a third step we perform a general principal axes' transformation and find the principal refractive indices n_1, n_2, and n_3 as well as the associated directions of vibrations of the $\boldsymbol{D}$-vectors. With this method one can achieve, without special measures, an accuracy of about 10^{-4} for the principal refractive indices, when the specimens are homogenous and adequately oriented and prepared. The dimensions of the prisms play an important role with respect to the quality of the preparation, particularly with soft crystals. The edge lengths of the prisms should not be essentially smaller than 6 mm and the prism angles should be selected to be as large as possible up to just about the limit for total reflection.

The determination of the indicatrix of crystals of higher symmetry is substantially easier. For example, with optical uniaxial crystals one needs only one prism, in which the entrance face of the light ray contains the respective principal symmetry axis.

4.3.6.3 **Plane-Parallel Plate between Polarizers at Perpendicular Incidence**

In the practical application of crystal optics, the plane-parallel plate is the most important standard preparation in polarization microscopy as well as in technical components. This also applies to the investigation of higher optical effects (electro-optics, piezo-optics). The transmission takes place almost exclusively in perpendicular incidence. Here we derive the intensity formula for the transmitted ray dependent on the position of the direction of vibration.

A linear polarized wave emerging from the first polarizer P_1 strikes the crystal plate. There it is split into components parallel to the directions of vibrations of the associated sectional ellipse. After passing through a thickness L both waves leave the plate and unify themselves to an elliptically polarized wave. This means, that we have a wave composed of two plane-polarized partial waves with the same propagation vector (and hence the same frequency), however, with different directions of vibration and phase (time difference of the maximum amplitude). A second polarizer P_2 transmits only that component, whose direction of vibration corresponds to the transmission direction of P_2. From the square of the amplitude of this component we then obtain a measure for the intensity of the transmitted light.

We introduce the following notations for the calculation (Fig. 4.15):

- Normal on the boundary surface: $\boldsymbol{e}_3$,
- Direction of vibration of the first polarizer P_1: $\boldsymbol{e}_1$,
- Directions of vibration of the sectional ellipse (directions of major and minor semiaxis): $\boldsymbol{e}'$ and $\boldsymbol{e}''$,
- Direction of vibration of the second polarizer P_2: $\boldsymbol{e}_1'$,
- Angle between $\boldsymbol{e}_1$ and $\boldsymbol{e}'$: ϕ,
- Angle between $\boldsymbol{e}_1$ and $\boldsymbol{e}_1'$: ψ.

We describe the incident wave by $\boldsymbol{D} = D_0 e^{2\pi i(\boldsymbol{k}\cdot\boldsymbol{x}-\nu t)}\boldsymbol{e}_1$, whereby $\boldsymbol{k} = k_3\boldsymbol{e}_3$.

When the wave enters the crystal ($x_3 = 0$), $\boldsymbol{D}$ it is split into components parallel to $\boldsymbol{e}'$ and $\boldsymbol{e}''$,

$$D_0\boldsymbol{e}_1 = D_0'\boldsymbol{e}' + D_0''\boldsymbol{e}'' \quad \text{with} \quad D_0' = D_0\cos\varphi, \quad D_0'' = -D_0\sin\varphi.$$

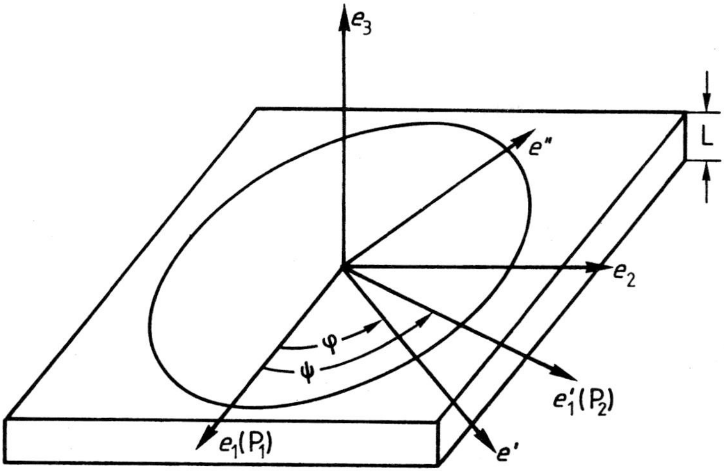

Figure 4.15 Plane-parallel plate between two polarizers P_1 and P_2.

After passing through a distance $x_3 = L$ both partial waves combine on the top side of the plate to

$$\boldsymbol{D} = D_0 \cos\varphi e^{2\pi i(k_3' L - \nu t)} \boldsymbol{e}' - D_0 \sin\varphi e^{2\pi i(k_3'' L - \nu t)} \boldsymbol{e}''.$$

After reentering in air or vacuum, respectively, we have

$$\boldsymbol{D} = D_0 \cos\varphi e^{2\pi i M} \boldsymbol{e}' - D_0 \sin\varphi e^{2\pi i(M+d)} \boldsymbol{e}'' \quad \text{with}$$
$$M = L/\lambda' + (x_3 - L)/\lambda_0 - \nu t$$

and $d = L(n'' - n')/\lambda_0$, whereby λ_0 means the wavelength in air or vacuum ($n = \lambda_0/\lambda$!), respectively.

Because

$$\boldsymbol{e}' = \cos(\psi - \varphi)\boldsymbol{e}_1' - \sin(\psi - \varphi)\boldsymbol{e}_2',$$
$$\boldsymbol{e}'' = \sin(\psi - \varphi)\boldsymbol{e}_1' + \cos(\psi - \varphi)\boldsymbol{e}_2' \quad (\text{with } \boldsymbol{e}_2' \cdot \boldsymbol{e}_1' = 0)$$

we obtain for the amplitude of the wave transmitted through the second polarizer

$$D_{\|P_2} = D_0 \cos\varphi \cos(\psi - \varphi) e^{2\pi i M} - D_0 \sin\varphi \sin(\psi - \varphi) e^{2\pi i(M+d)}.$$

If we ignore reflection losses at both boundary surfaces, we obtain for the time average of the transmitted intensity I the following value:

$$I = I_0 \cos^2\varphi \cos^2(\psi - \varphi)(1 - \tan\varphi \tan(\psi - \varphi) e^{2\pi i d})(1 - \tan\varphi \tan(\psi - \phi) e^{-2\pi i d}).$$

Here we have used the fact that the time average of the intensity is proportional to the square of the magnitude of the complex amplitude of $\boldsymbol{D}$, hence $I = qD_{\|P_2}\bar{D}_{\|P_2}$, and I_0, the intensity of the primary wave, is to be correspondingly represented by $I_0 = qD_0\bar{D}_0$. q is a proportionality factor.

With the help of the relations

$$\begin{aligned} e^{is} + e^{-is} &= 2\cos s, \quad \cos s = 1 - 2\sin^2(s/2), \\ \sin(u+v) &= \sin u \cos v + \sin v \cos u \quad \text{and} \\ \cos(u+v) &= \cos u \cos v - \sin u \sin v \end{aligned}$$

we can write the intensity formula in a much simpler form:

$$I = I_0(\cos^2\psi + \sin 2\varphi \sin 2(\psi - \varphi)\sin^2 \pi d).$$

The quantity $d = L(n'' - n')/\lambda_0$ is called the optical path difference, measured in wavelengths (for air or vacuum, respectively). Of special practical importance is the case of crossed polarizers $\psi = 90°$.

Here

$$I = I_0 \sin^2 2\varphi \sin^2 \pi d.$$

This relationship is the basis of polarization microscopy as well as of the quantitative measurement of the optical path difference and its variation under external conditions. If $\boldsymbol{e}'$ is parallel or perpendicular to $\boldsymbol{e}_1$ or $\boldsymbol{e}_1' = \boldsymbol{e}_2$, the directions of vibrations of the polarizers, then I vanishes completely ($\varphi = 0$ or $90°$; extinction position). In the so-called diagonal position $\varphi = 45°$ (and odd multiples thereof), I takes on a maximum. The factor $\sin^2 \pi d$ causes a strong dependence on the wavelength of the primary radiation. If one works with white light, then those regions of the spectrum are attenuated or extinguished for which $\pi d = m\pi$ (m integer), hence $d = m$ or $\lambda_0 = L(n'' - n')/m$. As a consequence, one observes the characteristic color phenomena, from which the experienced polarization microscopist, at least in the range of small values of m, can draw quantitative conclusions about the existing path difference.

If one places an arrangement of several crystal plates with parallel directions of vibration between the polarizers, the path differences d_i add according to $d = \sum_i d_i$. The application of the so-called compensators is based on this fact. Compensators are crystal plates whose effective path difference can be varied within certain limits by changing the active thickness (quartz-wedge compensator, Fig. 4.16) or by rotating a plate perpendicular to the transmission direction (rotatory compensator).

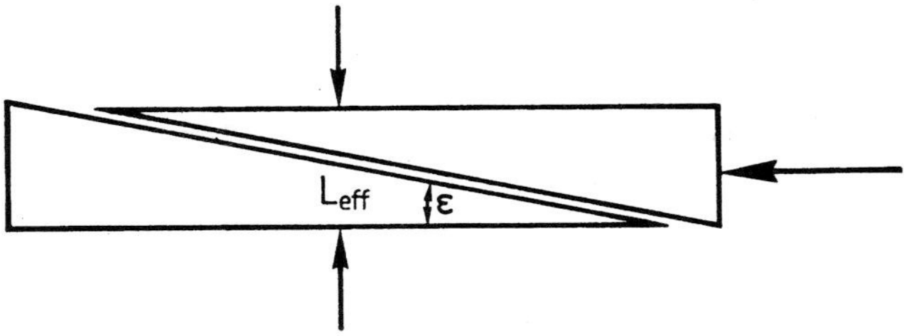

Figure 4.16 Quartz-wedge compensator. The effective thickness L_{eff} can be varied by translation of the wedges. In order to achieve high resolution the edge angle ε should be chosen to be sufficiently small. The wedges possess the same orientation. The optic axis is oriented within the outside faces of the plate.

4.3.6.4 **Directions of Optic Isotropy: Optic Axes, Optic Character**

In each crystal there exist directions having a circle as the sectional ellipse. The associated directions of vibration of the $\boldsymbol{D}$-vectors are, in this case, not fixed by the sectional ellipse. The refractive indices of all waves with this propagation direction are the same. We call these directions of isotropy of wave propagation *optic axes*. We consider the case of an arbitrary indicatrix in the principal axes' system:

$$\sum_i x_i^2/n_i^2 = 1 \quad \text{with} \quad n_1 < n_2 < n_3.$$

If we vary the position of the propagation directions perpendicular to $\boldsymbol{e}_2$, then one principal axis, respectively, of the sectional ellipse remains unaltered, namely that parallel to $\boldsymbol{e}_2$ with a length of n_2. The other principal axis changes its position between the values n_1 and n_3. Since n_2 lies between these values, there exists one direction in which $n' = n'' = n_2$.

Let the given propagation directions form an angle V with the direction $\boldsymbol{e}_3$ (Fig. 4.17). We thus have $\cos V = x_1/n_2$.

The condition for the circular section is

$$x_1^2 + x_3^2 = n_2^2 \quad \text{and} \quad x_1^2/n_1^2 + x_3^2/n_3^2 = 1.$$

After eliminating x_3 we find

$$\cos V = \frac{x_1}{n_2} = \pm\frac{n_1}{n_2}\sqrt{\frac{n_3^2 - n_2^2}{n_3^2 - n_1^2}} = \pm\sqrt{\frac{\dfrac{n_3^2}{n_2^2} - 1}{\dfrac{n_3^2}{n_1^2} - 1}}.$$

The angle $2V$ between both these equivalent optical axes is called the *axial angle*. We call the bisector of the smaller angle between the optical axes the *acute*

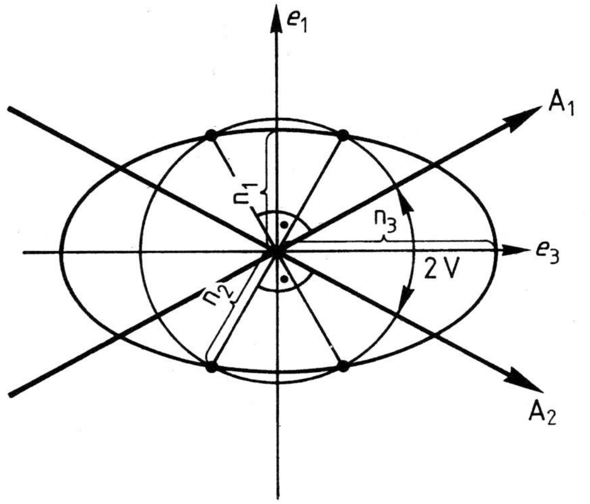

Figure 4.17 Construction of the optic axes A_1 and A_2 in the plane perpendicular to the medium semiaxis e_2 (axial plane). In the present case the acute bisectrix runs parallel to e_3 (optically positive, if $n_3 > n_1$).

bisectrix, the other bisector is called the *obtuse bisectrix*. The plane spanned by both axes is called the axial plane, their normal (parallel e_2) is called the optical binormal. We can now define the *optic character* in general:

optically positive, if acute bisectrix parallel to largest semiaxis n_3 (largest refractive index),

optically negative, if acute bisectrix parallel to shortest semiaxis n_1.

In crystals with a three-, four-, or six-fold principal axis, the indicatrix takes on the form of an ellipsoid of revolution. Both optical axes converge to one optical axis. Such crystals are therefore called optical uniaxial as opposed to the orthorhombic, monoclinic, and triclinic crystals which are referred to as optical biaxial.

The optic character is an important and easily accessible criterion for the diagnosic work. If a crystal plate, cut approximately perpendicular to an optic axis and aligned between crossed polarizers, is irradiated by a divergent beam of light, one can observe simultaneously the change of the path difference as a function of the deviation of the transmission direction from the optic axis (conoscopic imaging). With optic uniaxial crystals, the directions of identical path difference lie in cones around the optic axis and are identified by concentric rings around the piercing point of the optic axis. If one works with white light, one observes color phenomena which correspond to the progress of interference colors of a plane-parallel plate with increasing path difference. For this reason the curves are termed isochromates. Furthermore, the directions,

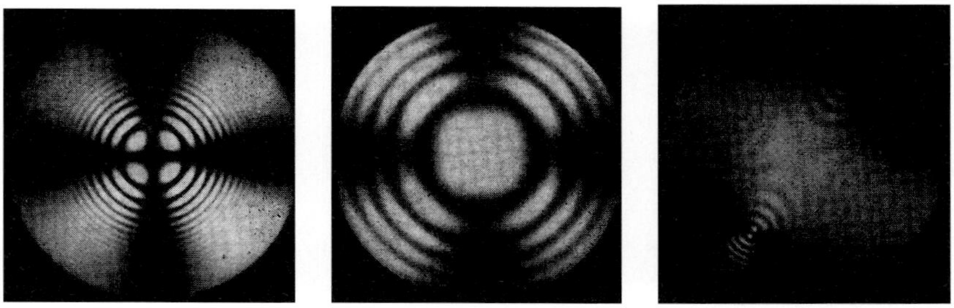

Figure 4.18 Conoscopic images around optical axes taken under crossed polarizers. (a) uniaxial, not optically active ($C(NH_2)_3Al(SO_4)_2 \cdot 6H_2O$); (b) uniaxial, optical active (cesium tartrate); (c) biaxial, not optical active (calcium formiate, diagonal position).

in which the principal axes of the sectional ellipse run parallel to the directions of vibration of the polarizers, show extinction. These directions converge in the image field to dark lines called isogyres. In optical biaxial crystals, the isochromates and the isogyres deviate in a characteristic manner from the image of uniaxial crystals (Fig. 4.18). If one now places an auxiliary crystal, with a path difference of around a quarter of a wavelength, in the ray path, one observes in different directions addition or subtraction of the path difference and thus a characteristic change of the interference colors. One can immediately recognize the optical character from the increase or decrease of the effective path difference in the vicinity of an optical axis, dependent on the position of the major semiaxis of the sectional ellipse of the auxiliary crystal.

The characteristic axis images can also be used for a fast and convenient determination of the orientation of large crystals. For this purpose, one places the crystal to be investigated in a cuvette with a liquid possessing a refractive index roughly coincident with that of the crystal. Two polarizers are then placed in front and behind the cuvette in crossed position. With divergent white light coming directly from a light bulb in front of the first polarizer, the position of the optical axis(es) can be located, with high accuracy, within a few minutes by turning the crystal.

The optical axes play a role in the determination of the directions of vibration of the $\boldsymbol{D}$-vectors for an arbitrary propagation direction g with the aid of Fresnel's construction in a stereographic projection. We draw the optical axes A_1 and A_2 as well as the propagation vector g from the center of the indicatrix. The normals L_1 and L_2 on the planes spanned by A_1 and g as well as A_2 and g intercept the indicatrix in points possessing a distance n_2 from the center. This means, both normals lie symmetric to one of the semiaxes of the sectional ellipse (perpendicular to g). The bisectors of the normals L_1 and

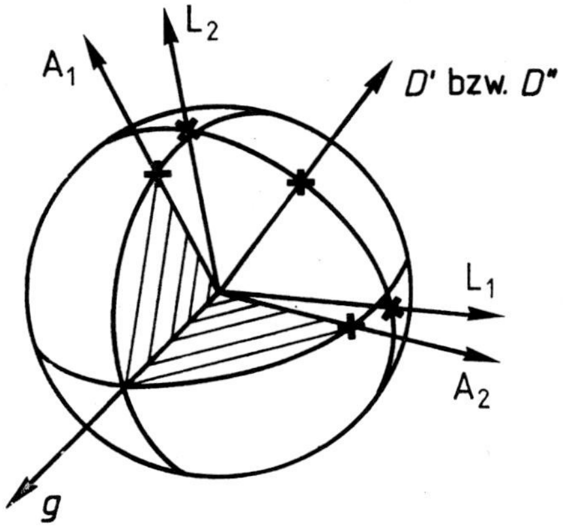

Figure 4.19 Fresnel's construction of the directions of vibration $\boldsymbol{D}'$ and $\boldsymbol{D}''$, respectively, from the position of the propagation vector g and the optical axes A_1 and A_2. The construction can be performed conveniently by the aid of the stereographic projection.

L_2 then deliver the associated directions of vibration (Fig. 4.19). In uniaxial crystals, the position of the directions of vibrations is particularly easy to recognize. The $\boldsymbol{D}$-vector of the extraordinary wave vibrates in the plane spanned by $\boldsymbol{g}$ and the optical axis (principal section), and the $\boldsymbol{D}$-vector of the ordinary wave vibrates perpendicular to the principal plane.

4.3.6.5 **Sénarmont Compensator for the Analysis of Elliptically Polarized Light**

Any number of waves with a fixed frequency ν and a common direction of propagation $\boldsymbol{g}$, but with different directions of vibration and arbitrary phase combine in isotropic media to form an elliptically polarized wave. One can imagine this wave as composed of two linear polarized waves with mutually perpendicular directions of vibration and with a phase difference of 90°. This is recognized as follows. Let arbitrary waves $\boldsymbol{D}_j = \boldsymbol{D}_{j0} e^{2\pi i(\boldsymbol{k}\cdot\boldsymbol{x}-\nu t+\alpha_j)}$ with $\boldsymbol{D}_j \cdot \boldsymbol{k} = 0$ unite to the resultant $\boldsymbol{D} = \sum_j \boldsymbol{D}_j$. In a Cartesian reference system let $\boldsymbol{D}_{j0}$ be resolved into components

$$\boldsymbol{D}_{j0} = D_{j0}\cos\varphi_j \boldsymbol{e}_{\mathrm{I}} + D_{j0}\sin\varphi_j \boldsymbol{e}_{\mathrm{II}} \qquad (\boldsymbol{k} \parallel \boldsymbol{e}_{\mathrm{III}}).$$

We then have

$$\boldsymbol{D} = e^{2\pi i(\boldsymbol{k}\cdot\boldsymbol{x}-\nu t)}(\boldsymbol{D}_{\mathrm{I}0} e^{2\pi i\varphi_{\mathrm{I}}} + \boldsymbol{D}_{\mathrm{II}0} e^{2\pi i\varphi_{\mathrm{II}}}) = e^{2\pi i A}(\boldsymbol{D}_{\mathrm{I}0} + \boldsymbol{D}_{\mathrm{II}0} e^{2\pi i\delta}) = \boldsymbol{D}_{\mathrm{I}} + \boldsymbol{D}_{\mathrm{II}},$$

where

$$\boldsymbol{D}_{\mathrm{I}0}e^{2\pi i\varphi_{\mathrm{I}}} = \sum_j D_{j0}\cos\varphi_j e^{2\pi i\alpha_j}\boldsymbol{e}_{\mathrm{I}},$$

$$\boldsymbol{D}_{\mathrm{II}0}e^{2\pi i\varphi_{\mathrm{II}}} = \sum_j D_{j0}\sin\varphi_j e^{2\pi i\alpha_j}\boldsymbol{e}_{\mathrm{II}},$$

$$A = (\boldsymbol{k}\cdot\boldsymbol{x} - \nu t + \varphi_{\mathrm{I}}) \quad \text{and} \quad \delta = (\varphi_{\mathrm{II}} - \varphi_{\mathrm{I}})$$

The resultant of $\boldsymbol{D}$ to a certain value of A is obtained from the real part of $\boldsymbol{D}$ in the directions $\boldsymbol{e}_{\mathrm{I}}$ and $\boldsymbol{e}_{\mathrm{II}}$. Thus

$$\mathrm{Re}(D_{\mathrm{I}}) = D_{\mathrm{I}} = D_{\mathrm{I}0}\cos 2\pi A \quad \text{and}$$
$$\mathrm{Re}(D_{\mathrm{II}}) = D_{\mathrm{II}} = D_{\mathrm{II}0}\cos 2\pi(A+\delta).$$

Elimination of A using

$$\cos 2\pi(A+\delta) = \cos 2\pi A\cos 2\pi\delta - \sin 2\pi A\sin 2\pi\delta$$

gives

$$D_{\mathrm{II}}/D_{\mathrm{II}0} - D_{\mathrm{I}}/D_{\mathrm{I}0}\cos 2\pi\delta = -\sqrt{1-(D_{\mathrm{I}}/D_{\mathrm{I}0})^2}\sin 2\pi\delta.$$

After squaring we have

$$D_{\mathrm{I}}^2/D_{\mathrm{I}0}^2 + D_{\mathrm{II}}^2/D_{\mathrm{II}0}^2 - 2D_{\mathrm{I}}D_{\mathrm{II}}\cos 2\pi\delta/D_{\mathrm{I}0}D_{\mathrm{II}0} = \sin^2 2\pi\delta.$$

This is the equation of an ellipse. The resultant of $\boldsymbol{D}$ passes through the ellipse once per vibration. The position of the principal axes of the ellipse and the extreme values of $\boldsymbol{D}$ are calculated with the help of a plane principal axes' transformation. The result for an angle of rotation ψ around $\boldsymbol{e}_{\mathrm{III}}$ is

$$\tan 2\psi = \frac{2D_{\mathrm{I}0}D_{\mathrm{II}0}\cos 2\pi\delta}{D_{\mathrm{I}0}^2 - D_{\mathrm{II}0}^2}.$$

The associated extreme values are $D'_{\mathrm{I}0}$ and $D''_{\mathrm{II}0}$. In the rotated reference system, the equation of the ellipse is $(D'_{\mathrm{I}}/D'_{\mathrm{I}0})^2 + (D'_{\mathrm{II}}/D'_{\mathrm{II}0})^2 = 1$. This means, the elliptic polarized wave can also be represented as a superposition of two linear polarized waves with mutually perpendicular directions of vibration. These exhibit a phase difference of 90°, as one recognizes from the equivalence of the equation of the ellipse in the principal axes' system with the parameter representation

$$D'_{\mathrm{I}} = D'_{\mathrm{I}0}\cos 2\pi A' \quad \text{and} \quad D''_{\mathrm{II}} = D'_{\mathrm{II}0}\sin 2\pi A'.$$

We use the Sénarmont compensator to analyze elliptic polarized light. This consists of a crystal plate with a path difference of $\lambda_0/4$ (e.g., a cleavage lamella of mica or a quartz plate of corresponding thickness) and a polarizer, both of which can be rotated independently around their common normal. In order to analyze the incident elliptic polarized light, one rotates the $\lambda_0/4$-plate and the direction of vibration of the polarizer until complete extinction occurs. From the position of the directions of vibration of the semiaxes of the $\lambda_0/4$-plate and the direction of vibration of the polarizer one obtains the amplitude ratio D'_{I0}/D'_{II0} and the position of the principal axes.

We explain the relation first for elliptically polarized light emerging from a double refracting crystal plate exposed to a linear polarized wave with a direction of vibration $\boldsymbol{e}_1$, which bisects the principal axes of the sectional ellipse (diagonal position). After leaving the plate, both waves have amplitudes

$$\boldsymbol{D}' = D_0 \frac{1}{\sqrt{2}} \cos 2\pi A \boldsymbol{e}' \quad \text{and} \quad \boldsymbol{D}'' = D_0 \frac{1}{\sqrt{2}} \cos 2\pi (A + d) \boldsymbol{e}''$$

with $d = L(n'' - n')/\lambda_0$.

D_0 is the amplitude of the incident wave. We now split both waves into components parallel and perpendicular to the direction of vibration of the polarizer according to

$$\begin{aligned} D_1 &= \frac{D_0}{2} \cos 2\pi A + \frac{D_0}{2} \cos 2\pi (A + d) \\ D_2 &= -\frac{D_0}{2} \cos 2\pi A + \frac{D_0}{2} \cos 2\pi (A + d). \end{aligned}$$

With the aid of

$$\begin{aligned} \cos u + \cos v &= 2 \cos \left(\frac{u + v}{2}\right) \cos \left(\frac{u - v}{2}\right) \quad \text{and} \\ \cos u - \cos v &= -2 \sin \left(\frac{u + v}{2}\right) \sin \left(\frac{u - v}{2}\right) \end{aligned}$$

we obtain

$$\begin{aligned} D_1 &= D_0 \cos(\pi d) \cos 2\pi (A + d/2) \\ D_2 &= -D_0 \sin(\pi d) \sin 2\pi (A + d/2). \end{aligned}$$

This is the normal representation of an elliptically polarized wave (partial amplitudes vibrate mutually perpendicular and have a phase difference of 90°). If we now place the $\lambda_0/4$-plate in the ray path such that its directions of vibration correspond to those of D_1 and D_2, hence parallel and perpendicular to the direction of vibration of the polarizer, the phase difference changes by a further 90°. If the direction of vibration of the slower wave of the compensator lies parallel to $\boldsymbol{e}_1(n'')$, the phase difference of both partial waves increases to 180°. In the converse case, the phase difference vanishes ($\boldsymbol{e}_1(n')$). In each case, we obtain a plane-polarized wave as a superposition of both partial waves. The position of the resultant

$$D_{\text{res}} = \mp D_0 \cos(\pi d)\boldsymbol{e}_1 - D_0 \sin(\pi d)\boldsymbol{e}_2$$

($-$ for n'' of the compensator parallel to $\boldsymbol{e}_1$, $+$ for n' perpendicular to $\boldsymbol{e}_1$, $n'' > n'$) can be determined with the help of the second rotatable polarizator, which is rotated to the extinction position. The angle ψ between resultant and $\boldsymbol{e}_1$ is given by

$$\tan\psi = \varepsilon = \left(\frac{D_2}{D_1}\right)_{\text{res}} = \sin(\pi d)/\cos(\pi d) = \tan\pi d,$$

where ε specifies the eccentricity of the ellipse. Hence, one can not only analyze the elliptically polarized light emerging from the crystal plate but also measure the path difference of the given crystal plate, if it lies in the range of up to one wavelength. This method finds its most important application in the measurement of induced changes of the path difference, such as, e.g., in electro-optic or piezo-optic effects (see Section 4.4.2). In order to improve the accuracy, further devices of similar type have been developed. With the aid of a left-quartz right-quartz double plate after Nakamura, the extinction position can be most precisely fixed.

Circular polarized light is present, when $|\varepsilon| = 1$, i.e., when both partial waves possess the same amplitude. One specifies elliptically polarized light as left or right rotating depending on the sense of rotation of the resultant on the ellipse. When facing in the propagation direction of the wave, one observes right-rotating light, if the sense of rotation is counterclockwise, otherwise left-rotating light. Often also the terms laev- and dextro-rotating are used.

4.3.6.6 **Absorption**

In isotropic absorbing substances one notes that the relative change in intensity $\Delta I/I$ of a light ray is proportional to the infinitesimal thickness Δx through which it passes, thus $\Delta I/I = -\mu\Delta x$. μ is the absorption coefficient. On integrating one finds $I = I_0 e^{-\mu L}$, that is, an exponential decline with thickness L (*Lambert's absorption law*).

How can we describe this behavior in the material constants of wave propagation? If one introduces, instead of the refractive power n, which up to now was considered as a real quantity, a complex refractive index $n = n_0 + i n_0 \kappa$, where n_0 and κ are real quantities, a wave propagating in the direction $\boldsymbol{e}_3$ attains the form

$$\boldsymbol{D} = \boldsymbol{D}_0 e^{2\pi i(n x_3/\lambda_0 - \nu t)} = \boldsymbol{D}_0 e^{-2\pi n_0 \kappa x_3/\lambda_0} e^{2\pi i(n_0 x_3/\lambda_0 - \nu t)}.$$

The quantity $4\pi n_0\kappa/\lambda_0$ then corresponds to the absorption coefficient mentioned above ($I \sim |\boldsymbol{D}|^2$).

The phenomenological description of absorption in the framework of Maxwell's equations is provided by the relation between refractive index and the dielectric constant $\epsilon_{\text{rel},jj} = n_j^2$ (Maxwell's relation, see Section 4.3.6). As was shown in 4.3.3, there exists a relationship between the real part of the electric conductivity and the imaginary part of the dielectric constant. Let $\epsilon_{ij} = \epsilon'_{ij} + i\epsilon''_{ij}$. If we set

$$\epsilon_{\text{rel},ij} = n_{ij}^2, \quad \text{with} \quad n_{ij} = n_{ij0} + i n_{ij0}\kappa_{ij},$$

we get by multiplying

$$\epsilon'_{\text{rel},ij} = n_{ij0}^2(1 - \kappa_{ij}^2) \quad \text{and} \quad \epsilon''_{\text{rel},ij} = 2\kappa_{ij} n_{ij0}^2.$$

The derivation of the basic equations for the propagation of plane waves runs analogously to the procedure for nonabsorbing media, whereby all quantities may now be complex. We will not display this here in detail, but will mention the most important results. The basic law of crystal optics takes the following form: In an absorbing medium, two elliptically polarized waves with $\boldsymbol{D}$-vectors perpendicular to the propagation direction can propagate in any direction. The major semiaxes of the elliptically polarized waves are mutually perpendicular; the ellipses possess the same eccentricity and their $\boldsymbol{D}$-vector moves in the same sense.

A derivation of this law is found, e.g., in Szivessy (1928). For the quantitative determination of the optical properties of absorbing crystals, two reference surfaces must be considered, the indicatrix, which reflects the real part of the dielectric tensor and a second surface for the absorption coefficient, which is also to be thought of as a second-order surface. In case the position of the indicatrix is fixed by symmetry conditions, then the absorption surface is also fixed by symmetry. One has to assign to each of the three principal refractive indices a principal absorption coefficient, with the help of which one can describe the complete optical behavior. In monoclinic and triclinic crystals, only one or none of the principal axes of both reference surfaces, respectively, coincide.

With weak absorption, as, e.g., in most crystals with low electric conductivity, the law of crystal optics is valid, to a sufficient degree, for nonabsorbing media. This means that for each direction of propagation, there exist two linear polarized vibrations with $\boldsymbol{D}$-vectors lying parallel to the principal axes of the sectional ellipse. However, the absorption coefficients of both waves are, in general, different. Such a crystal plate shows, in linear polarized white light, color phenomena dependent on the position of the direction of vibration of the polarizer. This phenomenon is called pleochroism. In many crystals, e.g., in certain varieties of tourmaline, the absorption for one of the two directions of vibration is so strong that such a crystal plate can be directly used for the production of polarized light. Commercial polarization foils contain strongly pleochroitic, parallel oriented crystals, which already in thin films give rise to almost complete absorption in one of the two directions of vibration.

Fresnel's formulae can also be directly adopted for absorbing media when one allows complex indices of refraction. The law of refraction for light waves passing through the boundary surface from a nonabsorbing medium I into an absorbing medium II takes the form

$$\frac{\sin \alpha_{\mathrm{I}}}{\sin \alpha_{\mathrm{II}}} = \frac{n_{\mathrm{II}}}{n_{\mathrm{I}}} = \frac{n_{\mathrm{II}0}(1 + i\kappa_{\mathrm{II}})}{n_{\mathrm{I}}}.$$

In the case of perpendicular incidence, we obtain from the amplitude ratio with $n_{\mathrm{I}} = 1$ (vacuum)

$$\frac{E_{\mathrm{Ir}}}{E_{\mathrm{I}}} = \frac{n_{\mathrm{II}} - 1}{n_{\mathrm{II}} + 1}$$

the reflectance

$$R = \left| \frac{n_{\mathrm{II}} - 1}{n_{\mathrm{II}} + 1} \right|^2 = \frac{(n_{\mathrm{II}0} - 1)^2 + n_{\mathrm{II}0}^2 \kappa_{\mathrm{II}}^2}{(n_{\mathrm{II}0} + 1)^2 + n_{\mathrm{II}0}^2 \kappa_{\mathrm{II}}^2}.$$

In strongly absorbing media ($\kappa \gg n_{\mathrm{II}0}$) R approaches the value 1 (metallic reflection).

If the medium II is anisotropic, one observes, in general, even with perpendicular incidence, a dependence of the reflectance on the direction of vibration of the incident linear polarized light. This phenomenon is called reflection pleochroism. With oblique incidence such effects are even more prominent. One can determine the optic constants of such crystals from an analysis of the reflected light, which in absorbing media is usually elliptically polarized.

4.3.6.7 **Optical Activity**

In certain acentric crystals, as e.g., quartz or $LiIO_3$, one observes a rotation of the direction of vibration of the $\boldsymbol{D}$-vector when a wave propagates in a certain direction. The rotation is proportional to the thickness of the medium traversed. This effect, which can also occur in cubic crystals and in liquids, is called *optical activity*. The phenomenological description of optical activity is achieved with the help of an extension of the relationship between $\boldsymbol{E}$ and $\boldsymbol{D}$ according to

$$D_i = \epsilon_{ij} E_j + g_{ijk} \frac{\partial E_j}{\partial x_k}.$$

What must be the nature of the tensor $\{g_{ijk}\}$ so that, corresponding to observation, it does not change the energy content of an electromagnetic wave

$$\boldsymbol{E} = \boldsymbol{E}_0 e^{2\pi i(\boldsymbol{k}\cdot\boldsymbol{x} - \nu t)}$$

by a reversal in propagation direction?

The electric energy density is calculated according to

$$W_{\text{el}} = \sum_i \int E_i dD_i = \int \sum_{i,j} E_i \epsilon_{ij} dE_j + \int \sum_{i,j} E_i 2\pi i g_{ijk} k_k dE_j$$

with

$$\frac{dE_j}{dx_k} = 2\pi i k_k E_j.$$

The second term vanishes when $g_{ijk} = -g_{jik}$. $\{g_{ijk}\}$ must, therefore, be antisymmetric in the first two indices, if the energy density is to be independent of the direction of the propagation vector. If one substitutes g_{ijk} by γ_{lk} according to $\frac{2\pi}{\lambda} g_{ijk} = \gamma_{lk}$ or $-\gamma_{lk}$ with $l \neq i, j$ and i, j, l cyclic or anticyclic in 1, 2, 3, respectively, one arrives at a clearer representation

$$g_{ijk} \frac{dE_j}{dx_k} = g_{ijk} \cdot 2\pi i k_k E_j = -i(\boldsymbol{G} \times \boldsymbol{E})_i,$$

whereby $\boldsymbol{G}$, the *gyration vector*, possesses the components $G_i = \gamma_{ij} g_j$. g is as previously, the unit vector in the direction $\boldsymbol{k}$. Thus we have $D_i = \epsilon_{ij} E_j - i(\boldsymbol{G} \times \boldsymbol{E})_i$. While $\{g_{ijk}\}$ represents, in both first index positions, an antisymmetric polar tensor, the tensor γ_{ij} derived from this as well as the vector $\boldsymbol{G}$ are pseudo tensors. $\boldsymbol{G} \times \boldsymbol{E}$ delivers a polar vector because $\boldsymbol{D}$ and $\boldsymbol{E}$ are polar; hence $\boldsymbol{G}$ is a pseudo tensor. From symmetry, the prescribed form of the tensor in the point symmetry groups is found in Table 4.2. Recall that the existence of an inversion center lets all components γ_{ij} vanish, in agreement with experimental findings.

We will now discuss at least some essentials of the propagation of plane electromagnetic waves in optically active crystals in order to be prepared for an understanding of higher order optical effects in a later section.

We get the index equation as follows. We proceed from the relation derived from Maxwell's equations

$$\frac{v^2}{c^2} D_j = E_j - (\boldsymbol{g} \cdot \boldsymbol{E}) g_j.$$

With

$$\frac{v^2}{c^2} = n^{-2}, \quad \epsilon_{\text{rel},jj} = n_j^2 \quad \text{and} \quad D_i = \epsilon_{ij} E_j - i(\boldsymbol{G} \times \boldsymbol{E})_i.$$

The following equations are then valid in the principal axes' system of the real dielectric tensor:

$$E_j(n_j^2 - n^2(1 - g_j^2)) + E_{j+1}(n^2 g_j g_{j+1} + iG_{j+2}) \\ + E_{j+2}(n^2 g_{j+2} g_j - iG_{j+1}) = 0 \quad \text{for} \quad j = 1, 2, 3.$$

This homogenous system of equations has a solution for $\boldsymbol{E} \neq 0$ only when its determinant vanishes. From this condition, and after appropriate collection, we obtain the index equation for optically active crystals

$$n^4\left(\sum_j n_j^2 g_j^2\right) - n^2\left(\sum_j n_j^2 n_{j+1}^2(1-g_{j+2}^2) - (\boldsymbol{g}\times\boldsymbol{G})^2\right) + n_1^2 n_2^2 n_3^2 - \sum_j n_j^2 G_j^2 = 0.$$

The coefficient for n^6 cancels out. A solution with given direction of propagation $\boldsymbol{g}$ requires the knowledge of the principal refractive indices and of the gyration vector. In the form

$$n^4 - n^2\frac{\sum_j n_j^2 n_{j+1}^2(1-g_{j+2}^2)}{\sum_j n_j^2 g_j^2} + \frac{n_1^2 n_2^2 n_3^2}{\sum_j n_j^2 g_j^2} = \frac{\sum_j n_j^2 G_j^2 - n^2(\boldsymbol{g}\times\boldsymbol{G})^2}{\sum_j n_j^2 g_j^2} = q^2$$

the left-hand side corresponds to the index equation without activity (see Section 4.3.6.1; with $x_i = n g_i$). Let the solutions be n_0' and n_0''. Hence the index equation has the form $(n^2 - n_0'^2)(n^2 - n_0''^2) = q^2$. From this, one can find approximations for n' and n'', both possible refractive indices for the direction of propagation $\boldsymbol{g}$, when one writes in q^2 for n^2 an approximate value, e.g., $n_0'^2$ or $n_0''^2$. Experiments have shown that in almost all crystals investigated so far, it suffices to calculate q^2 under the assumption $n_1 = n_2 = n_3 = n$. Then q^2 becomes

$$q_0^2 = (\boldsymbol{g}\cdot\boldsymbol{G})^2 \quad \text{because} \quad \boldsymbol{G}^2 - (\boldsymbol{g}\times\boldsymbol{G})^2 = (\boldsymbol{g}\cdot\boldsymbol{G})^2.$$

As a result, for n' and n'' we have

$$n'^2, n''^2 = \frac{1}{2}\left\{n_0'^2 + n_0''^2 \pm \sqrt{(n_0'^2 - n_0''^2)^2 + 4q_0^2}\right\},$$

where

$$n' > n'' \quad \text{and} \quad n_0' > n_0''.$$

If one enters n' and n'' into the basic equations and keeps the condition $n_1 = n_2 = n_3 = \bar{n}$ for the terms connected with the gyration vector, whereby $\bar{n}$ is a mean refractive index, one finds the following approximation law for the propagation of plane waves in optically active nonabsorbing crystals:

Two elliptically polarized transverse waves of the same ellipticity, but with opposite sense of rotation can propagate in any arbitrary direction $\boldsymbol{g}$. The associated refractive indices are n' and n''. The major semiaxes of both ellipses of vibration are mutually perpendicular and lie parallel to the directions of vibration that would exist without optical activity in the given direction of propagation (see, e.g., Szivessy, p. 811).

Table 4.4 Optic properties of optical active crystals of trigonal, tetragonal and hexagonal symmetry.

PSG 3, 4, 6 ($n_1 = n_2$)	PSG 32, 42, 62 ($n_1 = n_2$)
$\gamma_{11} = \gamma_{22}, \gamma_{33}, \gamma_{12} = -\gamma_{21}$	$\gamma_{11} = \gamma_{22}, \gamma_{33}$
$g = e_3$: $G_k = \gamma_{k3} g_3$ $G_1 = G_2 = 0, G_3 = \gamma_{33}$ $g \cdot G = \gamma_{33};\ g \times G = 0$ Index equation (for both groups): $n^4 - 2n_1^2 n^2 + n_1^4 - \gamma_{33}^2 = 0,$ $n'^2 = n_1^2 + \gamma_{33},\ n''^2 = n_1^2 - \gamma_{33}$ Basic equations: $D_1 = \epsilon_{11} E_1 + i\gamma_{33} E_2$ $D_2 = \epsilon_{11} E_2 - i\gamma_{33} E_1$ $D_3 = \epsilon_{33} E_3$	 $G_1 = G_2 = 0, G_3 = \gamma_{33}$ $g \cdot G = \gamma_{33};\ g \times G = 0$
$g = e_1$: $G_1 = \gamma_{11}, G_2 = \gamma_{21}, G_3 = 0$ $g \times G = \gamma_{21} e_3$ $g \cdot G = \gamma_{11}$ Index equation: $n^4 - n^2 \left\{ n_1^2 + n_3^2 - \frac{1}{n_1^2} \gamma_{12}^2 \right\}$ $+ n_1^2 n_3^2 - (\gamma_{11}^2 + \gamma_{12}^2) = 0$ With $n = n_1$ we have in both cases n'^2 resp. $n''^2 = (n_1^2 + n_3^2)/2 +$ resp. $- \frac{1}{2}\sqrt{(n_1^2 - n_3^2)^2 + 4\gamma_{11}^2}$ Basic equations: $D_1 = \epsilon_{11} E_1 - i\gamma_{21} E_3$ $D_2 = \epsilon_{11} E_2 + i\gamma_{11} E_3$ $D_3 = \epsilon_{33} E_3 - i\gamma_{11} E_2 + i\gamma_{21} E_1$	 $G_1 = \gamma_{11}, G_2 = G_3 = 0$ $g \times G = 0$ $g \cdot G = \gamma_{11}$ $n^4 - n^2(n_1^2 + n_3^2) + n_1^2 n_3^2 - \gamma_{11}^2 = 0$

The $\boldsymbol{D}$-vectors of these waves combine according to $\boldsymbol{D} = \boldsymbol{D}' + iu\boldsymbol{D}''$ for n' and $\boldsymbol{D} = \boldsymbol{D}'' + iu\boldsymbol{D}'$ for n'', whereby $\boldsymbol{D}'$ and $\boldsymbol{D}''$, respectively, run parallel to the semiaxes of the sectional ellipse for missing optical activity and $|\boldsymbol{D}'| = |\boldsymbol{D}''|$. The ellipticity u is dependent on the optic constants. In very rare cases of extremely large double refraction, one of course expects a certain deviation from this situation.

As an example, we now discuss the propagation of light parallel and perpendicular to the optic axis in crystals of the PSG 3, 4, 6, 32, 42, and 62. According to Table 4.4, the components $\gamma_{11} = \gamma_{22}$ and γ_{33} as well as $\gamma_{12} = -\gamma_{21}$ exist for the case of pure rotation axes. In the PSG 32, 42, and 62, $\gamma_{12} = \gamma_{21} = 0$. Table 4.4 lists the relationships for both groups.

Besides the three material equations, we have available the following relationships $D_j = n^2 E_j - n^2(\boldsymbol{g} \cdot \boldsymbol{E}) g_j$ from the Maxwell equations to determine the components of the $\boldsymbol{D}$-vectors. For the case $\boldsymbol{g} = \boldsymbol{e}_3$ we have $D_3 = 0$ because $\operatorname{div} \boldsymbol{D} = \boldsymbol{k} \cdot \boldsymbol{D} = 0$. If we introduce for D_1 and D_2 the values of the material equations, we get

$$D_1 = \epsilon_{11}E_1 + i\gamma_{33}E_2 = n^2E_1$$
$$D_2 = \epsilon_{11}E_2 - i\gamma_{33}E_1 = n^2E_2.$$

Hence, with $\epsilon_{11} = n_1^2$ and the relationship $n^2 - n_1^2 = \pm\gamma_{33}$ we have: $E_1 = iE_2$ and $D_1 = iD_2$ for n' as well as $E_2 = iE_1$ and $D_1 = -iD_2$ for n''.

The resulting waves $\boldsymbol{D}' = D_1\boldsymbol{e}_1 - iD_1\boldsymbol{e}_2$ and $\boldsymbol{D}'' = D_1\boldsymbol{e}_1 + iD_1\boldsymbol{e}_2$ represent two circular polarized waves, of which the first is left rotating and the second right rotating, when $\boldsymbol{g}$ points in the positive $\boldsymbol{e}_3$-direction.

If linear polarized light enters a plate cut perpendicular to the optic axis in perpendicular incidence, two circular polarized waves of opposite sense of rotation originate in the crystal and propagate with a velocity corresponding to the refractive indices n' and n''. After traversing the thickness L they immerse in the vacuum and interfere to a linear polarized wave, whose direction of vibration with respect to the primary wave is rotated by an angle φ.

The explanation is as follows: Let the primary wave be given by

$$\boldsymbol{D} = D_0 e^{2\pi i(k_3x_3 - \nu t)}\boldsymbol{e}_1.$$

When entering the plate ($x_3 = 0$), it splits up into two circular polarized waves

$$\boldsymbol{D}_+ = (D_0/2)e^{2\pi i(k_3'x_3 - \nu t)}\boldsymbol{e}_1 - (iD_0/2)e^{2\pi i(k_3'x_3 - \nu t)}\boldsymbol{e}_2$$

and

$$\boldsymbol{D}_- = (iD_0/2)e^{2\pi i(k_3''x_3 - \nu t)}\boldsymbol{e}_2 + (D_0/2)e^{2\pi i(k_3''x_3 - \nu t)}\boldsymbol{e}_1.$$

For $x_3 = 0$, the necessary condition $\boldsymbol{D} = \boldsymbol{D}_+ + \boldsymbol{D}_-$ is fulfilled.

After traversing the thickness L, both circular polarized waves propagate with the same velocity.

The components vibrating in $\boldsymbol{e}_1$ and $\boldsymbol{e}_2$, respectively, are combined to give

$$D_1 = (D_0/2)e^{2\pi iA} + (D_0/2)e^{2\pi i(A+d)}$$
$$D_2 = (iD_0/2)e^{2\pi i(A+d)} - (iD_0/2)e^{2\pi iA},$$

where

$$A = k_3'L + k_3(x_3 - L) - \nu t \quad \text{for} \quad x_3 > L \quad \text{and}$$
$$d = (k_3'' - k_3')L.$$

The physically effective contents of D_1 and D_2 are the real parts

$$\mathrm{Re}(D_1) = D_0/2(\cos 2\pi A + \cos 2\pi(A + d)),$$
$$\mathrm{Re}(D_2) = D_0/2(\sin 2\pi A - \sin 2\pi(A + d)).$$

With the help of the relations

$$\cos u + \cos v = 2\cos\frac{u+v}{2}\cdot\cos\frac{u-v}{2} \quad \text{and}$$
$$\sin u - \sin v = 2\sin\frac{u-v}{2}\cdot\cos\frac{u+v}{2}$$

one gets

$$\mathrm{Re}(D_1) = D_0\cos\pi d\cdot\cos 2\pi\left(A+\frac{d}{2}\right) \qquad \text{and}$$
$$\mathrm{Re}(D_2) = D_0\sin\pi d\cdot\cos 2\pi\left(A+\frac{d}{2}\right).$$

Both components together form a linear polarized wave with the direction of vibration

$$\boldsymbol{e} = \cos\varphi\,\boldsymbol{e}_1 + \sin\varphi\,\boldsymbol{e}_2.$$

We have

$$\varphi = \pi(k_3'' - k_3')L = \frac{\pi(n''-n')L}{\lambda_0}.$$

φ can be easily measured from the extinction position of a rotatable polarizer placed behind the crystal plate. With φ one also has γ_{33} because $n'^2 - n''^2 = 2\gamma_{33}$, hence

$$\gamma_{33} = (n''-n')(n''+n')/2 = \varphi\lambda_0(n'+n'')/2\pi L.$$

For $(n'+n'')$ we can write $2n_1$ as an approximation. The negative angle of rotation per mm thickness is termed the specific rotation: $\alpha = -180°|n' - n''|/\lambda_0$ (unit, degree per mm).

The derived relations are also valid for cubic crystals of the PSG 23 and 43 as well as for optically active isotropic substances.

For the case $\boldsymbol{g} = \boldsymbol{e}_1$ we assume that γ_{12} vanishes. Since $\boldsymbol{k}\cdot\boldsymbol{D} = 0$, $D_1 = 0$, and because $D_1 = \epsilon_{11}E_1$, $E_1 = 0$ and $\boldsymbol{g}\cdot\boldsymbol{E} = 0$. If one introduces in the dynamic basic equations $D_j = n^2E_j - n^2(\boldsymbol{g}\cdot\boldsymbol{E})g_j$ for D_j the values of the material equations, one gets

$$D_2 = \epsilon_{11}E_2 + i\gamma_{11}E_3 = n^2E_2$$
$$D_3 = \epsilon_{33}E_3 - i\gamma_{11}E_2 = n^2E_3.$$

From these one also finds here a fixed ratio of E_3/E_2, namely

$$E_3/E_2 = (n^2 - n_1^2)/i\gamma_{11} = -i\gamma_{11}/(n^2 - n_3^2),$$

where $\epsilon_{11} = n_1^2$ and $\epsilon_{33} = n_3^2$. The first expression follows from the first equation, and the second from the second. Hence,

$$D_2/D_3 = i(n^2 - n_3^2)/\gamma_{11} = iu.$$

With the known values for n' or n'' (see Table 4.4) we obtain the following results for both amplitude ratios

$$(D_2/D_3)' \quad \text{or} \quad (D_2/D_3)'' = i\{(n_1^2 - n_3^2) \pm \sqrt{(n_1^2 - n_3^2)^2 + 4\gamma_{11}^2}\}/2\gamma_{11}$$
$$= iu' \quad \text{or} \quad iu'',$$

respectively. One recognizes immediately that $u' = -1/u''$, i.e., there exists two elliptically polarized waves with the same ellipticity, but opposite sense of rotation. We mention at this point that the general proof for an arbitrary direction of propagation and arbitrary symmetry proceeds in an entirely analogous manner. From div $\boldsymbol{D} = 0$ it always follows that $\boldsymbol{g} \cdot \boldsymbol{D} = 0$, i.e., $\boldsymbol{D}$ has no components along the direction of propagation. The material equations are then to be considered only for both components in the plane perpendicular to $\boldsymbol{g}$ with corresponding results as above.

The determination of γ_{11} is possible by analysing the elliptically polarized light emerging from a plane-parallel plate when a linear polarized wave enters the plate in perpendicular incidence. The plate normal runs parallel to $\boldsymbol{e}_1$.

Let the primary wave be represented by

$$\boldsymbol{D} = D_0 e^{2\pi i(k_1 x_1 - \nu t)} \boldsymbol{e}_2.$$

Thus the two waves

$$\boldsymbol{D}' = \frac{D_0}{1+u''^2} e^{2\pi i(k_1' x_1 - \nu t)} \boldsymbol{e}_2 + \frac{iu'' D_0}{1+u''^2} e^{2\pi i(k_1' x_1 - \nu t)} \boldsymbol{e}_3,$$

$$\boldsymbol{D}'' = -\frac{iu'' D_0}{1+u''^2} e^{2\pi i(k_1'' x_1 - \nu t)} \boldsymbol{e}_3 + \frac{u''^2 D_0}{1+u''^2} e^{2\pi i(k_1'' x_1 - \nu t)} \boldsymbol{e}_2$$

propagate within the crystal ($x_1 \geq 0$).

After traversing the thickness L, the two elliptically polarized waves superimpose to a wave whose ellipticity and position of the major semiaxis is to be measured according to the methods discussed in Section 4.3.6.5. These quantities contain the value γ_{11} as well as the refractive indices n' and n''.

The resulting elliptically polarized wave is composed of the real parts of the vibrating components in $\boldsymbol{e}_2$ and $\boldsymbol{e}_3$:

$$\mathrm{Re}(D_2) = \frac{D_0}{1+u''^2}(\cos 2\pi A + u''^2 \cos 2\pi(A+d))$$

$$\mathrm{Re}(D_3) = \frac{u'' D_0}{1+u''^2}(\sin 2\pi(A+d) - \sin 2\pi A) \quad \text{with}$$

$$A = k_1(x_1 - L) + k_1' L - \nu t \quad \text{for} \quad x_1 > L \quad \text{and}$$
$$d = (k_1'' - k_1')L = (n'' - n')L/\lambda_0.$$

With the relation

$$\cos u + B\cos(u+v) = \sqrt{1 + B^2 + 2B\cos v}\cos(u+\delta),$$

where

$$\tan\delta = \frac{B\sin v}{1 + B\cos v}$$

and the relation used above for $(\sin u - \sin v)$ one finds

$$\mathrm{Re}(D_2) = \frac{D_0}{1+u''^2}\sqrt{1 + u''^4 + 2u''^2\cos 2\pi d}\cos 2\pi(A+\delta)$$

$$\mathrm{Re}(D_3) = \frac{2u'' D_0}{1+u''^2}\sin\pi d\cdot\cos 2\pi(A + d/2) \quad \text{with}$$

$$\tan 2\pi\delta = \frac{u''^2\sin 2\pi d}{1 + u''^2\cos 2\pi d}.$$

With $u'' = 1$ these values change into the formula derived previously for the superposition of two circular polarized waves.

Finally we note that one can also have formally effects of *second-order optical activity* in all centrosymmetric PSG except 4/m3, which are to be described by a fourth rank tensor $\{g_{ijkl}\}$ given by

$$D_i = \epsilon_{ij}E_j + g_{ijk}\frac{\partial E_j}{\partial x_k} + g_{ijkl}\frac{\partial^2 E_j}{\partial x_k \partial x_l}$$

with $g_{ijkl} = -g_{jikl}$ (Haussühl, 1990).

4.3.6.8 Double refracting, optically active, and absorbing crystals

Interference phenomena occurring, in general, in a plane-parallel plate can be calculated approximately with the aid of a model constructed of thin, only double refracting, only optically active, and only absorbing layers stacked alternately. When entering the next respective layer, both linearly polarized or both elliptically polarized waves must then be split according to the new directions of vibration. Using a computer, this process can be repeated in quasi-infinitesimal steps. In practice, it has been shown that in crystals with weak double refraction (e.g., $|n' - n''| < 0.05$) a few triple layers per mm of traversed thickness give sufficient agreement with experimental findings.

Other methods, as, e.g., the visually clear Poincaré representation or an elegant matrix method suggested by Jones (1948) are described in detail in an overview article by Ramachandran and Ramaseshan (1967).

4.3.6.9 Dispersion

The dependence of the refractive indices, the absorption coefficients and the components of the gyration tensor on the frequency of the wave is called *dispersion*. In an extended sense, the dependence of inducing quantities, such as temperature, mechanical stress, electric and magnetic fields, also fall in the domain of dispersion. We shall discuss these effects (electro-optics, piezo-optics, magneto-optics) partly in more detail in later sections. The phenomena of dispersion can only be described in a satisfactory manner using atomistic models of matter. In the classical picture, optical phenomena are attributed to the interaction of the electric field with the oscillators present in the matter. On the one hand, this concerns the states of the electrons responsible for the optical behavior close to the appropriate eigenfrequencies and on the other hand, the mutual oscillations of the lattice particles. The latter determine in many substances the optical properties in the infrared spectral region. As with all interactions in oscillating systems, the difference between the frequency ν of the exciting wave and the eigenfrequencies ν_k play a decisive role. From simple model calculations one can show that the interaction of the individual oscillators can be described to a good approximation by the following relation for the principal refractive indices:

$$n_j(\nu)^2 = 1 + \sum_k \frac{(e^2/m)A_{jk}}{\nu^2 - \nu_k^2},$$

where the coefficients A_{jk} are proportional to the number of the kth oscillators with eigenfrequency ν_k. e is the electron charge, and m the mass of the electron. In crystals with a simple structure, dispersion, in a not too broad spectral range, can often be well described by two dispersion oscillators. In many applications, dispersion formulae, as e.g., the *Sellmeier equation*

$$n^2 = B_1 + B_2/(\lambda_0^2 - B_3) - B_4\lambda_0^2$$

have proven useful. Their approximate validity confirms the relation mentioned above. $\lambda_0 = c/\nu$ is the vacuum wavelength. The functions $n_j(\nu)$ are obtained by a computer fit of the coefficients B_i to the measured values.

The *Lorentz–Lorenz formula* gives excellent results for the estimation of a mean refractive index, especially in ionic crystals. The formula is obtained from the Clausius–Mosotti relation (see Section 4.3.3), when one replaces ϵ_{rel} by n^2:

$$\frac{n^2 - 1}{n^2 + 2} = \frac{\rho}{M}R.$$

R is proportional to the mean effective polarizability for the respective frequency and is called molar refraction. Experimentally one finds that R is composed of quasi-additive and quasi-invariant contributions of the lattice particles, the so-called ionic refractions.

4.3.7 Electrical Conductivity

The well-known statement of Ohm's law $U = IR$, where U represents the electric voltage in a piece of conducting wire, R the electric resistance and I the current density, must be written in a generalized form when we want to apply it to crystals. We imagine a linear relationship between the vector of the electric current density $\boldsymbol{I}$ and the vector of the electric field strength given by

$$I_i = s_{ij} E_j.$$

The components s_{ij} are material constants independent of the form of the crystal. They specify the *electrical conductivity tensor*, and are to be considered as reciprocal resistances. In experiments we measure the current density in units of charge per mm^2 cross-section persecond. The Coulomb (Symbol C) is mainly used as the unit of charge. Thus the unit of current density is Ampére mm^{-2}. The electric field strength is given in Volt mm^{-1}. The components s_{ij} then have the dimension Ampére $\text{Volt}^{-1}\,\text{mm}^{-1} = \text{Ohm}^{-1}\,\text{mm}^{-1}$. The scarce experimental findings so far on conducting crystals of low symmetry indicate that the electrical conductivity tensor is symmetric, hence $s_{ij} = s_{ji}$. This condition is also required by the *Onsager principle* that is based on the reversibility of processes at atomic scale. A short and secure explanation cannot be given here, so that in the following we will assume the correctness of Onsager's principle (see e.g., Landau-Lifschitz, 1968, p. 374). In crystals, normally no charge creation takes place so that $\operatorname{div} \boldsymbol{I} = 0$. The electric field strength is connected to the electrostatic potential by $E_j = -(\operatorname{grad} U)_j$.

Thus the general potential equation for conducting crystals is

$$s_{ij} \frac{\partial^2 U}{\partial x_i \partial x_j} = 0 \quad \text{(summed over } i, j = 1, 2, 3).$$

The measurement of electrical conductivity is naturally best achieved with the aid of longitudinal effects on thin crystal plates. One measures the electric voltage U applied on the metallized surfaces of the plate and the current density I. The electric field is perpendicular to the surface of the plate. In the case of very low conductivity it is important to take care that the currents over the edges of the plates are kept small or do not enter into the results of the measurement. In order to prevent the interfering effects of heating up the probe during the measurement, it is recommended to work only with extremely small electric power. Since highly accurate commercial current- and voltage-measuring instruments are available also for extremely small currents up to an order of magnitude of 10^{-14} Ampére, such measurements do not cause any principal difficulties. It has been shown, however, that the values of conductivity exhibit strong scattering, even in crystals of very high purity and good quality (low density of lattice defects). The reason is the special role of lattice imperfections for the conduction mechanisms. In metallic conducting crystals lattice imperfections lead to a scattering of the moving charge carriers and thus to a reduction in conductivity, while impurities and lattice imperfections in distinctly weak conductors as, e.g., in most ionic crystals usually result in a strong increase in conductivity. Insights on the type of lattice imperfections can be gained from measurements of the temperature dependence of electric conductivity. Such investigations are thus suitable, within a certain scope, for the characterization of lattice defects.

The anisotropy effects of electrical conductivity are usually extremely small. One observes, for example, in the trigonal crystals of bismuth a ratio s_{11}/s_{33} of about 1.29 at 273 K. In the layer structure of graphite one finds a far better conductivity in the planes perpendicular to the sixfold axis than in those parallel to the sixfold axis. Aside from these layer structures, a distinct anisotropy also exists in structures with conductivity along certain lattice directions similar to a bundle of mutually isolated metal threads. Such a cable-type and uniaxial conductivity, as observed, for example, in certain platinum compounds such as $K_2Pt(CN)_4Br_{0.3} \cdot 3.2H_2O$ or in $LiIO_3$, where a s_{33}/s_{11} of about 1000 appears, cannot always be adequately described with our picture of a second-rank tensor because practically only current threads exist parallel to the unique cable direction. This is represented for a probe of hexagonal $LiIO_3$ cut at 45° to the sixfold axis (Fig. 4.20). If the probe is long enough, the conductivity along the sixfold axis is no longer effective, because the current threads can no longer connect both electrodes. A further important phenomenon is the formation of layers of extremely reduced conductivity at the boundary of two different types of media. These so-called depletion layers play an important role in

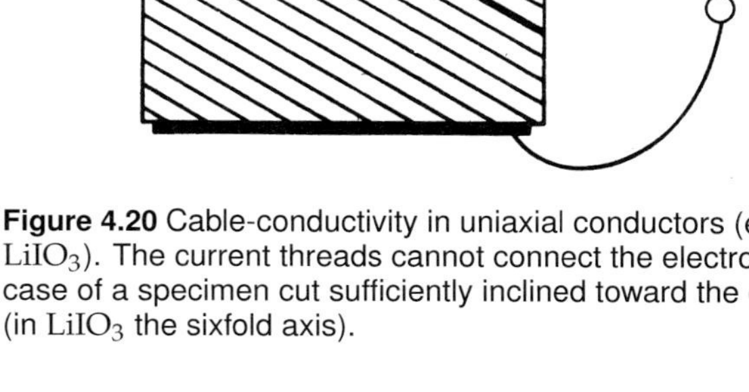

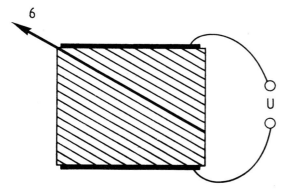

Figure 4.20 Cable-conductivity in uniaxial conductors (example $LiIO_3$). The current threads cannot connect the electrodes in the case of a specimen cut sufficiently inclined toward the cable axis (in $LiIO_3$ the sixfold axis).

electronic devices, where they are created in a great variety of forms for specific functions (transistor properties). In crystals with polar symmetry, the formation of depletion layers in a unique preferential direction, as for example, in hexagonal $LiIO_3$, can generate highly differing values of the direct current conductivity along the unique direction and its counterdirection. Such effects can be explained by a change of the medium close to the electrodes during current passage. In ionic crystals such effects result often by the depletion of an ion type, for example, lithium ions in the region of the cathode. These phenomena of conductivity cannot be explained by our tensorial approach.

Further effects in connection with electrical conductivity, such as the deviation from Ohm's law, the influence of a magnetic field, or external mechanical strain will be discussed in later sections.

4.3.8
Thermal Conductivity

The transport of thermal energy obeys laws analogous to the transport of electric charge. The thermal current density appears instead of the electric current density and the electric field $\boldsymbol{E} = -\operatorname{grad} U$ is replaced by the temperature gradient $\operatorname{grad} T = (\partial T/\partial x_i)\boldsymbol{e}_i$, where we consider the temperature T as a function of position.

The *thermal conductivity* is then represented by

$$Q_i = -\lambda_{ij}(\operatorname{grad} T)_j.$$

The vector $\boldsymbol{Q}$ specifies the flow of heat per second through a cross-section of $1\,\mathrm{mm}^2$ (units Joule $\mathrm{mm}^{-2}\,\mathrm{s}^{-1}$). Here we also assume the validity of Onsager's principle, which requires that the tensor $\{\lambda_{ij}\}$ is symmetric, hence $\lambda_{ij} = \lambda_{ji}$. The measurement should be carried out preferably on a heat plate conductometer, if high accuracy is required and moreover when strong anisotropy effects are expected. This equipment is constructed analogously to the case of electrical conductivity (Fig. 4.21). Again only longitudinal effects are measured. The probe, in the form of a plate, is placed between two metal plates of high thermal conductivity (e.g., copper), which because of their good conducting properties maintain an almost equal temperature over their complete volume. A small heater is mounted on one of the plates. The heat flows through the probe into the second metal plate, that is cooled either by the uniform flow of a coolant or by a Peltier element. If one succeeds in directing the heat loss in the heated plate almost exclusively over the probe, then the applied electrical energy is equal to the total amount of heat transported. After equilibrium, the temperatures T_1 and T_2 produce a temperature gradient $(T_1 - T_2)/D$ in the probe of thickness D. Hence the longitudinal effect along the plate normals is

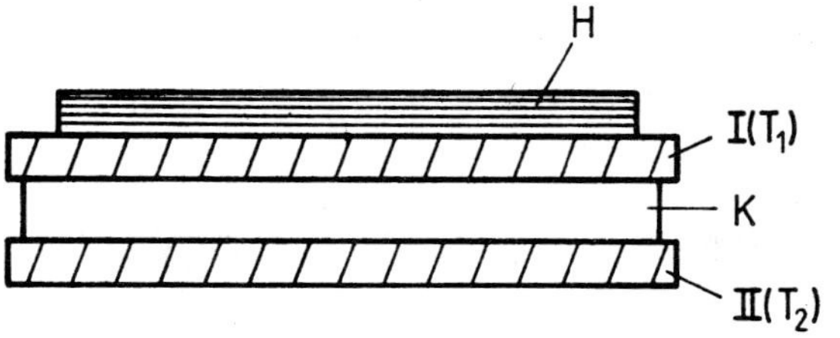

Figure 4.21 Thermal conductometer. H heating plate, K crystal. The plates $I(T_1)$ and $II(T_2)$ are contacted by a thin film of a liquid with high thermal conductivity.

given by

$$Q'_1 = -\lambda'_{11}(T_1 - T_2)/D = W_{el}/F,$$

where W_{ep} is the applied electrical power and F is the cross-section of the probe.

Suitable measures must be taken to suppress heat losses. Furthermore, just as in the case of electrical conductivity one must ensure a quasi-loss-free transfer of heat from the plates to the probe and out of the probe. Thin films of oil as a contact material have proven suitable in this respect. If one performs measurements on plates of the same orientation but with different thicknesses and plots $Q'_1/(T_1 - T_2)$ as a function of $1/D$, the slope of the curve gives a reliable value for the longitudinal effect λ'_{11}.

Another method is based on the measurement of the propagation velocity of heat pulses along a cylindrical rod. For crystals this method is only useful in special cases due to the large lengths required.

The interpretation of thermal conductivity proceeds from the scattering of acoustic waves while propagating through a crystal. The heat content determines the number of thermally generated phonons. According to a simple theory of Debye, which should be familiar to the reader from basic courses in physics, the dependence of heat conductivity on the mean free wavelength Λ of the phonons, the specific heat C_p, and the mean sound velocity $\bar{v}$ of the phonons in a given direction is described by the relationship:

$$\lambda' = \tfrac{1}{3} C_p \Lambda \bar{v}.$$

Here we recognize the important relation between thermal conductivity and acoustic properties. The latter will be discussed in Section 4.5.1.

4.3.9 Mass Conductivity

The phenomenon of *mass conductivity*, which appears for example, during filtration through a porous layer, the migration of oil through porous rocks, or the passage of gas through a thin membrane is completely analogous to electrical and thermal conductivity. The driving force is the gradient of the hydrostatic pressure, in other words, the pressure difference $(p_1 - p_2)/D$ per unit length. The mass current density vector S, which describes the transported mass in units of time and surface area is connected to the pressure gradient through

$$S_i = -q_{ij}(\text{grad}\, p)_j.$$

The mass permeability tensor $\{q_{ij}\}$ depends not only on the material but also on the liquid or gas to be transported.

Mass conductivity measurements are made, in principle, under the same aspects discussed for the measurement of thermal conductivity, i.e., it is appropriate to measure the longitudinal effect of a flow running perpendicularly through a plane-parallel plate. In single crystals this property has not yet been quantitatively investigated so far. Crystals with very large lattice particles or with certain vacancies or channel structures, as for example, the zeolites or chelates, which are sometimes used as "molecular sieves," possess an anisotropic permeability. The diffusion of liquids or gases in crystals under the influence of a pressure gradient also belongs to the field of mass conductivity.

4.3.10
Deformation Tensor

The mechanical change of shape of a medium is called deformation. This is described by the deformation tensor, also called strain tensor, whereby the rigid displacements and rotations accompanying the deformation shall be left out of consideration. We measure the deformation through the displacement experienced by two neighboring points P_1 and P_2. Let both points have the coordinates (x_1, x_2, x_3) and $(x_1 + \Delta x_1, x_2 + \Delta x_2, x_3 + \Delta x_3)$ in a fixed undeformed coordinate system. Deformation carries over both points to the points P_1' and P_2' with the coordinates

$$(x_1 + \xi_1, x_2 + \xi_2, x_3 + \xi_3) \quad \text{and}$$
$$(x_1 + \Delta x_1 + \xi_1 + \Delta\xi_1, x_2 + \Delta x_2 + \xi_2 + \Delta\xi_2, x_3 + \Delta x_3 + \xi_3 + \Delta\xi_3).$$

. Before deformation, the mutual position of both points is described by the vector $\Delta \boldsymbol{x}$ and after deformation by the vector $\Delta \boldsymbol{x} + \Delta \boldsymbol{\xi}$ (Fig. 4.22). The vector

$$\Delta \xi = (\Delta\xi_1, \Delta\xi_2, \Delta\xi_3)$$

is called the displacement vector.

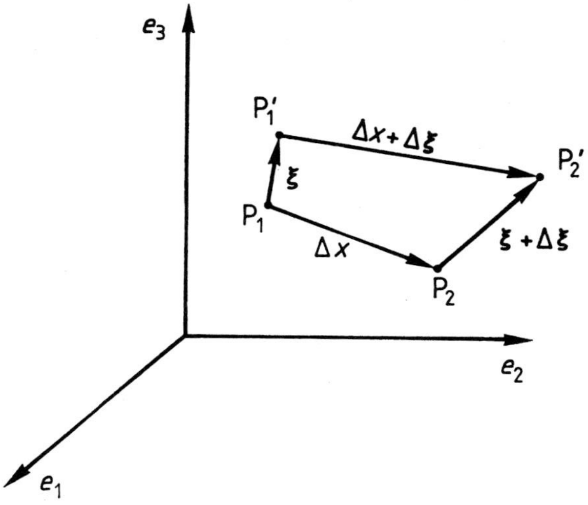

Figure 4.22 Positions of points P_1 and P_2 before and after deformation (P_1' and P_2').

We now consider the components of the vector $\Delta\boldsymbol{\xi}$ expanded in a Taylor series in components of $\Delta\boldsymbol{x}$, thus

$$\xi_i(\Delta\boldsymbol{x}) - \xi_i(0) = \Delta\xi_i = \frac{\partial\xi_i}{\partial x_j}\Delta x_j + \frac{1}{2}\frac{\partial^2\xi_i}{\partial x_j \partial x_k}\Delta x_j \Delta x_k + \cdots$$

The first term of the expansion is sufficient for many purposes:

$$\Delta\xi_i = \frac{\partial\xi_i}{\partial x_j}\Delta x_j.$$

The quantities $\frac{\partial\xi_i}{\partial x_j}$ are the components of the *displacement tensor*—a second rank tensor. The tensor property is recognized from the interconnection of the vectors $\Delta\boldsymbol{\xi}$ and $\Delta\boldsymbol{x}$.

We now resolve the displacement tensor

$$\frac{\partial\xi_i}{\partial x_j} = \varepsilon_{ij} + r_{ij} \quad \text{with} \quad \varepsilon_{ij} = \frac{1}{2}\left(\frac{\partial\xi_i}{\partial x_j} + \frac{\partial\xi_j}{\partial x_i}\right) \quad \text{and} \quad r_{ij} = \frac{1}{2}\left(\frac{\partial\xi_i}{\partial x_j} - \frac{\partial\xi_j}{\partial x_i}\right).$$

The symmetric part $\{\varepsilon_{ij}\}$ is called the *deformation tensor*, and the antisymmetric part $\{r_{ij}\}$ the *rotation tensor*. The tensor $\{r_{ij}\}$ gives rise to a rigid rotation. If $\boldsymbol{u}$ is a rotation vector whose length is proportional to the angle of rotation and whose axis runs parallel to the rotation axis, whereby the rotation appears clockwise along the direction of $\boldsymbol{u}$, we have for sufficiently small angles to a reasonable approximation

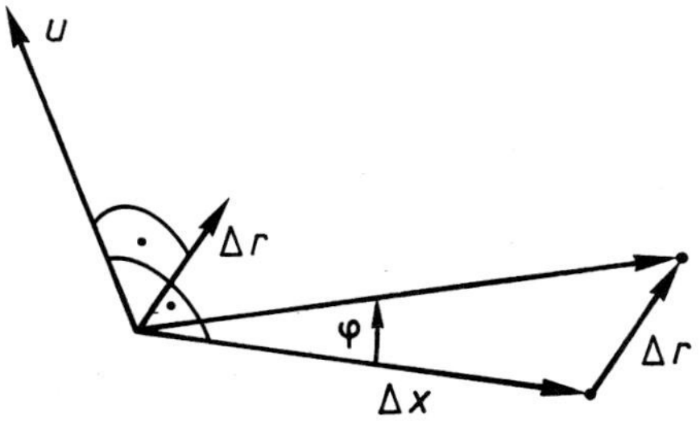

Figure 4.23 Rotary part Δr of the displacement vector.

$$\boldsymbol{u} \times \Delta \boldsymbol{x} = \begin{vmatrix} \boldsymbol{e}_1 & \boldsymbol{e}_2 & \boldsymbol{e}_3 \\ u_1 & u_2 & u_3 \\ \Delta x_1 & \Delta x_2 & \Delta x_3 \end{vmatrix} = \Delta \boldsymbol{r}.$$

Hence $\Delta \boldsymbol{r}$ is perpendicular to the axis of rotation and its length is proportional to the angle of rotation and to the distance of the end point of $\Delta \boldsymbol{x}$ from the axis of rotation (Fig. 4.23). This corresponds to a rigid rotation about the axis $\boldsymbol{u}$, where $\boldsymbol{u}$ is an axial vector (pseudo vector), because the vector product of $\boldsymbol{u}$ with a polar vector $\Delta \boldsymbol{x}$ generates a polar vector $\Delta \boldsymbol{r}$. Setting $\boldsymbol{u} = -(r_{23}, r_{31}, r_{12})$ gives $(\boldsymbol{u} \times \Delta \boldsymbol{x})_i = r_{ij} \Delta x_j = \Delta r_i$, thus the asserted property of $\{r_{ij}\}$ is proven. In the following, we will mainly be concerned with the deformation tensor $\{\varepsilon_{ij}\}$. For some applications it is useful to introduce the so-called Lagrangian deformation tensor instead of $\{\varepsilon_{ij}\}$. This is obtained when we consider the difference of the squares of the distance of P_1 and P_2 after and before the deformation.

We have

$$\begin{aligned}(\Delta \boldsymbol{x} + \Delta \boldsymbol{\xi})^2 - (\Delta \boldsymbol{x})^2 &= 2 \Delta \boldsymbol{x} \cdot \Delta \boldsymbol{\xi} + (\Delta \boldsymbol{\xi})^2 \\ &= 2 \Delta x_i \frac{\partial \xi_i}{\partial x_j} \Delta x_j + \frac{\partial \xi_k}{\partial x_i} \frac{\partial \xi_k}{\partial x_j} \Delta x_i \Delta x_j,\end{aligned}$$

when we again content ourselves with the first approximation. Thus

$$\begin{aligned}\frac{1}{2}((\Delta \boldsymbol{x} + \Delta \boldsymbol{\xi})^2 - (\Delta \boldsymbol{x})^2) &= \frac{1}{2} \left\{ \left(\frac{\partial \xi_i}{\partial x_j} + \frac{\partial \xi_j}{\partial x_i} \right) + \frac{\partial \xi_k}{\partial x_i} \frac{\partial \xi_k}{\partial x_j} \right\} \Delta x_i \Delta x_j \\ &= \eta_{ij} \Delta x_i \Delta x_j.\end{aligned}$$

$\{\eta_{ij}\}$ is the *Lagrangian deformation tensor*, which according to $\eta_{ij} = \varepsilon_{ij} + \zeta_{ij}$ contains, apart from the usual deformation tensor, the tensor formed from the product $\frac{\partial \xi_k}{\partial x_i} \frac{\partial \xi_k}{\partial x_j}$ and hence provides a next step in the approximation of finite deformations. We will return to this point in particular when discussing nonlinear elastic phenomena.

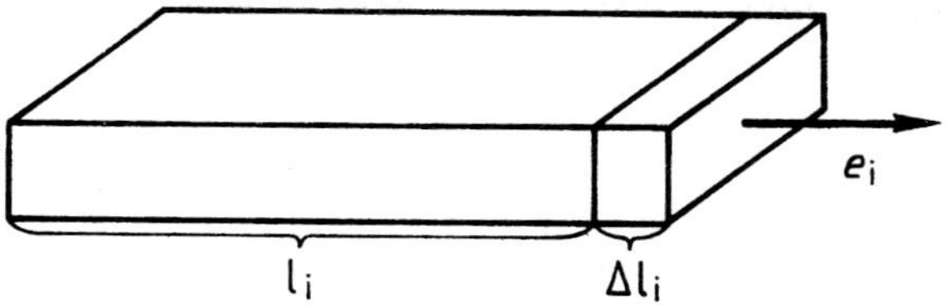

Figure 4.24 Longitudinal effect of the strain tensor along e_i: $\Delta l_i / l_i = \varepsilon_{ii}$.

The components of the deformation tensor are easily accessible to a physical interpretation. The longitudinal components ε_{ii} describe the relative change in length in the direction e_i occurring during the deformation. Since

$$\Delta\xi_i = \varepsilon_{ii}\Delta x_i + \frac{\partial \xi_i}{\partial x_j}\Delta x_j,$$

where $j \neq i$. The longitudinal effect is

$$\frac{\partial \xi_i}{\partial x_i} = \frac{\Delta l_i}{l_i} = \varepsilon_{ii},$$

when we specify the difference of the coordinates in direction e_i by l_i and their change by Δl_i (Fig. 4.24).

The components ε_{ij} ($i \neq j$) are called shear components because they directly indicate the shear of a volume element. Consider two points P_1 and P_2 lying on the coordinate axes belonging to e_1 or e_2, hence possessing the coordinates $(\Delta x_1, 0, 0)$ and $(0, \Delta x_2, 0)$ (Fig. 4.25). The deformation carries the

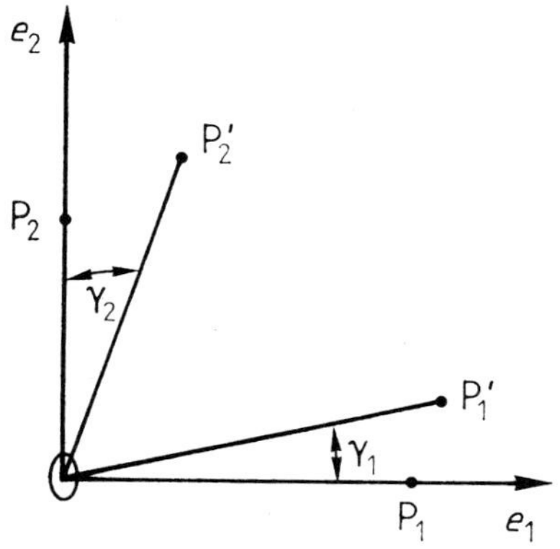

Figure 4.25 Interpretation of the component ε_{ij} ($i \neq j$) as shear component.

points over into

$$P_1' : \left\{ \left(1+\frac{\partial \xi_1}{\partial x_1}\right)\Delta x_1, \frac{\partial \xi_2}{\partial x_1}\Delta x_1, \frac{\partial \xi_3}{\partial x_1}\Delta x_1 \right\} \quad \text{and}$$
$$P_2' : \left\{ \frac{\partial \xi_1}{\partial x_2}\Delta x_2, \left(1+\frac{\partial \xi_2}{\partial x_2}\right)\Delta x_2, \frac{\partial \xi_3}{\partial x_2}\Delta x_2 \right\}.$$

The angles γ_1 and γ_2 which are enclosed by the sections $0P_1'$ and $0P_2'$ and the coordinate axes $\boldsymbol{e}_1$ or $\boldsymbol{e}_2$ are approximately given by $\left(\frac{\partial \xi_i}{\partial x_j} \ll 1\right)$:

$$\tan\gamma_1 = \frac{\partial \xi_2}{\partial x_1}\Delta x_1 \Big/ \left(1+\frac{\partial \xi_1}{\partial x_1}\right)\Delta x_1 \approx \frac{\partial \xi_2}{\partial x_1}$$
$$\tan\gamma_2 = \frac{\partial \xi_1}{\partial x_2}\Delta x_2 \Big/ \left(1+\frac{\partial \xi_2}{\partial x_2}\right)\Delta x_2 \approx \frac{\partial \xi_1}{\partial x_2}.$$

The difference γ_{12} of the angles $\sphericalangle(P_1\,0\,P_2)$ and $\sphericalangle(P_1'\,0\,P_2')$ is called the shear in the x_1, x_2 plane. If the angles are sufficiently small we can replace the tangent by the argument and obtain as an approximation

$$\gamma_{12} = \gamma_1 + \gamma_2 \approx \tan\gamma_1 + \tan\gamma_2 \approx \frac{\partial \xi_2}{\partial x_1} + \frac{\partial \xi_1}{\partial x_2} = 2\varepsilon_{12}.$$

The same is true in all other coordinate planes. In general $\gamma_{ij} \approx 2\varepsilon_{ij}$. Hence γ_{ij} describes, to a first approximation, the change in the angle of a right angle formed by the sides parallel $\boldsymbol{e}_i$ and $\boldsymbol{e}_j$.

In total we have six independent tensor components, three longitudinal and three transversal ones. However, in the principal axes' representation only three longitudinal components appear, i.e., the general deformation can be described, in the framework of the approximation discussed, by three mutually perpendicular longitudinal deformations. In the directions of the principal axes $l_i + \Delta l_i = \Delta x_i' = \Delta x_i + \varepsilon_{ii}\Delta x_i = (1+\varepsilon_{ii})\Delta x_i$. A sphere $\sum_i (\Delta x_i)^2 = 1$ thus becomes the deformation ellipsoid

$$\sum_i \frac{(\Delta x_i')^2}{(1+\varepsilon_{ii})^2} = 1.$$

The relative change in volume associated with the deformation, to a first approximation, is given by

$$\frac{V'-V}{V} = \frac{\Delta V}{V} \approx \varepsilon_{11} + \varepsilon_{22} + \varepsilon_{33} \quad \text{with}$$
$$V = \Delta x_1 \Delta x_2 \Delta x_3 \quad \text{and} \quad V' = \Delta x_1' \Delta x_2' \Delta x_3'.$$

Since the sum of the principal components is an invariant, this relationship is valid in any arbitrary reference system.

We will become more familiar with deformations induced by thermal, mechanical, electrical, or magnetic processes in the chapters to come.

4.3.11
Thermal Expansion

Thermal expansion belongs to the anharmonic effects. It is caused by the change of the mean center-of-mass position of the lattice particles resulting from an increase in the amplitude of thermal vibrations with increasing temperature. The phenomenological description is given by

$$\varepsilon_{ij} = \alpha_{ij}\Delta T + \beta_{ij}(\Delta T)^2 + \gamma_{ij}(\Delta T)^3 + \cdots$$

In a small temperature range of a few K the first term normally suffices when the deformation is to be represented with an accuracy of around 1%. Otherwise higher terms must be added.

The tensors of thermal expansion $\{\alpha_{ij}\}$, $\{\beta_{ij}\}$, $\{\gamma_{ij}\}$, and so on are all of second rank. With the aid of longitudinal effects they can be measured directly from the change of the linear dimensions within a small temperature interval. The following methods have proven suited for precision measurements of small changes in length:

(a) Inductive dilatometer(Fig. 4.26). The probe in the form of a parallelepiped or cylinder carries a rod of extremely low thermal expansion (quartz glass; invar), on the top of which a cylindrical permanent magnet (ferrite rod) is attached. The magnet immerses in a coil. The inductivity of the coil depends on the depth of immersion. Hence a change in length results in a change of inductance and thus a change in the resonant frequency of the oscillating

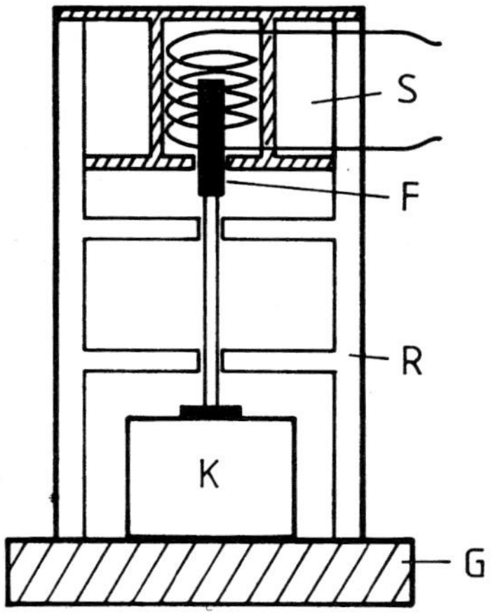

Figure 4.26 Inductive dilatometer. G ground plate, K crystal, R distance holder ring, F ferrite rod, S HF coil.

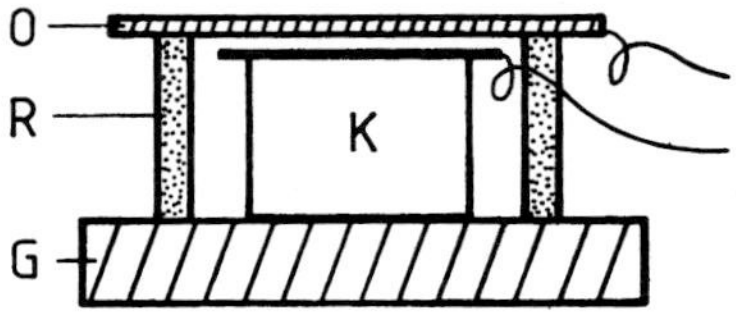

Figure 4.27 Capacitive dilatometer. G ground plate, K crystal with electrode, R distance ring made from invar or quartz glass, O upper copper electrode.

circuit coupled to the coil. The unit is calibrated by measuring the change in frequency as a function of immersion depth. In commercial devices the change in frequency, over a wide range, is proportional to immersion depth. These dilatometers possess the advantage that the requirements in respect to quality and plane-parallelism of the probe are low.

(b) Capacitive dilatometer (Fig. 4.27). One side of the probe, machined plane-parallel, carries one plate of a plate capacitor. The other plate of the capacitor is mounted at a fixed distance from the probe plate by the aid of a quartz glass ring. A change in probe thickness leads to a change in capacitance, which is measured to high accuracy by the change in the resonant frequency of an oscillating circuit in a similar manner as with the inductive dilatometer.

(c) Optical interference dilatometer(Fizeau interferometer; Fig. 4.28). In this method, the change in length is measured directly from the change of the optical path between the surface of the probe and a reference plate.

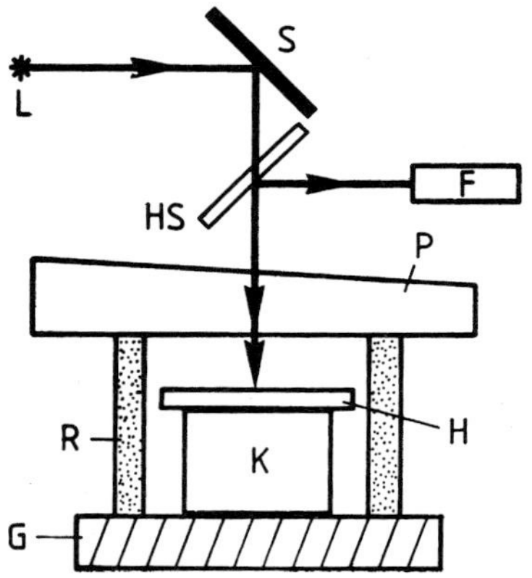

Figure 4.28 Optical interference dilatometer (Fizeau interferometer). G ground plate, K crystal with one-sided polished auxiliary plate H, distance ring R, slightly wedge-shaped reference plate P, semipermeable mirror HS, metallic mirror M, light source L, telescope or photo-detector F. The waves reflected at the bottom-side of P and the top-side of H generate the interference pattern, which can be directly observed above P. The parts H, R and P are made of quartz glass.

The crystal K and the support for the reference plate are mounted on a flat metal plate. Polishing the probe, which must have plane-parallel faces perpendicular to the direction of measurement, is not required when one places a thin plate of quartz glass, plane-polished on one side, on the probe. The ring carrying the reference plate should preferably be made of quartz glass, too. The optical assembly is mounted in a temperature controlled cell.

The beam of light reflected on the top surface of the auxiliary plate and on the lower surface of the reference plate generates an interference pattern which can be observed with a telescope (adjusted approximately to the lower surface of the reference plate). The number N of interference fringes passing an engraved mark on the reference plate within a small temperature range ΔT provides a direct measure of the change in length. A correction must be made for the change in the optical path resulting from the quartz support and from the change of refractive power in air in the small gap between both plates.

We then have

$$\alpha'_{11} = \frac{\varepsilon'_{11}}{\Delta T} = \frac{\Delta D}{D\Delta T} = \frac{N'\lambda_0}{2D\Delta T},$$

where λ_0 is the wavelength of light, D is the thickness of the probe along $\boldsymbol{e}'_1$ and N' is the corrected number of fringes. The correction is calculated with the help of the known expansion coefficients of air and quartz glass (see annex Table 12.10).

Visual observation with the simple interferometer enables a reading accuracy of about 1/20 of a fringe. Photometric methods enable a substantially increased accuracy. The reproducibility of such measurements is often affected by the plastic deformation of the probe due to the unavoidable inhomogenous temperature distribution in the probe during heating and cooling.

(d) Strain gauge. This method uses the change in electrical resistance of a wire as a function of mechanical strain to measure the deformation of rigid bodies. The strain gauge is cemented on a side face of the probe in the prescribed measurement direction. Normally strain gauges cannot be used for further measurements on other probes, because when removed from a probe, the resulting mechanical strain leads to a change in the specific resistance per unit strain as given by the manufacturer. Since this information only represents mean values of a production series, one can only have limited confidence in the measured values.

(e) X-ray precision measurements of the temperature dependence of the lattice constants. The methods of the precise determination of lattice constants are described in detail in the literature. For large crystals the Bragg method or the method of Bond (simple goniometer) is especially suitable. For small crystals, the multicircle diffractometer and for powder specimens, the Guinier method are suitable. One must pay particular attention to the exact temperature setting of the crystal. As an example for the evaluation, let us consider the gen-

eral case of a triclinic crystal, where the temperature dependence $\partial\theta/\partial T$ of the diffraction angle θ was measured for the following reflections: (100), (010), (001), (110), (101) and (011). Of course any reflections can be used additionally. From Bragg's condition $2d\sin\theta = \lambda$ (λ is the wavelength of the x-rays) we obtain for the temperature dependence of the lattice constant d

$$\frac{1}{d}\frac{\partial d}{\partial T} = -\cot\theta\frac{\partial\theta}{\partial T}.$$

The d-value of a reflection $\boldsymbol{h} = h_i\boldsymbol{a}_i^* = (h_1h_2h_3)$ is given by $d = 1/|\boldsymbol{h}|$. h_i are the Miller indices and $\boldsymbol{a}_i^*$ the vectors of the reciprocal system (Section 1.5.3). Hence with the temperature dependence of a d-value we also know the longitudinal deformation (longitudinal effect) along $\boldsymbol{h}$. To determine the tensors of thermal expansion in the crystal-physical system we only need to calculate the unit vectors in the direction of the $\boldsymbol{h}$ vectors in the crystal-physical system. For this purpose we use the formulae derived in Section 2.2.

For example

$$\frac{\boldsymbol{h}_{100}}{|\boldsymbol{h}_{100}|} = \frac{a_3}{a_1^*a_2^*V}\boldsymbol{e}_1 + \cos\alpha_3^*\boldsymbol{e}_2.$$

The measurement perpendicular to (010) gives us directly $\alpha_{22} = \partial d/d\partial T$. For the longitudinal effect in all other directions $\boldsymbol{e}_1' = u_{1j}\boldsymbol{e}_j$ we have

$$\frac{\partial d}{d\partial T} = \alpha_{11}' = u_{1i}u_{1j}\alpha_{ij}.$$

With our six selected reflections we still have available five further linear equations for the determination of the remaining α_{ij}. It is appropriate to include additional measurements of the temperature dependence of other d-values and to apply a least-squares procedure to the overdetermined system.

As an example we consider triclinic lithium hydrogen oxalate hydrate. Table 4.5 presents the values measured with an optical interference dilatometer. The measurements were performed in the temperature interval from 20 to −20°C.

Using the formula for the longitudinal effect one gets α_{11}, α_{22} and α_{33} directly from measurement nos. 1, 2, and 3. With these and from no. 4 we get α_{12} and from no. 6 α_{13}. From no. 8 one calculates finally α_{23}. The nos. 5 and 7 are used for control purposes. The tensor of thermal expansion is then, in units of 10^{-6}/K

$$\alpha_{11} = 23.7; \qquad \alpha_{22} = 19.2; \qquad \alpha_{33} = 60.9;$$
$$\alpha_{12} = -6.0; \qquad \alpha_{13} = -40.5; \qquad \alpha_{23} = 18.5.$$

The principal axes' transformation yields the principal values

$$\lambda_1 = 91.76, \qquad \lambda_2 = 15.60, \qquad \lambda_3 = -3.56$$

Table 4.5 Measurement of the thermal expansion of triclinic $LiHC_2O_4 \cdot H_2O$ at 273 K.

Nr.	u_{11}	u_{12}	u_{13}	Direction of measurement	Longitudinal effect α'_{11} $[10^{-6}K^{-1}]$
1	1	0	0	$[100]'$	23.7
2	0	1	0	$[010]'$	19.2
3	0	0	1	$[001]'$	60.9
4	0.714	0.700	0	$\approx [110]'$	15.6
5	−0.700	0.714	0	$\approx [\bar{1}10]'$	27.5
6	−0.465	0	0.885		85.6
7	0.885	0	0.465	$[101]$	−2.1
8	−0.329	$\sqrt{2}/2$	0.626		73.2

with the associated principal axes' directions (eigenvectors) (see Section 4.3.2)

$$e'_1 = 0.5112e_1 - 0.2518e_2 - 0.8218e_3,$$
$$e'_2 = -0.3505e_1 - 0.9341e_2 + 0.0681e_3,$$
$$e'_3 = 0.7849e_1 - 0.2529e_2 + 0.5657e_3.$$

Accordingly, the crystal possesses an unusually strong anisotropy. With increasing temperature the crystal contracts most strongly in a direction approximately parallel to $[101]$, hence it lies almost in the cleavage plane $(\bar{1}01)$. One observes the maximum thermal expansion nearly perpendicular to the cleavage plane. This property is essentially responsible for the unusually strong tendency for crack formation parallel to the cleavage plane. Consequently, when working with the crystal (grinding and polishing) one must pay special attention to ensure that no large temperature gradients are created in the probe.

In particular, it should be mentioned that the principal axes of the quadries of the dielectric constants and the thermal expansion almost coincide. The dielectric maximum or minimum corresponds to a minimum or maximum, respectively, of the thermal expansion. Here we observe a close correlation of two physical properties which otherwise do not exhibit a direct relationship.

The case of negative expansion is rather rarely observed. As examples for negative expansion we mention the orthorhombic crystal species calcium formate and ammonium hydrogen phthalate. In these crystals, the quadric has the form of a hyperboloid. These crystals can be used to prepare samples which show practically no thermal expansion in one direction within a certain temperature interval (see Exercise 15).

The tensors $\{\beta_{ij}\}$, $\{\gamma_{ij}\}$, and so on, introduced above, can be derived from the temperature dependence of the tensor components

$$\alpha_{ij}(\Delta T) = \alpha_{ij}(0) + \frac{\partial \alpha_{ij}}{\partial T}\Delta T + \frac{1}{2}\frac{\partial^2 \alpha_{ij}}{\partial T^2}(\Delta T)^2 + \cdots$$

(e.g., $\beta_{ij} = \partial \alpha_{ij}/\partial T$).

We will take up the relation between thermal expansion and elastic properties when discussing the Grüneisen tensor (see Exercise 29).

4.3.12
Linear Compressibility at Hydrostatic Pressure

By analogy to thermal expansion, one can describe a deformation produced by a change in the hydrostatic pressure Δp, by a second-rank tensor:

$$\varepsilon_{ij} = l_{ij}\Delta p,$$

where $\{l_{ij}\}$ is the *tensor of linear compressibility*. This tensor represents a second-rank tensor invariant of the elasticity tensor. We will return to this relationship later (see Section 4.5.3).

In principle, linear compressibility can be measured with methods similar to those employed for thermal expansion, where instead of a thermal measurement cell a pressure cell is used. If one works, for example, with an optical interferometer, then the change in the optical path resulting from a change in pressure in the empty cell must be carefully taken into account. If the required accuracy is not too high, the use of a strain gauge is recommended. A far more convenient and accurate method is the determination of the linear compressibility with the help of the measurement of sound velocities (see Section 4.5.6). X-ray methods for the determination of lattice constants have proven successful for the investigation of linear compressibility subjected to extreme pressures and temperatures.

4.3.13
Mechanical Stress Tensor

We consider the entirety of the external forces acting on a volume element of a body. Imagine the volume element to have the form of a parallelepiped with edges parallel to the basic vectors of a Cartesian reference system (Fig. 4.29). The external forces then act on the boundary faces of this parallelepiped. We resolve the forces in components acting perpendicular and tangential to the faces. If the forces are distributed homogenously over the faces, it is useful to introduce the so-called stresses, i.e., the forces per unit area. Hence stress = force/area. Positive stress means tensile force. The following nine mutually independent stresses can appear: three *normal stresses* σ_{ii} and six *shear stresses* σ_{ij} $(i \neq j), (i, j = 1, 2, 3)$.

The first index gives the direction of the force (stress), the second the direction of the normal on the face on which the stress acts. For example σ_{22} means the stress acting in the direction of the basic vector e_2 and exerted across the face perpendicular to e_2. The stress σ_{32} points in the direction e_3 and acts across the face perpendicular to e_2. If one goes on to infinitesimal dimensions

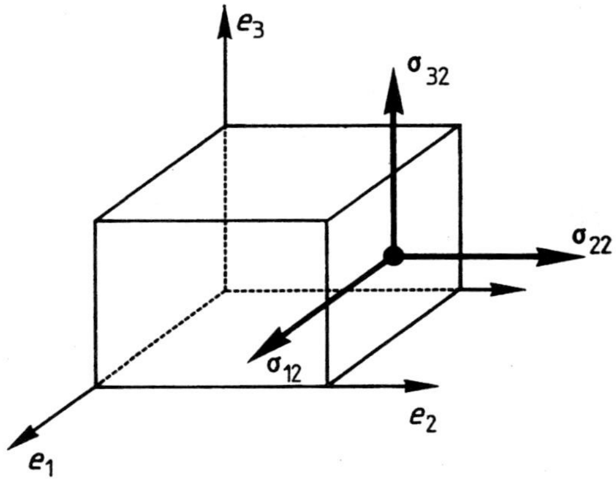

Figure 4.29 Definition of the stress tensor.

of the volume element, one can describe the state of stress at each point of a body by specifying these nine stresses. The quantities σ_{ij} are components of a second-rank tensor, which we will prove as follows:

Let the force acting on an arbitrary test triangle $A_1A_2A_3$ with the normals $f = f_i e_i$ and given state of stress be $P = P_i e_i$ (Fig. 4.30). We then have $P_i = \sigma_{ij} q_j$, where q_j is the area of the triangle $0A_iA_k$ (= projection of the triangle $A_1A_2A_3$ on the plane perpendicular to e_j). Let the length of f be equal to the

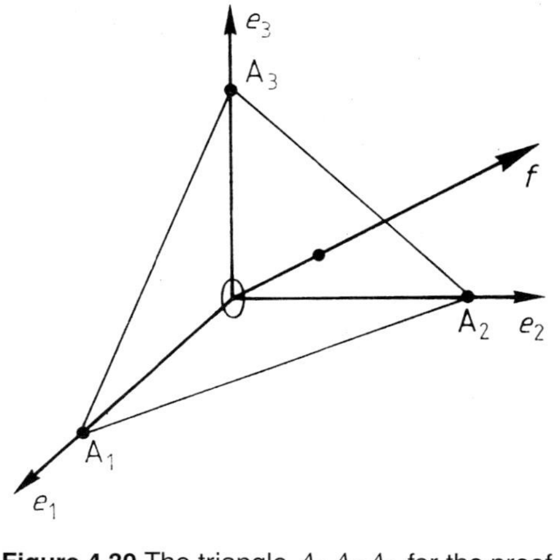

Figure 4.30 The triangle $A_1A_2A_3$ for the proof of the tensor character of the stress tensor.

area of the triangle $A_1A_2A_3$. Then $f_i = q_i$. This results from

$$\begin{aligned} \boldsymbol{f} &= (d_2\boldsymbol{e}_2 - d_1\boldsymbol{e}_1) \times (d_3\boldsymbol{e}_3 - d_2\boldsymbol{e}_2)/2 \\ &= (d_2d_3\boldsymbol{e}_1 + d_3d_1\boldsymbol{e}_2 + d_1d_2\boldsymbol{e}_3)/2 \quad = \quad q_i\boldsymbol{e}_i, \end{aligned}$$

where we set $0A_i = d_i$ for the intercepts. The same result is obtained in a more elegant way with the help of Gauss's theorem

$$\int_V \operatorname{div} \boldsymbol{u} dV = \int_{\text{surface}} \boldsymbol{u} \cdot d\boldsymbol{f}.$$

If $\boldsymbol{u}$ is a constant vector, the left side vanishes, and hence

$$\int_{\text{surface}} \boldsymbol{u} \cdot d\boldsymbol{f} = 0.$$

Applying this to the tetrahedron $0A_1A_2A_3$ gives $f_i = q_i$, when all face normals point outward of the tetrahedron. Hence we have

$$P_i = \sigma_{ij} f_j.$$

Since P_i and also f_j are components of a vector, the quantities σ_{ij}, according to the theorem on tensor operations, form a second-rank tensor, the *stress tensor*.

We now examine the symmetry of the stress tensor. Consider the torque $\boldsymbol{M}$ exerted on the test parallelepiped of Fig. 4.29. A force $\boldsymbol{K}$ acting on a lever arm, represented by the vector $\boldsymbol{x}$, gives rise to the torque $\boldsymbol{M} = \boldsymbol{x} \times \boldsymbol{K}$. $\boldsymbol{M}$ is an axial vector, which as we have seen in Section 4.3.10, can be represented by an antisymmetric second-rank tensor analogous to the rotation vector. Let the parallelepiped have edge lengths x_i along the basic vectors $\boldsymbol{e}_i$. We then get, for example,

$$M_1\boldsymbol{e}_1 = x_2\boldsymbol{e}_2 \times \boldsymbol{K}_3 + x_3\boldsymbol{e}_3 \times \boldsymbol{K}_2,$$

where $\boldsymbol{K}_2$ or $\boldsymbol{K}_3$ are the tangential forces acting along $\boldsymbol{e}_2$ or $\boldsymbol{e}_3$, respectively,

$$\boldsymbol{K}_3 = \sigma_{32}x_1x_3\boldsymbol{e}_3, \qquad \boldsymbol{K}_2 = \sigma_{23}x_1x_2\boldsymbol{e}_2.$$

Hence

$$M_1 = V(\sigma_{32} - \sigma_{23}).$$

$V = x_1x_2x_3$ is the volume of the parallelepiped. In general

$$M_i = -V(\sigma_{jk} - \sigma_{kj}), \quad i, j, k \text{ cyclic in } 1, 2, 3.$$

If the torque is represented as an antisymmetric tensor $\{M^*_{ij}\}$, then $M_i = M^*_{jk} = -M^*_{kj}$.

All torques must vanish, if the volume element is in equilibrium, i.e., $\sigma_{ij} = \sigma_{ji}$; the stress tensor is symmetric in this case. The antisymmetric part of the stress tensor establishes the torque. The question of whether, in a static experiment, a torque appears or not can be clearly decided. In dynamic processes, for example, the propagation of elastic waves, the existence of torques and corresponding rotational motions must be carefully checked. If nothing is explicitly said, we assume in the following that the stress tensor is always symmetric. Hence it normally possesses six independent components. In the principal axes' representation only the three principal stresses σ'_{ii} appear. These are the longitudinal components in the directions of the principal axes (normal stresses). This means that each state of stress, that does not generate a torque, is to be represented by three mutually perpendicular normal stresses. Shear stresses do not appear in a parallelepiped cut parallel to the principal axes' directions of the stress tensor!

The most important examples of stress tensors in the principal axes' representation are:

a) Uniaxial tension parallel to $\boldsymbol{e}_1$. The stress tensor only contains the components σ_{ii}. We have $\sigma_{ii} = K_i/q_i$, where K_i acts in the direction $\boldsymbol{e}_i$ on the face perpendicular to $\boldsymbol{e}_i$ with cross-section q_i.

b) Biaxial tension parallel to $\boldsymbol{e}_i$ and $\boldsymbol{e}_j$. Here we only have the two principal stresses σ_{ii} and σ_{jj}. This state of stress appears when we apply a tension on two faces of a rectangular parallelepiped or put it in a biaxial clamping device.

c) Triaxial tension in three mutually perpendicular directions. This is realizable by clamping a rectangular parallelepiped in an assembly with three independent uniaxial pressure generators arranged perpendicular to one another, where only the components σ_{11}, σ_{22}, and σ_{33} appear. An important special case is hydrostatic pressure p, where $\sigma_{11} = \sigma_{22} = \sigma_{33} = -p$.

d) Pure shear stress. We assume $\sigma_{12} = \sigma_{21} \neq 0$, all other $\sigma_{ij} = 0$. The associated principal axes' representation is obtained through a plane principal axes' transformation by rotating about $\boldsymbol{e}_3$. According to Section 4.3.2 we have for the angle φ of rotation

$$\tan 2\varphi = \frac{2\sigma_{12}}{\sigma_{11} - \sigma_{22}}, \quad \text{hence} \quad 2\varphi = 90^\circ, \quad \varphi = 45^\circ.$$

Thus $\sigma'_{11} = \sigma_{12}$ and $\sigma'_{22} = -\sigma_{12}$, all remaining σ'_{ij} vanish. That is, the state of a pure shear stress σ_{12} is equivalent to a longitudinal stress (tension) of magnitude $\sigma'_{11} = \sigma_{12}$ along $(\boldsymbol{e}_1 + \boldsymbol{e}_2)$ and perpendicular to this a longitudinal stress of opposite sign (compressive stress) $\sigma'_{22} = -\sigma_{12}$

along $(e_1 - e_2)$. These directions run parallel to the diagonals of the face of the cube perpendicular to e_3 (Fig. 4.29).

The stress tensor, as an inducing quantity, is easy to realize in static experiments.

4.4 Third-Rank Tensors

First of all we try to gain an overview of the form of the tensors in the 32 PSG. As already discussed in the general introduction, all polar tensors of uneven rank vanish with the existence of an inversion center. Hence we only need to consider the following PSG (non-Laue groups): 1, 2, m, 22, mm, 3, 32, 3m, 4, 42, 4m, $\bar{4}$, $\bar{4}2$, 6, 62, 6m, $\bar{6}$, $\bar{6}2$, 23, 43, and $\bar{4}3$. In the case of pseudo tensors, the Laue groups require a more detailed inspection.

At first we consider a two-fold symmetry axis parallel to e_i or a symmetry plane perpendicular to e_i. These are represented by the following transformation matrices (see also Section 3.8.2):

$$\begin{aligned} 2 \parallel e_i: &\quad u_{ii} = 1, \quad u_{jj} = u_{kk} = -1, \quad u_{ij} = 0 \text{ for } i \neq j, \\ \bar{2} \parallel e_i: &\quad u_{ii} = -1, \quad u_{jj} = u_{kk} = 1, \quad u_{ij} = 0 \text{ for } i \neq j. \end{aligned}$$

In both cases $t'_{ijk} = (-1)^q t_{ijk}$. In the first case, q is equal to the number of indices j and k together and in the second, equal to the number of index i. With uneven q, the respective tensor component vanishes. Therefore, in the case of a two-fold axis parallel e_i, only those components exist, in which the indices j and k together occur an even number of times, i.e., the index i occurs an odd number of times. In the case of a symmetry plane perpendicular to e_i the complementary existence condition is valid: the index i occurs an even number of times. Similar as with second-rank tensors, it is useful for an overview to write the components in the form of a matrix. This matrix looks like

$$(t_{ijk}) = \begin{pmatrix} t_{111} & \underline{t_{112}} & t_{113} & \underline{t_{121}} & t_{122} & \underline{t_{123}} & t_{131} & \underline{t_{132}} & t_{133} \\ \underline{t_{211}} & t_{212} & \underline{t_{213}} & t_{221} & \underline{t_{222}} & t_{223} & \underline{t_{231}} & t_{232} & \underline{t_{233}} \\ t_{311} & \underline{t_{312}} & t_{313} & \underline{t_{321}} & t_{322} & \underline{t_{323}} & t_{331} & \underline{t_{332}} & t_{333} \end{pmatrix}.$$

Even with tensors of higher rank, the overview is made easier, when the components are arranged in three rows, where the first index in the first row is 1, in the second row 2 and in the third row 3. The further indices in each row are to be selected in the sequence of the scheme for a tensor of the next lowest rank.

The 13 components existing in the case of a two-fold axis parallel to e_2 are underlined; the remaining 14 components exist in the case of $\bar{2} \equiv \text{m}$ parallel to e_2 (symmetry plane perpendicular to e_2!). A third-rank pseudo tensor with m $\parallel e_2$ is of the same type as a polar tensor with $2 \parallel e_2$!

For all PSG with the subgroup 22 it is immediately recognizable that only those components exist in which each index occurs an odd number of times, hence t_{123}, t_{132}, t_{231}, t_{213}, t_{312}, and t_{321}. In the PSG mm in the standard setting (e_3 parallel to the intersection edge of the symmetry planes) only such components can appear, in which the indices 1 or 2 occur an even number of times, hence t_{113}, t_{131}, t_{223}, t_{232}, t_{311}, t_{322}, and t_{333}.

The effect of a three-fold axis parallel e_3 requires a more detailed inspection of the transformation behavior. At this point we present a complete derivation of the conditions in order to show how, in analogous cases of higher rank tensors, one proceeds in practice. The rotation about a three-fold axis is represented by

$$R_{\pm 3\parallel e_3} = \begin{pmatrix} -1/2 & \pm\sqrt{3}/2 & 0 \\ \mp\sqrt{3}/2 & -1/2 & 0 \\ 0 & 0 & 1 \end{pmatrix}$$

$+$ means clockwise rotation, $-$ anticlockwise rotation, when one looks in the direction e_3. The simultaneous calculation for both directions of rotation yields two conditions for each operation, because the sign-dependent parts must be considered separately.

We obtain

$$t'_{111} = t_{111} = -\frac{1}{8}t_{111} \pm \frac{\sqrt{3}}{8}(t_{112} + t_{121}) - \frac{3}{8}t_{122} \pm \frac{\sqrt{3}}{8}t_{211} - \frac{3}{8}(t_{212} + t_{221}) \pm \frac{3\sqrt{3}}{8}t_{222}.$$

The parts coupled with the change of sign must vanish, hence

$$t_{112} + t_{121} + t_{211} + 3t_{222} = 0. \qquad (4.1)$$

The remainder gives

$$3t_{111} + t_{122} + t_{212} + t_{221} = 0. \qquad (4.2)$$

Further conditions result from

$$t'_{122} = t_{122} = -\frac{3}{8}t_{111} \mp \frac{\sqrt{3}}{8}(t_{112} + t_{121}) - \frac{1}{8}t_{122} \pm \frac{3\sqrt{3}}{8}t_{211} + \frac{3}{8}(t_{212} + t_{221}) \pm \frac{\sqrt{3}}{8}t_{222}.$$

From which we obtain

$$-t_{112} - t_{121} + 3t_{211} + t_{222} = 0 \qquad (4.3)$$

and

$$t_{111} + 3t_{122} - t_{212} - t_{221} = 0. \tag{4.4}$$

From (4.1) and (4.3) one finds

$$t_{211} = -t_{222} = \tfrac{1}{2}(t_{112} + t_{121}),$$

and from (4.2) and (4.4)

$$t_{122} = -t_{111} = \tfrac{1}{2}(t_{212} + t_{221}).$$

From

$$t'_{112} = t_{112} = \mp\frac{\sqrt{3}}{8}t_{111} - \frac{1}{8}t_{112} + \frac{3}{8}t_{121} \pm \frac{\sqrt{3}}{8}t_{122} + \frac{3}{8}t_{211} \pm \frac{\sqrt{3}}{8}t_{212} \mp \frac{3\sqrt{3}}{8}t_{221} - \frac{3}{8}t_{222}$$

we get the conditions

$$-t_{111} + t_{122} + t_{212} - 3t_{221} = 0 \tag{4.5}$$

and

$$-3t_{112} + t_{121} + t_{211} - t_{222} = 0. \tag{4.6}$$

Similarly from

$$t'_{212} = t_{212} = -\frac{3}{8}t_{111} \mp \frac{\sqrt{3}}{8}t_{112} \pm \frac{3\sqrt{3}}{8}t_{121} + \frac{3}{8}t_{122} \mp \frac{\sqrt{3}}{8}t_{211} - \frac{1}{8}t_{212} + \frac{3}{8}t_{221} \pm \frac{\sqrt{3}}{8}t_{222}$$

we get the conditions

$$-t_{112} + 3t_{121} - t_{211} + t_{222} = 0 \tag{4.7}$$

and

$$t_{111} - t_{122} + 3t_{212} - t_{221} = 0. \tag{4.8}$$

With (4.5) and (4.8) as well as (4.6) and (4.7) one finally finds

$$t_{212} = t_{221} = t_{122} = -t_{111}$$

and

$$t_{112} = t_{121} = t_{211} = -t_{222}.$$

We see that these relations are permutable in the indices 1 and 2, as we expect on the basis of the equivalence of e_1 and e_2 with respect to the three-fold axis. This relationship is also of great value for similar questions. Thus the transformation of t_{121}, t_{211}, t_{221}, and t_{222} does not lead to new conditions.

We now turn our attention to those components carrying the index 3 once or twice.

$$t'_{113} = t_{113} = \frac{1}{4}t_{113} \mp \frac{\sqrt{3}}{4}t_{123} \mp \frac{\sqrt{3}}{4}t_{213} + \frac{3}{4}t_{223}.$$

Consequently $t_{113} = t_{223}$ and $t_{123} = -t_{213}$. The analogous transformation for t_{131} and t_{311} leads to $t_{131} = t_{232}$, $t_{132} = -t_{231}$, $t_{311} = t_{322}$, and $t_{312} = -t_{321}$.

$$t'_{133} = t_{133} = -\frac{1}{2}t_{333} \pm \frac{\sqrt{3}}{2}t_{233} \quad \text{gives} \quad t_{133} = t_{233} = 0;$$

likewise one finds $t_{313} = t_{323} = t_{331} = t_{322} = 0$.

Finally we have $t'_{333} = t_{333}$ (no condition!).

Therefore, in the PSG 3 there exist the following nine independent components of a third-rank polar tensor:

$$\begin{aligned}
&t_{111} = -t_{122} = -t_{212} = -t_{221}\\
&t_{222} = -t_{211} = -t_{121} = -t_{112}\\
&t_{113} = t_{223}, \quad t_{123} = -t_{213}, \quad t_{131} = t_{232},\\
&t_{132} = -t_{231}, \quad t_{311} = t_{322}, \quad t_{312} = -t_{321}, \quad t_{333}.
\end{aligned}$$

With these we can now immediately determine the form of the third-rank tensors in the remaining trigonal and hexagonal PSG by taking into account the additional symmetry elements (see Table 4.6).

Finally the effect of 4 and $\bar{4}$ has to be checked. Since 4 as well as $\bar{4}$ contain a two-fold axis, it suffices to consider only such tensor components containing the index 3 an odd number of times (4 or $\bar{4} \parallel e_3$!).

With

$$\boldsymbol{R}_{\pm 4\parallel e_3} = \begin{pmatrix} 0 & \pm 1 & 0 \\ \mp 1 & 0 & 0 \\ 0 & 0 & 1 \end{pmatrix} \quad \text{and} \quad \boldsymbol{R}_{\pm \bar{4}\parallel e_3} = \begin{pmatrix} 0 & \mp 1 & 0 \\ \pm 1 & 0 & 0 \\ 0 & 0 & -1 \end{pmatrix}$$

one obtains for PSG 4 the seven independent components $t_{113} = t_{223}$, $t_{131} = t_{232}$, $t_{311} = t_{322}$, $t_{123} = -t_{213}$, $t_{132} = -t_{231}$, $t_{312} = -t_{321}$, t_{333} and for PSG $\bar{4}$ the six independent components $t_{113} = -t_{223}$, $t_{131} = -t_{232}$, $t_{311} = -t_{322}$, $t_{123} = t_{213}$, $t_{132} = t_{231}$, $t_{312} = t_{321}$.

Table 4.6 gives the form of the third-rank tensors in the remaining tetragonal and cubic PSG. The effect of the other symmetry elements results from the rules previously discussed. With the cubic PSG, the three-fold axis demands $t_{123} = t_{231} = t_{312}$ etc. (cyclic permutation!).

Table 4.6 Independent components of polar tensors in acentric PSG of the trigonal, tetragonal, hexagonal, and cubic systems. All otherwise existing components result from the relations valid in PSG 3 and 4, respectively (see text). The PSG of lower symmetry are discussed in the text.

PSG	Positions of the symmetry elements	Conditions for existence	Independent t_{ijk}
3	$3 \parallel e_3$	–	$t_{111}, t_{222}, t_{113}, t_{123}, t_{131}, t_{132}, t_{311}, t_{312}, t_{333}$
32	$2 \parallel e_1$	Index 1 odd-numbered	$t_{111}, t_{123}, t_{132}, t_{312}$
3m	$\mathrm{m} = \bar{2} \parallel e_1$	Index 1 even-numbered	$t_{222}, t_{113}, t_{131}, t_{311}, t_{333}$
4	$4 \parallel e_3, 2 \parallel e_3$	Index 3 odd-numbered	$t_{113}, t_{131}, t_{311}, t_{123}, t_{132}, t_{312}, t_{333}$
$\bar{4}$	$\bar{4} \parallel e_3, 2 \parallel e_3$	Index 3 once	$t_{113}, t_{131}, t_{311}, t_{123}, t_{132}, t_{312}$
6	$6 \parallel e_3, 2 \parallel e_3$	Index 3 odd-numbered	$t_{113}, t_{123}, t_{131}, t_{132}, t_{311}, t_{312}, t_{333}$
62	$2 \parallel e_1$	Indices 3 and 1 odd-numbered	$t_{123}, t_{132}, t_{312}$
6m	$\mathrm{m} = \bar{2} \parallel e_1$	Index 3 odd-numbered, Index 1 even-numbered	$t_{113}, t_{131}, t_{311}, t_{333}$
$\bar{6} \equiv 3/\mathrm{m}$	$\mathrm{m} = \bar{2} \parallel e_3$	Index 3 even-numbered	t_{111}, t_{222}
$\bar{6}2 \equiv 3/\mathrm{m}2$	$\mathrm{m} = \bar{2} \parallel e_3, 2 \parallel e_1$	Index 3 even-numbered, Index 1 odd-numbered	t_{111}
42	$2 \parallel e_1$	Index 1 odd-numbered	$t_{123}, t_{231}, t_{312}$
4m	$\mathrm{m} = \bar{2} \parallel e_1$	Index 1 even-numbered	$t_{113}, t_{131}, t_{311}, t_{333}$
$\bar{4}2$	$2 \parallel e_1$	Index 1 odd-numbered	$t_{123}, t_{231}, t_{312}$
23	$2 \parallel e_i$	All indices odd-numbered and cyclic	t_{123}, t_{132}
$\bar{4}3$	$\bar{4} \parallel e_i$	All indices odd-numbered and cyclic	$t_{123} = t_{213}$
43	$4 \parallel e_i$	All indices odd-numbered and cyclic	$t_{123} = -t_{213}$

Furthermore, symmetry reduction leads to the result that third-rank pseudo tensors can exist in all PSG.

The quadric $F = t_{ijk}x_i x_j x_k$ contains ten coefficients that can be conveniently determined with the help of longitudinal effects. Measurements in the directions $\boldsymbol{e}_i$ directly give the components t_{111}, t_{222} and t_{333}. The longitudinal effect in the bisectors of the Cartesian basic vectors $\boldsymbol{e}'_1 = \sqrt{2}/2(\boldsymbol{e}_i \pm \boldsymbol{e}_j)$ is

$$t'_{111\pm} = \frac{\sqrt{2}}{4}t_{iii} \pm \frac{\sqrt{2}}{4}t_{jjj} \pm \frac{\sqrt{2}}{4}(t_{iij} + t_{iji} + t_{jii}) + \frac{\sqrt{2}}{4}(t_{jji} + t_{jij} + t_{ijj}).$$

From $(t'_{111+} + t'_{111-})$ and $(t'_{111+} - t'_{111-})$ one gets the six coefficients $(t_{iij} + t_{iji} + t_{jii})$ with $i, j = 1, 2, 3$. From a further measurement, e.g., in the direction of one of the space diagonals $\boldsymbol{e}'_1 = \frac{\sqrt{3}}{3}(\boldsymbol{e}_1 \pm \boldsymbol{e}_2 \pm \boldsymbol{e}_3)$, one can also determine the tenth coefficient $(t_{123} + t_{132} + t_{231} + t_{213} + t_{312} + t_{321})$ taking into account the already known coefficients. To improve accuracy, it is often advisable to perform measurements in other directions, too.

In any case, we can acquire only a part of the tensor components with the help of longitudinal effects. We need the measurement of transversal effects to completely determine the tensor.

We will address the problem of extreme values when discussing concrete examples of these tensors. A principal axes' transformation, whereby all components with mixed indices vanish, as is the case with second-rank tensors, does not exist here.

4.4.1
Piezoelectric Tensor

In the pyroelectric effect we observe an electric polarization $\boldsymbol{P}$, proportional to the temperature difference within a small temperature range. The application of mechanical stress, expressed by the stress tensor $\{\sigma_{ij}\}$, can also give rise to an electric polarization. This phenomenon is known as the *piezoelectric effect*. For a sufficiently small, homogenous test volume, one has

$$\Delta P_i = d_{ijk}\sigma_{jk} + d_{ijklm}\sigma_{jk}\sigma_{lm} + \cdots$$

The tensor $\{d_{ijk}\}$ represents the *linear piezoelectric effect* and $\{d_{ijklm}\}$ the *quadratic piezoelectric effect* and so on. Normally the linear effect is sufficient for the description of an experiment. Because the stress tensor is symmetric, there is no possibility to distinguish between the components d_{ijk} and d_{ikj}. Therefore, we must assume that the tensor $\{d_{ijk}\}$ is symmetric in the second and third index positions, hence $d_{ijk} = d_{ikj}$. This condition reduces the number of independent components of the third-rank tensor from 27 to 18. In the light of the overview gained in the previous section, the form of the piezoelectric tensor can be immediately read. Table 4.7 presents the results for all

Table 4.7 Independent components of third-rank tensors in the acentric PSG under interchangeability of the second and third order index position (example piezoelectric tensor). Z is the number of independent components.

PSG	Z	Independent components
1	18	$d_{ijk} = d_{ikj}\ (i,j,k = 1,2,3)$
$2 \parallel e_2$	8	$d_{112}, d_{123}, d_{211}, d_{213}, d_{222}, d_{233}, d_{312}, d_{323}$
$\bar{2} \parallel e_2$	10	$d_{111}, d_{113}, d_{122}, d_{133}, d_{212}, d_{223}, d_{311}, d_{313}, d_{322}, d_{333}$
22	3	$d_{123}, d_{231}, d_{312}$
mm2, $2 \parallel e_3$	5	$d_{113}, d_{223}, d_{311}, d_{322}, d_{333}$
$3 \parallel e_3$	6	$d_{111} = -d_{122} = -d_{212}, d_{222} = -d_{211} = -d_{121}, d_{113} = d_{223},$ $d_{123} = -d_{213}, d_{311} = d_{322}, d_{333}$
32, $2 \parallel e_1$	2	$d_{111} = -d_{122} = -d_{212}, d_{123} = -d_{213}$
3m, $\bar{2} \parallel e_1$	4	$d_{113} = d_{223}, d_{222} = -d_{211} = -d_{121}, d_{311} = d_{322}, d_{333}$
$4 \parallel e_3$	4	$d_{113} = d_{223}, d_{123} = -d_{213}, d_{311} = d_{322}, d_{333}$
42, $2 \parallel e_1$	1	$d_{123} = -d_{213}$
4m, $\bar{2} \parallel e_1$	3	$d_{113} = d_{223}, d_{311} = d_{322}, d_{333}$
$\bar{4}$	4	$d_{113} = -d_{223}, d_{123} = d_{213}, d_{311} = -d_{322}, d_{312}$
$\bar{4}2$, $2 \parallel e_1$	2	$d_{123} = d_{213}, d_{312}$
$6 \parallel e_3$	4	$d_{113} = d_{223}, d_{123} = -d_{213}, d_{311} = d_{322}, d_{333}$
62, $2 \parallel e_1$	1	$d_{123} = -d_{213}$
6m, $\bar{2} \parallel e_1$	3	$d_{113} = d_{223}, d_{311} = d_{322}, d_{333}$
$\bar{6}$	2	$d_{111} = -d_{122} = -d_{212}, d_{222} = -d_{211} = -d_{121}$
$\bar{6}2$, $2 \parallel e_1$	1	$d_{111} = -d_{122} = -d_{212}$
23	1	$d_{123} = d_{231} = d_{312} = d_{213} = d_{132} = d_{321}$
43	0	–
$\bar{4}3$	1	$d_{123} = d_{231} = d_{312} = d_{213} = d_{132} = d_{321}$
$\infty \parallel e_3$	4	$d_{113} = d_{223}, d_{123} = -d_{213}, d_{311} = d_{322}, d_{333}$
$\infty 2$	1	$d_{123} = -d_{213}$
∞m	3	$d_{113} = d_{223}, d_{311} = d_{322}, d_{333}$
$\bar{\infty}, \bar{\infty}2$	0	–

piezoelectric PSG. Of special interest is the fact that PSG 43 shows no piezoelectric effect, because the conditions $d_{123} = -d_{213}$ and $d_{123} = d_{231} = d_{213}$ can only be fulfilled with $d_{ijk} = 0$. The two other cubic piezoelectric PSG to be considered, namely 23 and $\bar{4}3$, possess only one independent component $d_{123} = d_{231} = d_{312} = d_{132} = d_{213} = d_{321}$.

4.4.1.1 **Static and Quasistatic Methods of Measurement**

Piezoelectric effects can best be observed and measured as longitudinal and transversal effects under the application of uniaxial pressure. In the quantitative determination of the piezoelectric tensor one must pay attention that mechanical stresses are homogenously transferred over the test object so that the influence of boundary effects can be neglected and also that a simple relationship can be assumed between inducing stress and charge generation. The easiest accessible method is the longitudinal effect (Fig. 4.31). At first we will study a probe in the form of a rectangular parallelepiped. If we apply a uniaxial pressure σ'_{11} in the direction e'_1, one observes a change in charge ΔQ on the

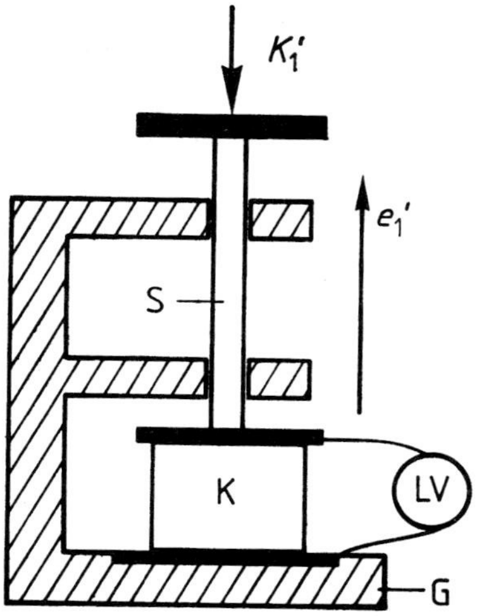

Figure 4.31 Measurement of the longitudinal piezoelectric effect. G ground plate, K crystal, S pressure bar, and LV charge amplifier.

faces cut perpendicular to e_1'. The latter must be made electrically conducting by vapor deposition of a metal or with conducting paste. We then have

$$\Delta P_1' = K\frac{\Delta Q}{x_2' x_3'} = d_{111}' \frac{KK_1'}{x_2' x_3'}, \quad \text{hence} \quad \Delta Q = d_{111}' K_1',$$

where ΔQ means the charge difference appearing on the face in question, which can be measured with a commercial charge amplifier, and K_1' is the uniaxial acting force. K is the constant of the system of measurement used. Most important is that the dimensions of the probe do not come into play. As long as the state of stress distributes itself to a certain extent homogenously over the cross-section, one observes only minor deviations from the theoretical value, even when the cross-section of the probe along the direction of pressure varies considerably. The measurement of the longitudinal effect can be miniaturized, without special precautions, down to probe dimensions 0.1 mm thick (plates) and 0.5 mm long (thin needle), where one can still obtain reliable values for the constant d_{111}'. Thus according to the evaluation of longitudinal effects described in the previous section, the quadric $F = t_{ijk}x_i x_j x_k$ with its ten coefficients can be completely determined.

Including transversal effects it is now possible to completely determine the piezoelectric tensor also in the case of triclinic crystals. We use the same pressure cell as with the longitudinal effect and also apply a uniaxial stress σ_{11}' parallel e_1'. The metallized pair of faces, however, are now arranged parallel to the direction of pressure, that is, perpendicular to e_2' or e_3' (Fig. 4.32). Thus

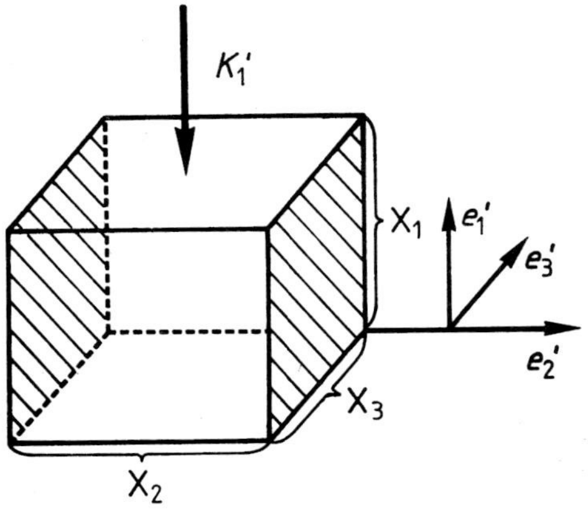

Figure 4.32 Parallelepipedic specimen for the measurement of the piezoelectric transversal effect d'_{211} (uniaxial stress parallel e'_1, surfaces of the electrodes perpendicular to e'_2).

for the generation of charge on the face perpendicular to e'_2 we have

$$\Delta P'_2 = K\frac{\Delta Q}{x'_1 x'_3} = d'_{211}\frac{KK'_1}{x'_2 x'_3}, \quad \text{hence} \quad \Delta Q = d'_{211}K'_1\frac{x'_1}{x'_2}.$$

Accordingly, with constant force K'_1, the observed change in charge depends on the ratio of the length x'_1 of the specimen to the separation of the electrodes x'_2. The dimension x'_3 does not matter. For a high accuracy of the measurement as well as for the practical generation of electric charges with the help of the transversal effect the specimen should be prepared as long as possible in the direction of pressure and as thin as possible in the direction of the normals on the surface of the electrodes. The measurements require that the specimens possess nearly the form of a rectangular parallelepiped. Moreover, geometrically similar specimens produce the same change in charge ΔQ under identical load conditions.

If we conduct such measurements with uniaxial stress direction parallel to e_i and electrode normals parallel to e_j, we directly obtain the six components d_{jii} $(i, j = 1, 2, 3)$. Thus we can now determine the components d_{iij} using the coefficients $(d_{jii} + d_{iij} + d_{iji}) = (d_{jii} + 2d_{iij})$ available from the quadric. The three missing components d_{123}, d_{231}, and d_{312} are obtained from measurements of transversal effects on 45° cuts, in other words, parallelepipeds bounded by a pair of faces perpendicular to $e'_1 = e_i$ and two pair of faces with the normals $e'_2 = \sqrt{2}/2(e_j + e_k)$ and $e'_3 = \sqrt{2}/2(-e_j + e_k)$, $(i, j, k$ cyclic in $1, 2, 3)$. For this orientation and with uniaxial stress parallel to e'_2 or e'_3 we have

$$d'_{122} = \tfrac{1}{2}d_{ijj} + \tfrac{1}{2}d_{ikk} + d_{ijk} \quad \text{or} \quad d'_{133} = \tfrac{1}{2}d_{ijj} + \tfrac{1}{2}d_{ikk} - d_{ijk}.$$

Thus

$$d_{ijk} = (d'_{122} - d'_{133})/2.$$

The procedure outlined above now provides a measurement scheme for the complete determination of the piezoelectric tensor.

To check the measurements and to increase accuracy, it is also useful to determine the transversal effects d'_{311}, d'_{322}, d'_{233}, and d'_{211} on the 45° cuts for which the following relations, to be proved by transformation, are valid:

$$\begin{aligned}
d'_{211} &= \frac{\sqrt{2}}{2}d_{jii} + \frac{\sqrt{2}}{2}d_{kii},\\
d'_{311} &= -\frac{\sqrt{2}}{2}d_{jii} + \frac{\sqrt{2}}{2}d_{kii},\\
d'_{233} &= \frac{\sqrt{2}}{4}(d_{jjj} + d_{kkk} - 2d_{jkj} + d_{jkk} - 2d_{kjk} + d_{kjj}), \quad \text{and}\\
d'_{322} &= \frac{\sqrt{2}}{4}(-d_{jjj} + d_{kkk} - 2d_{jkj} - d_{jkk} + 2d_{kjk} + d_{kjj}).
\end{aligned}$$

Of interest are the relations

$$\begin{aligned}
d'_{211} + d'_{311} &= \sqrt{2}d_{kii},\\
d'_{211} - d'_{311} &= \sqrt{2}d_{jii},\\
d'_{233} + d'_{322} &= \frac{\sqrt{2}}{2}(d_{kkk} + d_{kjj} - 2d_{jkj}), \quad \text{and}\\
d'_{233} - d'_{322} &= \frac{\sqrt{2}}{2}(d_{jjj} + d_{jkk} - 2d_{kjk}).
\end{aligned}$$

From the principal cut (edges parallel to $\boldsymbol{e}_i$) and the three 45° cuts we obtain 36 (=4×9) measurement results, which allow a reliable determination of the piezoelectric tensor.

One can achieve a higher measurement accuracy by employing a dynamic pressure generator, as for example, a spring connected to a harmonically vibrating loud speaker membrane or connected to a motor-driven eccentric. With the help of a phase-sensitive amplification of the piezoelectric signal it is possible to attain a hundred-fold increase in the sensitivity of the method. In any case, one must pay particular attention to the fact that just as with the pyroelectric effect, the surface conductivity remains small or does not cause a reduction in the true piezoelectric signal. For the measurement of electrical charges, compensation methods have proven to be of great advantage. In these methods, the charge appearing is immediately compensated by an opposing charge generated by the compensator so that only very small charge differences and hence small electric voltages are present at the crystal. The complete compensation charge is then measured. Also in the case of a certain volume conductivity of the specimen it is important to eliminate and measure the charge as promptly as possible.

In crystals of higher symmetry, such as α-quartz, lithium niobate, potassium bromate, lithium iodate, KH_2PO_4 or in cubic crystals like the isotype group of zinc blende-type, just to mention a few examples, the determination of the tensor is comparably unproblematic. These examples are briefly explained below. The experimental values are presented in Table 12.14 (annex).

(a) α-Quartz (32)

The following components are present: $d_{111} = -d_{122} = -d_{212} = -d_{221}$ and $d_{123} = d_{132} = -d_{231} = -d_{213}$. Measuring the longitudinal effect along the direction of the two-fold axis (e_1) gives us immediately d_{111}. As a control, a transversal measurement with uniaxial stress in the direction e_2 (perpendicular to e_1 and e_3) and charge generation on the pair of faces perpendicular to e_1 is useful. In doing so, the constant to be measured d_{122} is identical to $-d_{111}$. For the determination of d_{123} we appropriately employ the transversal effect on a 45° cut with the edges parallel to $e'_1 = e_1$, $e'_2 = \frac{\sqrt{2}}{2}(e_2 + e_3)$ and $e'_3 = \frac{\sqrt{2}}{2}(-e_2 + e_3)$. With uniaxial stress along e'_2 or e'_3 we obtain

$$d'_{122} = \tfrac{1}{2}d_{122} + d_{123}$$

or

$$d'_{133} = \tfrac{1}{2}d_{122} - d_{123}$$

and hence two independent values for d_{123}.

At present, α-quartz is used more than all other crystals together in the application of the piezoelectric effect. We will discuss the construction of sound generators, frequency stabilizers, frequency filters etc. later. Because of their good mechanical stability, quartz crystals are also employed preferably in devices for the measurement of mechanical stress and deformation as well as for electrically controlled precision feeds.

b) Lithium niobate ($LiNbO_3$) and $KBrO_3$ (3m)

Here, the following components exist: $d_{222} = -d_{112} = -d_{121} = -d_{211}$, $d_{113} = d_{223} = d_{131} = d_{232}$, $d_{311} = d_{322}$, d_{333}.

d_{222} and d_{333} are found directly from the longitudinal effects along e_2 and e_3, respectively, whereby e_1 runs perpendicular to the mirror plane. From the transversal effect with uniaxial stress in the direction e_1 and charge generation on the pair of faces perpendicular to the three-fold axis (e_3) one obtains d_{311}. The still missing component d_{113} results from the longitudinal effect with uniaxial stress parallel to

$$e_1' = \frac{\sqrt{2}}{2}(\mp e_1 + e_3).$$

Owing to their relatively large coefficients d_{333}, both crystals are especially suitable for the production of piezoelectric stress sensors and sound generators. Why, in spite of their superiority with respect to piezoelectric effects, the preferential application of α-quartz is not in danger will be discussed later.

c) Lithium iodate (6)

The following components exist: $d_{113} = d_{223} = d_{131} = d_{232}$, $d_{311} = d_{322}$, $d_{123} = d_{132} = -d_{213} = -d_{231}$, and d_{333}. From the longitudinal effect along the six-fold axis one directly obtains d_{333}. The component d_{311} is directly measurable from the transversal effect with uniaxial stress along $\boldsymbol{e}_1$ (this direction can be chosen arbitrarily perpendicular to the six-fold axis, since the piezoelectric tensor possesses cylindrical symmetry in all hexagonal PSG!) and charge generation on a pair of faces perpendicular to the six-fold axis. d_{113} and d_{123}, as described under (a) or (b), are to be determined on 45° cuts.

Lithium iodate possesses unusually large longitudinal- and transversal effects, which can be used to generate sound waves. We will discuss the influence of the extreme anisotropic electric conductivity in Section 4.5.5.

d) Sphalerite type ($\bar{4}3$)

Only one independent constant exists $d_{123} = d_{231} = d_{312} = d_{132} = d_{213} = d_{321}$. Its most reliable determination is via the longitudinal effect along the space diagonals of the Cartesian reference system, hence along

$$e_1' = \frac{\sqrt{3}}{3}(\pm e_1 \pm e_2 \pm e_3).$$

Here we have

$$d_{111}' = \pm\left(\frac{\sqrt{3}}{3}\right)^3 6d_{123} = \pm\frac{2}{3}\sqrt{3}d_{123}.$$

The sign is equal to the product of the signs of the components of $\boldsymbol{e}_1'$. In addition, one has the transversal effects d_{122}' and d_{133}' on a 45° cut with the edges $e_1' = e_1$, $e_2' = \frac{\sqrt{2}}{2}(e_2 + e_3)$, $e_3' = \frac{\sqrt{2}}{2}(-e_2 + e_3)$, which yields $d_{122}' = d_{123}$ and $d_{133}' = -d_{123}$ directly, whereby $\boldsymbol{e}_2'$ or $\boldsymbol{e}_3'$ specify the direction of the uniaxial stress and $\boldsymbol{e}_1$ denotes the normal on the charge generating pair of faces.

If these crystals form tetrahedra $\{111\}$, corresponding to the associated point symmetry group, then one can measure the constant d_{123} without further preparation.

4.4.1.2 **Extreme Values**

For many applications it is important to know the directions of the maximum or minimum effects. As an example, we consider the extreme values of the longitudinal effect in the point symmetry groups containing the symmetry group 22 as a subgroup (22, 42, 62, $\bar{4}2$, 23, $\bar{4}3$). The longitudinal effect along the direction $e'_1 = u_{1i}e_i$ is

$$d'_{111} = u_{1i}u_{1j}u_{1k}d_{ijk} = u_{11}u_{12}u_{13} \cdot 2(d_{123} + d_{231} + d_{312}) = u_{11}u_{12}u_{13} \cdot 2d.$$

In the PSG 42 and 62 $d = 0$ because $d_{123} = -d_{231}$ and $d_{312} = 0$. We proceed according to the well-known prescription and form the auxiliary function $H = d'_{111} - \lambda \sum_i u_{1i}^2$, which under the constraint $\sum_i u_{1i}^2 = 1$ allows the calculation of the extreme value of d'_{111}. λ is a factor still to be determined. An extreme value of H and hence of d'_{111} is present, when

$$\frac{\partial H}{\partial u_{1i}} = 2u_{1j}u_{1k}d - 2\lambda u_{1i} = 0.$$

If we divide $u_{1j}u_{1k}d = \lambda u_{1i}$ by the corresponding equation $u_{1i}u_{1k}d = \lambda u_{1j}$, we obtain

$$\frac{u_{1j}}{u_{1i}} = \frac{u_{1i}}{u_{1j}} \quad \text{or} \quad u_{1i}^2 = u_{1j}^2 = u_{1k}^2 = \frac{1}{3} \quad \text{because} \quad \sum_i u_{1i}^2 = 1.$$

Hence extreme values appear in the directions of the space diagonals of the Cartesian reference system

$$e'_1 = \frac{\sqrt{3}}{3}(\pm e_1 \pm e_2 \pm e_3).$$

The extreme values are

$$\frac{\pm 2\sqrt{3}}{9}d.$$

In the cubic PSG $d'_{111} = \pm \frac{2}{\sqrt{3}} d_{123}$.

For certain applications of the piezoelectric effect it is desirable that pure longitudinal effects appear, i.e., that the associated transversal effects completely vanish. We consider this question for crystals of the minimum symmetry 22. The conditions are

$$d'_{211} = d'_{311} = 0.$$

We have

$$\begin{aligned} d'_{k11} &= 2u_{k1}u_{12}u_{13}d_{123} + 2u_{k2}u_{13}u_{11}d_{231} + 2u_{k3}u_{11}u_{12}d_{312} \\ &= 2e'_k \cdot (u_{12}u_{13}d_{123}e_1 + u_{13}u_{11}d_{231}e_2 + u_{11}u_{12}d_{312}e_3) = 0 \end{aligned}$$

for $k = 2$ or 3. Since $e'_k \cdot e'_1 = 0$ we find from the above equation that the components u_{1i} of $e'_1 = u_{1i}e_i$ have the following form:

$$u_{11} = qu_{12}u_{13}d_{123}, \quad u_{12} = qu_{13}u_{11}d_{231}, \quad \text{and} \quad u_{13} = qu_{11}u_{12}d_{312}$$

with an arbitrary factor q. If we multiply these expressions one after the other with u_{11} or u_{12} and u_{13} we obtain

$$u_{11}^2 = qu_{11}u_{12}u_{13}d_{123},$$
$$u_{12}^2 = qu_{11}u_{12}u_{13}d_{231},$$
$$u_{13}^2 = qu_{11}u_{12}u_{13}d_{321}.$$

Hence the ratio

$$u_{11}^2 : u_{12}^2 : u_{13}^2 = d_{123} : d_{231} : d_{312} \quad \text{must be satisfied.}$$

From this we see that all d_{ijk} must have the same sign. For cubic crystals, we obtain the result that pure longitudinal effects appear in the directions of the space diagonals of the Cartesian (and crystallographic) reference system.

Finally we take a look at the situation of the PSG 32, in which also α-quartz crystallizes. For the longitudinal effect we obtain

$$d'_{111} = u_{11}^3 d_{111} + u_{11}u_{12}^2 d_{122} + u_{11}u_{12}^2 \cdot 2d_{212} + u_{11}u_{12}u_{13} \cdot 2(d_{123} + d_{231}).$$

The last term vanishes because $d_{123} = -d_{231}$. With $d_{122} = d_{212} = -d_{111}$ we get $d'_{111} = d_{111}(u_{11}^3 - 3u_{11}u_{12}^2)$. From

$$H = d'_{111} - \lambda \sum_i u_{1i}^2 \quad \text{and} \quad \frac{\partial H}{\partial u_{1i}} = 0$$

we find $u_{13} = 0$. Accordingly, the extreme values lie in the plane perpendicular to the three-fold axis. With $u_{11}^2 + u_{12}^2 = 1$ and $u_{11} = \cos\varphi$ we obtain

$$d'_{111} = d_{111}(4u_{11}^3 - 3u_{11}) = d_{111}(4\cos^3\varphi - 3\cos\varphi) = d_{111}\cos 3\varphi.$$

It is now sufficient to study the behavior of

$$\frac{\partial d'_{111}}{\partial \varphi} = -3d_{111}\sin 3\varphi = 0.$$

We obtain $\varphi = m\pi/3$ with integer m. An extreme value appears in the direction e_1 ($m = 0$), which is repeated with alternating sign after each $\pi/3$ rotation. This rapid change of the extreme values is the reason why d_{111} in crystals of the PSG 32 are relatively small. The variation of d'_{111} in the plane perpendicular to the three-fold axis is shown in Fig. 4.33.

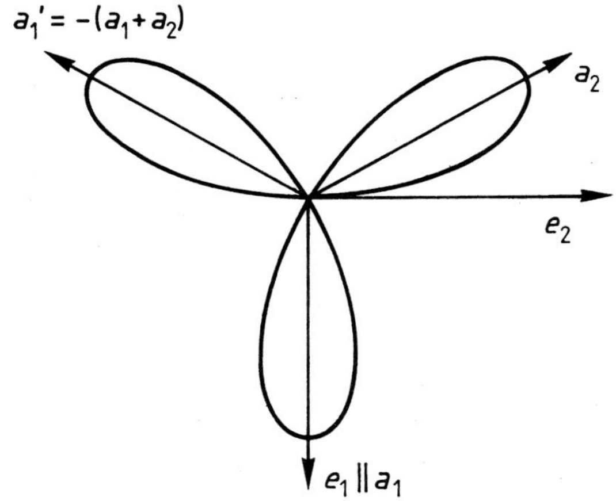

Figure 4.33 Variation of the longitudinal effect d'_{111} in crystals belonging to PSG 32 along directions perpendicular to the threefold axis (e.g., α-quartz).

4.4.1.3 **Converse Piezoelectric Effect (First-Order Electrostriction)**

A type of reversal of the usual piezoelectric effects appears when a crystal in an external electric field experiences a deformation. For a small homogenous volume element, the relation is written as

$$\varepsilon_{ij} = \hat{d}_{ijk} E_k.$$

$\{\varepsilon_{ij}\}$ is the deformation tensor and $\{\hat{d}_{ijk}\}$ the *first-order electrostriction tensor*. Because of the symmetry of the deformation tensor we must also assume the interchangeability of the first two index positions, hence $\hat{d}_{ijk} = \hat{d}_{jik}$. Otherwise all the aspects discussed for the piezoelectric tensor are valid when one takes into consideration that the interchangeability of the indices is different. As we shall see in Section 5, thermodynamical relations demand that the components of the piezoelectric and electrostrictive tensors take on the same numerical value. That is, $\hat{d}_{ijk} = d_{kij}$, where, however, certain boundary conditions are to be observed. Thus one can measure the piezoelectric tensor also with the help of electrostriction. In particular, the measurement of both effects on the same specimen provides very good control potential. The measurement of the deformations appearing in the converse piezoelectric effect is preferably performed using optical interferometry methods of high resolution. Figure 4.34 shows the scheme of such an arrangement (*Michelson interferometer*). Again, the specimen has preferably the form of a rectangular parallelepiped. It is furnished on one side with an optical mirror. A laser beam strikes a semi-transparent optical plate at 45° incidence so that the beam is split. Let one beam, e.g., that passing straight through, strike the specimen in vertical inci-

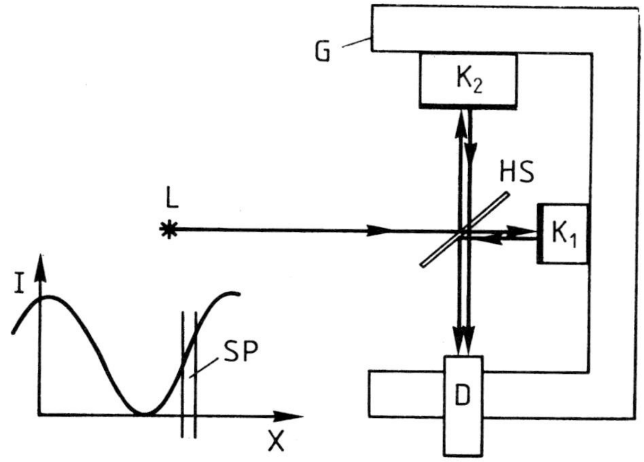

Figure 4.34 Scheme of a Michelson interferometer. G ground plate, L light source, HS semitransparent mirror, K_1 and K_2 crystal and auxiliary crystal, respectively, with electrodes (and polishing), D detector. Maximum resolution is realized, if the detector slit SP is placed at the position of the steepest intensity variation in the interference pattern (between maximum and minimum of intensity).

dence. This beam is reflected and deflected by 90° at the beam splitter. The second beam strikes the auxiliary crystal of the known piezoelectric effect, also furnished with a mirror, where it is reflected. The beam passing straight through the splitter interferes with the other beam to build a pattern of interference fringes. By carefully adjusting one of the crystals one can set the interfringe distance to the desired width. With the help of a narrow slit in front of the detector (photomultiplier or photodiode) one can find a position of the strongest increase in intensity. This is located between a maximum and an adjacent minimum of the fringes. This is the setting of the highest sensitivity for the detection of fringe pattern displacement. If in one of the two crystals a homogenous mechanical deformation is produced along the direction of the wave normal of the beam, one observes a displacement of the fringes corresponding to twice the deformation path. An especially high accuracy is realizable, when the fringe displacement is compensated with the help of the second crystal by producing an equally large deformation, for example by means of the converse piezoelectric effect. If one works with an electric alternating field, where the amplitude can be set independently at each crystal, one can achieve, with the help of a lock-in amplifier, an accuracy an order of magnitude higher than required for the measurement of the normal converse first-order piezoelectric effect. The lock-in technique largely suppresses a substantial part of disturbing phenomena (vibration, inhomogeneous heating, and detector intrinsic noise). For this reason such an arrangement is also suitable to measure second- and higher order effects. We will return to this in Section 4.5.3.

Here we again measure longitudinal effects with the electric field in the direction of the beam and transversal effects with the electric field perpendicular to the beam direction. The complete determination of the first-order electrostriction tensor can be carried out using a strategy similar to that outlined above.

Of course, other types of interferometers can be used for purposes of this kind. Somewhat more versatile than the interferometer described here is a device where the light beam is transmitted through the specimen. Changes in the optical path are measured. These, however, contain a fraction arising from a change in the refractive index under the influence of the electric field (electrooptical effect, Section 4.4.2).

This method requires a substantially higher optical quality of the crystal and also a rather perfect preparation (plane-parallelism of the surfaces through which the light passes, surface finish).

The measurement of electrostriction with the help of strain gauges, comfortable as it is, hardly achieves the desired accuracy except in the case of very large effects. In contrast, the highly precise determination of lattice constants offers a competitive alternative to the optical interference methods (see Section 4.3.11).

We will become acquainted with other methods to measure piezoelectric effects in connection with the propagation of sound waves in piezoelectric crystals, where a coupling appears between the elastic and piezoelectric constants. This electromechanical coupling is the basis for the construction of sound generators and resonators.

A method to qualitatively test whether a substance possesses piezoelectric properties was reported by Giebe and Scheibe. The method works on powder samples and allows certain conclusions concerning the order of magnitude of the effects. The powder is immersed in a high-frequency electric field. Individual grains, excited to acoustic resonances at a certain frequency depending on their dimensions, give weak interfering signals which are amplified and detected.

4.4.2 First-Order Electro-Optical Tensor

In Section 4.3.6 we became familiar with the indicatrix $a_{ij}x_ix_j = 1$ which we employed to describe the propagation of electromagnetic waves in non-absorbing and optically non-active crystals. The refractive indices n_{ij} are related to the "polarization constants" a_{ij} according to $a_{ij} = n_{ij}^{-2}$ (see Section 4.3.6). An external electric field $\boldsymbol{E}$ can give rise to a change in the polarisation constants

$$a_{ij} - a_{ij}^0 = \Delta a_{ij} = r_{ijk}E_k + r_{ijkl}E_kE_l + \dots$$

The tensor $\{r_{ijk}\}$ is called the ***first-order electro-optical tensor***. It is of the same form as the converse piezoelectric tensor $\{\hat{d}_{ijk}\}$ because of the interchangeability of the first two index positions. Hence, we can directly take over the set of independent tensor components.

The concept of the measurement of longitudinal- and transversal effects with respect to the optical properties of crystals cannot be directly realized, because the laws of light propagation do not allow arbitrary directions of vibration in the propagation directions. In a simple measurement arrangement let us assume that a light wave strikes the surface of a rectangular parallelepiped at vertical incidence. The possible directions of vibration in the crystal, without external field, are deduced from the sectional ellipse of the original indicatrix. With the electric field applied either parallel or perpendicular to the ray direction (Fig. 4.35), we expect a change in the indicatrix, which will be expressed in two phenomena; firstly, a new sectional ellipse is created with altered principal axes directions and secondly, the lengths of the principal axes are altered, i.e., a change results in the refractive indices, hence in the velocity of light. In the orthorhombic system, the directions of the principal axes remain unchanged, to a first approximation, due to symmetry. The same is valid for the distinct principal axis parallel to the twofold axis or perpendicular to the symmetry plane in monoclinic crystals and for the principal axis parallel

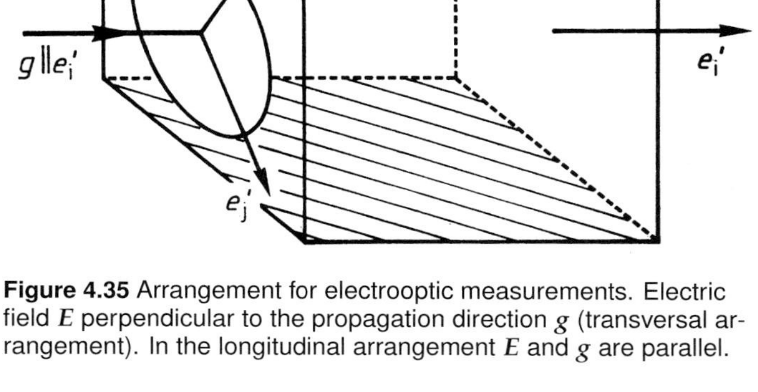

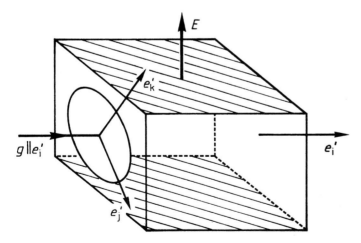

Figure 4.35 Arrangement for electrooptic measurements. Electric field $\boldsymbol{E}$ perpendicular to the propagation direction $\boldsymbol{g}$ (transversal arrangement). In the longitudinal arrangement $\boldsymbol{E}$ and $\boldsymbol{g}$ are parallel.

to the principal symmetry axis in trigonal, tetragonal and hexagonal crystals. In cubic and in optically uniaxial crystals—except in certain singular directions of the electric field—the degeneracy of the indicatrix is canceled, i.e., the indicatrix takes on the form of a triaxial ellipsoid. We will now examine to what extent we can actually expect a measurable rotation of the indicatrix compared with the initial situation.

In the following, we designate the transmission direction (vertical incidence) with e'_i and the two directions of vibration with e'_j and e'_k (directions of the semiaxes of the sectional ellipse). The sectional ellipse in the plane perpendicular to e'_i is

$$a'_{jj}x'^2_j + a'_{kk}x'^2_k = 1.$$

After the change $\{\Delta a_{ij}\}$ (calculated in the dotted system!) we have

$$(a'_{jj} + \Delta a'_{jj})x'^2_j + (a'_{kk} + \Delta a'_{kk})x'^2_k + 2\Delta a'_{jk}x'_j x'_k = 1$$

(no summation here).

We now perform a rotation in the new principal axes system of the sectional ellipse and obtain for the angle of rotation (see Section 4.3.2)

$$\tan 2\varphi = \frac{2\Delta a'_{jk}}{a'_{jj} + \Delta a'_{jj} - a'_{kk} - \Delta a'_{kk}}.$$

By tensor transformation we can calculate the quantities $\Delta a'_{jk}$ from the tensor components in the crystal–physical reference system according to $\Delta a'_{jk} = r'_{jkl}E'_l$.

The quantities a'_{jj} and a'_{kk} are known ($a'_{jj} = n'^{-2}_{jj}$, $a'_{kk} = n'^{-2}_{kk}$); furthermore,

$$\Delta a'_{jj} = -2\Delta n'_{jj}/n'^3_{jj} \quad \text{and} \quad \Delta a'_{kk} = -2\Delta n'_{kk}/n'^3_{kk} \quad (\text{because}\, a_{ij} = n^{-2}_{ij}).$$

Hence, the magnitude of the angle of rotation depends decisively on the difference of refractive indices and naturally, on the quantity $\Delta a'_{jk}$. Experience shows that the angle of rotation remains virtually unobservable by a double refraction $(n'_{jj} - n'_{kk}) > 0.03$, i.e., the position of the principal axes of the sectional ellipse is virtually unchanged. If, however, the sectional ellipse has the form of a quasicircle ($n'_{jj} \approx n'_{kk}$) or a circle, then the rotation or the position of the sectional ellipse due to the field must be taken into account.

In all other cases, the approximation of the vanishing angle of rotation leads to a substantial simplification of the measurement procedure. Let the optical path difference of the two light waves passing through a crystal of thickness L and the refractive indices n'_{jj} and n'_{kk} in the direction e'_i be given by $G = L(n'_{jj} - n'_{kk})$. For a change, we have

$$\Delta G = L\Delta(n'_{jj} - n'_{kk}) + (n'_{jj} - n'_{kk})\Delta L.$$

If we write for $\Delta n'$ the values $-n'^3 \Delta a'/2$ and for ΔL the induced quantity from the converse piezoelectric effect

$$\Delta L = L\varepsilon'_{ii} = L\hat{d}'_{iil} E'_l,$$

we get

$$\Delta G = -\frac{L}{2}(n'^3_{jj} r'_{jjl} E'_l - n'^3_{kk} r'_{kkl} E'_l) + (n'_{jj} - n'_{kk}) L\hat{d}'_{iil} E'_l.$$

(with fixed j and k; summation over l). Accordingly, to determine the components r_{ijk} the knowledge of the piezoelectric tensor is also required, because the electric field causes a change of the geometric path length L. When the effects are sufficiently large, the measurements can be made with the help of a Sénarmont compensator (see Section 4.3.6.5). Higher accuracy is achieved when one works with an alternating electric field that induces an oscillation of the direction of vibration of the linear polarized light behind the Sénarmont compensator. A Faraday cell placed in the ray path behind the Sénarmont compensator can be used to compensate this oscillation by producing an opposing magnetooptical Faraday effect (see Section 4.4.5). The current in the coil of the calibrated Faraday cell required for this purpose is a measure of the rotation and thus for the path difference originating in the crystal.

With the aid of these so-called relative measurements, performed in different directions, the components r_{ijk} in many PSG, in particular in those of higher symmetry, can be completely determined. In triclinic and monoclinic crystals it is, however, also necessary to add so-called absolute measurements, where only the change in the optical path of one of the waves propagating in the direction e'_i is measured separately. High measurement accuracy is not only achieved with a *Jamin interferometer* (Fig. 4.36), where light is transmitted

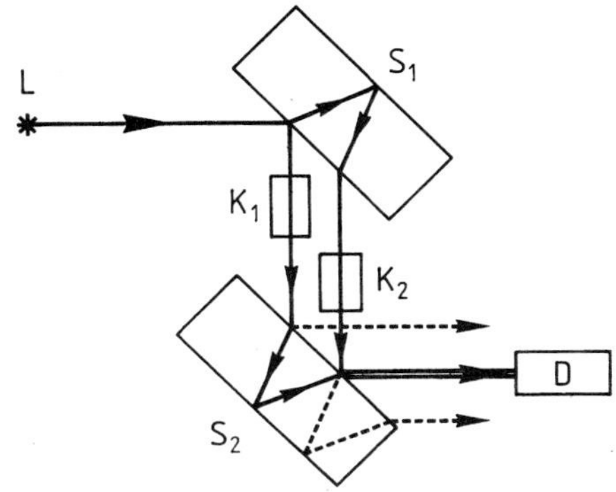

Figure 4.36 Scheme of a Jamin interferometer. L light source, S_1 and S_2 are glass plates with metallized reverse side, K_1 and K_2 are specimen and auxiliary crystal, respectively, D detector.

through the crystal only once, but also with the previously described Michelson interferometer. One now places a crystal with known electrooptical effect in the second beam to enable a compensation measurement. The above formula for ΔG is also true for the absolute measurement when instead of n'_{kk} or n'_{jj} for the refractive index of air ($n \approx 1$) the value 0 is written for the second beam and for r'_{kkl} or r'_{jjl}. The accuracy is increased by orders of magnitude, when one works with an electric alternating field employing a lock-in amplifier, whereby the detector, as described in Section 4.4.1.3, adjusted in a region of largest changes in intensity (between a maximum and an adjacent mimimum of the interference pattern). The difference of the absolute measurements for both directions of vibration $\boldsymbol{e}'_j$ and $\boldsymbol{e}'_k$ must naturally agree with relative measurement in the respective direction. Thus one also has a complete control of the measurements. Except on lithium hydrogen oxalate-hydrate (Ramadan, 1982) no measurements of the complete electrooptical tensor of a triclinc crystal have been reported. The strategy for such a task is not essentially more complicated as with the piezoelectric effect. For a better overview it is appropriate to leave the crystal–physical reference system and select the Cartesian reference system of the semi-axes of the indicatrix instead. In this system, where we specify its basic vectors with $\boldsymbol{e}_i^0$, the above formula derived for the path difference takes on an essentially simpler form, when we also note the electrooptical tensor in the optical reference system. In particular, one can, if the rotation of the indicatrix is negligible, directly determine longitudinal effects and transversal effects with the help of absolute measurements, at least for the optical principal directions. In this way, one can immediately acquire the following nine components: r^0_{iii}, r^0_{iij} $(i, j = 1, 2, 3)$. If, for example, we apply the electric field in the direction $\boldsymbol{e}_2^0$ and irradiate also in the direction $\boldsymbol{e}_2^0$, we obtain the transversal effects r^0_{112} or r^0_{332} for the directions of vibration $\boldsymbol{e}_1^0$ or $\boldsymbol{e}_3^0$. If we transmit in the direction $\boldsymbol{e}_1^0$, we obtain the longitudinal effect r^0_{222} and again the transversal effect r^0_{332} for the directions of vibration $\boldsymbol{e}_2^0$ or $\boldsymbol{e}_3^0$. Transmission in the direction $\boldsymbol{e}_3^0$ again yields r^0_{112} and r^0_{222}. Further measurements on 45°-cuts give the possibility of measuring the remaining nine components. With the help of the transformation $\boldsymbol{e}_i = u_{ij}\boldsymbol{e}_j^0$, which presents the basic vectors of the crystal–physical system in the crystal–optical system, we obtain the components r_{ijk} from the components r^0_{ijk} according to

$$r_{ijk} = u_{ii^*} u_{jj^*} u_{kk^*} r^0_{i^*j^*k^*}.$$

In any case it is necessary to correct for the converse piezoelectrical effect beforehand.

If the sectional ellipse rotates when an electric field is applied, then the formula for the path difference given above must be modified. The new position of the semiaxes of the sectional ellipse, expressed through the angle of rotation φ, results in an additional equation containing the components r_{ijk}. While the

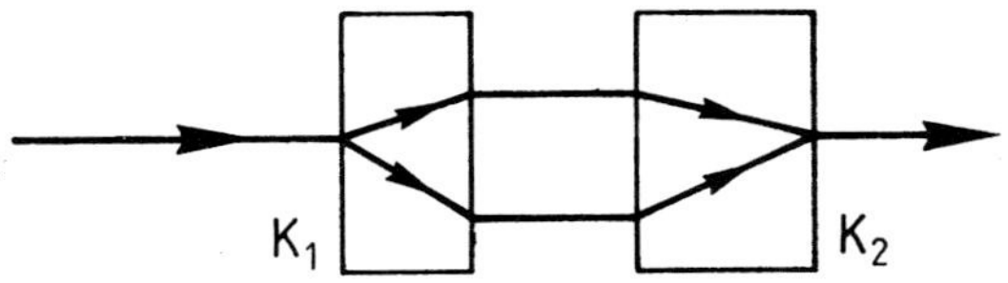

Figure 4.37 Compensation of the ray double refraction of plane-parallel plates. K_1 crystal (specimen), K_2 auxiliary crystal with known ray double refraction.

piezoelectric correction remains, one must enter into the first term the calculated change of the refractive indices for the new directions of vibration. We will return to this type of special cases later.

At this point it is appropriate to make a few comments on the performance of the measurements. Often the crystals possess such a strong double refraction that when making relative measurements the waves are separated in space and cannot interfere. In such a situation one places a second specimen in the ray path to compensate for the double refraction. For the routine measurements of such effects it is useful to prepare a set of plates of known double refraction (ray separation at vertical incidence). For this purpose, it is best to use crystals of high optical quality and good stability of the surface finish, for example, orthorhombic calcium formate. A suitable compensation plate is then selected for each measurement and placed with the correct orientation in the ray path (Fig. 4.37).

A certain problem appears when measuring the effect, if ray direction and electric field are parallel to each other. The specimen must then be equipped with transparent electrodes. Here, optical glass plates with a thin conducting film of SnO_2 have proven good. The plates are prepared, for example, by spraying them with a solution of $SnCl_4$ in HCl and methanol. At about 450°C a conducting transparent film of SnO_2 is formed by oxidation, which is also mechanically very stable. The glass plates are attached with paraffin oil or some other immersion liquid to the finely ground faces of the specimen. In this manner one can almost always avoid polishing the specimen.

In most PSG the electrooptical effects are also superimposed by effects of optical activity. This, however, is only noticeable when a sectional ellipse in the form of a circle or quasi-circle is present, as for example, in cubic crystals of the PSG 23.

We now discuss a few typical examples. First we consider crystals with the subgroup 22 (PSG 22, 42, $\bar{4}2$, 62, 23 and $\bar{4}3$). No more than three independent components exist r_{123}, r_{231} and r_{312}. We have $\Delta a_{12} = r_{123}E_3$, $\Delta a_{23} = r_{231}E_1$ and $\Delta a_{31} = r_{312}E_2$ and $\Delta a_{ii} = 0$. Hence, the sum of the principal refractive indices remains constant because $\sum_i \Delta a_{ii} = 0$. Furthermore, no measurable effect occurs in the direction of the semiaxes of the indicatrix. We must therefore employ 45°-cuts. We apply the electric field parallel to e_1' at the 45°-cut with the edges parallel $e_1' = e_i$, $e_2' = \frac{\sqrt{2}}{2}(e_j + e_k)$, $e_3' = \frac{\sqrt{2}}{2}(-e_j + e_k)$. Using the transformation matrix

$$\begin{pmatrix} 1 & 0 & 0 \\ 0 & \frac{\sqrt{2}}{2} & \frac{\sqrt{2}}{2} \\ 0 & -\frac{\sqrt{2}}{2} & \frac{\sqrt{2}}{2} \end{pmatrix}$$

one finds for $i = 1$, with transmission along e_3', $\Delta a_{11}' = 0$ and $\Delta a_{22}' = \Delta a_{23} = r_{231}E_1$. Moreover, $\hat{d}_{331}' = -\hat{d}_{231}$ and

$$n_{22}'^2 = \frac{2n_{22}^2 n_{33}^2}{n_{22}^2 + n_{33}^2}.$$

One obtains analogous values for $i = 2$ or 3. Hence, for arbitrary i, the change in the path difference is given by

$$\Delta G = \frac{L}{2}(n_{22}')^3 r_{jki} E_i - (n_{ii} - n_{22}') L \hat{d}_{jki} E_i$$

(do not sum over i!).

Thus we directly obtain the components r_{jki} ($i \neq j \neq k \neq i$).

The peculiar situation of a circular sectional ellipse may be explained on crystals of the PSG $\bar{4}2$ as an example. Let the electric field and the ray direction be parallel to the fourfold axis, therefore parallel to e_3. Then $\Delta a_{12} = \Delta a_{21} = r_{123}E_3$. After the field is applied, the sectional ellipse takes the form

$$a_{11}x_1^2 + a_{11}x_2^2 + 2\Delta a_{12}x_1x_2 = 1.$$

The principal axes transformation gives $\tan 2\varphi = \infty$, hence $\varphi = 45°$. The new directions of vibration are

$$e_1' = \frac{\sqrt{2}}{2}(e_1 + e_2) \quad \text{and} \quad e_2' = \frac{\sqrt{2}}{2}(-e_1 + e_2).$$

Consequently $a_{11}' = a_{11} + \Delta a_{12}$ and $a_{22}' = a_{11} - \Delta a_{12}$. The path difference is then

$$\Delta G = -Ln_{11}^3 r_{123} E_3 = n_{11}^3 r_{123} U.$$

The electrostrictive part vanishes since $\hat{d}_{333} = 0$. In this case, with fixed electrical voltage U, ΔG does not depend on the length L of the transmission path because $E_3 = -U/L$. If the light passes through the same specimen in the direction e_1' and the electric field persists parallel e_3, a change

$$\Delta G = -\frac{L}{2}n_{11}^3 r_{123} E_3 + (n_{11} - n_{33}) L \hat{d}_{123} E_3$$

in the path difference occurs, whereby the electrooptical part, at the same field strength, is only half as large as in the case of transmission in the direction of the electric field.

We now want to get an overview of the obtainable electrooptical effects in different transmission and field directions in cubic crystals. We search for the directions of extreme changes of refractive indices in a given field direction. We proceed in the same manner as discussed in Section 4.3.2 (principal axes transformation). The auxiliary function

$$H=(a_{11}(x_1^2+x_2^2+x_3^2)+2r_{123}x_1x_2E_3+2r_{231}x_2x_3E_1+2r_{312}x_3x_1E_2-1)-\lambda\sum_i x_i^2$$

also attains an extreme value, when the length of the radius vector of the indicatrix, which is equal to the associated refractive index, becomes extreme. For Δa_{ij} we write the quantity $r_{ijk}E_k$, respectively. Extreme values appear, when $\partial H/\partial x_i = 0$ for $i = 1,2,3$, thus, for example,

$$\begin{aligned}\frac{\partial H}{\partial x_1} &= (a_{11}-\lambda)2x_1 + 2rx_2E_3 + 2rx_3E_2 &= 0\\ \frac{\partial H}{\partial x_2} &= (a_{11}-\lambda)2x_2 + 2rx_1E_3 + 2rx_3E_1 &= 0\end{aligned}$$

with $r_{123} = r_{231} = r_{312} = r$. Multiplying the first equation by x_2, the second by x_1, and subtracting gives $E_3(x_2^2 - x_1^2) = x_3(x_1E_1 - x_2E_2)$. The corresponding expressions are obtained by forming the difference

$$\left(x_i\frac{\partial H}{\partial x_j} - x_j\frac{\partial H}{\partial x_i}\right) = 0.$$

In general, it is true that $E_k(x_i^2 - x_j^2) = -x_k(x_iE_i - x_jE_j)$, where i, j, k are cyclic in $1,2,3$.

We now consider some concrete examples. First let $\boldsymbol{E}$ be parallel to $[110]$, thus

$$\boldsymbol{E} = \frac{\sqrt{2}|\boldsymbol{E}|}{2}(\boldsymbol{e}_1 + \boldsymbol{e}_2).$$

This results from the above equation with $E_3 = 0$: $x_3 = 0$ or $x_1 = x_2$ (because $E_1 = E_2$). With $x_3 = 0$ one gets $x_1^2 = -x_1x_2$ and $x_2^2 = -x_1x_2$, hence $x_2 = -x_1$ and $x_1 = 0$ or $x_2 = 0$.

Therefore, we have found three extreme directions

$$\boldsymbol{e}_1' = \frac{\sqrt{2}}{2}(\boldsymbol{e}_1 - \boldsymbol{e}_2) \parallel [1\bar{1}0], \quad \boldsymbol{e}_1'' = \boldsymbol{e}_2 \quad \text{and} \quad \boldsymbol{e}_1''' = \boldsymbol{e}_1.$$

The refractive indices do not change in these directions ($\Delta a_{11} = \Delta a_{22} = \Delta a_{12} = 0$ for $E_1 = E_2$ and $E_3 = 0$!).

The other condition $x_1 = x_2$ gives $E_1(x_1^2 - x_3^2) = -x_1^2 E_1$ with $E_1 = E_2$, hence $x_3 = \pm\sqrt{2}x_1$, and thus

$$e_2' = \frac{1}{2}(e_1 + e_2 - \sqrt{2}e_3),$$
$$e_3' = \frac{1}{2}(e_1 + e_2 + \sqrt{2}e_3).$$

Both these directions are perpendicular to e_1'. The associated principal refractive indices are calculated from the corresponding eigenvalues λ_i. These results, for example, from $\partial H/\partial x_i = 0$

$$\begin{aligned}\lambda_1 &= a_{11} \rightarrow \Delta a_{11}' = 0;\\ \lambda_2 &= a_{11} - r_{123}\sqrt{2}E_1 \rightarrow \Delta a_{22}' = -r_{123}\sqrt{2}E_1;\\ \lambda_3 &= a_{11} + r_{123}\sqrt{2}E_1 \rightarrow \Delta a_{33}' = r_{123}\sqrt{2}E_1.\end{aligned}$$

Transmitting light perpendicular to the electric field in the direction $e_1' \parallel [1\bar{1}0]$ produces a path difference $\Delta G = Ln^3 r_{123}\sqrt{2}E_1 = Ln^3 r_{123}|E|\,\mathrm{sign}\,E$. The electrostrictive part vanishes because $n_{11} = n_{22} = n_{33}$.

In cubic crystals the direction of the maximal longitudinal effect coincides with a principal axis direction of the indicatrix modified by the electric field. By analogy with the method discussed pveviously for the piezoelectric effect (see Section 4.4.1.2), we form the auxiliary function $H = r_{111}' - \lambda \sum_i x_i^2$ for the effect $\Delta a_{11}' = r_{111}' E_1'$ and calculate its extreme values. As before, for $e_1' = u_{1i} e_i$ we obtain the directions of the space diagonals of the cube, hence

$$u_{1i} = \pm\frac{1}{\sqrt{3}}.$$

With this we find

$$\Delta a_{11}' = r_{111}' E_1' = 6 r_{123}\left(\frac{1}{\sqrt{3}}\right)^3 |E|\,\mathrm{sign}\,E.$$

In the plane perpendicular to e_1', thus perpendicular to a threefold axis, the indicatrix forms a circular section. We have $\Delta a_{22}' = \Delta a_{33}' = -\Delta a_{11}'/2$ because $\sum_i \Delta a_{ii} = 0$. Hence, a transmission in all directions perpendicular to the electric field (E along the threefold axis) produces a path difference

$$\Delta G = -\frac{\sqrt{3}Ln^3}{2} r_{123}|E|\,\mathrm{sign}\,E \quad \text{with} \quad \Delta a_{11}' - \Delta a_{22}' = \frac{3}{2}\Delta a_{11}'.$$

Table 4.8 presents the most important arrangements of field and transmission directions for cubic crystals of the PSG $\bar{4}3$.

Crystals of the PSG 23 usually exhibit such a strong optical activity that the formulae derived in Section 4.3.6.7 for the interplay of optical activity and

Table 4.8 Electro-optical effect in distinct directions of cubic crystals; g propagation vector, e_2' and e_3' directions of vibration, L length of transmission.

$\mathbf{E} \parallel$	$g \parallel e_1' \parallel$	$e_2' \parallel$	$e_3' \parallel$	ΔG
[001]	[001]	[110]	$[\bar{1}10]$	$-Ln^3 r_{123}\lvert\mathbf{E}\rvert \operatorname{sign} \mathbf{E}$
[001]	$[\bar{1}10]$	[001]	[110]	$\frac{L}{2} n^3 r_{123}\lvert\mathbf{E}\rvert \operatorname{sign} \mathbf{E}$
[110]	$[1\bar{1}0]$	$[11-\sqrt{2}]$	$[11\sqrt{2}]$	$Ln^3 r_{123}\lvert\mathbf{E}\rvert \operatorname{sign} \mathbf{E}$
[111]	$[1\bar{1}0]$	[111]	$[11\bar{2}]$	$-\frac{\sqrt{3}L}{2} n^3 r_{123}\lvert\mathbf{E}\rvert \operatorname{sign} \mathbf{E}$

double refraction (here induced double refraction) must be used. For a first approximation one assumes that the dependence of the optical activity on the electric field strength is negligible. Under this presupposition the tensor components r_{123} of optically active crystals can be measured with methods similar to those outlined above.

The practical application of the electrooptical effect is limited mainly to the modulation of light. All other modulation methods cannot compete with the almost inertia-less control of the electrooptical modulator. This is especially true for cubic crystals because the electrostrictive part is missing. In this context, it is very convenient that the modulation, to a sufficient approximation, follows the frequency of the applied electric field. The specimen is located in diagonal position between crossed polarizers (see Section 4.3.6.3; $\varphi = 45°$). If one neglects reflection and absorption losses, the transmitted intensity is given by

$$I' = I_0 \sin^2 \pi d = I_0(1 - \cos 2\pi d)/2,$$

where $d = L(n'' - n')/\lambda_0 = G/\lambda_0$ is the measured path difference in units of the vacuum wavelength. Changing the path difference, for example, via the electrooptical effect, gives

$$I'' = I_0 \sin^2\{\pi(G + \Delta G)/\lambda_0\} = I_0(1 - \cos\{2\pi(G + \Delta G)/\lambda_0\})/2.$$

Thus we observe a modulated part

$$I'' - \frac{I_0}{2} = \Delta I = \frac{1}{2} I_0 \left\{ \sin \frac{2\pi}{\lambda_0} G \sin \frac{2\pi}{\lambda_0} \Delta G - \cos \frac{2\pi}{\lambda_0} G \cos \frac{2\pi}{\lambda_0} \Delta G \right\}.$$

If L is adjusted so that $L(n'' - n')$ is an odd multiple of a quarter of the vacuum wavelength, hence $G = (2m + 1)\lambda_0/4$, for m integer, we obtain

$$\Delta I = \pm \frac{I_0}{2} \sin \frac{2\pi}{\lambda_0} \Delta G.$$

For sufficiently small ΔG a quasi-sinusoidal modulation arises, when the electric field varies sinusoidally. This is approximately correct for very small ΔG

even with an arbitrary primary path difference G. In cubic crystals, because $G = 0$, the following is true:

$$\Delta I = -\frac{I_0}{2}\cos 2\pi\frac{\Delta G}{\lambda_0}.$$

A measure for the efficiency of the electrooptical effect is the half-wave voltage U_0 required to produce a path difference change of $\lambda_0/2$ and thus a maximum of the intensity I'' for the case $G = m\lambda_0$, m integer. The dimensions of the specimen must be given, for example, transmission length and thickness in the field direction both 10 mm. In the devices, half-wave voltages as small as possible are preferred. At present, only a few crystals are known exhibiting a half-wave voltage less than 100 V as, for example, $K(Nb, Ta)O_3$.

4.4.3 First-Order Nonlinear Electrical Conductivity (Deviation from Ohm's Law)

The dependence of current density on the electric field is for most substances linear up to very high field strengths. The possible deviations are described by a Taylor series of the type

$$I_i = s_{ij}E_j + s_{ijk}E_jE_k + s_{ijkl}E_jE_kE_l + \cdots.$$

The tensor $\{s_{ijk}\}$ contains the *first-order nonlinear conductivity*. Its behavior is completely analogous to the piezoelectric tensor $\{d_{ijk}\}$. The measurement is performed with the help of the longitudinal effect as with the measurement of Ohmic conductivity. If one applies an alternating electric field $E_1' = E_0 \sin\omega t$ to a plane-parallel plate with the normal $\boldsymbol{e}_1'$, one observes a current density

$$I_1' = s_{11}'E_1' + s_{111}'E_1'^2 = s_{11}'E_0\sin\omega t + s_{111}'E_0^2(1-\cos 2\omega t)/2$$

(with $\sin^2\omega t = (1-\cos 2\omega t)/2$). Hence there appears a direct current component $s_{111}'E_0^2/2$ and a frequency-doubled component $s_{111}'E_0^2(\cos 2\omega t)/2$, both of which can be measured with sensitive instruments. The frequency-doubled component can best be detected using a lock-in amplifier with compensation. Consequently, the heat produced in the specimen by the high field and current strengths and therefore the associated change in Ohmic conductivity have no influence on the measurement. The quadratic dependence of the effect on the field strength allows a reliable control of the measurement.

Centrosymmetrc crystals (almost all metals) and crystals of the PSG 43 do not possess a first-order nonlinear effect. Therefore, one can test, with the help of nonpiezoelectric cubic crystals, whether the primary voltage is free of components of the first harmonic. Here, the deviation from Ohm's law begins with the second-order effect (third power of the field strength), whereby, in the case of an alternating field, components with the triple frequency appear in the current density. In semiconductors, conduction to a large extent is mainly caused by defects. The symmetry properties of these defects do not always coincide with the space group symmetry or the site symmetry in the undisturbed lattice. In such cases, first-order effects can also be observed in centrosymmetric crystals.

4.4.4
Nonlinear Dielectric Susceptibilty

The dielectrical and optical properties discussed in Sections 4.3.3 and 4.3.6 are based on a linear relation between the electric displacement $\boldsymbol{D}$ and the electric field strength $\boldsymbol{E}$ according to $D_i = \epsilon_{ij}E_j$. When discussing higher order effects, this relationship must be supplemented by adding higher order terms, hence

$$\begin{aligned} D_i = {} & \epsilon_{ij}E_j + \epsilon_{ijk}E_jE_k + \epsilon_{ijkl}E_jE_kE_l + \cdots \\ & + g_{ijk}\frac{\partial E_j}{\partial x_k} + g_{ijkl}\frac{\partial^2 E_j}{\partial x_k \partial x_l} + \cdots \\ & + f_{ijkl}E_j\frac{\partial E_k}{\partial x_l} + \cdots . \end{aligned}$$

The top row corresponds to the normal Taylor series. With the first term in the second row we were able to describe in Section 4.3.6.7 the optical activity. Until recently, the prevailing opinion was that higher order dielectric effects were practically undetectable. In the meantime we know quite a number of higher order phenomena which are simple to realize experimentally. With the developement of coherent light sources (laser) it became possible to produce effects of frequency multiplication and frequency mixing in the field of optics. At low frequencies, hence in the usual dielectric range, one still requires special experimental techniques to detect higher order effects. In this section we will address the optical effects. Instead of the hitherto employed material equation $D_i = \epsilon_{ij}E_j$ we must use the equation above. Here we will consider only one more step in the series expansion in each case, for example, $\epsilon_{ijk}E_jE_k$ or $\epsilon_{ijkl}E_jE_kE_l$. The simplest and for most problems adequate model for nonlinear effects consists in allowing the existence of one or more plane waves propagating in a crystal according to the usual laws of crystal optics, hence without loss of energy. The directions of vibration of the $\boldsymbol{D}$-vector associated with each direction of propagation are fixed as well as the associated electric fields. If a nonlinear interaction of the components of the electric fields produces additional electric displacements, i.e., polarizations, then these generate secondary waves when migrating through the crystal, which differ in direction of vibration, direction of propagation, and frequency from the primary waves. If we let a linear polarized wave strike a crystal plate, two linear polarized waves

$\boldsymbol{D}'$ and $\boldsymbol{D}''$ propagate in the crystal with their propagation vectors running parallel only in the case of perpendicular incidence or optical isotropy. The associated $\boldsymbol{E}$-vectors $\boldsymbol{E}' = \boldsymbol{E}'_0 e^{2\pi i(\boldsymbol{k}'\cdot\boldsymbol{x}-\nu t)}$ and $\boldsymbol{E}'' = \boldsymbol{E}''_0 e^{2\pi i(\boldsymbol{k}''\cdot\boldsymbol{x}-\nu t)}$, cause in the volume elements, through which they pass, a nonlinear polarization $\boldsymbol{D}^{\mathrm{NL}}$:

$$D_i^{\mathrm{NL}} = \epsilon_{ijk}(E'_j E'_k + E'_j E''_k + E''_j E'_k + E''_j E''_k) \\ + \epsilon_{ijkl}(E'_j E'_k E'_l + E'_j E'_k E''_l + \cdots E''_j E''_k E''_l).$$

The quadratic part of the first term consists of the partial waves

$$D_i^{\mathrm{NL}}(2\nu) = \epsilon_{ijk} E'_{j0} E'_{k0} e^{2\pi i(2\boldsymbol{k}'_0\cdot\boldsymbol{x}-2\nu t)} \\ + \epsilon_{ijk}(E'_{j0} E''_{k0} + E''_{j0} E'_{k0}) e^{2\pi i((\boldsymbol{k}'_0+\boldsymbol{k}''_0)\cdot\boldsymbol{x}-2\nu t)} \\ + \epsilon_{ijk} E''_{j0} E''_{k0} e^{2\pi i(2\boldsymbol{k}''_0\cdot\boldsymbol{x}-2\nu t)}.$$

Here we assume that $\{\epsilon_{ijk}\}$ is virtually independent of frequency.

These three waves are firmly coupled with the fundamental wave. We call them bound waves. They possess twice the frequency of the fundamental wave. We therefore speak of the *frequency doubling* effect (or SHG: *second harmonic generation*). The generation of the first and third partial wave, each of them emanating from only one fundamental wave, is refered to as a *type-I-process*, the generation of the "mixed wave" as a *type-II-process*. Only those components of $\boldsymbol{D}^{\mathrm{NL}}(2\nu)$ compatible with the indicatrix for the double frequency and compatible with the associated directions of vibration allow the propagation of the so-called free waves. The intensity of these frequency-doubled waves is calculated from the Maxwell equations taking into consideration the previously discussed boundary conditions in Section 4.3.6.1. An important situation for the frequency doubling of laser radiation is realized, when the fundamental wave and the frequency-doubled free wave (harmonic) possess the same propagation velocity with respect to direction and magnitude. In this case the fundamental waves, on passing through the crystal, feed their frequency-doubled components in-phase to the harmonic. The intensity of the harmonic, as opposed to an arbitrary situation, increases by many orders of magnitude due to constructive interference and is thus made easily accessible to visual observation. This special situation is called *phase matching*. The conditions for type-I-processes are $n'(\nu) = n(2\nu)$ or $n''(\nu) = n(2\nu)$ and for type-II-processes $n'(\nu) + n''(\nu) = 2n(2\nu)$. Possible directions of phase matching are obtained from the intercept of the index surfaces for both frequencies ν and 2ν. A vector leading from the center of the index surfaces to an intercept point shows a possible phase matching direction. Under normal dispersion ($\partial n/\partial\nu > 0$), the associated directions of vibration are different. Only a specific anisotropy of refractive indices allows an intercept of the index surfaces at all. In cubic crystals no normal phase matching occurs due to the missing anisotropy. Nevertheless, phase matching in a very restricted frequency range is conceivable in the case of anomalous dispersion ($\partial n/\partial \nu < 0$). In optical uniaxial crystals, the directions of vibration of the fundamental and of the free second harmonic waves are mutually perpendicular in the case of phase matching. Biaxial crystals offer a higher probability for the existence of phase matching directions due to the wider variation of anisotropy. According to Hobden (1967) one can distinguish between a total of 13 different types of phase matching (Hobden classes).

Since the anisotropy as well as the dispersion of the refractive indices are very small, a very good knowledge of the refractive indices or vacuum wavelengths is required to determine the directions of phase matching. We will return to the calculation of phase matching directions at the end of this section.

The effects of frequency doubling, as far as they are based on the existence of the tensor $\{\epsilon_{ijk}\}$, only appear in piezoelectric crystals, because the tensors $\{\epsilon_{ijk}\}$ and $\{d_{ijk}\}$ belong to the same symmetry type (see Table 4.7). Similarly with the piezoelectric effect, there also exists a powder test, with which one can determine to a high degree of probability, whether a crystal species possesses an inversion center or not (here an exception is also the PSG 43 with vanishing third-rank tensors). This so-called SHG test is carried out as follows: the specimen in the form of a powder is applied as a thin layer on a polished glass plate. If the preparation is irradiated with a strong laser beam, one can expect that several crystal grains always are oriented in a favorable position for frequency doubling. The light emerging from the preparation, from which the primary radiation was removed by suitable filters, is analyzed with the help of photon counting techniques, if necessary, using energy-discrimination methods. Decisive is, whether the number of measured photons with twice the energy (frequency) surmounts the background value. On comparing with specimens, which show no effect, for example corundum powder, this statement is in almost all cases investigated so far unambiguously answered. The method can be made substantially more sensitive than required for routine investigations by employing pulse techniques coupled with a synchronously running counter.

A number of methods are available to measure the tensor $\{\epsilon_{ijk}\}$ of which in particular the *wedge method* (Jerphagnon, 1968) and the method of *Maker interferences* deserve special attention (Maker et al., 1962). In the wedge method, the intensity of the second harmonic wave at perpendicular incidence is measured on a slightly wedge-shaped crystal (Fig. 4.38). If one neglects the effect of double refraction and absorption losses, one expects an intensity of the second harmonic varying periodically with the transmission length through the crystal. Up to the first maximum, only the constructive components add to

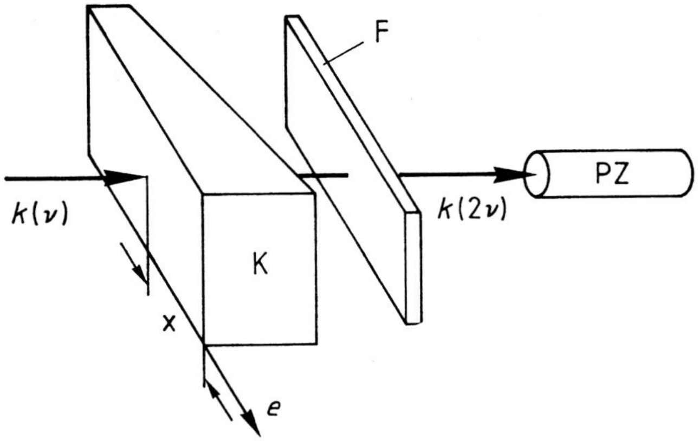

Figure 4.38 Wedge method. K crystal (specimen), F filter for suppressing the primary radiation, PZ photon counter. The intensity of the frequency-doubled radiation (second harmonic) varies periodically with the shift x of the wedge-shaped specimen along the direction $\boldsymbol{e}$.

the harmonic. The associated transmission thickness

$$L_c = \frac{\lambda_0}{4(n(2\nu) - n(\nu))}$$

is termed the coherence length for the given direction of propagation, whereby λ_0 refers to the vacuum wavelength of the fundamental wave. With increasing thickness further contributions of the generating wave become more and more out of phase with the harmonic and finally effect complete destruction of the harmonic. After this minimum a maximum of the second harmonic again builds up and so on. Thus if one shifts the specimen perpendicular to the finely bundled fundamental wave, one observes a periodic change in the intensity. The intensity of the maxima is proportional to the square of the product of coherence length, effective nonlinear coefficient ϵ_{eff} and intensity of the fundamental wave. ϵ_{eff} is calculated as a function of the components ϵ_{ijk} for the given direction of vibration of the fundamental wave by means of tensor transformation. The calibration of the intensity of the harmonic can be performed with the aid of a standard specimen, for example made of KH_2PO_4. By varying the transmission direction and the direction of vibration of the fundamental wave, it is possible, in principle, to determine all tensor components. The problem of the relative sign of tensor components is solvable by measuring those effective nonlinear coefficients, which contain sums or differences of the components ϵ_{ijk}.

In the Maker interference method, the intensity of the harmonic is investigated on a thin plane-parallel plate as a function of the angle of incidence

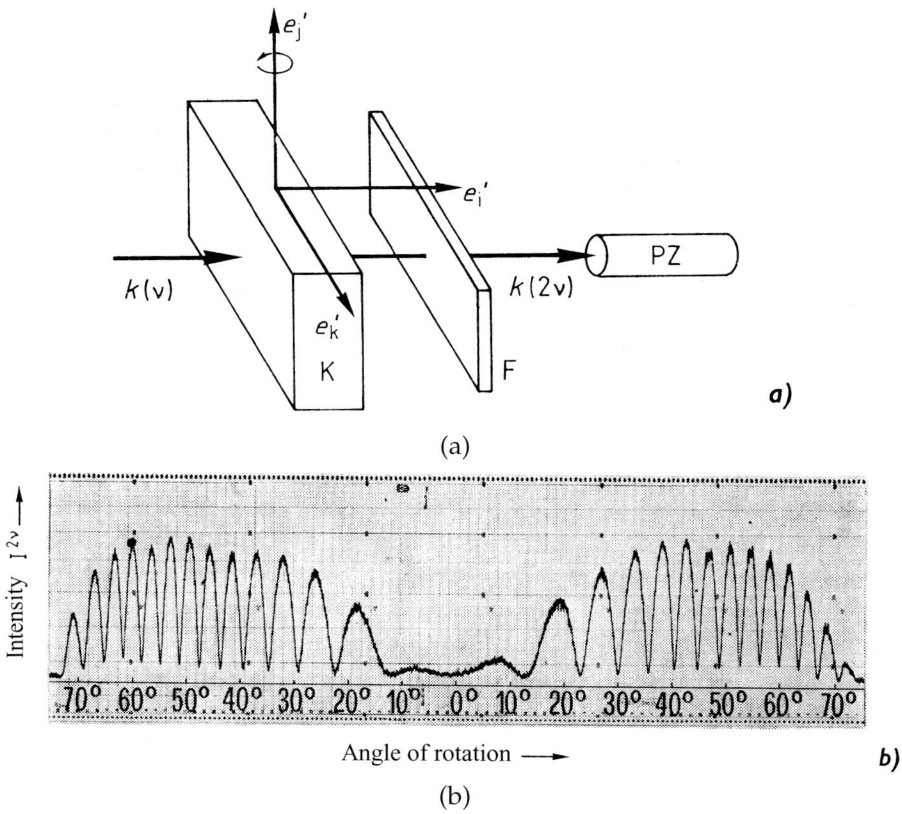

Figure 4.39 (a) Method of Maker interferences. K plane-parallel specimen, rotatable around the axis e_j', F filter for suppressing the primary radiation, PZ photon counter. (b) Maker interferences obtained on a thin (010)-plate of $LiHC_2O_4 \cdot H_2O$. Axis of rotation and direction of vibration of the fundamental wave parallel to the shortest semi-axis of the indicatrix, direction of vibration of the frequency-doubled wave (second harmonic) perpendicular to the axis of rotation.

and the direction of vibration of the fundamental wave (Fig. 4.39). Maxima appear also here, when an odd multiple of the coherence length is realized in the given direction. In contrast to the wedge method, the effective nonlinear coefficient changes from maximum to maximum. Furthermore, differences in reflection losses appear as a function of the angle of incidence, which must be taken into consideration. As before, a standard specimen is used for the calibration of the intensity. From the envelope of the maxima, which can even be determined with good accuracy on specimens of moderate optical quality, one can extract the effective nonlinear coefficient for the direction of perpendicular incidence. Important is that in the case of negligible double refraction the envelope is independent of the specimen thickness. An evaluation of the

Table 4.9 Arrangement for the measurement of non-linear optical effects on thin plates.

No.	Plate normal and vector g	$D(\nu)$ (fundamental wave)	$D(2\nu)$ (2nd harmonic)	Eff. coeff. for perp. incidence	Type
1	e_i^0	e_j^0	e_j^0	ϵ_{jjj}	I
2	e_i^0	e_j^0	e_k^0	ϵ_{kjj}	I
3	e_i^0	$\frac{\sqrt{2}}{2}(e_j^0 \pm e_k^0)^*$	e_j^0	$\epsilon_{jjj}, \epsilon_{jkk}, \epsilon_{jjk}$	II**
4	$\frac{\sqrt{2}}{2}(e_i^0 + e_j^0)$	$\frac{\sqrt{2}}{2}(-e_i^0 + e_j^0)$	e_k^0	$\frac{1}{2}(\epsilon_{kii} - 2\epsilon_{kij} + \epsilon_{kjj})$	I

* before entering the specimen
**) the effective coefficient is a function of ϵ_{jjj}, ϵ_{jkk}, ϵ_{jjk}, n_{jj} and n_{kk}.

complete range of the envelope curve and also of the whole interference curve with the help of computer-supported matching of the intensity to the observations can lead to a substantial improvement in accuracy. The progression of the intensity is calculated by variation of the components ϵ_{ijk}.

A derivation of formulae for the evaluation of Maker interferences on optically uniaxial crystals was given by Jerphagon & Kurtz (1970). The corresponding formulae for biaxial crystals were derived by Bechthold (1977). In the meantime, the latter have also been used for the determination of the complete tensor $\{\epsilon_{ijk}\}$ on triclinic lithium hydrogen oxalate-monohydrate.

A substantial simplification of the measurements results, when one assumes the validity of the so-called *Kleinman rule* (1962). This rule states that nonabsorbing dispersion-free crystals possess a totally symmetric tensor $\{\epsilon_{ijk}\}$. Experimental observations confirm, in fact, that only small deviations from total symmetry appear. For crystals of the PSG 42 and 62, the Kleinman rule would have the consequence that absolutely no frequency doubling effect would be observed. On the few examples of these PSG investigated so far, weak effects of frequency doubling could be observed, so that one cannot assume a severe validity of the Kleinman rule.

Measurements with the wedge method and with the method of Maker interferences are preferentially carried out on plates cut parallel to the principal planes of the indicatrix.

The following arrangements have proven useful and are also suitable for the complete determination of the tensor of triclinic crystals (see Table 4.9). Let the principal axes of the indicatrix be specified by e_i^0. Except in triclinic and monoclinic systems, the optical principal axes coincide anyway with the basic vectors e_i of the crystal–physical system. The indices i, j, k should be different.

Variation of the indices results in a total of 21 different arrangements already at perpendicular incidence. To record Maker interferences one turns the plate, during transmision, about the direction of $D(\nu)$ or $D(2\nu)$ taken as the axis of rotation. From the progression of the envelope of the Maker interferences one can obtain information on the relative sign of the components ϵ_{ijk}.

In order not to jeopardize the validity of the simple model, the plate thickness should not be too large so that a noticeable spacial separation of the volume, through which the fundamental and the harmonic pass, does not occur as a consequence of ray double refraction.

From the above table we recognize immediately that all crystals with 22 as a subgroup, hence also those of PSG $\bar{4}2$ and the cubic crystals of PSG 23 and $\bar{4}3$, show no effect in the arrangements 1 to 3. In these cases we work with arrangement 4, whereby the orientation of the 45°-cuts, in the case of the cubic crystals, is referred to the crystal–physical system.

There exists another way of determining the sign, which, however, yet has not been used routinely. Namely, one obtains, apart from the frequency-doubled radiation via nonlinear processes, also a contribution of constant polarization. Let us consider, for example, a type-I-process of the kind $D_1 = \epsilon_{12}E_2 + \epsilon_{122}E_2^2$ with $\boldsymbol{E} = E_2\boldsymbol{e}_2 = E_{20}\cos(2\pi(\boldsymbol{k}\cdot\boldsymbol{x} - \nu t))\boldsymbol{e}_2$. Because $\cos^2 u = (1+\cos 2u)/2$ we obtain a constant contribution $D_{10} = E_{20}^2\epsilon_{122}/2$, which is measurable in the case of sufficiently high field strength. In the example discussed the specimen has to be equipped with a conducting pair of faces perpendicular to $\boldsymbol{e}_1$, on which one observes a charge density similar to that detected in pyroelectric or piezoelectric effects. The sign of ϵ_{122} corresponds directly to the sign of the measured polarization.

Apart from collinear processes discussed here, where the fundamental and the harmonic show approximately the same propagation direction, the possibility exists of generating a new radiation, when two laser beams meet at a finite angle. In the volume covered by both waves, nonlinear frequency mixing effects occur, where waves appear with sum and difference frequencies of both fundamental waves. Even with collinear superposition of two parallel laser beams of different frequencies, such sum and difference frequencies are observable, as one can immediately derive the general relationship between $\boldsymbol{D}$ and $\boldsymbol{E}$.

Furthermore, under certain conditions, the so-called threshold processes can produce rays with half frequencies and arbitrary frequencies within a certain frequency range. These first appear, just as with starting a laser, from a definite energy density onwards. The frequency of the generated rays depends on the direction of the fundamental in the crystal and can be varied, to an extent, by rotating the crystal (parametric oscillator; see Bloembergen, 1965).

Finally, we discuss some questions concerning the practical application of frequency doubling under phase matching conditions. In optical uniaxial crystals the phase matching angle can be easily calculated from the index surfaces, because they are rotationally symmetric for both frequencies. The

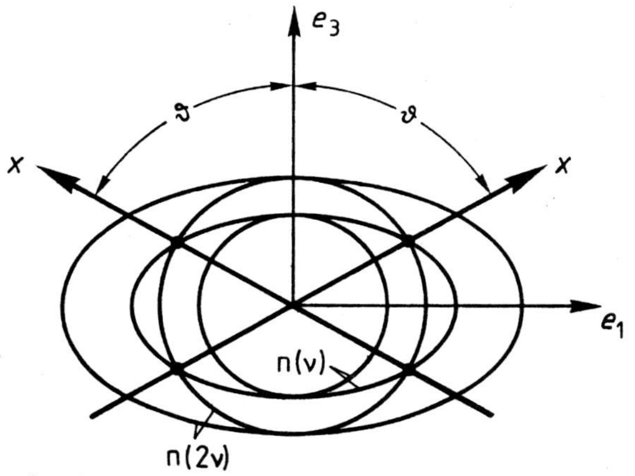

Figure 4.40 Construction of the direction of phase matching of fundamental wave with frequency ν and second harmonic with frequency 2ν for the case of an optically positive uniaxial crystal ($n_1 = n_2 < n_3$). The ordinary wave of the doubled frequency (harmonic), which is represented in the index-surface by a sphere with radius $n(2\nu)$ takes on the same propagation velocity as the extraordinary fundamental wave along all directions within a cone around the optical axis parallel e_3.

equation of the index surface is (see Section 4.3.6.1)

$$\sum_i x_i^2 = n_1^2 \quad \text{and} \quad (x_1^2 + x_2^2)/n_3^2 + x_3^2/n_1^2 = 1.$$

Due to rotational symmetry it suffices to consider a principal cut with $x_2 = 0$. In the case of normal dispersion ($n(2\nu) > n(\nu)$) in optical positive crystals ($n_3 > n_1$), the refractive index of ordinary waves for the double frequency ($n_1(2\nu)$) can coincide with the extraordinary wave of the fundamental frequency $n_e(\nu)$ (Section 4.40). For optically negative crystals ($n_3 < n_1$) the reverse is true.

In the case of $n_3 > n_1$, $x_1^2 + x_3^2 = n_1^2(2\nu)$ and $x_1^2/n_3^2(\nu) + x_3^2/n_1^2(\nu) = 1$ intersect. ϑ is the angle between the phase matching direction and the optic axis. With $x_1 = n_1(2\nu)\sin\vartheta$ and $x_3 = n_1(2\nu)\cos\vartheta$ one finds

$$\sin^2\vartheta = \frac{1/n_1^2(\nu) - 1/n_1^2(2\nu)}{1/n_1^2(\nu) - 1/n_3^2(\nu)}.$$

Correspondingly, for $n_3 < n_1$ we get

$$\sin^2\vartheta = \frac{1/n_1^2(\nu) - 1/n_1^2(2\nu)}{1/n_3^2(2\nu) - 1/n_1^2(2\nu)}.$$

Normally, the range of the angle $\Delta\vartheta$ of the cone about the optic axis, in which phase matching is observable, is extremely small, namely of the order of a few arc minutes. In rare cases, the index surfaces intersect just about tangentially, i.e., the refractive indices differ very little over a large angular range. Such a situation is called noncritical phase matching, where $\vartheta \approx 90^\circ$. This means that for all partial waves of a divergent laser beam, there exists large coherence lengths and thus a high rate of conversion of the energy of the fundamental into the frequency-doubled harmonic.

The phase matching directions for type-II-processes are calculated in an analogous way.

To judge whether a crystal is suitable as a technical frequency doubler, the following criteria should be considered:

1. good transparency in the frequency range between ν and 2ν,
2. existence of quasinoncritical phase matching directions,
3. large effective nonlinear coefficients,
4. high irradiation strength (since the efficiency of frequency conversion grows in proportion to the square of the intensity of the fundamental wave, it is advantageous to work with very high laser power!) and
5. capability to manufacture crystals of sufficient size and optical quality.

At present, irradiation strength and manufacturing capability can hardly be predicted. Crystals with high mechanical strength as, for example, $LiNbO_3$ tend to form optical defects under weak irradiation. Others, like lithium formate-hydrate are mechanically weak in comparison, however, show a good irradiation strength. A certain role is surely played by the regeneration capability of the crystal and the existence of certain primary defects, in particular chemical impurities and micro-inclusions in the crystal.

Further effects of nonlinear optical properties as, for example, the effects of frequency tripling and frequency quadrupling, brought about by the tensors $\{\epsilon_{ijkl}\}$ and $\{\epsilon_{ijklm}\}$ will only be mentioned here. Since $\{\epsilon_{ijkl}\}$ is a fourth-rank tensor, it exists in all materials. Therefore, with sufficiently high intensity of the coherent fundamental wave, one can observe in all materials a radiation of triple frequency arising from the product $E_j E_k E_l$. While frequency doubling effects are observable in phase matching directions of crystals already at very weak laser powers of the fundamental wave in an amount of a fraction of a milliwatt, one requires for the generation of frequency tripling and higher effects far higher powers, even with phase matching. In other words, the search for crystals with high irradiation strength as well as strong effects of this kind is still in progress. Fields of application are, for example, laser systems based on the generation of stimulated Raman scattering (see e.g., Kaminskii et al., 2000).

4.4.5 Faraday Effect

If a cubic crystal (electrically nonconducting) or some other optically isotropic medium is placed in a magnetic field, one observes a rotation of the direction of vibration of a linear polarized wave similar as in optical activity. However, while in optical activity the rotation per mm optical path is the same in all directions, one finds here a change in the sense of rotation, when the propagation direction is reversed. This magnetooptical effect, named after its discoverer *Faraday effect*, is proportional to the magnetic field strength. It emerges that the influence of the magnetic field on the propagation of light also appears in anisotropic crystals. To describe the phenomena, we demand, just as with optical activity, that no additional energy terms arise, i.e., we attach an additional term to the electric displacement that is proportional to the magnetic field $\boldsymbol{H}$ and satisfies the condition $q''_{ij} = -q''_{ji}$

$$D_i = \epsilon_{ij}E_j + z_{ijk}E_jH_k = (\epsilon_{ij} + z_{ijk}H_k)E_j.$$

The analogous relationship in optical activity is

$$D_i = \epsilon_{ij}E_j + g_{ijk}\frac{\partial E_j}{\partial x_k} = \epsilon_{ij}E_j + 2\pi i g_{ijk}k_kE_j,$$

for the case that $\boldsymbol{E}$ represents a plane wave $\boldsymbol{E} = \boldsymbol{E}_0 e^{2\pi i(\boldsymbol{k}\cdot\boldsymbol{x}-\nu t}$ (see Section 4.3.6.7).

Corresponding to the condition $g_{ijk} = -g_{jik}$ for optical activity, here we must also demand that z_{ijk} are antisymmetric in the first two indices and furthermore, are purely imaginary. If we substitute z_{ijk} by $i\zeta_{lk}$ or $-i\zeta_{lk}$ with $l \neq i, j$ and i, j, l cyclic or anticyclic in 1,2,3 and introduce the vector $\boldsymbol{Z}$ with the components $Z_l = \zeta_{lk}H_k$, we obtain

$$D_i = \epsilon_{ij}E_j - i(\boldsymbol{Z} \times \boldsymbol{E})_i.$$

This form corresponds to the description of optical activity. Now, with respect to tensor properties, $\{z_{ijk}\}$ behaves as a third-rank axial tensor in crystals with rotation–inversion axes. This is different to the polar tensor $\{g_{ijk}\}$ which we used in the description of optical activity. The tensor $\{z_{ijk}\}$ makes no distinction between rotation axes and rotation–inversion axes. It exists in all point symmetry groups. The experimentally found difference to optical activity also comes to light. If we substitute $\boldsymbol{k}$ by $-\boldsymbol{k}$, hence reverse the direction of propagation, the vector $\boldsymbol{G}$ changes its sign in optical activity, while the vector $\boldsymbol{Z}$ remains unchanged. In optical activity, if we look in the propagation direction, the direction of $\boldsymbol{G}$ reverses when the viewing direction is reversed, so that the sense of rotation of the $\boldsymbol{D}$-vector of an elliptically polarized wave is conserved, in contrast to the Faraday effect. We can describe the modification of light propagation by the Faraday effect directly by adopting the corresponding relationships for optical activity (see Section 4.3.6.7). In particular it is true that linear polarized waves in a crystal without magnetic field become elliptically polarized waves, as soon as a magnetic field acts. We expect pure circular polarized waves, if the magnetic field lies in the direction of the optic axis in optically uniaxial crystals.

An application of the Faraday effect is realized in the Faraday modulator, a cylindrical specimen made from an isotropic material inserted in a current-carrying cylindrical magnetic coil. If an alternating current flowing through the coil generates a magnetic field of the same frequency, then the direction of vibration of a linear polarized wave propagating along the coil axis in the material oscillates at the same cycle about the position without magnetic field. The Faraday cell is thus a useful aid in optical precision measurements with modulation techniques (see Section 4.4.2).

In conducting materials the magnetic field gives rise to far higher additional effects due to the direct interaction with the charge carriers, which we can only refer to here (Voigt effect, magnetoband absorption).

In ferromagnetic and other magnetically ordered materials, the associated magnetic point symmetry groups are to be applied to derive the independent components of the tensor $\{z_{ijk}\}$.

Meanwhile the Faraday effect has been investigated on numerous crystals. A measurement arrangement has proven successful, where the specimen (a thin crystal plate) placed in a magnetic field, is continuously rotated about an axis perpendicular to the magnetic field and the change in the light emerging from the plate is continuously analyzed (Kaminsky, Haussühl, 1993). An important result of the investigations on ionic crystals is the quasiadditivity of the specific Faraday rotation (Verdet's constant) of the lattice particles (Haussühl, Effgen, 1988).

4.4.6
Hall Effect

As an example for a more complex relationship we discuss here one of the many interesting magnetic effects. The *Hall effect* is observed in electric conductors including semiconductors. Let a plate, through which an electric current density $\boldsymbol{I}$ is flowing, be simultaneously immersed in a magnetic field $\boldsymbol{H}$ with a component transverse to the current. The charge carriers experience a force deflecting them perpendicular to $\boldsymbol{I}$ and $\boldsymbol{H}$ (Lorentz force). Thus in the surface elements perpendicular to $\boldsymbol{I}$ and $\boldsymbol{H}$ an electric polarization develops and builds up an electric field. Figure 4.41 shows a schematic arrangement for the measurement of the effect. To a first approximation the field strength is

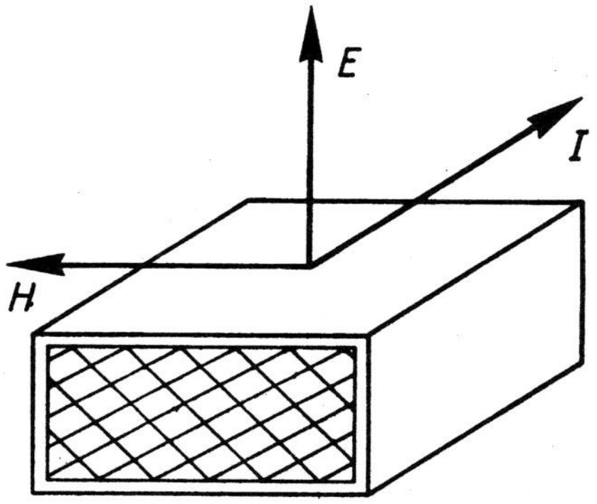

Figure 4.41 Scheme for the measurement of the Hall effect on a rectangular parallelepipedron. The surfaces perpendicular to the current density vector $\boldsymbol{I}$ and the electric field $\boldsymbol{E}$, respectively, are metallized. Between these surfaces a sufficiently wide isolating gap has to be installed.

proportional to current density and magnetic field strength

$$E_i = k_{ijk} I_j H_k.$$

Thus we are dealing with a third-rank axial tensor, which as we have seen, can exist in all PSGs. Since $\boldsymbol{I}$ as well as $\boldsymbol{H}$ change their sign with time reversal, the Hall tensor $\{k_{ijk}\}$ just as $\boldsymbol{E}$ is not affected by time reversal. In crystals with magnetic order the respective magnetic symmetry group must be considered for symmetry reduction. Furthermore, it should be noted that the Onsager relation for transport processes with time reversal is valid, which here takes of the form $k_{ijk} = -k_{jik}$. Hence, the transformation properties are completely analogous to those of the Faraday effect $\{z_{ijk}\}$ (see Section 4.4.5). For example, cubic crystals possess only one component, namely k_{123}

$$k_{123} = k_{231} = k_{312} = -k_{213} = -k_{321} = -k_{132}.$$

In crystals with 22 as a subgroup only three independent components appear, namely, k_{123}, k_{231} and k_{312}.

Introducing a vector $\boldsymbol{R}$ with the components $R_l = r_{lk} H_k$ and the condition $r_{lk} = k_{ijk}$ ($l \neq i, j$ and i, j, l cyclic in 1, 2, 3), the Hall effect can also be described by a vector product $E_i = (\boldsymbol{I} \times \boldsymbol{R})_i$, where $\boldsymbol{R}$ is an axial vector and $\{r_{lk}\}$ a polar tensor.

The measurement of the Hall effect is most simple, even in the case of triclinic crystals. Maximally nine independent components k_{ijk} exist. If we select the principal axes system of the tensor of electrical conductivity as the reference system, we can preset the direction of the electric current vector $\boldsymbol{I} \parallel e_2'$ on a thin plate cut according to the Cartesian axes of this system. Similarly, we can set the external magnetic field parallel to the plate normal e'_3. We then directly measure the coefficients k'_{123} perpendicular to I and H according to

$$E'_1 = k'_{123} I'_2 H'_3.$$

All coefficients can be determined by interchanging the directions of I, H, and E. Hence, we require three such plates.

The Hall effect allows important statements concerning charge carriers in conductors. From its sign one can directly deduce the sign of the charge (positive Hall coefficient means positive charge carriers). Moreover, the absolute magnitude of the Hall coefficient provides a measure for the mobility of the charge carriers.

We mention here a qualitative interpretation of the Hall effect for an isotropic medium sufficient for many applications. The field strength E generated by the Hall effect perpendicular to I and H must compensate the Lorentz force K acting on the charge carriers, which move with the mean velocity v in the direction I. Hence, $K = ev \times H = eE$ and consequently $E = v \times H$, where e is the charge of a carrier. The total current is then $I_{\mathrm{tot}} = Qenv$. $Q = db$ is the cross-section, d is the thickness of the plate parallel to the field and b is the width of the plate = separation of the electrodes, where the electric voltage $U = b|E|\mathrm{sign}E$ is measured. n means the number of charge carriers per volume element. Hence,

$$E = \frac{I_{\mathrm{tot}} \times H}{Qen} \quad \text{or with} \quad I = I_{\mathrm{tot}}/Q$$

$$E = \frac{I \times H}{en}.$$

Thus the Hall coefficient corresponds to the quantity $1/en$.

A specimen exhibiting a large Hall effect, the so-called Hall generator (e.g., made from InAs or InSb) can be used to directly measure the magnetic field strength.

4.5 Fourth-Rank Tensors

The symmetry reduction for the general fourth-rank tensor with its $3^4 = 81$ components would take up too much space here. Thus we limit ourselves to the two important types, namely those tensors with permutability of the first and second as well as the third and fourth position (type A) and those tensors with the additional permutability of both first positions with both last positions (type B). We are thus dealing with the conditions

$$\text{Type A}: \; t_{ijkl} = t_{jikl} = t_{ijlk} = t_{jilk},$$
$$\text{Type B}: \; t_{ijkl} = t_{klij} \quad \text{(in addition to the conditions of type A).}$$

The most important representatives for type A are: the piezooptical tensor, the second-order electrooptical effect (Kerr effect), electrostriction and second-order magnetostriction. The following relationships are valid:

$$\Delta a_{ij} = q_{ijkl}\sigma_{kl}, \qquad \Delta a_{ij} = r_{ijkl}E_k E_l,$$
$$\varepsilon_{ij} = \hat{d}_{ijkl}E_k E_l, \qquad \varepsilon_{ij} = f_{ijkl}H_k H_l.$$

The permutability is a consequence of the interconnection of two second-rank symmetric tensors (Δa_{ij}, σ_{ij}, E_iE_j, H_iH_j). The most significant representative of type B is the elasticity tensor, which links the deformation tensor and the mechanical stress tensor according to $\sigma_{ij} = c_{ijkl}\varepsilon_{kl}$ or $\varepsilon_{ij} = s_{ijkl}\sigma_{kl}$. Here, the pairwise permutability is based on the reversibility of mechanical deformation work, which we will consider in detail.

In any case a careful check is required to determine the influence of symmetry and physical aspects. Since a symmetric second-rank tensor possesses at most six independent components, the maximum number of independent components in type A reduces to $6 \times 6 = 36$ and in type B to $(6 \cdot 5)/2 + 6 = 21$. Because the intrinsic symmetry of all the even-rank tensors possesses an inversion center, the number of distinguishable tensors based on crystallographic symmetry are at most equal to the number of the Laue groups, hence 11 ($\bar{1}$, 2/m, 2/mm, $\bar{3}$, $\bar{3}$2, 4/m, 4/mm, 6/m, 6/mm, m3, 4/m3). Only these 11 cases are to be investigated for polar tensors.

If a twofold axis parallel to e_2 exists, then

$$t_{ijkl} = (-1)^q t_{ijkl},$$

where q is equal to the number of indices 2, i.e., the index 2 only is allowed to occur an even number of times. It is appropriate to arrange the fourth-rank tensors of type A and B in the form of 6×6-matrices according to the following scheme (for monoclinic crystals $2 \parallel e_2$):

	11	22	33	23	31	12
11	1111	1122	1133	–	1131	–
22	2211	2222	2233	–	2231	–
33	3311	3322	3333	–	3331	–
23	–	–	–	2323	–	2312
31	3111	3122	3133	–	3131	–
12	–	–	–	1223	–	1212

Hence, in type A there exists 20 independent components and 13 independent components in type B.

In the PSG 22 each index is allowed to appear only an even number of times. Thus for orthorhombic crystals only the following scheme persists with 12 independent components for type A and 9 for type B:

$$\begin{pmatrix} 1111 & 1122 & 1133 & & & \\ 2211 & 2222 & 2233 & & 0 & \\ 3311 & 3322 & 3333 & & & \\ & & & 2323 & & \\ & 0 & & & 3131 & \\ & & & & & 1212 \end{pmatrix}.$$

As with second- and third-rank tensors, the operation of the threefold axis requires the most effort also here. Again we have

$$3 \parallel e_3, \quad \text{hence} \quad R_{\pm 3\parallel e_3} = \begin{pmatrix} -\frac{1}{2} & \pm\frac{\sqrt{3}}{2} & 0 \\ \mp\frac{\sqrt{3}}{2} & -\frac{1}{2} & 0 \\ 0 & 0 & 1 \end{pmatrix}.$$

In order to arrange the symmetry reduction clearly one must proceed systematically. In a symmetry operation the respective tensor components transform in certain subsets among themselves. We will again return to this aspect latter. For the moment, let us consider the special case of index 3 which is directly assigned to the direction of the axis of rotation. There exist five subsets of tensor components, arranged according to the number of index 3

- missing (9 components of type $t_{1111}, t_{1112}, t_{1122}$ etc.),
- once (12 components of type $t_{1113}, t_{1123}, t_{1223}$ etc.),
- twice (10 components of type $t_{1133}, t_{1313}, t_{1323}$ etc.),
- three times (4 components $t_{1333}, t_{2333}, t_{3313}, t_{3323}$),
- four times (1 component t_{3333}).

We begin with the first subset and find for type A:

$$t'_{1111} = t_{1111} = \frac{1}{16}(t_{1111} + 9t_{2222} + 3t_{1122} + 3t_{2211} + 12t_{1212}) \mp \frac{\sqrt{3}}{8}(t_{1112} + t_{1211} + 3t_{2212} + 3t_{1222}), \tag{4.9}$$

$$t'_{2222} = t_{2222} = \frac{1}{16}(9t_{1111} + t_{2222} + 3t_{1122} + 3t_{2211} + 12t_{1212}) \pm \frac{\sqrt{3}}{8}(3t_{1112} + 3t_{1211} + t_{2212} + t_{1222}), \tag{4.10}$$

$$t'_{1122} = t_{1122} = \frac{1}{16}(3t_{1111} + 3t_{2222} + t_{1122} + 9t_{2211} - 12t_{1212}) \pm \frac{\sqrt{3}}{8}(t_{1112} - 3t_{1211} + 3t_{2212} - t_{1222}), \tag{4.11}$$

$$t'_{2211} = t_{2211} = \frac{1}{16}(3t_{1111} + 3t_{2222} + 9t_{1122} + t_{2211} - 12t_{1212}) \pm \frac{\sqrt{3}}{8}(-3t_{1112} + t_{1211} - t_{2212} + 3t_{1222}). \tag{4.12}$$

Correspondingly, there exist further equations for t'_{1112}, t'_{1211}, t'_{2212}, t'_{1222} and t'_{1212}, which, however, do not result in new conditions. From these four double equations we obtain eight conditions. The terms with alternating signs must vanish. Hence, from (4.9) or (4.10) we have

$$t_{1112} + t_{1211} + 3t_{2212} + 3t_{1222} = 0$$

and

$$3t_{1112} + 3t_{1211} + t_{2212} + t_{1222} = 0$$

and thus

$$t_{1211} = -t_{1112} \quad \text{and} \quad t_{2212} = -t_{1222}.$$

Similarly (4.11) and (4.12) yield

$$t_{1112} - 3t_{1211} + 3t_{2212} - t_{1222} = 0 \quad \text{and}$$
$$-3t_{1112} + t_{1211} - t_{2212} + 3t_{1222} = 0,$$

hence together with the above result

$$t_{1112} = -t_{2212} = -t_{1211} = t_{1222}.$$

From (4.9) or (4.10) we find

$$5t_{1111} = 3t_{2222} + t_{1122} + t_{2211} + 4t_{1212} \quad \text{and}$$
$$5t_{2222} = 3t_{1111} + t_{1122} - t_{2211} + 4t_{1212}.$$

From the difference and the sum of these conditions we get $t_{1111} = t_{2222}$ or $t_{1212} = (t_{1111} - t_{1122}/2 - t_{2211}/2)/2$. From (4.11) and (4.9) we get $t_{1122} = t_{2211}$ as well as the important relation $t_{1212} = (t_{1111} - t_{1122})/2$.

The remaining conditions give us nothing new. In total, the subset with missing index 3 furnishes the following three independent components t_{1111}, t_{1122} and t_{1112}.

In the subset with one index 3, symmetry transformation gives the following result:

$$t'_{1113} = t_{1113} = -\frac{1}{8}(t_{1113} + 6t_{1223} + 3t_{2213}) \pm \frac{\sqrt{3}}{8}(3t_{2223} + 2t_{1213} + t_{1123}), \tag{4.13}$$

$$t'_{2223} = t_{2223} = -\frac{1}{8}(t_{2223} + 6t_{1213} + 3t_{1123}) \mp \frac{\sqrt{3}}{8}(3t_{1113} + 2t_{1223} + t_{2213}), \tag{4.14}$$

$$t'_{1123} = t_{1123} = \frac{1}{8}(-3t_{2223} + 6t_{1213} - t_{1123}) \pm \frac{\sqrt{3}}{8}(-t_{1113} + 2t_{1223} - t_{2213}), \tag{4.15}$$

$$t'_{1213} = t_{1213} = \frac{1}{8}(-3t_{2223} + 2t_{1213} + 3t_{1123}) \pm \frac{\sqrt{3}}{8}(-t_{1113} - 2t_{1223} + t_{2213}). \tag{4.16}$$

The conditions for the components with interchanged first and second index pairs have the same coefficients. Further conditions, for example, for t'_{2123} bring nothing new.

From (4.13) we have:

$$3t_{2223} + 2t_{1213} + t_{1123} = 0 \quad \text{and} \tag{4.17}$$

$$3t_{1113} + 2t_{1223} + t_{2213} = 0; \tag{4.18}$$

from (4.15)

$$-3t_{2213} + 2t_{1223} - t_{1113} = 0 \quad \text{and} \tag{4.19}$$

$$3t_{1123} + t_{2223} - 2t_{1213} = 0; \tag{4.20}$$

from (4.16)

$$-t_{1113} - 2t_{1223} + t_{2213} = 0 \quad \text{and} \tag{4.21}$$

$$2t_{1213} + t_{2223} - t_{1123} = 0. \tag{4.22}$$

From (4.17) and (4.20) respectively from (4.18) and (4.19) one obtains $t_{2223} = -t_{1123}$ and $t_{1123} = t_{1213}$, respectively, $t_{1113} = -t_{2213}$ and $t_{1113} = -t_{1223}$.

The pairwise interchange of the indices (1. pair against 2. pair) yields analogous equations with the same coefficients. Hence, in this subset we have the following four independent components with one index 3:

$$t_{1113} = -t_{2213} = -t_{1223}, \quad t_{2223} = -t_{1123} = -t_{1213},$$
$$t_{1311} = -t_{1322} = -t_{2312}, \quad t_{2322} = -t_{2311} = -t_{1312}.$$

For the subset with twice index 3 we have

$$t'_{1133} = t_{1133} = \frac{1}{4}(t_{1133} + 3t_{2233} \pm \sqrt{3}2t_{1233}),$$
$$t'_{1313} = t_{1313} = \frac{1}{4}(t_{1313} + 3t_{2323} \pm \sqrt{3}t_{2313} \pm \sqrt{3}t_{1323}).$$

This gives

$$t_{1133} = t_{2233}, \quad t_{1313} = t_{2323}, \quad t_{1233} = 0 \quad \text{and} \quad t_{2313} = -t_{1323}.$$

Furthermore, after interchanging the index pairs we get $t_{3311} = t_{3322}$ and $t_{3312} = 0$. In total there also exist four independent components.

The components with index 3 three times all vanish because $t'_{1333} = -\frac{1}{2}t_{1333} \pm \frac{\sqrt{3}}{2}t_{2333}$ and so on. Finally, only $t'_{3333} = t_{3333}$ remains. Accordingly, a PSG with threefold rotation axis or rotation–inversion axis (PSG 3 and $\bar{3}$) possesses the following type A tensor with a total of 12 independent components:

	11	22	33	23	31	12
11	1111	1122	1133	1123	1131	1112
22	1122	1111	1133	−1123	−1131	−1112
33	3311	3311	3333	–	–	–
23	2311	−2311	–	3131	2313	−1311
31	3111	−3111	–	−2313	3131	2311
12	−1112	1112	–	−1131	1123	1212*)

*) $t_{1212} = \frac{1}{2}(t_{1111} - t_{1122})$

In the Laue class $\bar{3}$m we place the twofold axis parallel e_1 and note the condition "index 1 even number of times." Starting from the above table, we obtain in the PSG 32, 3m and $\bar{3}$m the following eight independent components for type A: t_{1111}, t_{1122}, t_{1133}, t_{1123}, t_{3311}, t_{3333}, t_{2311} and t_{1313}. Which components appear at all is read directly from the above scheme.

For the symmetry reduction resulting from a fourfold axis parallel e_3 we only need to consider those components where the index 3 appears an even number of times because a fourfold axis simultaneously represents a twofold axis (2 is a subgroup of 4!). With

$$R_{\pm 4\|e_3} = \begin{pmatrix} 0 & \pm 1 & 0 \\ \mp 1 & 0 & 0 \\ 0 & 0 & 1 \end{pmatrix}$$

we obtain the following conditions:

$$\begin{aligned} &t_{1111}=t_{2222}, \quad t_{1122}=t_{2211}, \quad t_{1212}=t_{2121}, \quad t_{1121}=-t_{2212}, \quad t_{1211}=-t_{2122}, \\ &t_{1133}=t_{2233}, \quad t_{3311}=t_{3322}, \quad t_{1313}=t_{2323}, \quad t_{1323}=-t_{2313}, \quad t_{1233}=-t_{2133}=0, \\ &t_{3312}=-t_{3321}=0, \quad t_{3333}=t_{3333}. \end{aligned}$$

Hence a type A tensor has the following form (10 independent components) for the Laue class 4/m (PSG 4, $\bar{4}$ and 4/m):

	11	22	33	23	31	12
11	1111	1122	1133	–	–	1112
22	1122	1111	1133	–	–	−1112
33	3311	3311	3333	–	–	–
23	–	–	–	3131	2331	–
31	–	–	–	−2331	3131	–
12	1211	−1211	–	–	–	1212

In the Laue class 4/mm the twofold axis parallel e_i ($i = 1, 2, 3$) requires that all indices are present an even number of times. Thus from the above scheme t_{1112}, t_{1211} and t_{2331} vanish, so that only seven independent components remain.

We calculate the effect of a sixfold rotation axis or rotation–inversion axis by adding to the result of the threefold axis the condition of a twofold axis parallel e_3 (index 3 only an even number of times). For the Laue class 6/m (PSG 6, $\bar{6}$ and 6/m) this leads to the following eight independent tensor components: t_{1111}, t_{1122}, t_{1133}, t_{1112}, t_{3311}, t_{3333}, t_{3131}, t_{2313}. Otherwise, the scheme for the threefold axis is valid. The Laue class 6/mm (PSG 62, 6m, $\bar{6}$m and 6/mm) includes the condition "index 1 an even number of times" because of the twofold axis parallel e_1, so that only the following six independent components exist:

$$t_{1111}, t_{1122}, t_{1133}, t_{3311}, t_{3333}, t_{3131}.$$

We achieve the symmetry reduction in both cubic Laue classes in the shortest way by applying the operation of the threefold axis in the direction of the space diagonals of the Cartesian reference system to the result of orthorhombic or tetragonal crystals. We have

$$R_{3\|e_1+e_2+e_3} = \begin{pmatrix} 0 & 1 & 0 \\ 0 & 0 & 1 \\ 1 & 0 & 0, \end{pmatrix}$$

i.e. the transformation interchanges the indices cyclically. Hence, in the Laue class m3 (PSG 23 and m3) there exists the following four independent components:

$$t_{1111} = t_{2222} = t_{3333}, \quad t_{1122} = t_{2233} = t_{3311},$$
$$t_{2211} = t_{3322} = t_{1133}, \quad \text{and} \quad t_{1212} = t_{2323} = t_{3131}.$$

In the Laue class 4/m3 (PSG 43, $\bar{4}3$ and 4/m3) t_{1122} equals t_{2211}, so that only three independent components exist.

In the case of the cylindrical symmetries ∞/m and ∞/mm we have the situation of the Laue class 6/m and 6/mm, respectively. The isotropy, derived from the conditions of the Laue class 4/m3 under the inclusion of cylindrical symmetry, demands that

$$t_{1212} = (t_{1111} - t_{1122})/2,$$

so that only two independent components exist.

The corresponding type B tensors can be immediately derived from the complete formulae for type A tensors given above by allowing the permutation of the index pairs (see Table 4.10).

Tensors with other intrinsic symmetries, for example, the permutability of three index positions, as in the case of nonlinear dielectric susceptibility or the permutability of indices of only one pair, as with the electrically induced piezoelectric effect, can be easily reduced by employing an analogous method as above.

4.5.1
Elasticity Tensor

Mechanical stresses and deformations are, within the limits of *Hooke's law*, proportional to one another. For sufficiently small stresses and deformations we can describe the relationship of these quantities, when we neglect infinitesimal torques or rotations, by:

$$\varepsilon_{ij} = s_{ijkl}\sigma_{kl} \quad \text{or} \quad \sigma_{ij} = c_{ijkl}\varepsilon_{kl}.$$

In the first case we assume that the mechanical stresses are inducing quantities, which represent the normal situation in elastostatics. We will see later that the reverse is an even more favorable representation in dynamic processes. Furthermore, we assume that the stresses and deformations are of such a nature that no plastic, i.e., permanent deformations appear. We call this limit for the deformations or stresses the critical limit with the corresponding critical deformation or critical stress. We will return to this point in Section 6.1.1. Moreover, we assume that the deformations are small enough to guarantee

Table 4.10 Fourth-rank tensors of type B (permutability of indices within the first and second pair and pairwise; example elasticity tensor).

$\bar{1}$ (Z = 21):

1111	1122	1133	1123	1131	1112
1122	2222	2233	2223	2231	2212
1133	2233	3333	3323	3331	3312
1123	2223	3323	2323	2331	1223
1131	2231	3331	2331	3131	1231
1112	2212	3312	1223	1231	1212

2/m (Z = 13):

1111	1122	1133	0	1131	0
1122	2222	2233	0	2231	0
1133	2233	3333	0	3331	0
0	0	0	2323	0	1223
1131	2231	3331	0	3131	0
0	0	0	1223	0	1212

2/mm (Z = 9):

1111	1122	1133	0	0	0
1122	2222	2233	0	0	0
1133	2233	3333	0	0	0
0	0	0	2323	0	0
0	0	0	0	3131	0
0	0	0	0	0	1212

$\bar{3}$ (Z = 7)

1111	1122	1133	1123	1131	0
1122	1111	1133	−1123	−1131	0
1133	1133	3333	0	0	0
1123	−1123	0	3131	0	−1131
1131	−1131	0	0	3131	1123
0	0	0	−1131	1123	1212*

$\bar{3}$m (Z = 6):

1111	1122	1133	1123	0	0
1122	1111	1133	−1123	0	0
1133	1133	3333	0	0	0
1123	−1123	0	3131	0	0
0	0	0	0	3131	1123
0	0	0	0	1123	1212*

4/m (Z = 7):

1111	1122	1133	0	0	1112
1122	1111	1133	0	0	−1112
1133	1133	3333	0	0	0
0	0	0	3131	0	0
0	0	0	0	3131	0
1112	−1112	0	0	0	1212

4/mm (Z = 6):

1111	1122	1133	0	0	0
1122	1111	1133	0	0	0
1133	1133	3333	0	0	0
0	0	0	3131	0	0
0	0	0	0	3131	0
0	0	0	0	0	1212

6/m, 6/mm, ∞/m, ∞/mm (Z = 5):

1111	1122	1133	0	0	0
1122	1111	1133	0	0	0
1133	1133	3333	0	0	0
0	0	0	3131	0	0
0	0	0	0	3131	0
0	0	0	0	0	1212*

m3, 4/m3 (Z = 3):

1111	1122	1122	0	0	0
1122	1111	1122	0	0	0
1122	1122	1111	0	0	0
0	0	0	1212	0	0
0	0	0	0	1212	0
0	0	0	0	0	1212

∞/m∞ (Isotropy, Z = 2):

1111	1122	1122	0	0	0
1122	1111	1122	0	0	0
1122	1122	1111	0	0	0
0	0	0	1212*	0	0
0	0	0	0	1212*	0
0	0	0	0	0	1212*

* $t_{1212} = \frac{1}{2}(t_{1111} - t_{1122})$

the validity of Hooke's law. We will discuss the nonlinear elastic properties observed far before the limit to plastic deformation is reached in Section 4.6.3. While we can still work, in this section, with the ordinary deformation tensor, we will employ the Lagrangian deformation tensor $\{\eta_{ij}\}$ for the relationship between stress and deformation in the case of nonlinear elastic properties.

An important simplification arises from the permutability of the first with the second index pair. The proof is analogous to that of the symmetry of the dielectric tensor, namely from the reversibility of the deformation work. This is per unit volume

$$\Delta W = \sigma_{ij}\Delta\varepsilon_{ij}, \quad \frac{\partial W}{\partial \varepsilon_{ij}} = \sigma_{ij} c_{ijkl}\varepsilon_{kl}.$$

Reversibility demands that ΔW forms a complete differential, hence

$$\frac{\partial^2 W}{\partial\varepsilon_{ij}\partial\varepsilon_{kl}} = \frac{\partial^2 W}{\partial\varepsilon_{kl}\partial\varepsilon_{ij}}$$

and thus $c_{ijkl} = c_{klij}$ is true.

The permutability of the indices within the first and second pair is a consequence of the symmetry of the stress tensor and the deformation tensor. The same is true for the s-tensor $s_{ijkl} = s_{klij}$ and so on, as one recognizes from the calculation of $\varepsilon_{ij} = s_{ijkl}\sigma_{kl}$ from $\sigma_{ij} = c_{ijkl}\varepsilon_{kl}$.

Voigt already introduced an abbreviation, which we also use here, with two indices instead of the four, according to the following scheme: $ii \to i$, $ij \to 9 - i - j$ for $i \neq j$ (thus, for example, $c_{1122} \to c_{12}$ or $c_{1232} \to c_{64}$). One must never forget that this abbreviation does not lead to a second-rank tensor. In all operations involving tensor transformations, one must, in any case, return to the four indices representation. We call the quantities c_{ij}, following common usage, the elastic constants and s_{ij} the elastic coefficients. Many authors prefer the terms stiffnesses for c_{ij} and compliances for s_{ij}.

If the s-tensor or the c-tensor is known, then one can be calculated from the other by matix inversion. Because

$$\varepsilon_{ij} = s_{ijkl}\sigma_{kl} = s_{ijkl}c_{klmn}\varepsilon_{mn}, \quad \text{i.e.,} \quad s_{ijkl}c_{klmn} = \delta_{im}\delta_{jn}.$$

In Voigt notation, one has

$$s_{qp} = \frac{(-1)^{p+q}A_{pq}}{D},$$

where A_{pq} is the subdeterminant of the matrix (c_{pq}) after eliminating the pth row and the qth column and D is the determinant of (c_{pq}). When converting from s_{pq} to s_{ijkl} it is absolutely essential to note the following multiplicity rule resulting from the permutability of the indices:

$$s_{ij} \to s_{iijj}, \quad s_{i,9-k-l} \to 2s_{iikl}, \quad s_{9-i-j,9-k-l} \to 4s_{ijkl}.$$

As an example, we consider the relations for cubic crystals. We have:

$$s_{1111} = s_{11} = \frac{c_{11} + c_{12}}{(c_{11} - c_{12})(c_{11} + 2c_{12})},$$

$$s_{1122} = s_{12} = \frac{-c_{12}}{(c_{11} - c_{12})(c_{11} + 2c_{12})}, \quad 4s_{1212} = s_{66} = 1/c_{66}.$$

With the first attempts to describe anisotropic elasticity in the eighteenth century a discussion began concerning the question, whether the elasticity tensor is totally symmetric or not (rari-constant or multiconstant theory). We can express the deviation from total symmetry by the six relationships $c_{iijk} - c_{ijik} = g_{mn}(-1)^{mn(m-n)/2}$, where $m \neq i, j$ and $n \neq i, k$. We are dealing with a second-rank tensor invariant with the components g_{mn}, as one can immediately recognize from the representation $g_{mn} = \frac{1}{2} e_{mik} e_{njl} c_{ijkl}$, where $\{e_{mik}\}$ is the Levi-Civitá tensor (see Section 3.7).

The relations $c_{iijk} - c_{ijik} = 0$ are called *Cauchy relations*. The tensor $\{g_{mn}\}$ is a second-rank tensor invariant of the elasticity tensor. It describes the deviations from the Cauchy relations. The prerequisites for the validity of the Cauchy relations are given as (see Leibfried, 1955):

1. pure central forces exist between the lattice particles,
2. each particle possesses a central symmetry and thus is located in a symmetry center of the lattice,
3. no anharmonicity (pure Hooke's law),
4. no thermal energy content,
5. no initial stress.

There does not exist one single substance for which the Cauchy relations are fulfilled, although some crystals, as for example, NaCl, exhibit only a small deviation. From the experimental data one can recognize that the deviations from the Cauchy relations allow statements concerning bonding properties of the particles. Crystals with predominantly ionic bonds show, to a large extent, positive components g_{mn}, those with directional bonding, in particular, with strongly covalent bonds, negative g_{mn} (Haussühl, 1967). Interestingly, an asphericity of the electron distribution of the ions gives rise to positive g_{mn}. That is, the deviation from the Cauchy relations may give hints on atomistic bonding details. We will become acquainted with other physically interesting invariants of the elasticity tensor in the following.

The tensor $\{g_{mn}\}$ represents the deviation of the elasticity tensor from its totally symmetric part. Accordingly, Hehl and Itin (2002) were able to show rather elegantly that$\{g_{mn}\}$ constitutes, in a group-theoretical sense, an irreducible piece of the elasticity tensor.

4.5.2
Elastostatics

Here we consider the deformations appearing in equilibrium with external forces or the required mechanical stresses in equilibrium with a given deformation. The time-dependent processes until equilibrium is achieved will be

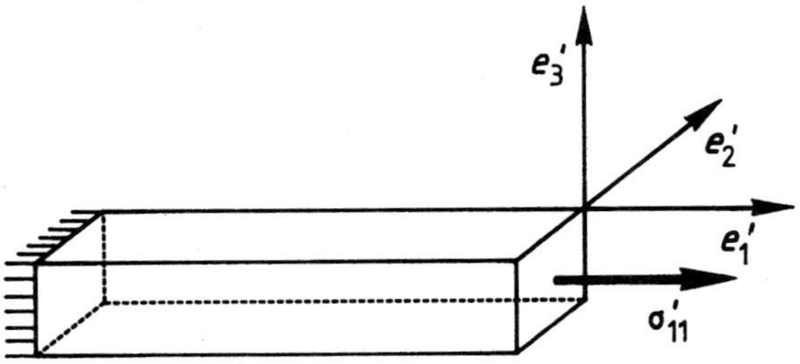

Figure 4.42 Measurement of Young's modulus on a thin bar by application of uniaxial stress σ'_{11}.

neglected. We then have

$$\frac{\partial \sigma_{ij}}{\partial t} = 0 \quad \text{and} \quad \frac{\partial \varepsilon_{ij}}{\partial t} = 0 \quad (t \text{ time}).$$

All quantities before and after the deformation are described in the same fixed reference system. Furthermore, we assume the validity of Hooke's law.

Whereas the solution of general problems of elastostatics often requires considerable mathematical efforts, as, for example, the problem of the bending of a curved bar with variable cross-section or the torsion of a thin rod of nonuniform cross-section, we will treat here only problems that are relevant for practical measurments in crystals. A decisive role is played by the boundary conditions to which the specimen is exposed. These must be formulated as simply as possible in order to directly apply the fundamental elastostatic equation $\varepsilon_{ij} = s_{ijkl}\sigma_{kl}$. In particular, arrangements are preferred that lead to a homogoneous deformation of the complete crystal, that is, where

$$\frac{\partial \varepsilon_{ij}}{\partial x_k} = 0.$$

As a very important and experimentally easily accessibile example we first investigate the longitudinal deformation of a bar with fixed cross-section (according to form and size) under uniaxial stress along the bar axis e'_1 (Fig. 4.42). We have

$$\varepsilon'_{11} = \left(\frac{\Delta l_1}{l_1}\right)' = s'_{1111}\sigma'_{11}.$$

$(\Delta l_1/l_1)'$ is the relative change in length along e'_1. The boundary conditions are $\sigma'_{11} \neq 0$, otherwise $\sigma'_{ij} = 0$. The material property $1/s'_{1111}$ is called *Young's modulus*. For an arbitrary direction $e'_1 = u_{1i}e_i$ one has

$$s'_{1111} = u_{1i}u_{1j}u_{1k}u_{1l}s_{ijkl}.$$

The right-hand side contains the following 15 coefficients:

$$s_{iiii}\ (3),\quad 4s_{iiij}\ (6),\quad 2(s_{iijj}+2s_{ijij})\ (3)\quad \text{and}\quad 4(s_{iijk}+2s_{ijik})\ (3)\qquad \text{with}\quad i=1,2,3.$$

Accordingly, from such longitudinal measurements, even when we select many different directions e_1', we obtain only the coefficients of the quadric $F = s_{ijkl}x_ix_jx_kx_l$.

If we also include measurements of the longitudinal deformation occurring perpendicular to the bar axis, the so-called *lateral contraction*

$$\varepsilon_{22}' = \left(\frac{\Delta l_2}{l_2}\right)' = s_{2211}'\sigma_{11}' \quad \text{and} \quad \varepsilon_{33}' = \left(\frac{\Delta l_3}{l_3}\right)' = s_{3311}'\sigma_{11}'$$

we can, in principle, determine all 21 independent tensor component s_{ijkl} of a triclinic crystal.

The measurements are performed as in the case of thermal expansion with optical methods employing inductive or capacitive path sensors as well as with the help of strain gauges. The uniaxial stress is transferred to the crystal, for example, by applying weights, or by springs tightened by a motor-driven feed. An increase in measurement accuracy of the signal is achieved using a lock-in amplifier, when one works with a periodic load. One can also employ X-ray methods to determine the change in the lattice constants under uniaxial stress resulting from the deformation as with measurements of thermal expansion (see Section 4.3.11). The complete determination of the elasticity tensor, using these methods, has only been successful for a few highly symmetric crystals. For isotropic substances (building materials, ceramics, glasses, plastics), where sufficiently large specimens can be prepared, the static method is of major importance, in particular in understanding the elastic behavior beyond Hooke's law up to plastic deformation and fracture.

Since the qualitative measurement, even on small crystals with dimensions of a few mm, delivers at least the order of magnitude of the deformation and in particular its sign, such tests, for example, with the help of strain gauges, can serve to supplement and check other methods.

A variant of the measurement of Young's modulus is the bending test of thin rods with homogeneous cross-section (*beam bending*), where Young's modulus is the effective material constant for the beam axis (longitudinal dilatation or compression with bending on the convex or concave side of the rod, respectively). For practical measurements, this method, used by Voigt and coworkers (1884) in thousands of measurements to determine the elasiticity tensor of NaCl and KCl, is today rarely taken into consideration because of the low accuracy. This also applies to variants such as bending of thin plates, as employed, for example, by Coromilas for the measurement of some elastic coefficients of gypsum and mica (1877).

The ratio of longitudinal transverse contraction and dilatation, in our example $\varepsilon'_{22}/\varepsilon'_{11}$, for arbitrary beam directions e'_1 and directions e'_2 perpendicular to these, is called *Poisson's ratio*. It plays, among other things, an important role in the evaluation of the stiffness of composite materials.

4.5.3
Linear Compressibility Under Hydrostatic Pressure

If we expose an arbitrarily shaped specimen to a hydrostatic pressure p ($\sigma_{ij} = -p\delta_{ij}$), we observe a change in the dimensions, described by the longitudinal effect

$$\varepsilon'_{11} = \sum_k s'_{11kk}\sigma_{kk}$$

($\sigma'_{kk} = \sigma_{kk}$). With $e'_1 = u_{1i}e_i$ we have

$$\varepsilon'_{11} = \sum_k u_{1i}u_{1j}s_{ijkk}(-p) = -pu_{1i}u_{1j}S_{ij}.$$

The quantities $S_{ij} = s_{ijkl}\delta_{kl}$ form a second-rank tensor invariant (contraction of the s-tensor). This experimentally easily accessible tensor property, provides, for many applications, sufficient information concerning the elastic behavior of materials. The measurement is performed conveniently by attaching strain gauges to the side faces of a parallelepipedic specimen. However, one must make a correction for the pressure dependence of the electrical resistance of the strain gauges.

From the sum $\sum \varepsilon'_{ii}$ of the principal deformations we obtain the *volume compressibility*

$$K = -\frac{d\log V}{dp} \approx \sum_i \varepsilon'_{ii}/p = -\frac{\Delta V}{Vp},$$

where p is sufficiently small and V is an arbitrary volume of the specimen. A precise measurement of the volume compressibility is even possible with small crystal grains suspended in a nonsolving liquid in a pressure cell to measure the change in volume ΔV after pressure loading. The measured total compressibility of the suspension arises from the fractional volumina V_i/V and the compressibilities K_i of the ith components of the suspension

$$K = \sum_i K_i V_i/V,$$

as one can easily verify. In the case of two components, when the compressibility of one of them is known as well as the fractional volumina, one can determine the compressibility of the other component. At higher hydrostatic pressures the X-ray determination of the lattice constants has been proven to be very successful for the determination of the compressibility. (see Section 4.3.11).

For K we obtain from above

$$K = \sum_{i,k} s_{iikk}.$$

For cubic crystals one has

$$K = 3(s_{11} + 2s_{12}) = \frac{3}{c_{11} + 2c_{12}};$$

and for trigonal, tetragonal, and hexagonal crystals

$$K = (s_{11} + s_{22} + s_{33} + 2s_{12} + 2s_{23} + 2s_{31}) = \frac{c_{11} + c_{12} + 2c_{33} - 4c_{31}}{c_{33}(c_{11} + c_{12}) - 2c_{31}^2}$$

(matrix inversion of the c-tensor, see above).

K is a scalar invariant of the elasticity tensor, thus independent of the reference system. For crude considerations concerning stiffness behavior, the reciprocal compressibility is often a very useful measure. Moreover, just this elastic quantity is easily accessible to calculation from simple lattice models.

The tensor of the linear compressibility $\{S_{ij}\}$ is also involved in experiments, where crystals are exposed to a general stress described by the stress tensor $\{\sigma_{ij}\}$. According to Hooke's law we expect deformations $\varepsilon_{kl} = s_{klij}\sigma_{ij}$, which lead to a variation of the metric of the crystal (Lattice constants a_i and angles α_i). From these variations, which are accessible by high-resolution X-ray or neutron diffraction techniques, the resulting relative change of the unit cell volume $\varepsilon = \Delta V/V = \varepsilon_{11} + \varepsilon_{22} + \varepsilon_{33}$ can be derived. From Hooke's law we obtain $\varepsilon = \varepsilon_{11} + \varepsilon_{22} + \varepsilon_{33} = \sum s_{kkij}\sigma_{ij} = S_{ij}\sigma_{ij}$ (summation over $k, i, j = 1, 2, 3$). We recognize that the tensor connecting the components ε_{ij} of the deformation tensor with the hydrostatic pressure p ($\sigma_{ij} = -p$ for $i = j$ and 0 otherwise) and the tensor connecting the relative variation of the volume $\varepsilon = \Delta V/V$ with the components σ_{ij} of the stress tensor are identical. This is due to the interchangeability of the first and second pair of indices ($s_{ijkl} = s_{klij}$). In the field of X-ray studies S_{ij} are called X-ray elasticity factors and bear the symbol F_{ij}.

4.5.4 Torsion Modulus

If one succeeds in producing thin crystal rods, fibers or films, then one can apply the known methods to measure the *torsion modulus* of metal fibers also to crystals. The requirement is a constant cross-section of the fiber or rod and a certain minimum length of a few mm. The specimen can be statically twisted and the restoring force measured with a torsion balance. Low-frequency dynamic measurements, where the specimen is furnished with a large external mass to reduce the frequency of oscillation, can enable a substantial increase in measuring accuracy. Since in the meantime one can manufacture thin hairlike crystals (so-called "whiskers") from many materials, this method is of certain significance, when larger crystals are not available. The same applies to the torsion of strips or thin plates.

Concerning the mathematical treatment of problems with complex boundary conditions, we refer to numerous literature on this subject (for example, Love, 1926, 1944; Voigt, 1928).

4.5.5
Elastodynamic

With the development of ultrasound techniques the dynamic methods, in particular, for precision measurements on crystals, have largely superseded the static methods. The dynamic processes are characterized by the fact that at least for one component σ_{ij} the following is true:

$$\frac{\partial \sigma_{ij}}{\partial x_k} \neq 0 \quad \text{and} \quad \frac{\partial \sigma_{ij}}{\partial t} \neq 0;$$

hence for at least one ε_{ij}

$$\frac{\partial \varepsilon_{ij}}{\partial x_k} \neq 0 \quad \text{and} \quad \frac{\partial \varepsilon_{ij}}{\partial t} \neq 0$$

is fulfilled.

To begin with, let us study the forces acting on a volume of a parallelepipedic specimen of dimensions Δx_i in the directions $\boldsymbol{e}_i$ ($i = 1, 2, 3$) (Fig. 4.43).

$$\{\sigma_{ii}(\boldsymbol{x}) - \sigma_{ii}(\boldsymbol{x}_0)\}\Delta x_j \Delta x_k + \{\sigma_{ij}(\boldsymbol{x}) - \sigma_{ij}(\boldsymbol{x}_0)\}\Delta x_i \Delta x_k \\ + \{\sigma_{ik}(\boldsymbol{x}) - \sigma_{ik}(\boldsymbol{x}_0)\}\Delta x_i \Delta x_j \quad \text{with} \quad i \neq j \neq k \neq i.$$

Here $\boldsymbol{x} - \boldsymbol{x}_0 = \Delta x_i \boldsymbol{e}_i$. These forces are equal to the product of the mass of the specimen volume

$$\Delta m = \rho \Delta x_1 \Delta x_2 \Delta x_3$$

and the ith component of the acceleration $\partial^2 \xi_i / \partial t^2$, where $\boldsymbol{\xi}$ is the displacement vector and ρ is the density. If one expands

$$\sigma_{ij}(\boldsymbol{x}) = \sigma_{ij}(\boldsymbol{x}_0) + \frac{\partial \sigma_{ij}}{\partial x_q} \Delta x_q + \cdots \qquad \text{(here one must sum over } q\text{)}$$

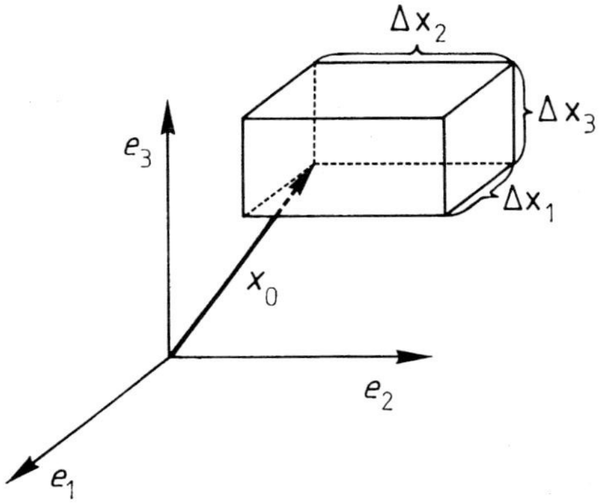

Figure 4.43 Parallelepipedic specimen.

and contends oneself with the first differential quotient, one gets

$$\frac{\partial \sigma_{ii}}{\partial x_q}\Delta x_j \Delta x_k \Delta x_q + \frac{\partial \sigma_{ij}}{\partial x_q}\Delta x_i \Delta x_k \Delta x_q + \frac{\partial \sigma_{ik}}{\partial x_q}\Delta x_i \Delta x_j \Delta x_q = \rho \Delta x_1 \Delta x_2 \Delta x_3 \frac{\partial^2 \xi_i}{\partial t^2}.$$

After dividing by $\Delta x_1 \Delta x_2 \Delta x_3$, one has for $i = 1, 2, 3$

$$\frac{\partial \sigma_{ii}}{\partial x_q}\frac{\Delta x_q}{\Delta x_i} + \frac{\partial \sigma_{ij}}{\partial x_q}\frac{\Delta x_q}{\Delta x_j} + \frac{\partial \sigma_{ik}}{\partial x_q}\frac{\Delta x_q}{\Delta x_k} = \rho \frac{\partial^2 \xi_i}{\partial t^2}.$$

Since Δx_i shall be freely selectable, independent of one another, we must demand that $\partial \sigma_{ij}/\partial x_q = 0$ for $q \neq j$.

This results in the elastodynamic basic equations

$$\frac{\partial \sigma_{ij}}{\partial x_j} = \rho \frac{\partial^2 \xi_i}{\partial t^2} \quad \text{(summing over } j = 1, 2, 3 \text{ at fixed } i\text{, respectively)},$$

which we can also write as $\mathrm{Div}\, \sigma_{ij} = \rho \partial^2 \xi_i / \partial t^2 \boldsymbol{e}_i$. The operation Div is the so-called vector divergence which, by differentiation and summation, generates the vector $\frac{\partial \sigma_{ij}}{\partial x_j}\boldsymbol{e}_i$ from the tensor $\{\sigma_{ij}\}$.

If we now introduce the linear relationship of Hooke's law $\sigma_{ij} = c_{ijkl}\varepsilon_{kl}$ with

$$\varepsilon_{kl} = \frac{1}{2}\left(\frac{\partial \xi_k}{\partial x_l} + \frac{\partial \xi_l}{\partial x_k}\right), \quad \text{we obtain}$$

$$\frac{\partial \sigma_{ij}}{\partial x_j} = \frac{1}{2} c_{ijkl}\left(\frac{\partial^2 \xi_k}{\partial x_l \partial x_j} + \frac{\partial^2 \xi_l}{\partial x_k \partial x_j}\right).$$

Because $c_{ijkl} = c_{ijlk}$, the following elastodynamic basic equations result for deformations within the limits of Hooke's law:

$$\frac{\partial \sigma_{ij}}{\partial x_j} = c_{ijkl}\frac{\partial^2 \xi_k}{\partial x_j \partial x_l} = \rho \frac{\partial^2 \xi_i}{\partial t^2}.$$

Now we will consider the propagation of plane waves, where especially simple laws exist just as in optics. Hence we write $\boldsymbol{\xi} = \boldsymbol{\xi}_0 e^{2\pi i(\boldsymbol{k}\cdot\boldsymbol{x}-\nu t)}$. $\boldsymbol{k}$ is the propagation vector with magnitude $|\boldsymbol{k}| = 1/\lambda$ as before; λ is the wavelength of the plane wave and ν the frequency. With $\boldsymbol{k}\cdot\boldsymbol{x} = k_i x_i$ and

$$\frac{\partial^2 \xi_i}{\partial t^2} = -4\pi^2\nu^2\xi_i \quad \text{and} \quad \frac{\partial^2 \xi_k}{\partial x_j \partial x_l} = -4\pi^2 k_j k_l \xi_k$$

we obtain $\rho\nu^2\xi_i - c_{ijkl}k_j k_l \xi_k = 0$.

Setting $\boldsymbol{k} = |\boldsymbol{k}|\boldsymbol{g}$ with $\boldsymbol{g} = g_i \boldsymbol{e}_i$ and noting that $\nu/|\boldsymbol{k}| = \nu\lambda = v$ signifies the propagation velocity, the elastodynamic basic equations for plane waves, assuming the validity of Hooke's law, take the form

$$(-\rho v^2 \delta_{ik} + c_{ijkl} g_j g_l)\xi_k = A_{ik}\xi_k = 0$$

for $i = 1, 2, 3$ ($\delta_{ik} = 1$ for $i = k$, otherwise 0). Here ξ_k denote the components of the displacement vector, whose position with respect to the propagation vector is to be calculated from the system of equations, when the components c_{ijkl} are known and the propagation direction $\boldsymbol{g}$ is given. The system only yields a solutions for $\boldsymbol{\xi} \neq 0$ when its determinant vanishes, hence

$$|A_{ik}| = |-\rho v^2 \delta_{ik} + c_{ijkl} g_j g_l| = 0.$$

Since we will use these determinants, the so-called *Christoffel determinants*, repeatedly, let us write them out

$$\begin{vmatrix} -\rho v^2 + c_{1j1l}g_j g_l & c_{1j2l}g_j g_l & c_{1j3l}g_j g_l \\ c_{2j1l}g_j g_l & -\rho v^2 + c_{2j2l}g_j g_l & c_{2j3l}g_j g_l \\ c_{3j1l}g_j g_l & c_{3j2l}g_j g_l & -\rho v^2 + c_{3j3l}g_j g_l \end{vmatrix} = 0.$$

Thus we obtain a third-order equation in v^2 for each given propagation direction $\boldsymbol{g}$, i.e., in general, three different values for v^2. The propagation velocity of the associated wave is the same in a direction and in its opposite direction. We now show that the velocities v', v'', and v''' belonging to the displacement vectors $\boldsymbol{\xi}'$, $\boldsymbol{\xi}''$, and $\boldsymbol{\xi}'''$ are mutually perpendicular, i.e., $\boldsymbol{\xi}'\cdot\boldsymbol{\xi}'' = 0$, where $\boldsymbol{\xi}'$ and $\boldsymbol{\xi}''$ are two of the three displacement vectors.

The following basic equations are valid for both velocities v' and v'',

$$-\rho v'^2 \xi'_i + c_{ijkl} g_j g_l \xi'_k = 0$$
$$-\rho v''^2 \xi''_i + c_{ijkl} g_j g_l \xi''_k = 0.$$

Multiplying the first equation by ξ''_i, the second by ξ'_i, forming the difference and summing over i gives us

$$-\rho(v'^2 - v''^2)\xi'_i\xi''_i + c_{ijkl} g_j g_l (\xi'_k \xi''_i - \xi'_i \xi''_k) = 0.$$

The second term vanishes because $c_{ijkl} = c_{klij}$. Therefore,

$$(v'^2 - v''^2)\xi'_i\xi''_i = 0, \quad \text{hence} \quad \boldsymbol{\xi}' \cdot \boldsymbol{\xi}'' = 0, \quad \text{in the case} \quad v' \neq v''.$$

Accordingly, the three displacement vectors are mutually perpendicular, as stated, and form a Cartesian reference system, which also in the case of triclinic crystals is produced by nature for each arbitrary propagation direction. The special case $v' = v''$ (degeneracy) appears in crystals only in distinct singular directions. Here, we can only ascertain that the displacement vectors for v' and v'' are perpendicular to the displacement vector $\boldsymbol{\xi}'''$ belonging to v'''. The case $v' = v'' = v'''$ does not exist. We can easily calculate the position of the displacement vectors, when we write the basic equations in the form $A_{ik}\xi_k = 0$ with

$$A_{ik} = -\rho v^2 \delta_{ik} + c_{ijkl} g_j g_l,$$

where A_{ik}, for fixed $i = 1, 2, 3$, form the components of a vector perpendicular to the associated $\boldsymbol{\xi}$ (for each fixed v, which first must be calculated from the determinant). Thus, $\boldsymbol{\xi}$ runs parallel to the vector product of any two arbitrary vectors of the three vectors $\boldsymbol{A}_i = A_{ik}\boldsymbol{e}_k$, hence $\boldsymbol{\xi} = q\boldsymbol{A}_i \times \boldsymbol{A}_j$, where q is an arbitrary constant.

In general, the propagation direction $\boldsymbol{g}$ and the displacement vector $\boldsymbol{\xi}$, according to the above, form an angle ζ given by

$$\cos\zeta = \frac{[\boldsymbol{g}, \boldsymbol{A}_i, \boldsymbol{A}_j]}{|\boldsymbol{A}_i \times \boldsymbol{A}_j|}.$$

If $\zeta = 0$, we speak of a pure *longitudinal wave*, as also appearing, for example, in liquids and gases; for the case $\zeta = 90°$ we have a pure *transverse wave* analogous to the $\boldsymbol{D}$-wave in optics. Here, we are also dealing with singular cases. In an arbitrary direction, we expect a *combination wave* composed of a longitudinal and a transversal component. Both propagate simultaneously in the same direction with the same velocity.

An interesting relationship exists between the extreme values of the longitudinal effect c'_{1111} in a direction $\boldsymbol{e}'_1 = u_{1j}\boldsymbol{e}_j$ and the character of the wave. We consider the auxiliary function $H = c'_{1111} + \lambda \sum_i u_{1i}^2$, where $\sum_i u_{1i}^2 = 1$. Extreme values of H appear, when $\partial H/\partial u_{1i} = 0$, hence,

$$4c_{ijkl}u_{1j}u_{1k}u_{1l} + 2\lambda u_{1i} = 0 \quad \text{with} \quad c'_{1111} = u_{1i}u_{1j}u_{1k}u_{1l}c_{ijkl}.$$

The factor 4 in the first term stems from the permutability of the indices. If we multiply this equation with u_{1i} and sum over i, we find $\lambda = -2c'_{1111}$. Thus

$$-c'_{1111}u_{1i} + c_{ijkl}u_{1j}u_{1k}u_{1l} = 0.$$

The elastodynamic basic equation is completely analogous to

$$-\rho v^2 \xi_i + c_{ijkl} g_j g_l \xi_k = 0.$$

Multiplying the first equation by ξ_i, the second by u_{1i}, summing over i and forming the difference gives

$$(-c'_{1111} + \rho v^2) u_{1i} \xi_i + c_{ijkl}(u_{1j} u_{1k} u_{1l} \xi_i - g_j g_l u_{1i} \xi_k) = 0.$$

The second term vanishes, when we make $e'_1 = g$, i.e., the propagation direction lies in the direction of an extreme value of the longitudinal effect. If $e'_1 \cdot \xi \neq 0$, we have $\rho v^2 = c'_{1111}$. If we now compare the initial equations and note that $g_i = u_{1i}$, we recognize immediately that ξ_i must equal $q g_i$; q is again an arbitrary constant. This means that a pure longitudinal wave runs in the direction of an extreme value of the longitudinal effect. Because of the orthogonality of the three displacement vectors, both waves perpendicular to the propagation direction are polarized and thus represent pure transverse waves. As a supplementary comment, let us note that according to Neumann's principle, an extreme value for c'_{1111} must be present along rotation axes or rotation–inversion axes with $n \geq 2$. Accordingly, pure longitudinal waves exist in these directions. Even in triclinic crystals there exists at least three different directions with pure longitudinal waves.

As just discussed, the condition of the solvability of the elastodynamic basic equations leads to a third-order equation in ρv^2. The coefficients of the powers of ρv^2 depend on c_{ijkl} as well as on the propagation direction g. The coefficient for the term $(\rho v^2)^2$ possesses a special significance in so far as, according to the fundamental theorem of algebra, it is equal to the sum of the roots, hence, equal to

$$\rho(v'^2 + v''^2 + v'''^2).$$

From the determinant $|A_{ik}|$, one obtains, for these coefficients, the value $c_{ijkl} g_j g_l \delta_{ik} = \sum_{i=1,2,3} c_{ijil} g_j g_l$. The quantities $E_{jl} = c_{ijkl} \delta_{ik}$ represent the components of a second-rank tensor (contraction of the tensor $\{c_{ijkl}\}$). The sum of the squares of the propagation velocities, multiplied by the density, is accordingly, for each arbitrary direction $e'_1 = g_{1i} e_i$ (with $g_{1i} = g_i$) equal to the longitudinal effect of the tensor $\{E_{jl}\}$. We call the tensor $\{E_{jl}\}$ the *dynamic elasticity*. It represents the directional dependence of the averaged squares of the sound velocity and is thus a useful measure for the elastic anisotropy. This relation plays a helpful role as a control in the practical determination of the elastic constants.

At a first glance, the impression exists that the basic equations derived here adequately describe the elastic behavior for all materials. This is, however, incorrect. Of great practical importance are the additional interactions occurring in piezoelectric and piezomagnetic crystals (crystals with ordered magnetic structure). The piezoelectric effect causes, via the stress components σ_{ij}, an electric displacement. The elastic waves are thus accompanied by electric waves. The simultaneous elastic, piezoelectric, and dielectric interactions are described to a first approximation, under negligibly small pyroelectric effect, by the following equations:

$$\varepsilon_{ij} = s^E_{ijkl}\sigma_{kl} + \hat{d}_{ijm}E_m = s^E_{ijkl}\sigma_{kl} + d_{mij}E_m,$$
$$D_i = d_{ikl}\sigma_{kl} + \epsilon^\sigma_{im}E_m \quad (\hat{d}_{ijm} = d_{mij}, \quad \text{see Section 4.4.1.3}).$$

In the case of a elastomagnetic interaction, a corresponding term of the magnetostriction and the associated relation for the magnetization would be added. As we have seen in the derivation of the elastodynamic basic equations, it may be easy to use the deformation as an inducing quantity. Instead of the above equations we then have with $\hat{e}_{ijm} = -e_{mij} = c_{ijkl}\hat{d}_{klm}$

$$\sigma_{ij} = c^E_{ijkl}\varepsilon_{kl} + \hat{e}_{ijm}E_m = c^E_{ijkl}\varepsilon_{kl} - e_{mij}E_m$$
$$D_i = e_{ikl}\varepsilon_{kl} + \epsilon^\varepsilon_{im}E_m.$$

The upper indices E and ε mean: at constant electric field and constant deformation state, respectively. The reason for replacing the quantities $-e_{mij}$ by the coefficients $\hat{e}_{ijm}$ in the same manner as $\hat{d}_{ijm} = d_{mij}$ (apart from the sign), will be explained in Section 5. Moreover, the Maxwell equations must be fulfilled.

As before $\operatorname{Div}\sigma_{ij} = \rho\partial^2\xi_i/\partial t^2 \boldsymbol{e}_i$. Further, in nonconducting crystals we have $\operatorname{div}\boldsymbol{D} = 0$. We can describe the electric field coupled to the elastic wave, to sufficient approximation, by $\boldsymbol{E} = -\operatorname{grad}\phi$, where the potential ϕ possesses the form $\phi = \phi_0 e^{2\pi i(\boldsymbol{k}\cdot\boldsymbol{x} - vt)}$ analogous to the elastic wave $\boldsymbol{\xi} = \boldsymbol{\xi}_0 e^{2\pi i(\boldsymbol{k}\cdot\boldsymbol{x} - vt)}$.

This means $\boldsymbol{E} \parallel \boldsymbol{g}$ and $\operatorname{rot}\boldsymbol{E} = 0$ and hence $\boldsymbol{B} = \text{const.}$ and $\boldsymbol{D} = \text{const.}$ The justification for this statement was given by Hutson & White (1962) on the basis of the exact relationship derived by Kyame (1949). These authors showed that the relative deviations from the exact solutions for v^2 are proportional to the square of the ratio of sound- to light velocity in the crystal, hence in the order of magnitude of at most 10^{-8}.

Thus,

$$\operatorname{Div}\sigma_{ij} = \{c^E_{ijkl}g_j g_l \xi_k + e_{mij}g_m g_j \phi\}(-4\pi^2 \boldsymbol{k}^2)\boldsymbol{e}_i.$$

$\operatorname{div}\boldsymbol{D} = 0$ gives $\boldsymbol{g}\cdot\boldsymbol{D} = 0$, if $\boldsymbol{D} = \boldsymbol{D}_0 e^{2\pi i(\boldsymbol{k}\cdot\boldsymbol{x} - vt)}$, hence $e_{nkl}g_n g_l \xi_k - \epsilon^\varepsilon_{rs}g_r g_s \phi = 0$ or

$$\phi = \frac{e_{nkl}g_n g_l \xi_k}{\epsilon^\varepsilon_{rs}g_r g_s}.$$

The denominator $\epsilon^{\varepsilon}_{rs} g_r g_s$ represents the longitudinal effect of the dielectric tensor in the propagation direction g. Thus, for piezoelectric crystals, we obtain the following basic equations ($i = 1, 2, 3$)

$$\left\{ -\rho v^2 \delta_{ik} + \left(c^E_{ijkl} + \frac{e_{mij} e_{nkl} g_m g_n}{\epsilon^{\varepsilon}_{rs} g_r g_s} \right) g_j g_l \right\} \xi_k = 0.$$

New indices were used in some positions, when this was required for the independence of the summation.

The difference compared to the basic equations for nonpiezoelectric materials is that instead of c^E_{ijkl} the quantities

$$c^D_{ijkl}(g) = c^E_{ijkl} + \frac{e_{mij} e_{nkl}}{\epsilon^{\varepsilon}_{rs} g_r g_s} g_m g_n$$

appear. That c^D_{ijkl} actually takes this form in the case $D = 0$, results from the relation $D_i = e_{ikl}\varepsilon_{kl} + \epsilon^{\varepsilon}_{im} E_m = 0$, after multiplying the ith equation by g_i, summing over i and introducing the resulting expression for ϕ in the equation for σ_{ij}.

This is consistent with the condition $D = \text{const.}$, which follows from the Maxwell equations owing to $\operatorname{rot} H = 0$. Hence, for piezoelectric and nonpiezoelectric crystals we can write the basic equation in the general form $(-\rho v^2 \delta_{ik} + c^D_{ijkl} g_j g_l)\xi_k = 0$. Since $c^D_{ijkl} = c^D_{klij}$, all relations derived from the ordinary elastodynamic basic equations are equally valid. In particular, even in piezoelectric crystals, the displacement vectors ξ', ξ'' and ξ''' associated with a propagation direction g are mutually orthogonal. If one succeeds to measure the components c^D_{ijkl} and c^E_{ijkl} or their difference, one obtains statements on piezoelectric quantities important not only for the determination of the piezoelectric tensor but also for the technical application of such crystals. The quantity $(c^D_{ijkl} - c^E_{ijkl})/c^D_{ijkl}$ is a measure of the fraction of electrical energy to the total energy of the elastic wave. This and similarly defined quantities are called *coupling factors*. We will return to these later.

In piezoelectric crystals possessing a certain electrical conductivity, as for example, LiO_3 or semiconductors of the GaAs-type or CdS-type, an attenuation of elastic waves occurs because the accompanying electric field breaks down due to the conductivity (Hutson & White, 1962). The elastic wave experiences thereby continuously a loss of energy. In the relationships just derived, this phenomenon can be taken into account by introducing complex dielectric constants. In pyroelectric crystals, the relationships derived above must be modified by adding pyroelectric terms (see Section 5).

Before we turn to the procedures for the measurement of propagation velocities of elastic waves, a few remarks are necessary on the validity of the derived relationships. From our approach it is clear that only small deformations are allowed in order to remain within the limits of Hooke's law (linear elasticity). We will discuss higher order effects in Section 4.6.3. Moreover, we have neglected all types of mechanical damping (scattering and excitation processes) by assuming real components c_{ijkl}, without, however, giving extra emphasis to this aspect. We have also not considered that similar effects as found in optical activity, may exist between the components of the stress and the deformation tensor. Such effects have actually been observed in certain crystals (*acoustic activity*). We will also return to this point. The greatest problem in all these measurements is, however, the approach of plane waves, in which we have assumed that they propagate completely independently in all directions. This strictly applies only to infinitely extended bodies. One must therefore carefully consider under which circumstances the relationships gained above may be taken as a sufficient approximation.

To what extent one can regard a crystal as approximately infinite depends on the ratio B of the specimen dimensions perpendicular to the wave normals and the wavelength λ. Experimentally it has been found that from $B/\lambda > 20$ on, the specimen dimensions have no influence on the observed propagation velocities exceeding 0.5% of the value for an infinite crystal. This is also confirmed by simple model calculations. For the practical measurements of propagation velocities it is therefore appropriate to adjust the wavelength by selecting the frequency to fulfil the above condition. In contrast to crystal optics in the visible spectrum, one observes over a wide frequency range, no substantial dependence of the propagation velocity of elastic waves on frequency. In most materials a weak dispersion begins first above 1000 MHz. The energy of quanta with a frequency of 1000 MHz is usually by far too small to excite states of elastic oscillations in lattice particles.

In some measurement methods the specimen is externally irradiated with sound waves (for example, pulse-echo; forced oscillations). It should then be noted that just as in optics, the sound wave and the associated elastic energy flux density, represented by the ray vector s, can encompass a finite angle, leading to a sideways drift of the sound ray. We obtain the energy flux density vector from the total elastic energy W per unit volume as follows: We have

$$W = \frac{1}{2}\int_V \left(\rho \frac{\partial \xi_i}{\partial t}\frac{\partial \xi_i}{\partial t} + \sigma_{ij}\varepsilon_{ij}\right) dV;$$

the first terms correspond to the kinetic energy constant and the second to the familiar elastic deformation work. The time change of this total energy is

$$\frac{\partial W}{\partial t} = \int_V \left(\rho \frac{\partial \xi_i}{\partial t}\frac{\partial^2 \xi_i}{\partial t^2} + \frac{\partial}{2\partial t}(\sigma_{ij}\varepsilon_{ij})\right) dV.$$

With $\sigma_{ij} = c_{ijkl}\varepsilon_{kl}$ we get

$$\frac{\partial}{\partial t}(\sigma_{ij}\varepsilon_{ij}) = 2\sigma_{ij}\frac{\partial \varepsilon_{ij}}{\partial t}.$$

Inserting the dynamic basic equation

$$\frac{\partial \sigma_{ij}}{\partial x_j} = \rho \frac{\partial^2 \xi_i}{\partial t^2}$$

in the first term results in

$$\frac{\partial W}{\partial t} = \int_V \left(\frac{\partial \xi_i}{\partial t} \frac{\partial \sigma_{ij}}{\partial x_j} + \sigma_{ij} \frac{\partial \varepsilon_{ij}}{\partial t} \right) dV = \int_V \frac{\partial}{\partial x_j} \left(\sigma_{ij} \frac{\partial \xi_i}{\partial t} \right) dV.$$

From Gauss's theorem, the volume integral is equal to a surface integral given by $\int_V \operatorname{div} \boldsymbol{Q} dV = \int_O \boldsymbol{Q} \cdot d\boldsymbol{o} = \int_O Q_j n_j df$, where $d\boldsymbol{o}$ specifies a surface element of size df with the surface normal (unit vector) $\boldsymbol{n} = n_j \boldsymbol{e}_j$. Thus

$$\frac{\partial W}{\partial t} = \int_O \sigma_{ij} \frac{\partial \xi_i}{\partial t} n_j df.$$

If we now imagine the specimen volume as a rectangular parallelepiped with edges parallel to the Cartesian basic vectors, we recognize that $\sigma_{ij} \partial \xi_i / \partial t$ represents the jth component of the energy flux density vector. Hence, we have $s_j = \sigma_{ij} \partial \xi_i / \partial t$, whereby the sign is selected such that the components parallel to the propagation direction become positive. In the case of an elastic wave $\boldsymbol{\xi} = \boldsymbol{\xi}_0 e^{2\pi i (\boldsymbol{k} \cdot \boldsymbol{x} - \nu t)}$, $s_j = \frac{\omega^2}{v} c_{ijkl} \xi_i \xi_k g_l$, as one can easily confirm ($\omega = 2\pi\nu$).

For the time average we obtain

$$\bar{s}_j = \frac{\omega^2}{2v} c_{ijkl} \xi_{0i} \xi_{0k} g_l.$$

We can now calculate the angle between the ray vector $\boldsymbol{s}$ and propagation direction $\boldsymbol{g}$ with the help of $\boldsymbol{s} \cdot \boldsymbol{g}$. The important cases are those in which both vectors lie parallel to each other. This always applies to pure longitudinal waves. As proof let $\boldsymbol{e}_1'$ be parallel to the propagation direction. Then $g_1' = 1$, $g_2' = g_3' = \xi_2' = \xi_3' = 0$ and $\boldsymbol{\xi} = \xi_1' \boldsymbol{e}_1'$ parallel to the propagation direction. Thus $\bar{s}_1' = \frac{\omega^2}{2v} \xi_0^2 c_{1111}'$ and $\bar{s}_j' = \frac{\omega^2}{2v} \xi_0^2 c_{1j11}'$ for $j = 2$ and 3. From the basic equation $(-\rho v^2 \delta_{ik} + c_{ijkl} g_j g_l) \xi_k = 0$ one obtaines in the primed system for pure longitudinal waves

$$-\rho v^2 g_i' + c_{ijkl}' g_j' g_k' g_l' = 0$$

and especially for $g_1' = 1$, $g_2' = g_3' = 0$

$$g_1' = \frac{c_{1111}'}{\rho v^2} = 1, \quad g_2' = \frac{c_{2111}'}{\rho v^2} = 0 \quad \text{and} \quad g_3' = \frac{c_{3111}'}{\rho v^2} = 0.$$

Therefore, $c_{2111}' = c_{3111}' = 0$ and hence $\bar{s}_2' = \bar{s}_3' = 0$.

If transverse waves with $\boldsymbol{\xi} \parallel \boldsymbol{e}_1$ run along an n-fold rotation- or rotation-inversion axis parallel to $\boldsymbol{e}_3$, hence $g_3 = 1$, $g_1 = g_2 = \xi_2 = \xi_3 = 0$, one finds $\bar{s_j} = c_{1j13}\xi_0^2 \frac{\omega^2}{2v}$. If n or $\bar{n}$ is even, the index 3 may only occur an even number of times (2 or $\bar{2}$ is contained in n or $\bar{n}$, respectively!). This means $\bar{s_1} = \bar{s_2} = 0$ and $\bar{s_3} = c_{1313}\xi_0^2 \frac{\omega^2}{2v}$.

In general, with a threefold axis, the energy flow runs askew to the propagation direction. This causes the so-called internal conic refraction, an analogous phenomenon known in optics, which here, however, can also appear in cubic crystals (see Exercise 28).

4.5.6 Dynamic Measurement Methods

Dynamic methods to measure elastic properties have achieved a special significance in solid-state research with the development of ultrasound technology and precision frequency measurements. In the following, we discuss the most important methods.

These are

1. Pulse-echo methods,
2. Schaefer–Bergmann method,
3. Resonances of plates and rods,
4. Brillouin scattering,
5. Neutron scattering,
6. X-ray scattering (thermal scattering).

1. Pulse-Echo Methods and Related Methods

The specimen in the form of a plane-parallel plate of thickness L in the direction of the face normals is irradiated with short ultrasonic pulses of a few microseconds duration (about 20 wave trains at 10 MHz). The ultrasonic generator, usually a thin quartz plate excited to mechanical oscillations by a high-frequency electric field, is fixed to one side of the specimen by a thin film of an appropriate resin, for example (Fig. 4.44). The ultrasonic pulse is reflected at the boundary and returns to the generator, which in the meantime is electronically switched to act as a receiver. It transforms, via the piezoelectric effect, a part of the elastic energy of the incoming echo into a voltage pulse, which can be observed on an oscilloscope with time-proportional x-deflection. Repeated reflections of the same pulse generate a sequence of quasiequidistant echos in time, when the propagation direction remains unchanged. The time difference of m consecutive echos corresponds to the delay time of the sound wave

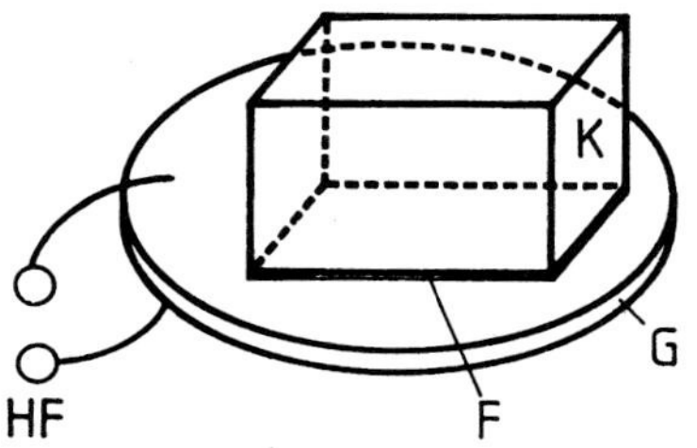

Figure 4.44 Pulse-echo method. The crystal K (a plane-parallel plate) is pasted on the ultrasonic generator G by an oil film F or a highly viscous material (necessary in the case of transverse waves). In the intervals between the pulses the generator works as a piezoelectric ultrasound probe.

through the $2m$-fold specimen length and a certain path length covered by the sound wave in the generator or receiver until reaching the maximum of the electric pulse. The required correction can be experimentally determined, for example, by measurements on specimens of the same orientation but different thicknesses. Let Δt be the corrected delay time. The velocity of the elastic wave is then $v = 2mL/\Delta t$. The delay time Δt can be directly read from the scale of the oscilloscope after calibration using standards with known sound velocity or determined by the aid of electronic measuring equipment. One repeats the pulses after a few milliseconds to hold the oscillogram still.

To generate longitudinal waves or waves with large longitudinal components we work with a so-called thickness resonator, for example, a quartz plate cut perpendicular to the direction e_1 of the longitudinal piezoelectric effect (perpendicular to a twofold axis; X-cut). Paraffin oil and at lower temperatures, a mixture of low boiling hydrocarbons have proven suitable as a cementing liquid. Transverse waves, for example, can be generated with quartz plates, whose normals are perpendicular to e_1 (general Y-cut). The displacement vector of these waves then lies parallel to e_1 (in the face of the plate) when an electric field with a component E_2 is applied. The transfer of the transverse waves to the specimen can only be achieved, with good efficiency, with the help of high-strength adhesives or cementing with a highly viscous liquid. Standard cementing materials are beeswax with paraffin oil additive for lower temperatures, benzophenone for temperatures around 20° and high-strength dental cement for higher temperatures. We will discuss below, in more detail, the quantitative treatment of the oscillating piezoelectric plate.

With commercial equipment, primarily conceived to test materials for cracks and other inhomogeneities, sound velocities can be measured to an accuracy of about 0.5%, when the dimensions of the specimens are at least ten times the wavelength and the propagation direction remains sufficiently sharp for all echos. In this regard, it is essential that the surface of the generator emitting the ultrasonic pulse is sufficiently large.

For precision measurements, the simple pulse-echo method is less suitable, in particular with crystals of low symmetry, because pure wave types are only possible in a few distinct directions. Hence two or all three combination waves are simultaneously excited. The selection of echos is easy in principle, but inaccuracies creep in when trying to determine the position of the maxima of the voltage pulses triggered by the echos.

We illustrate a special advantage, in comparison to most other methods, in more detail. This is the measurement of the attenuation of the sound waves which one can read directly from the decay of the maxima of the voltage pulses of consecutive echos. To describe the attenuation it is convenient to introduce an individual attenuation coefficient α for each wave.

Then

$$\xi = \xi_0 e^{2\pi i(\boldsymbol{k}\cdot\boldsymbol{x}-\nu t)} e^{-\alpha(\boldsymbol{g}\cdot\boldsymbol{x})}.$$

This behavior can also be formally described with a complex propagation vector $\boldsymbol{k} = \boldsymbol{k}' + i\boldsymbol{k}''$, where $\boldsymbol{k}'' \parallel \boldsymbol{k}$ and $v = \nu/|\boldsymbol{k}'|$.

Hence

$$\alpha = 2\pi|\boldsymbol{k}''| \qquad (\text{because} \quad \alpha\boldsymbol{g} = 2\pi\boldsymbol{k}'').$$

The attenuation coefficients with respect to elastic constants are by no means to be represented as components of a tensor. Similar to optics, there also exists a formal description for elastic absorption with the aid of complex components of the elasticity tensor. Since, however, the major part of the attenuation is usually generated by inhomogeneities, such an approach is of limited use. One can gain certain insights concerning the perfection of crystals from attenuation phenomena.

The value α results from the amplitude ratio of two consecutive echos according to

$$\frac{|\xi_{0,n+1}|}{|\xi_{0,n}|} = e^{-Q} e^{-\alpha 2L}.$$

The factor e^{-Q} takes into account the losses due to reflection at the generator and on the other side of the specimen. e^{-Q} can be eliminated by measurements on two specimens of different lengths L and L'.

The exponential decay of the amplitudes of the echos is directly read from the envelope of the voltage maxima on the screen of the oscilloscope (Fig. 4.45). Deviations from plane-parallelism as well as a divergence of the sound beam lead to oscillations of the envelope, which can impair the accuracy of the measurement.

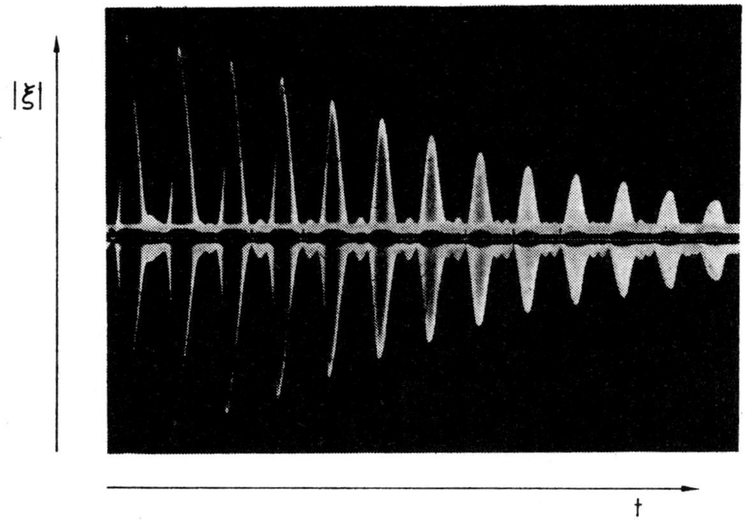

Figure 4.45 A sequence of pulse-echos of longitudinal waves in NaF. Propagation direction [100], frequency 5 MHz, length of specimen about 30 mm.

One can also directly record the sound pulse on the surface opposite to the generator-side with the help of a piezoelectric receiver (transmission).

Electronic processing of the time displacement of primary pulse and echo or transmission pulse enables a substantial improvement in the accuracy of measurement of small relative changes in sound velocity. Forgacs (1960) achieved an improvement using the time difference between start and arrival of the signal as a frequency-determining element for the repetition of the start pulse (sing-around-method). A second interesting method is the superposition of the echo or the transmission signal with the primary pulse. In this method one obtains, dependent on the frequency of the carrier wave of the pulse, sharply adjustable superposition profiles, which respond sensitively to small changes in the sound velocity (McSkimin, 1961). With these and related methods, one can, under suitable conditions, measure changes in sound velocity of the order of 10^{-8}. These methods are particularly suited for the measurement of changes in elastic properties under the influence of external parameters, as for example, hydrostatic or unidirectional pressure, temperature, electric and magnetic fields. The accuracy of the measurement of the absolute sound velocity cannot, however, be substantially increased by these methods, because the influence of the adhesive film and the generator or receiver cannot be directly eliminated.

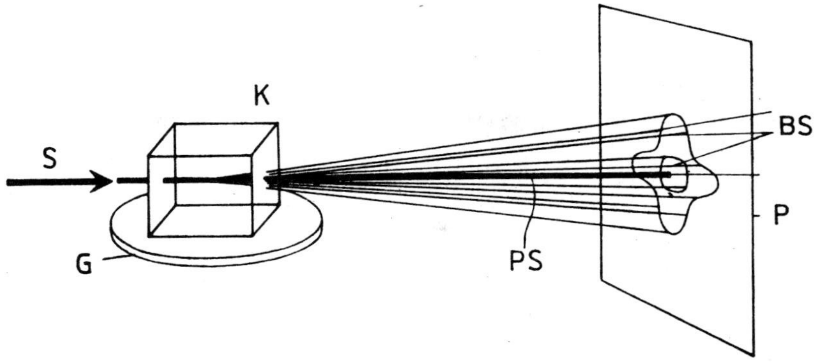

Figure 4.46 Schaefer–Bergmann method (diffraction of light by ultrasonic waves; S light beam, K crystal with ultrasound generator G, P photoplate or ground glass screen, PS primary beam, BS diffracted rays).

2. Schaefer–Bergmann Method

The periodic distortions accompanying a sound wave can cause a change in refractive indices due to the elastooptical effect $\Delta a_{ij} = p_{ijkl}\varepsilon_{kl}$. This results in the creation of a sine-shaped variation of the refractive index, that is, an optical grating, on which light can be diffracted. Now, if one irradiates a crystal, of no particular shape, with a sound wave of fixed frequency, a broad spectrum of sound waves of different propagation directions and displacement vectors are generated due to multiple reflections at the boundaries. A detailed analysis shows that in the case of small diffraction angles, only those sound waves can contribute to a noticeable diffraction of light, whose propagation vector is oriented approximately perpendicular to the propagation vector of the monochromatic light beam (Fig. 4.46). Hence, one furnishes the specimen with an optically transparent pair of faces approximately perpendicular to the face of the sound generator.

A monochromatic light beam entering the crystal approximately perpendicular to the pair of transparent faces, finds sound wave fields in nearly all propagation directions within these faces and is thus diffracted just about uniformly in all directions. The diffraction angle is taken from the formula for the optical ruled grating, as long as the diffraction angles are small enough and the optical path through the crystal is so short that the depth of the sound grating can be neglected. For optically quasiisotropic materials, a sufficient approximation is $2d \sin\vartheta = m\lambda$ analogous to the Bragg condition. 2ϑ is the diffraction angle, i.e., the angle of deflection of the diffracted wave from the direction of incidence, d is the acoustic wave length, λ the optical wavelength and m the order of the diffraction. Since, to a first approximation, only pure sinusoidal displacements occur, the higher orders cannot be interpreted by corresponding Fourier coefficients, as for example, in the case of X-ray diffraction. Rather, effects of multiple diffraction are present, that is, each strongly diffracted wave acts as a new primary wave. Since, in general, for each propagation direction, three different wave types exist with different propagation velocities, one obtains, outgoing from the crystal, three cones of diffracted rays which generate three curves on a photographic plate or screen behind the crystal. From these, the diffraction angles for the propagation directions perpendicular to the light beam and hence the propagation velocity of the ultrasonic waves can be determined. In the case of strong anisotropy, one obtains separate diffraction patterns for both directions of vibration of the incident light wave. These patterns show nearly the same diffraction angle in the case of weak ray double refraction. However, easily measurable deviations of around one percent occur in cases of strong double refraction, even if the diffraction angles are small (Küppers, 1971). The cause for the splitting is different refraction when the light wave enters and emerges as well as momentum conservation in the scattering process. Corresponding corrections are necessary, if a higher measurement accuracy is required. If R is the effective distance between the crystal and the photographic plate and $2r$ is the distance between the diffraction spots on the photographic plate belonging to the propagation vectors $\boldsymbol{g}$ and $-\boldsymbol{g}$, one obtains the angle of diffraction 2ϑ from

$$\tan 2\vartheta = r/R.$$

Assuming that the source of the diffracted ray lies in the geometrical center of the crystal, which is the case, when the specimen is uniformly irradiated by the primary ultrasonic wave. This takes into account the refraction of the diffracted ray at the rear of the crystal, one obtains for the effective distance $R = R_0 - L(1 - 1/n)/2$, where R_0 is the geometric distance crystal center-photographic plate, L the length of the crystal in the direction of the light wave and n the refractive index. Hence, one obtains for the velocity of the ultrasonic wave deflecting the light ray by the diffraction angle 2ϑ

$$v = \nu d = \nu \frac{\lambda}{2 \sin \vartheta} \qquad (m = 1),$$

where ν denotes the frequency of the ultrasonic wave. To achieve higher accuracy, it is recommended to make R_0, as well as the frequency ν of the ultrasonic wave as large as possible, for example, $R_0 = 5\,\text{m}$, ν around 20 MHz. Good sharpness of the diffraction spots is obtained, when the intense primary light beam travels through a pinhole (opening about 30 μm) and with the help of a field lens, through the vibrating crystal to the photographic plate to form there an image of the pinhole, enlarged by a factor of about five (Fig. 4.47). The third Laue condition is effective in long specimens and with large diffraction angles. That is, the intensity of the diffracted wave takes on its maximum under the

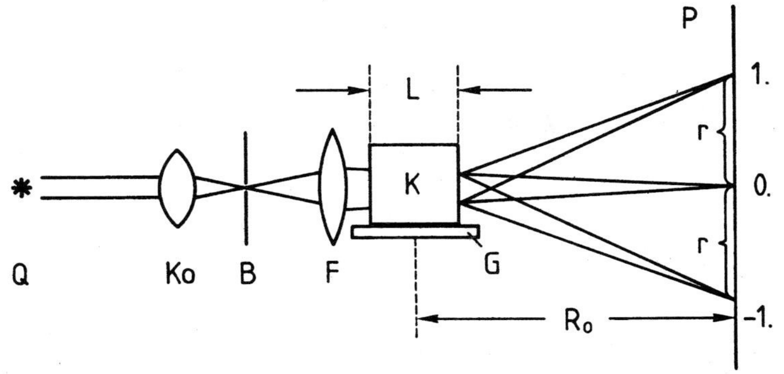

Figure 4.47 Arrangement of the Schaefer–Bergmann method. Q light source, Ko condenser, B fine pinhole, F field lens for imaging the pinhole through the crystal K on the photo plate or screen, respectively, G ultrasound generator, L length of the crystal, R_0 distance crystal—photo plate, r distance of the diffraction spots of first and minus first order from the primary spots 0.

Bragg condition (angle of incidence = angle of emergence, measured against the normals of the elastic wavefront). This means that one can only observe reflections with large diffraction angles (above about 30°) the Bragg condition is realized. Accordingly, with light waves of arbitrary incidence, one obtains, in such cases, only very few and weak reflections, if any at all. In order to obtain a diffraction spectrum of all propagation vectors perpendicular to the light ray it is thus necessary to keep the diffraction angles sufficiently small ($\vartheta < 3°$), i.e., the frequency of the ultrasound waves must be correspondingly low, in contrast to the above requirement for large diffraction angles in favor of higher accuracy. In specimens with arbitrary boundaries, a high proportion of reflected waves is created from the primary ultrasound wave, whose wave normals run outside the plane perpendicular to the light ray, and hence, do not contribute to diffraction. If one furnishes the crystal with a "gothic arch" type boundary as in Fig. 4.48, a major portion of the acoustic primary wave is transferred, via reflection, into partial waves with propagation directions within the plane. This results, in particular, in a more uniform distribution of the excited waves and to a substantial increase in the intensity of the diffracted light waves. The exposure times for making a diffraction photograph can thus be reduced to a fraction of a second. Furthermore, the intensity of the primary ultrasound wave can be minimized to largely avoid the interfering effects of crystal heating.

For the evaluation of Schaefer–Bergmann elastograms it is advantageous, when the plane of each effective propagation vector is chosen by taking into consideration the symmetry properties of the crystal, for example, in or-

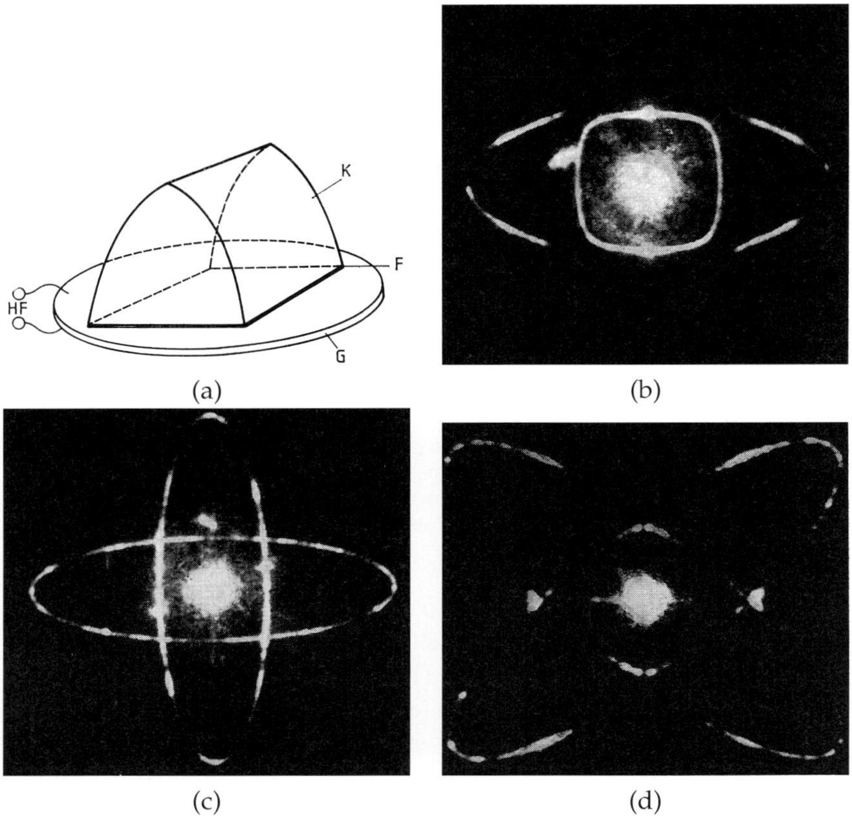

Figure 4.48 (a) "Gothic arch" type shape of the specimen for the generation of a wide spectrum of propagation vectors within the plane perpendicular to the light beam by reflection of the primary ultrasound wave (for symbols see Fig. 4.44). (b) Elastogram of KCl (PSG 4/m3), light beam along [100], primary ultrasound wave along [001]. (c) Elastogram of KH_2PO_4 (PSG $\bar{4}$m), light beam along [001], primary ultrasound wave along [100]. (d) Elastogram of calcium formate (PSG mmm), light beam along [001], primary ultrasound wave along [010]. The anisotropy of the diffraction angles allows to recognize the Laue symmetry in the direction of the optical transmission.

thorhombic crystals the planes (100), (010), and (001). In triclinic crystals the evaluation is simplified, when the light wave is incident in the direction of the Cartesian basic vectors e_i. Then namely, one g_i vanishes in each of the elastodynamic basic equations, so that only a fraction of the components c_{ijkl} is involved. From only three Schaefer–Bergmann elastograms of specimens with a 'gothic arch' shape one obtains the sound velocities for hundreds of independent directions, from which the complete elasticity tensor can be determined with the help of a suitable computer program.

This advantage is in confrontation with the disadvantage of low measurement accuracy. In the case of first-order diffraction, one normally achieves an accuracy of around 0.2% in determining the velocity of sound. The reason lies in a mutual coupling of the elastic waves, which, similar to coupled pendula, can influence the effective restoring forces (these correspond to the elastic constants). If one works with plane-parallel plates instead of the "gothic arch" shape, standing waves form between these plates at certain frequencies (resonances), generating a strongly pronounced optical grating in a single propagation direction. This produces multiple reflections of very high order, so that the diffraction angle can be measured to a relative accuracy of up to 10^{-5} thanks to the larger values for $2mr$. Moreover, by specifying the direction of vibration of the generator, one can excite quite definite wave types. Furthermore, with this boundary, an elastic primary wave of substantially lower power will suffice. By adding such precision measurements for some distinct directions, one can determine the elasticity tensor, even of triclinic crystals, to good accuracy. Indeed, the elasticity tensor of a triclinic crystal, namely $CuSO_4 \cdot 5H_2O$, was completely determined by this procedure (Haussühl & Siegert, 1969).

The Schaefer–Bergmann method is of course not suited for strongly absorbing crystals. Since the intensity of the diffracted rays depends on the elastooptical constants, the method of light diffraction on ultrasound waves can also be used to measure elastooptical properties (see Section 4.5.9.2).

3a. Resonances of Plates and Rods

We observe the simplest vibrating forms on plane-parallel plates and parallelepipedic rods thanks to the clear boundary conditions. Firstly, we consider a plane-parallel plate of thickness L with unlimited sides, suspended free of external stress. The possible forms of vibration must satisfy the differential equation

$$\frac{\partial \sigma_{ij}}{\partial x_j} = \frac{1}{2} c_{ijkl} \left(\frac{\partial^2 \xi_k}{\partial x_j \partial x_l} + \frac{\partial^2 \xi_l}{\partial x_j \partial x_k} \right) = \rho \frac{\partial^2 \xi_i}{\partial t^2}$$

as well as the boundary conditions "no external stresses." We select $\boldsymbol{e}_3$ parallel to the plate normal, and the origin in the center of the plate. Thus $\sigma_{3i}(L/2) = \sigma_{3i}(-L/2) = 0$ for $i = 1, 2, 3$. We need not consider the other boundary conditions, because we are only searching for solutions independent of the position coordinates x_1 and x_2, i.e., those that are homogeneous within each cross-section perpendicular to $\boldsymbol{e}_3$.

Assuming an antisymmetric solution for a standing wave

$$\begin{aligned} \xi &= \xi_0 \sin(2\pi \boldsymbol{k} \cdot \boldsymbol{x}) \cos \omega t \\ &= \xi_0 \sin(2\pi k_3 x_3) \cos \omega t \end{aligned}$$

in the above equation, gives

$$(-\rho v^2\delta_{ik} + c_{i3k3})\xi_k = 0,$$

where $\omega/2\pi k_3 = v$ is the vave velocity.

From the condition for the solution (Det. $= 0$), we obtain the three velocities v', v'' and v''', as for an infinite crystal where the wave propagates in the direction $\boldsymbol{e}_3$. The associated displacement vectors are calculated according to the rules discussed in Section 4.5.12. A general solution for the freely vibrating plate can be obtained by a linear combination of the three special solutions according to

$$\boldsymbol{\xi} = (\boldsymbol{\xi}_0' \sin(2\pi k_3' x_3) + \boldsymbol{\xi}_0'' \sin(2\pi k_3'' x_3) + \boldsymbol{\xi}_0''' \cdots) \cos \omega t.$$

The boundary conditions $\sigma_{3i}(\pm L/2) = 0$ lead to

$$c_{i3k3}\xi_{0k}' 2\pi k_3' \cos(2\pi k_3' L/2) + c_{i3k3}\xi_{0k}'' 2\pi k_3'' \cos(2\pi k_3'' L/2) + \xi_{0k}''' \ldots) = 0$$

with $i = 1, 2, 3$. This system of equations has only solutions with $\boldsymbol{\xi} \neq 0$, when its determinant vanishes. We must have $\cos(\pi k_3 L) = 0$ for one of the three values $k_3 = k_3', k_3''$ or k_3'''. This means $k_3 L = (2n-1)/2$, where n is integer. If, for example, the solution with v'' is excited, we must have $k_3'' = 1/\lambda'' = (2n-1)/2L$, i.e., an odd multiple of half the wavelength must equal the plate thickness.

We obtain a similar result for a symmetric solution of a standing wave $\boldsymbol{\xi} = \boldsymbol{\xi}_0 \cos(2\pi k_3 x_3) \cos \omega t$, however, with the condition, that now an integer multiple of the wavelength is equal to the plate thickness. Both results, the antisymmetric and the symmetric, together lead to the result that the characteristic vibrations of the infinite plane plate always appear, when an integer multiple of half the wavelength of the associated wave of the infinite crystal is equal to the plate thickness. The associated resonance frequencies are integer multiples of the fundamental frequency $\nu_1 = v/2L$, when no dispersion is present, which, as previously discussed, virtually always applies approximately below 1000 MHz. If one succeeds to excite the plate to its characteristic vibrations and measures the associated resonance frequencies, one has the possibility to determine the wave velocity v from the difference of these frequencies and the plate thickness according to $v = 2L(\nu_n - \nu_m)/(n-m)$, as one can directly read from the resonance condition ($v = \nu\lambda$). ν_n or ν_m specify the nth or mth resonance frequency.

An advantageous property of the resonances of thick plates consists in the fact, that the three states of vibration can be measured independent of one another, in other words, without coupling. For freely vibrating plane-parallel plates of piezoelectric crystals, the resonance condition $L = m\lambda/2$ (m integer) is also valid. Instead of c_{ijkl}, however, the components c_{ijkl}^D derived above must employed. The proof is carried out analogously, whereby

$$\sigma_{ij} = c^{E}_{ijkl}\varepsilon_{kl} - e_{mij}E_m, \qquad D_i = e_{ikl}\varepsilon_{kl} + \epsilon^{\varepsilon}_{im}E_m$$

with $E_m = -(\mathrm{grad}\,\phi)_m$ and $\boldsymbol{D} \cdot \boldsymbol{g} = 0$. Corresponding conditions apply to magnetostrictive crystals.

A change in the boundary conditions, as for example, in the case of forced vibrations, leads to a modified resonance condition, to which we will come later. The simplest way to excite resonances is by using a sound generator, which in order to transfer the vibrations must be brought in contact with one of the faces of the crystal. The vibrations can be transferred via an air gap, a thin wire, a rigid connection between generator and crystal, or directly by cementing the generator on the specimen. In particular, the latter is necessary, when exciting sound waves with strong transverse components. In excitation via air or wire contact, one observes hardly any interfering coupling between generator and specimen, i.e., the resonance frequencies of the crystal plate are nearly not influenced by the generator. This does not apply when the generator is firmly attached to the crystal. Rather, one observes resonances of the combined system. The resonance condition can be easily calculated, when the generator possesses also the form of a plane-parallel plate and generator and specimen posses a similar sound wave resistance ρv, i.e., when the sound wave can travel through the boundary surface between generator and specimen without strong reflection losses.[2] If, on the other hand, the sound wave resistance of both media is considerably different, one observes a strong coupling only close to the resonances of the generator. This arrangement has especially proven advantageous for the practical routine measurement of elastic properties. In a preferred version for investigations at very high frequencies (over 100 MHz) a thin piezoelectric generator, for example, made of hexagonal CdS is vapor-deposited on the crystal.

Electrostrictive crystals can be directly excited to vibrate by applying a high-frequency electric field to the metallized faces of the plate. Analogously magnetostrictive crystals can be excited by external magnetic fields. We will return to this in more detail in the next section. The resonances can also be excited in a wide frequency range with the help of the quadratic electrostrictive effect (see Section 4.5.10) or by the forces emanating from a high-frequency electric field of a plate capacitor, one plate of which is the metallized face of the

2) The intensity of the sound wave reflected at the boundary surface of two isotropic media I and II is given by (see, for example, Lord Raleigh, 1945)

$$\frac{I_{\mathrm{re}}}{I_0} = \frac{\left(1 - \frac{\rho_{\mathrm{I}} v_{\mathrm{I}} \cos\alpha_{\mathrm{II}}}{\rho_{\mathrm{II}} v_{\mathrm{II}} \cos\alpha_{\mathrm{I}}}\right)^2}{\left(1 + \frac{\rho_{\mathrm{I}} v_{\mathrm{I}} \cos\alpha_{\mathrm{II}}}{\rho_{\mathrm{II}} v_{\mathrm{II}} \cos\alpha_{\mathrm{I}}}\right)^2}.$$

specimen. Here, however, one only achieves small sound amplitudes, so that special measures are needed to detect the resonances.

In the state of resonance, the free faces of the plate vibrate with maximum amplitude, when the amplitude of the exciting wave remains quasiconstant over a broad frequency range. The most important resonance-detection methods are based on this property. In order not to affect the character of the free vibration, only a small part of the vibrational energy may be tapped for the detection. This, for example, can be accomplished with capacitive or piezoelectric transfer. We then talk of a capacitive or piezoelectric sensor. Another possibility consists in measuring the electric power of the generator as a function of the frequency of the field. In this case, one observes a weak feedback of the vibration state on the generator. Hence, one can determine the resonance frequencies very accurately with simple electronic measures.

A further possibility, which has proven preeminently useful in practice, is based on the diffraction of monochromatic light by standing acoustic waves in the resonance state, as with the Schaefer–Bergmann method. In the resonance state, only one strong wave exists, which is easy to observe via the diffraction effect, even when the acoustic power of the generator is weak. This simple method offers the special advantage, that from the angle of diffraction one obtains an independent statement on the velocity of sound and on the directly excited wave type. If one works with polarized light, whose direction of vibration is adjustable, one can observe the resonances at substantially lower acoustic powers. For this purpose, one places a second polarizer in the ray path, which eliminates a large part of the disturbing background radiation on the screen. Crystals with weak elasto-optical effects or insufficient optical transparency can also be investigated with this method. In this case one cements an auxiliary crystal, made of a material with an extremely large elasto-optical effect, as for example, RbI, on the face of the specimen plate opposite the generator. The auxiliary crystal is furnished with a plane parallel pair of faces for the transmission of the light beam. Furthermore, the auxiliary crystal should possess optical quality in order to produce sharp diffraction spots. Finally, one must ensure that the auxiliary crystal does not have a disturbing influence on the resonances of the crystal plate. This occurs through two measures: firstly, the auxiliary crystal is furnished with an irregular boundary on the face opposite to that where the sound enters (for example, "gothic arch" form as in the normal Schaefer–Bergmann method), and secondly, one selects an auxiliary crystal with a sound wave resistance ρv, sufficiently different from that of the specimen. The resonances are somewhat attenuated by the auxiliary crystal. The crucial thing, however, is that the resonance frequencies are not substantially influenced by the generator nor by the auxiliary crystal, when one remains outside the resonance area of the generators. One works with a set of generators covering a large frequency range from, for example,

Table 4.11 Resonance frequencies f in MHz of a plane-parallel plate of $LiHSeO_3$. Plate normal and propagation direction [001], thickness 7.353 mm, temperature 292 K. Only resonances of the order $(2m+1)$ were recorded. The measurement between the orders (80+1) and (104+1) are omitted.

m	f	m	f	m	f	m	f	m	f	m	f
1	10.534	19	15.140	37	19.760	55	24.381	73	28.990	115	39.767
3	11.047	21	15.665	39	20.272	57	24.888	75	29.500	117	40.282
5	11.560	23	16.180	41	20.786	59	25.400	77	30.013	119	40.790
7	12.070	25	16.693	43	21.300	61	25.915	79	30.525	121	41.305
9	12.582	27	17.209	45	21.810	63	26.430	(80–106)		123	41.815
11	13.095	29	17.720	47	22.320	65	26.942	107	37.713	125	42.331
13	13.607	31	18.223	49	22.830	67	27.452	109	38.226	127	42.841
15	14.120	33	18.735	51	23.344	69	27.962	111	38.735	129	43.358
17	14.630	35	19.248	53	23.870	71	28.475	113	39.250	131	43.872

10 to 30 MHz. One observes a strong coupling only close to the fundamental frequency of a generator, i.e., a frequency shift of the resonance compared to the free vibration. One recognizes this by comparison with the frequency of the resonances, recorded by another generator with another fundamental frequency. Investigations using piezoelectric crystals, whose resonances can be directly excited, confirm this important finding. One observes very weak coupling with transverse waves. If one utilizes only resonances lying outside the frequency range of the coupling with the generator, one obtains, with great accuracy, an equidistant frequency spectrum of the unbounded plan-parallel plate. As an example, we present the measurements on a crystal of $LiHSeO_3$ (PSG 222). Table 4.11 lists the measured fundamental frequencies (generator fundamental frequencies 8.24 and 40 MHz). The plate had a thichness of $L = 7.353$ mm. Hence, for the velocity of sound of the longitudinal wave along [001] we get the value

$$v = 2L\frac{\nu_n - \nu_m}{m - n} = 3.7713\,\mathrm{kms}^{-1}$$

with $m = 1$ and $n = 131$.

This value is in excellent agreement with the values obtained using the pulse-echo method or with the Schaefer–Bergmann method. Since both frequencies ν_m and ν_n were measured with an absolute error below 2 kHz, the value determined for v is at most subjected to an error of 0.05%, whereby $|\Delta L/L| < 0.03\%$ was assumed. This accuracy is sufficient for all normal requirements. The accuracy of the absolute determination of the velocity of sound is limited by the measurement of the thickness L. Only with special effort it is possible to determine the thickness with an error substantially below 0.01%. This also presupposes that the specimens are prepared plane-parallel with the corresponding accuracy. In the acquisition of relative changes of sound velocities, as, for example, occurring under the influence of temperature, hydrostatic or uniaxial pressure and electric or magnetic fields, one also achieves a substantially higher accuracy with this method, when one records the diffraction signals with a photomultiplier and employs the possibilities made available by electronic signal processing based on pulse controlled sound excitation and detection. A substantial improvement is also achieved using phase-sensitive lock-in techniques to amplify the signal during frequency-modulated sound excitation. Thus the accuracy of frequency measurements of resonances using well-prepared plates can be increased to give relative errors below 10^{-7} to 10^{-8} with comparatively little effort. This procedure requires, by far the least effort, and hence, must be given preference over other methods.

Finally, precision measurements of frequencies of fundamentals and low-order harmonics on rods, cylinders and plates must be mentioned, which similar to a tuning fork are mechanically excited. The specimens have to be arranged and excited in the way that those vibrational states occur, which are favored mainly by the respective boundary condition, and that damping by the arrangement is largely suppressed. This means that the specimen shall only lie with the nodes of vibration on the support and excitation shall occur on the expected antinodes. With the help of sensitive piezoelectric sensors, the vibrations of the specimen can be recorded by sound emission in air and the associated resonance frequencies determined with high accuracy. Naturally, optical interferometric methods can also be used to detect the vibrations. From longitudinal and flexural vibrations of rods one can very accurately determine dynamic Young's moduli and from transverse and torsional vibrations certain shear moduli, even on specimens with dimensions of a few mm.

3b. Forced Vibrations of Piezoelectric Crystals

We now consider two important examples of special specimen shapes, namely, the plane-parallel unbounded plate and the thin rod. Both can be experimentally realized to a good approximation. The thin plate plays an important role in the application of piezoelectric crystals for acoustic generators and detectors as well as in the manufacture of high-frequency generators and related devices such as frequency stabilizers and frequency selectors in communication technology. In both cases, a comparatively simple relation exists between the observed resonance frequencies and the material constants.

We first consider the plane-parallel plate. The basic equations are (see Section 4.5.12):

$$\sigma_{ij} = c^E_{ijkl}\varepsilon_{kl} - e_{mij}E_m, \qquad D_i = e_{ikl}\varepsilon_{kl} + \epsilon^{\varepsilon}_{im}E_m.$$

Moreover, we set $E_m = -(\operatorname{grad}\phi)_m$ and only allow plane waves

$$\xi = \xi_0 e^{2\pi i(\boldsymbol{k}\cdot\boldsymbol{x}-\nu t)} \quad \text{and} \quad \phi = \phi_0 e^{2\pi i(\boldsymbol{k}\cdot\boldsymbol{x}-\nu t)}.$$

As previously, for dynamic processes we have

$$\frac{\partial \sigma_{ij}}{\partial x_j} = \rho \frac{\partial^2 \xi_i}{\partial t^2}.$$

Hence, the general basic equation for the propagation of plane waves in the form derived above is valid

$$(-\rho v^2 \delta_{ik} + c^D_{ijkl} g_j g_l)\xi_k = 0$$

with

$$c^D_{ijkl} = c^E_{ijkl} + \frac{e_{mij} e_{nkl}}{\epsilon_{rs} g_r g_s} g_m g_n,$$

where g is the unit vector in the direction of the wave normals. We only consider vibrational states, which are homogeneous over the complete cross-section parallel to the faces of the plate; i.e., all derivatives $\partial/\partial x_1$ and $\partial/\partial x_2$ vanish, when $g = e_3$ is perpendicular to the faces of the plate. Let the plate be suspended force free, i.e., we have $\sigma_{3i}(x_3) = 0$ for $x_3 = \pm L/2$, hence,

$$\sigma_{31}(\pm L/2) = c^E_{31k3} \frac{\partial \xi_k}{\partial x_3} - e_{331} E_3 = 0$$
$$\sigma_{32}(\pm L/2) = c^E_{32k3} \frac{\partial \xi_k}{\partial x_3} - e_{332} E_3 = 0$$
$$\sigma_{33}(\pm L/2) = c^E_{33k3} \frac{\partial \xi_k}{\partial x_3} - e_{333} E_3 = 0.$$

Because $\boldsymbol{E} = -\operatorname{grad} \phi$ we have $\boldsymbol{E} \parallel \boldsymbol{g}$. These conditions fully correspond to the situation discussed above for free vibrations. Now, as a further decisive parameter we add the electric boundary condition. We imagine electrodes attached to the faces of the plate (for example, by deposition of a thin metallized film), to which we apply a voltage $\phi = \phi_0 \cos \omega t$, so that on both faces the boundary conditions

$$\phi(\pm L/2) = \pm \phi_0 \cos \omega t$$

are to be fulfilled.

To solve the problem, we must determine the three, in the direction $g = e_3$, possible sound velocities and the associated displacement vectors $\boldsymbol{\xi}^{(1)}$, $\boldsymbol{\xi}^{(2)}$ and $\boldsymbol{\xi}^{(3)}$ (previously specified by $\boldsymbol{\xi}', \boldsymbol{\xi}'', \boldsymbol{\xi}'''$) from the dynamic basic equations. Each solution is coupled to an electric potential ϕ, obtained from the relation

$$D_3 = e_{3k3} \frac{\partial \xi_k}{\partial x_3} + \epsilon^{\varepsilon}_{33} E_3 = \text{const}.$$

by integrating with respect to x_3 ($\boldsymbol{D}$ = const. was already derived as a secondary condition in Section 4.5.12). From $E_3 = -\partial\phi/\partial x_3$, one obtains by integration

$$e_{3k3}\xi_k - \epsilon^{\xi}_{33}\phi = qx_3 + q_0,$$

hence,

$$\phi = \frac{e_{3k3}}{\epsilon^{\xi}_{33}}\xi_k - q'x_3 - q'_0,$$

where q, q_0, q' and q'_0 are constants of integration possessing the same time dependence as ξ_k.

The electric boundary conditions then take on the form

$$\phi(\pm L/2) = \frac{e_{3k3}}{\epsilon^{\xi}_{33}}\xi_k(\pm L/2) \mp q'L/2 - q'_0 = \pm\phi_0 \cos\omega t.$$

The general solution consists of a superposition of the three waves in the form $\boldsymbol{\xi} = \sum_\mu C^{(\mu)}\boldsymbol{\xi}^{(\mu)}$, whereby the complex coefficients $C^{(\mu)}$ are so adjusted, that the boundary conditions are fulfilled. We now imagine $\boldsymbol{\xi}$ to be resolved in a symmetric and an antisymmetric part according to $\boldsymbol{\xi} = \boldsymbol{\xi}_s + \boldsymbol{\xi}_a$ with $\boldsymbol{\xi}_s(x_3) = \boldsymbol{\xi}_s(-x_3)$ and $\boldsymbol{\xi}_a(x_3) = -\boldsymbol{\xi}_a(-x_3)$. In the general approach for a plane wave, this means that $\boldsymbol{\xi}_s$ only contains terms of the form $\cos(2\pi k_3 x_3)$ and $\boldsymbol{\xi}_a$ only terms of the form $\sin(2\pi k_3 x_3)$.

Hence, by subtraction or addition of both boundary conditions, we obtain

$$q' = -\frac{2\phi_0}{L}\cos\omega t + \frac{2e_{3k3}}{L\epsilon^{\xi}_{33}}\xi_{a,k}(L/2)$$

and

$$q'_0 = 0.$$

This means, the electric boundary conditions are only fulfilled by the antisymmetric part of the solution. Conversely, the applied voltage can only excite antisymmetric forms of vibration. Symmetric vibrations can be excited through the quadratic electrostrictive effect (see Section 4.5.10), however, with by far lower amplitudes.

If we now eliminate E_3 in the mechanical boundary conditions $\sigma_{3i}(\pm L/2) = 0$ with the help of the above result for q' and consider only the antisymmetric solution

$$\boldsymbol{\xi}_a = \sum_\mu \boldsymbol{\xi}_0^{(\mu)} \sin(2\pi k_3^{(\mu)} x_3)\cos\omega t,$$

we obtain

$$\sum_{\mu}\left(c^{E}_{3ik3}+\frac{e_{33i}e_{3k3}}{\epsilon^{\varepsilon}_{33}}\right)2\pi k_3^{(\mu)}\xi_{0k}^{(\mu)}\cos(\pi k_3^{(\mu)}L)\cos\omega t+\frac{2e_{33i}}{L}\phi_0\cos\omega t$$
$$-\sum_{\mu}\frac{2e_{33i}e_{3k3}}{L\epsilon^{\varepsilon}_{33}}\xi_{0k}^{(\mu)}\sin(\pi k_3^{(\mu)}L)\cos\omega t=0,$$

hence,

$$\sum_{\mu}\left(c^{D}_{3ik3}\cdot\pi k_3^{(\mu)}\cos(\pi k_3^{(\mu)}L)-\frac{e_{33i}e_{3k3}}{L\epsilon^{\varepsilon}_{33}}\sin(\pi k_3^{(\mu)}L)\right)\xi_{0k}^{(\mu)}=-\frac{e_{33i}}{L}\phi_0$$
$$\text{for}\quad i=1,2,3.$$

This system of equations yields the contributions of the wave type $\boldsymbol{\xi}^{(\mu)}$ to the solution $\boldsymbol{\xi}_a$. The ratio $\xi_{01}^{(\mu)}:\xi_{02}^{(\mu)}:\xi_{03}^{(\mu)}$ is assumed to be known from the solution of the basic equations. Plate resonances appear, when at least one of the amplitudes $\xi^{(\mu)}$ becomes unlimited, i.e., when the determinant

$$\left|c^{D}_{3ik3}\pi k_3^{(\mu)}\cos(\pi k_3^{(\mu)}L)-\frac{e_{33i}e_{3k3}}{L\epsilon^{\varepsilon}_{33}}\sin(\pi k_3^{(\mu)}L)\right|$$

vanishes.

The practical utilization of this condition for the determination of material constants is only worthwhile in the case of pure longitudinal- or transverse waves.

For a pure longitudinal wave along $\boldsymbol{e}_3$ ($\xi_1=\xi_2=0$, $\xi_3\neq 0$), the boundary condition for $i=3$ is

$$\left(c^{D}_{3333}\cdot\pi k_3\cos(\pi k_3L)-\frac{e^2_{333}}{L\epsilon^{\varepsilon}_{33}}\sin(\pi k_3L)\right)\xi_{03}=-\frac{e_{333}}{L}\phi_0.$$

Hence, a resonance occurs when

$$\tan(\pi k_3L)=(\pi k_3L)\frac{\epsilon^{\varepsilon}_{33}c^{D}_{3333}}{e^2_{333}}=(\pi k_3L)/k_t^2,$$

where k_t^2 is the coupling factor $(c^{D}_{3333}-c^{E}_{3333})/c^{D}_{3333}=e^2_{333}/\epsilon^{\varepsilon}_{33}c^{D}_{3333}$, which we were previously already acquainted with. A pure transverse wave with displacement vector parallel $\boldsymbol{e}_1$ ($\xi_1\neq 0$, $\xi_2=\xi_3=0$) gives, for $i=1$, the boundary condition

$$\left(c^{D}_{3113}\cdot\pi k_3\cos(\pi k_3L)-\frac{e^2_{331}}{L\epsilon^{\varepsilon}_{33}}\sin(\pi k_3L)\right)\xi_{01}=-\frac{e_{331}}{L}\phi.$$

The analogous condition

$$\tan(\pi k_3 L) = (\pi k_3 L)/k_{35}^2$$

with the coupling factor $k_{35}^2 = e_{331}^2/\epsilon_{33}^{\xi} c_{3131}^{D}$ applies for diverging of ξ_{01}.

We obtain similar conditions for the case of other propagation and displacement directions of pure wave types.

The equation $\tan X = X/k^2$ with $X = \pi k_3 L$ represents a frequency condition. We can use the approximate condition $\tan X \to \infty$ for solutions with very high values of X/k^2. Here, $X = (2m-1)\pi/2$, hence, $L = (2m-1)\lambda/2$ or

$$\nu = \frac{(2m-1)}{2}\frac{v}{L} \quad \text{with} \quad k_3 = 1/\lambda$$

and m integer. This condition is identical with the resonance condition for the freely vibrating plate, where, however, only the odd multiples of half the wavelength (antisymmetric states of vibration) are allowed. Thus, from the higher order resonance frequencies we can determine, with sufficient accuracy, the propagation velocity of the associated wave. In the region of small and medium values of X/k^2 characteristic deviations of free resonances appear: The resonance frequencies are not multiples of the fundamental frequency. The situation is best seen by means of a graphical representation (Fig. 4.49). The line $Y = X/k^2$ intersects the tangent curves $Y = \tan X$ once in each interval between $(m-1)\pi < X \leq m\pi$. Since k^2 is smaller than one in each case, there exists for $m = 1$ also one solution. For practical utilization, the method proposed by Onoe, Thiersten & Meitzler (1963) has proven to be optimal. The solutions X_m of the equation $\tan X_m = X_m/k^2$ are calculated as a function of k^2 in the range $0 \leq k^2 < 1$, which can be carried out with a simple iteration method of the type

$$X_m^{(s+1)} = m\pi + \arctan(X_m^{(s)}/k^2).$$

The resulting ratios

$$\frac{X_n}{X_m} = \frac{\pi L/\lambda_n}{\pi L/\lambda_m} = \frac{\nu_n}{\nu_m}$$

are directly accessible to experiment. They depend monotonically on k^2, not, however, on the plate thickness L and the velocity of sound v. Therefore, from the measured ratios of resonance frequencies, one can determine the associated coupling factors by a comparison with the tabled values. A sufficiently large deviation from the ratio $(2n-1)/(2m-1)$ is only obtained with k-values above approximately 0.1 as well as for m and $n = 1, 2$ and 3. An extract of values sufficient for practical applications is presented in Table 4.12. With the

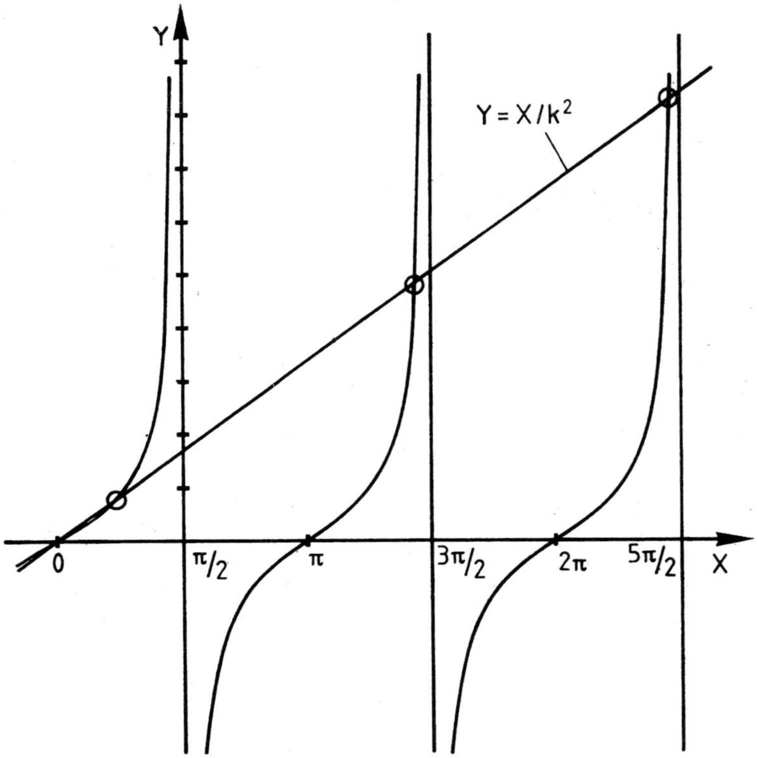

Figure 4.49 Intersection of $Y = \tan X$ with $Y = X/k^2$ ($k^2 < 1$). In each interval $(m-1)\pi < X \leq m\pi$ the line $Y = X/k^2$ intersects the curves $Y = \tan X$ just once.

help of these k^2-values and the sound velocities obtained from higher order resonance frequencies, one can determine the square of the piezoelectric coefficients e_{33i} according to

$$e_{33i}^2 = k_{3j}^2 \epsilon_{33}^{\varepsilon} c_{3i3i}^D$$

($j = 6 - i$), when the associated dielectric constant $\epsilon_{33}^{\varepsilon}$ is known, too. We can, to a sufficient approximation, insert a value for $\epsilon_{33}^{\varepsilon}$ measured in the range of higher frequencies, for example around 10 MHz. Of course, the relationships discussed here also apply approximately to the case of combination waves with strong longitudinal- or transverse components. The method of piezoelectric resonances of plane-parallel plates fails in the case of very small coupling coefficients, which, for the application, are usually of little interest. The resonance frequencies can be simply and very accurately identified by the minima of the ac resistance of the plate connected in series with an auxiliary resistor, when sweeping through a broad frequency range during electrical excitation.

Table 4.12 Ratio of resonance frequencies of quasi-unlimited plane plates as a function of the coupling coefficient k. Excitation of vibrations in an ac electric field parallel to the normals of the plate.

k	ν_2/ν_1	ν_3/ν_1	k	ν_2/ν_1	ν_3/ν_1	k	ν_2/ν_1	ν_3/ν_1
0	3	5	0.35	3.148	5.265	0.70	3.917	6.625
0.05	3.003	5.005	0.40	3.201	5.359	0.75	4.185	7.094
0.10	3.011	5.020	0.45	3.265	5.474	0.80	4.561	7.751
0.15	3.025	5.045	0.50	3.344	5.614	0.85	5.132	8.744
0.20	3.045	5.080	0.55	3.441	5.787	0.90	6.123	10.461
0.25	3.071	5.128	0.60	3.563	6.002	0.95	8.430	14.447
0.30	3.105	5.189	0.65	3.717	6.274			

If one is able to transmit light through the plate parallel to the plate surfaces, resonance frequencies can be observed by the diffraction of light as discussed above. With this method, resonances have been detected up to very high orders, in individual cases exceeding 1000.

The thin plane-parallel rod, which we now consider, is treated similarly to the plane-parallel plate. Let the length, width, and thickness extend in the directions $\boldsymbol{e}_1$, $\boldsymbol{e}_2$ and $\boldsymbol{e}_3$. Let $L_1 \gg L_2 > L_3$ specify the respective dimensions. If one applies an electric alternating field to the plane faces perpendicular to $\boldsymbol{e}_3$, one can excite vibrations in the rod, whose frequencies are fixed by the length L_1. Since the crystal is not exposed to any external forces

$$\sigma_{ij} = 0 \qquad (i, j = 1, 2, 3)$$

is true for all points on the surface. Inside the crystal, all σ_{2i} and σ_{3i} vanish to a good approximation, because the dimensions L_2 and L_3 are so small that no significant mechanical stresses can form between the side faces. This means, however, that only the stress component σ_{11} exists. It depends solely on x_1 because

$$\frac{\partial \sigma_{11}}{\partial x_2} \quad \text{and} \quad \frac{\partial \sigma_{11}}{\partial x_3}$$

are also approximately zero as a consequence of the small dimensions. The equation of motion for the rod is then

$$\frac{\partial \sigma_{11}}{\partial x_1} = \rho \frac{\partial^2 \xi_1}{\partial t^2}.$$

The mechanical deformation ε_{11} connected with σ_{11} is expressed by $\varepsilon_{11} = s^E_{1111}\sigma_{11} + d_{311}E_3$ hence,

$$\sigma_{11} = \frac{\varepsilon_{11}}{s^E_{1111}} - \frac{d_{311}}{s^E_{1111}} E_3.$$

We have, as a further material equation, to take notice of the relationship $D_3 = \epsilon_{33}^{\sigma} E_3 + d_{311}\sigma_{11}$, when ensuring that $E_1 = E_2 = 0$. The equation of motion then takes the form

$$\frac{1}{s_{1111}^E} \frac{\partial^2 \xi_1}{\partial x_1^2} = \rho \frac{\partial^2 \xi_1}{\partial t^2}$$

because

$$\frac{\partial E_3}{\partial x_1} = 0$$

(the broadsides perpendicular to e_3 are completely metallized!). Using the plane-wave approach $\xi_1 = \xi_{01} e^{2\pi i(k_1 x_1 - \nu t)}$ we get

$$\left(\rho\omega^2 - \frac{4\pi^2 k_1^2}{s_{1111}^E}\right)\xi_1 = 0, \quad \text{thus} \quad \rho v^2 = \frac{1}{s_{1111}^E}.$$

In contrast to the unlimited plate, the longitudinal component c_{1111}^D is not effective in the rod but rather Young's modulus $(1/s_{1111}^E)$. This result also applies to nonpiezoelectric crystals. We obtain the resonance condition, as before, from the boundary conditions. Again, we imagine the general solution as composed of a symmetric and an antisymmetric part given by

$$\xi_1 = \xi_s \cos(2\pi k_1 x_1) \cos\omega t + \xi_a \sin(2\pi k_1 x_1) \cos\omega t.$$

Inserting the boundary conditions $\sigma_{11}(\pm L/2) = 0$ in the relation

$$\varepsilon_{11} = s_{1111}^E \sigma_{11} + d_{311} E_3$$

gives

$$-\xi_{0s} \sin(\pm\pi k_1 L_1) + \xi_{0a} \cos(\pm\pi k_1 L_1) = d_{311} E_{03}/2\pi k_1,$$

hence,

$$\xi_{0s} = 0 \quad \text{and} \quad \xi_{0a} = d_{311} E_{03}/(2\pi k_1 \cos(\pi k_1 L_1)).$$

Accordingly, only the antisymmetric states of vibration are excited. Our solution is then

$$\xi_1 = \frac{d_{311} E_{03}}{2\pi k_1 \cos(\pi k_1 L_1)} \sin(2\pi k_1 x_1) \cos\omega t.$$

Resonances appear, when $\cos(\pi k_1 L_1) = 0$, hence when $\pi k_1 L_1 = (2m-1)\pi/2$ or

$$\nu_m = \frac{(2m-1)v}{2L_1}$$

just as with the unlimited plate.

The portion of alternating current flowing over the metallized surfaces is

$$I_3 = \int \frac{dD_3}{dt} dF,$$

where integration is carried out over the surface perpendicular to e_3. With $D_3 = D_{03} \cos \omega t$ one obtains

$$\frac{dD_3}{dt} = -\omega D_{03} \sin \omega t$$

and hence,

$$\begin{aligned} I_3 &= -\omega L_2 \int_{-L_1/2}^{+L_1/2} D_{03} \sin \omega t dx_1 \\ &= -\omega L_2 \int_{-L_1/2}^{+L_1/2} \left(\epsilon_{33}^{\xi} E_{03} + \frac{d_{311}}{s_{1111}^E} \frac{\partial \xi_{01}}{\partial x_1} \right) \sin \omega t dx_1 \end{aligned}$$

using $\epsilon_{33}^{\sigma} = \epsilon_{33}^{\xi} + d_{311}^2 / s_{1111}^E$. This relation is obtained by inserting the expression derived above for σ_{11} in $D_3 = \epsilon_{33}^{\sigma} E_3 + d_{311}\sigma_{11}$. The result is with ξ_1 as given above

$$I_3 = -\omega L_1 L_2 \left(\epsilon_{33}^{\xi} + \frac{d_{311}^2}{s_{1111}^E} \frac{\tan(\pi k_1 L_1)}{\pi k_1 L_1} \right) E_{03} \sin \omega t.$$

In the resonance state, the resistance is minimal because $\tan(\pi k_1 L_1) \to \infty$. However, there exists the possibility of an unlimited resistance when, namely, the expression in the brackets vanishes. For the respective frequency, the term "antiresonance frequency" has come into common usage. It must be

$$\tan X = -\frac{\epsilon_{33}^{\xi} s_{1111}^E}{d_{311}^2} X \quad \text{with} \quad X = \pi k_1 L_1.$$

This relation, apart from the sign, is analogous to the resonance condition of the unlimited plate. If one now introduces the effective coupling coefficient

$$\frac{\epsilon_{33}^{\sigma} s_{1111}^E}{d_{311}^2} = 1/k^2$$

one obtains, with

$$\epsilon_{33}^{\xi} = \epsilon_{33}^{\sigma} - \frac{d_{311}^2}{s_{1111}^E}$$

the antiresonance condition in the form

$$\tan X = X(k^2 - 1)/k^2 = -X/k'^2.$$

Table 4.13 Ratio of the first antiresonance frequency and the first resonance frequency of thin and slender rods as function of the coefficient $k'^2 = k^2/(1-k^2)$ at an excitation in an ac electric field along the rod axis.

k'^2	ν_{A1}/ν_1	k'^2	ν_{A1}/ν_1	k'^2	ν_{A1}/ν_1	k'^2	ν_{A1}/ν_1
0	1	0.50	1.0921	1.00	1.2916	1.50	1.4879
0.05	1.0010	0.55	1.1094	1.05	1.3126	1.55	1.5051
0.10	1.0040	0.60	1.1276	1.10	1.3335	1.60	1.5218
0.15	1.0094	0.65	1.1467	1.15	1.3541	1.70	1.5535
0.20	1.0160	0.70	1.1664	1.20	1.3745	1.80	1.5832
0.25	1.0247	0.75	1.1866	1.25	1.3945	1.90	1.6108
0.30	1.0352	0.80	1.2072	1.30	1.4141	2.00	1.6364
0.35	1.0473	0.85	1.2281	1.35	1.4332		
0.40	1.0609	0.90	1.2492	1.40	1.4520		
0.45	1.0760	0.95	1.2704	1.45	1.4702		

From a graphical representation, one sees that the antiresonance frequency ν_{Am} of mth order always lies higher than the resonance frequency ν_m. The impedance behavior as a function of the frequency of the electric field enables two sets of data to be gained, namely, the resonance frequencies ν_1, obtained with high accuracy from a higher resonance frequency according to $\nu_1 = \nu_m/(2m-1)$, and the associated antiresonance frequency, leading to a value for k'^2 or k^2. The ratio ν_{Am}/ν_m is largest for $m = 1$. Therefore, one confines oneself to the measurement of the first antiresonance frequency. Then, $\nu_{A1}/\nu_1 = X_{A1}/X_1 = X_{A1}/(\pi/2)$ because $X_1 = \pi/2$. This ratio only depends on k^2 and not on v and L_1. From a table of values $X_{A1}(k'^2)/X_1$, easily calculated by an iteration method of the type $X^{(s+1)} = \arctan(-X^{(s)}/k'^2)$, one can then extract k'^2 or k^2 and hence the piezoelectric coefficient d_{311}^2, when the dielectric constant ϵ_{33}^{ξ} is known. Table 4.13 presents the values $X_{A1}/X_1 = \nu_{A1}/\nu_1$ as a function of k'^2, which suffice for practical use. By conducting measurements on rods of different orientations, one can thus determine certain components of the piezoelectric and elastic tensors. In any case, for the complete measurement of the elasticity tensor other measurements must be included, because the vibrations discussed here are only coupled with the longitudinal effects $s_{1111}^{E\,\prime}$.

Other wave types can also be excited in the thin rod under discussion. Again let $L_1 \gg L_2 > L_3$. Whilst previously the length L_1 was the frequency determining factor, let it now be the width L_2.

It is convenient to intentionally arrange the dimension along e_1 irregularly in order to prevent the formation of the resonances discussed above. The electric field is again applied in the direction e_3. The boundary conditions are now: $\partial\sigma_{ij}/\partial x_1 = 0$, because the crystal is quasiunlimited in the direction e_1, $\sigma_{3i} = 0$ and $\partial\sigma_{ij}/\partial x_3 = 0$ in the whole crystal, because L_3 is very small, and $\sigma_{2i}(\pm L_2/2) = 0$.

We only consider states of vibration with propagation vector parallel e_2. Employing $\xi_i = \xi_{0i} e^{2\pi i(k_2 x_2 - \nu t)}$, we obtain from the elastodynamic basic equations, taking into consideration the above conditions

$$\rho \frac{\partial^2 \xi_1}{\partial t^2} = \frac{\partial \sigma_{12}}{\partial x_2} = c^E_{1212} \frac{\partial^2 \xi_1}{\partial x_2^2} + c^E_{1222} \frac{\partial^2 \xi_2}{\partial x_2^2}$$

and

$$\rho \frac{\partial^2 \xi_2}{\partial t^2} = \frac{\partial \sigma_{22}}{\partial x_2} = c^E_{2222} \frac{\partial^2 \xi_2}{\partial x_2^2} + c^E_{2212} \frac{\partial^2 \xi_1}{\partial x_2^2},$$

and hence,

$$(-\rho v^2 + c^E_{1212})\xi_1 + c^E_{1222}\xi_2 = 0,$$
$$c^E_{2212}\xi_1 + (-\rho v^2 + c^E_{2222})\xi_2 = 0.$$

This system has only solutions, when its determinant vanishes. We get the two solutions

$$\rho v^2 = \frac{c^E_{22} + c^E_{66}}{2} \pm \frac{1}{2}\sqrt{(c^E_{22} - c^E_{66})^2 + 4c^{E2}_{26}}$$

with

$$c_{2222} = c_{22}, \quad c_{1212} = c_{66}, \quad c_{2212} = c_{26}.$$

The boundary conditions

$$\sigma_{12}(\pm L_2/2) = c^E_{26} \frac{\partial \xi_2}{\partial x_2} + c^E_{66} \frac{\partial \xi_1}{\partial x_2} - e_{312} E_3 = 0,$$
$$\sigma_{22}(\pm L_2/2) = c^E_{22} \frac{\partial \xi_2}{\partial x_2} + c^E_{26} \frac{\partial \xi_1}{\partial x_2} - e_{322} E_3 = 0,$$
$$D_3 = \epsilon^\sigma_{33} E_3 + 2d_{312}\sigma_{12} + d_{322}\sigma_{22}$$

(all variables at the positions + or $-L_2/2$) demonstrate that the electric field must, in general, simultaneously excite both solutions, because for each solution the amplitude ratio ξ_{01}/ξ_{02} is fixed by the dynamic equations. Considerations similar to the previous situation lead to the conclusion that the resonance condition $\tan X = 0$ with $X = \pi \nu L_2 / v$ also applies, hence

$$\nu_m = \frac{(2m-1)}{2L_2} v$$
$$= \frac{(2m-1)}{2L_2} \sqrt{\frac{1}{2\rho}\left((c^E_{22} + c^E_{66}) \pm \sqrt{(c^E_{22} - c^E_{66})^2 + 4c^{E2}_{26}}\right)}.$$

Thus, further elastic data can be extracted, which together with the other results allow a complete determination of the elasticity tensor. Antiresonance frequencies appear, too, from which one can derive additional data for the determination of piezoelectric coefficients. Since the boundary conditions cannot be exactly realized, one must expect certain deviations of the thus obtained values from c_{ij}^{E} of the unlimited crystal.

Examples for the application of this method for the determination of elastic, dielectric and piezoelectric properties of the technologically important crystals are α-quartz, KH_2PO_4 and Rochelle salt, which were described in detail by Mason (1954). The special advantages of the investigation of quadratic plates were explained by Bechmann (1951).

3c. Resonant Ultrasound Spectroscopy (RUS)

We now come to a particularly effective method for the future experimental investigation of elastic properties, developed in the last decades. It presents the generalization of the determination of elastic properties from acoustic resonance frequencies of geometrically clearly defined test specimens, as discussed above for plane-parallel plates or rods. The problem of the calculation of acoustic resonances of homogeneous bodies, even in the case of anisotropic media, was solved by Demarest in 1969 for spheres and rectangular parallelepipeds. It turned out that the resonance frequencies of rectangular parallelepipeds, the so-called RPR spectrum ("rectangular parallelepiped resonances"), could be calculated with high accuracy from the density, orientation and dimensions of the parallelepiped and its elasticity tensor, however, with the aid of a powerful computer. The converse, namely the determination of the elasticity tensor from a sufficiently large region of the resonance spectrum of a rectangular parallelepipedic specimen with known density, orientation and dimensions could only be solved up to now with a least-squares-method. With this method one attempts to arrive at a data set of the elasticity tensor in iterative steps, whose calculated resonance spectrum agrees with the experimental data set. Before the calculations initial values for the components of the elasticity tensor must be selected most carefully. The first successful experiments were reported by Ohno (1976).

Rectangular parallelepipeds, whose edges run parallel to the basic vectors of a Cartesian reference system, in particular, that of the crystal-physical basic system, have proven best suited for the practical application of the method. In favorable cases, the dimensions of the specimens can be relatively small (under 1 mm). One must ensure that only minimal deviations from plane-parallelism (in the range of less than 0.5 per mille in the dimensions) and from orientation with respect to the Cartesian system (angular deviation less than 0.3°) occur. The larger the elastic anisotropy the more exact the orientation must be realized.

Commercial network analyzers are suitable for the measurement of the resonance frequencies. The specimens are fixed along a space diagonal, between a sound generator and a detector, such that one corner of the specimen lies on the generator and the opposite corner contacts the detector. The mechanical force on the specimen must be kept as low as possible to approximate the condition of a freely vibrating body. If one sweeps through a frequency interval with the generator, one expects in the case of a resonance deformation, which are communicated to the detector on the opposite corner (Fig. 4.50a). The inverse piezoelectric effect produces an electric signal at the detector, which can be recorded by a lock-in amplifier. Normally, i.e., at low acoustic damping, one obtains exceptionally sharp signals. For a comparatively unproblematic evaluation, it is advantageous, when the spectrum is recorded beginning with the lowest resonance frequency up to a specified maximum frequency without any gap. The number of the required resonances depends on the point symmetry group of the specimen and the desired accuracy. For example, with cubic crystals, about 30 resonances are sufficient, with triclinic crystals about six times more are needed to obtain confident results. If the experimental spectrum is incomplete, an evaluation with simple programs fails, since the computer-calculated resonance frequencies, continuously ordered according to rising frequency, do not agree with the sequence of the corresponding experimental frequencies. If, in the experimental sequence, one or the other resonance is missing, because, for example, the associated deformation only exhibited a small effective component at the detector, special computer programs must be employed to process the gaps in the spectrum. Such programs have been developed in the meanwhile. Anyway, additional efforts should be made to extract, by repeated measurements on the specimen in different settings, a complete spectrum within the specified frequency range. An example of such as spectrum is presented in Fig. 4.50b.

A short discussion of the fundamental principles of the method is appropriate. Naturally, the familiar basic equations previously discussed for the other methods apply. These we note here, however, for the case of piezoelectric crystals

$$\sigma_{ij} = c^E_{ijkl}\varepsilon_{kl} - e_{ijk}E_k \qquad \text{and} \qquad D_i = \epsilon^{\varepsilon}_{ij}E_j + e_{ijk}\varepsilon_{kl}.$$

As previously, $\{\sigma_{ij}\}$ designates the stress tensor, $\{\varepsilon_{kl}\}$ the deformation tensor, $\{e_{ijk}\}$ the piezoelectric e-tensor, $\{\epsilon^{\varepsilon}_{ij}\}$ the dielectric tensor at constant mechanical deformation, $\{D_i\}$ and $\{E_i\}$ the vectors of the electric displacement and the electric field strength (see Section 4.5.5). For each small specimen volume (Newton's axiom) we have

$$\frac{\partial \sigma_{ij}}{\partial x_j} = \rho \frac{\partial^2 \xi_i}{\partial t^2},$$

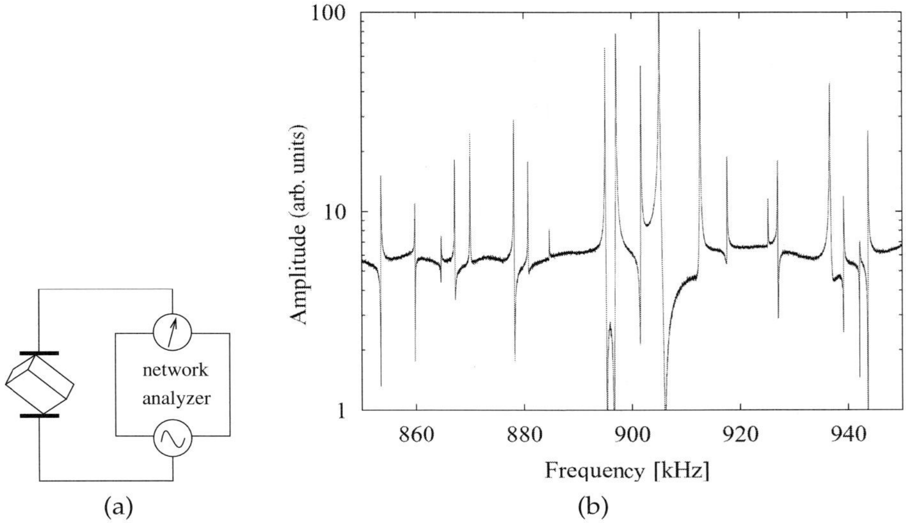

Figure 4.50 (a) Scheme of the RUS method. (b) Part of the RUS spectrum of a rectangular parallelepipedron of $La_3Ga_5SiO_{14}$ having edge lengths of about 6 mm.

where $\boldsymbol{\xi} = \xi_i \boldsymbol{e}_i$ is the displacement vector, ρ the density and t the time. For the freely vibrating parallelepiped, the mechanical stress at the edges must vanish. The calculation of the possible characteristic vibrations of the probe under these boundary conditions is performed with the Lagrange formalism

$$L = \int (\text{kinetic energy density} - \text{potential energy density})\, dV.$$

Integration is extended over the volume of the specimen. The first term follows from the product of the square of the sound velocity components with the density ($\sum_i \rho v_i^2/2$), the second, from the product of the components of the stress tensor with the components of the deformation tensor ($\sigma_{ij}\varepsilon_{ij}/2 = c^E_{ijkl}\varepsilon_{ij}\varepsilon_{kl}/2$ for nonpiezoelectric crystals). A harmonic time-dependence is assumed for the displacement vector, hence, $\xi = \xi_0 \cos(2\pi\nu t)$. According to the rules of variational calculus, the Lagrange function takes on extreme values in the case of stationary solutions, i.e., in the states of resonance. The method of Ritz, in which the components of the displacement vector and of the electric field strength are expanded with respect to suitable basic functions $\Phi(\boldsymbol{x})_s$ or $\Psi(\boldsymbol{x})_s$ in the configuration space (x_1, x_2, x_3), has proven a practical procedure. The expansion has the form

$$\xi_i(\boldsymbol{x}) = a_{is}\Phi_s(\boldsymbol{x}) \qquad \text{as well as} \qquad E_i = -b_s(\operatorname{grad}\Psi_s)_i.$$

In particular, such functions are preferably selected in the RPR-method, with which the boundary conditions can be adhered to without difficulty, for example, with Legendre polynomials. In the last years, however, it turned out that power functions of the type $\Phi(\boldsymbol{x})_s = x_1^\lambda x_2^\mu x_3^\nu$, after suitable normalization, are also capable to realize the boundary conditions in a straightforward way. The advantage of these power functions lies in the fact that they can be employed, in contrast to Legendre polynomials, with a multitude of simple specimen forms (spheres, ellipsoids, cylinders, cones etc.). If one selects certain limits for the positive exponents λ, μ, ν in the form $\lambda + \mu + \nu \leq Q$ with $Q \approx 14$, one obtains a correspondingly large number of coefficients a_{is} and b_s. For L to take on an extreme value, all derivatives of L with respect to these coefficients must vanish. This condition is equivalent to an eigenvalue problem, in which the eigenvalues correspond to the squares of resonance frequencies. Ultimately, we are dealing with the solution of an eigenvalue problem for which standard programs are available (for example, in the freely available LAPACK program library). If resonance frequencies are known, one can then determine the associated resonance states. Detailed presentations of these interrelationships are found in, for example, Leisure & Willis (1997) and Migliori & Sarrao (1997).

Fortunately, in the meantime, extended programs have been developed, which are also applicable to piezoelectric crystals. However, various supplements and improvements are still required, which we will, in part, comment more precisely below. Recently, Schreuer could show that the general approach, introduced above for the basic equations of piezoelectric crystals, is suitable for the quantitative description of the RUS spectrum. Conversely, from the resonance spectrum of a piezoelectric crystal, the components of its elasticity tensor as well as those of its piezoelectric tensor can be determined with high internal consistency (Schreuer, 2002). Only the dielectric constants are required as additional input. Thus a further important field of application arises from this new method.

Finally, let us point out certain advantages and disadvantages of the RUS method.

Advantages:

- Even in the case of triclinic crystals, one only needs a single specimen to completely determine the elasticity tensor. Its dimensions can be smaller than 1 mm provided that the required plane-parallelism is achieved, the dimensions are determined to sufficient accuracy and errors in orientation are minimized. This opens the possibility to investigate elastic properties of materials in abundance, which for lack of adequate crystal size could not be treated up till now with conventional methods.

- The influence of external conditions, for example, electric field, magnetic field, mechanical stress (including hydrostatic pressure) and temperature, can be directly investigated on such specimens. The associated effects arise from the shift of the resonance frequencies.

- Finally, phase transitions can be detected, especially second-order ones, from anomalies of the shift of resonance frequencies accompanied by a change in external parameters, above all, the temperature. Acoustic attenuation phenomena connected with these phase transitions can also be identified.

- Apart from the elasticity tensor, the piezoelectric tensor can be simultaneously determined without much additional effort, when the dielectric tensor is known.

- Already in the current state of development, data acquisition in the RUS method is largely automated, likewise the evaluation in the case of highly symmetric crystals and the selection of appropriate starting values for a least-squares method.

Disadvantages:

- The specimen must possess a very high quality, since even small inhomogeneities, especially surface defects, can sensitively interfere with the spectrum. Conversely, this circumstance can be used to investigate defects and their effects on elastic or piezoelectric quantities.

- With a parallelepiped, the high requirements with respect to plane-parallelism and the accuracy of orientation is only achieved with considerable effort in contrast, for example, with a plane-parallel plate with arbitrary side boundaries.

- With specimens of small dimensions, the danger exists that the boundary areas affected by the preparation falsify the spectrum as a consequence of mechanically induced inhomogeneities. Thus, especially with relatively soft crystals, the RUS method on very small specimens can be problematic.

- Experience in the evaluation of spectra of crystals of low symmetry has taught us that the starting data set for a least-squares method may only deviate a little from the true elasticity tensor (less than about 10%), in order to achieve convergence of the iteration. Hence, computing cycles with a wide variation of initial values are required to obtain a reliable solution. One must endeavor to make use of restrictions based on crystal-chemical, physical and mathematical reasoning to reduce computing time to an acceptable level.

- A barrier also arises from the necessity of a gap-free data set in the hitherto used programs. This problem is easily solved by modifying the programs used currently, however, at the cost of considerably higher computing time.

- Since the acoustic resonances extend over a large frequency interval, one must be prepared for certain dispersion effects leading to a limitation in the attainable accuracy. However, it is known that the maximum dispersion in the crystals investigated so far in the range between 0.1 and 2 MHz amounts to not much more than two percent.

- The individual components of the elasticity tensor cannot be directly determined. Rather, they appear first after the complete determination of the elasticity tensor. The same applies to the dependence of individual components on external parameters.

From these remarks it follows that after elimination of the mentioned difficulties, the RUS method will, in many areas, be superior to the other established methods for the investigation of elastic properties. Certain problems, for example, the change of elastic properties under uniaxial pressure, from which one can derive nonlinear elastic effects, presumably remain not suited for the RUS method.

4. Brillouin Scattering

In the Schaefer–Bergmann method, the diffraction of light on elastic waves entering the crystal is observed. In contrast, Brillouin scattering is caused by the thermally excited elastic waves of the lattice. These waves (phonons) represent a substantial part of the thermal energy of a crystal lattice. They propagate in all directions. Their frequency spectrum depends on the temperature and the bonding properties of the lattice. The highest frequencies, or the shortest wavelengths of these waves are bounded by two factors. Each photon possesses an energy quantum with the energy content $E = h\nu$, where h is Planck's constant and ν is the frequency of the photon. According to the equipartition theorem, each degree of freedom of the vibrating system will contribute, on average, the energy $kT/2$, where k is Boltzmann's constant and T temperature. This means that waves with frequencies higher than $kT/2h$ are comparatively weakly excited according to the requirements of a Boltzmann distribution $e^{-h\nu/kT}$. Furthermore, wavelengths of the thermally excited elastic waves are limited. They cannot be smaller than twice the interplanar spacing d of the lattice in the respective propagation direction ($\lambda \geq 2d$ or $1/\lambda = |\boldsymbol{k}| \leq |\boldsymbol{h}|/2$ with $|\boldsymbol{h}| = 1/d$, where $\boldsymbol{h}$ is the normal of the lattice plane concerned).

This condition corresponds to the highest resonance frequency of a thin plate bounded by parallel lattice planes. The $\boldsymbol{k}$-vectors originating from the origin of the reference system are accordingly bounded by a surface derived from the metric of the associated elementary cell (first Brillouin zone).

Naturally, light quanta can be diffracted by the thermally excited waves just as on artificially generated sound waves as in the case of the Schaefer–Bergmann method. However, the thermally excited waves are by no means monochromatic, so that only a few photons are found in a small frequency interval $\Delta\nu$. From the conservation of energy and momentum in an interaction between a phonon and a photon of the light wave, the possibility arises of measuring the frequency and wavelength of the phonon and thus the velocity of the elastic wave. The momentum of the incoming photons $h\boldsymbol{k}_0$ and the quasimomentum of the phonons $h\boldsymbol{k}_p$ sum to give the momentum of the scattered photon $h\boldsymbol{k}_e$: $\boldsymbol{k}_0 \pm \boldsymbol{k}_p = \boldsymbol{k}_e$ (momentum conservation). The energy is given by $E_0 \pm E_p = E_e$, hence, $\nu_0 \pm \nu_p = \nu_e$ and because $\nu_0\lambda_0 = c/n_0$ (n refractive index, c light velocity in vacuum) and $\nu_p\lambda_p = v$

$$\left|\frac{\nu_e - \nu_0}{\nu_0}\right| = \frac{v n_0 \lambda_0}{c\lambda_p} \qquad \text{(energy conservation).}$$

If one selects a suitable experimental arrangement so that $\lambda_0 \approx \lambda_p$, a crude estimation shows that

$$\left|\frac{\nu_e - \nu_0}{\nu_0}\right| \approx \frac{10^6 \cdot 1.5}{3 \cdot 10^{10}} = 10^{-4}/2,$$

i.e., the energy and hence, the magnitude of the momentum of incoming and scattered photons hardly differ. When $n_0 = n_e$, we can set $|\boldsymbol{k}_0| = |\boldsymbol{k}_e|$, thus giving Bragg's condition $2d \sin\vartheta = \lambda_0$ with $d = \lambda_p$. In the case of optical anisotropy, the following condition applies:

$$\frac{1}{\lambda_0^2} + \frac{1}{\lambda_e^2} - \frac{2}{\lambda_0'\lambda_e}\cos 2\vartheta = \frac{1}{\lambda_p^2},$$

obtained by taking the square of $\boldsymbol{k}_0 - \boldsymbol{k}_e = \boldsymbol{k}_p$. n_0 and n_e are the refractive indices belonging to $\boldsymbol{k}_0$ and $\boldsymbol{k}_e$, respectively, and not the refractive indices of the ordinary or extraordinary wave!

The experimental determination of the propagation velocity of elastic waves is best done, in particular, by selecting a fixed angle between the primary direction of photons of an intensive laser beam and the direction of observation of the scattered photons, preferably $2\vartheta = 90°$ (so called 90°-geometry, to suppress effects of refraction, Fig. 4.51). Then $\lambda_p = \lambda_0/\sqrt{2}$, in case $\lambda_0 = \lambda_e$. The associated frequencies are obtained from high-resolution spectroscopic measurements of the scattered light. Aside from photons of frequency ν_0 (Rayleigh scattering), one finds two lines with frequency $\nu_0 \pm \nu_p$, which can be measured, with not too much effort, to an accuracy of around 0.1%. Hence, one has $v = \nu_p\lambda_p$, the propagation velocity of the elastic wave in the given direction. According to past experimental findings, a noticeable dispersion of the propagation velocity of elastic waves first sets in at a wavelength smaller than about twenty times the lattice spacings, hence, roughly below about 50 nm. Since the wavelength of visible light, for example, in the green spectral region is about 500 nm one can assume that the measured v-values hardly deviate

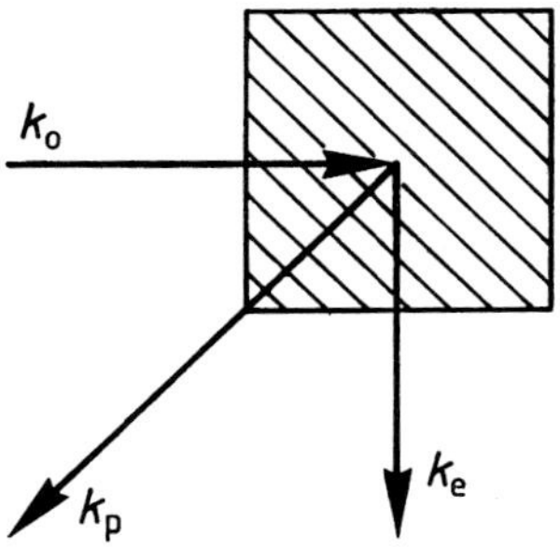

Figure 4.51 Brillouin scattering under 90°-geometry. The photons incoming along $\boldsymbol{k}_0$ appear to be reflected into the photons $\boldsymbol{k}_e$ by the phonons propagating along $\boldsymbol{k}_p$ (conservation of momentum and energy).

from the velocities observed with ultrasound methods at substantially lower frequencies. If one works with polarized laser light and takes into consideration the elastooptical effects of the elastic waves involved, one can, in favorable cases of higher symmetry, separately record the three possible wave types in each direction. Up to now, this method has been mainly applied to cubic crystals, in isolated cases also to orthorhombic and even to monoclinic crystals. Although the method does not achieve the precision of ultrasound measurements, it has gained particular importance because of the possibility of investigating relatively small crystals, most notably for the investigation of elastic properties under extreme pressure.

Related to the Brillouin scattering experiment just sketched are the so-called *stimulated Brillouin scattering* and its variants. If one fires an intensive laser beam (pulsed and focused) at a crystal that can withstand such a strong radiation exposure, the quadratic electrostrictive effect according to $\varepsilon_{ij} = d_{ijkl}E_kE_l$ produces a deformation, repeating itself periodically at a spacing of half the wavelength of the light wave, where $\boldsymbol{E}$ is the electric field of the light wave. Hence, the light wave generates phonons. These phonons can diffract the primary laser beam or even another laser beam entering at the correct scattering angle. We talk of stimulated Brillouin scattering in the case of back scattering of the primary laser light wave by these phonons. We then have $2\vartheta = 180°$, hence, $2\lambda_p = \lambda_0$. The difference in frequency of the scattered photons gives ν_p and thus the propagation velocity of the elastic waves. Whereas only a small fraction of the crystals tested so far could withstand the required high radiation exposure, more favorable conditions exist for the observation of diffraction, under oblique incidence, of a second laser beam on the phonons generated by the primary laser beam. In particular, this is also true when both pulsed light waves, time delayed, enter the crystal. In this way it is possible to derive the lifetime of the phonons concerned from the decay time of the scattered radiation (Eichler, 1977).

5. Neutron Scattering

With the development of intensive neutron sources, the scattering of neutrons has become one of the most attractive tools for the investigation of lattice dynamics. The scattering of neutrons on phonons also takes place under the conditions of momentum- or quasimomentum conservation and energy conservation. Consequently, the formal relationships are completely analogous to the Brillouin scattering just discussed. The experimental arrangement for the observation of neutron scattering is basically the same, albeit different in detail.

The first important problem is the generation of monochromatic neutrons. This occurs, for example, with the aid of time-of-flight spectrometers in the range of very slow neutrons or crystal monochromators, which reflect neutrons of discrete wavelengths through a fixed diffraction angle, similar to X-ray diffraction by constructive interference. Both methods are capable of continuously tuning the wavelength of the neutron wave over a wide range. The detection and frequency determination of the neutrons scattered by the crystal specimen is performed with time-of-flight spectrometers, crystal monochromators or scintillation detectors and related instruments. The effective $\boldsymbol{k}$-vectors of the phonons causing the scattering are fixed by adherence to a certain geometry for the primary and scattered neutron waves. The experimental evaluation, however, requires special efforts, since the wavelengths of the neutrons can change considerably in the scattering process. From the knowledge of λ_0, λ_e and from the angle 2ϑ, one obtains the wavelength λ_p of the phonons and from the change in energy of the neutrons, the frequency ν_p of the phonons and hence, the propagation velocity. Therefore, it is possible to measure the wave velocity and thus the dispersion of the phonons in a broad frequency- or wavelength range by rotating the crystal, varying the diffraction angle and the wavelength of the primary neutron wave. If one selects sufficiently short neutron wavelengths, then after a regular Bragg reflection at the lattice plane $\boldsymbol{h}$, a scattering by a phonon can follow. Then we have $\boldsymbol{k}_e = \boldsymbol{k}_0 + \boldsymbol{h} + \boldsymbol{k}_p$. The propagation vector $\boldsymbol{k}_e$ deviates from the Bragg direction. In this connection, one must not overlook the fact that multiple scattering processes can be superimposed on the neutron scattering discussed here, because each scattered neutron wave can act as a new primary wave. The low-frequency values obtained from the dispersion curves $\nu_p(\boldsymbol{k}_p)$ according to $v = \partial \nu_p / \partial |\boldsymbol{k}_p|$ at $|\boldsymbol{k}_p| = 0$ correspond to the acoustic sound velocities.

Figure 4.52 shows a phonon dispersion curve $\nu_p(\boldsymbol{k}_p)$. Since the length of the possible $\boldsymbol{k}$-vectors are limited by the lattice interplanar spacing ($\lambda/2 \geq d$), it is convenient to represent $\boldsymbol{k}_p$-vectors in the reciprocal lattice. The aforementioned first Brillouin zone mirrors this limitation of the $\boldsymbol{k}$-vectors. Specific symbols have been introduced for the notation of phonons of different propagation directions and lengths. An example is given in Fig. 4.53 for the cubic

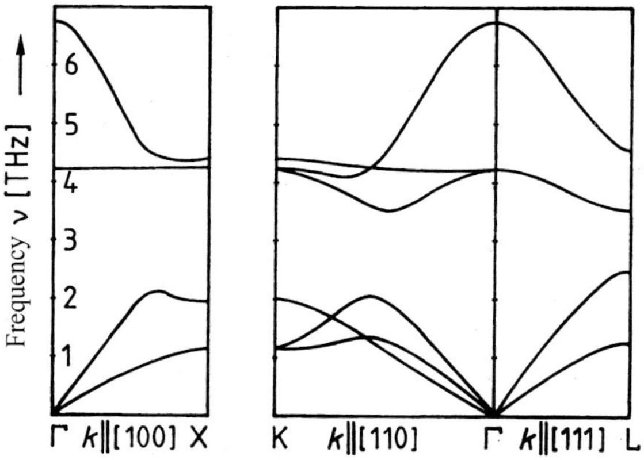

Figure 4.52 Phonon dispersion curves of cubic RbCN at 300 K (Ehrhardt et al., 1980). The frequency (∼ energy) of the phonons is recorded as a function of the length of the propagation vector $\boldsymbol{k}$. The curves starting from the Γ point represent the so-called acoustic branches. The linear region near Γ corresponds to the quasidispersion-free sound velocity. The other curves are called optical branches (see Fig. 4.53).

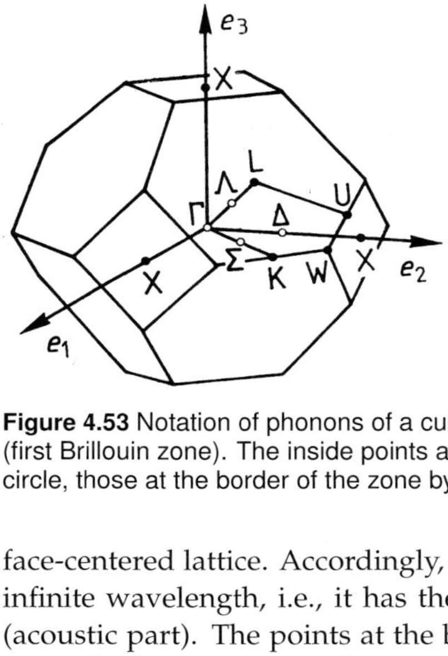

Figure 4.53 Notation of phonons of a cubic face-centered lattice (first Brillouin zone). The inside points are designated by an empty circle, those at the border of the zone by a full circle.

face-centered lattice. Accordingly, a phonon at the Γ-point is equivalent to an infinite wavelength, i.e., it has the propagation velocity for the frequency 0 (acoustic part). The points at the boundary of the first Brillouin zone belong to the highest resonance frequencies of the lattice in the propagation direction leading from the Γ-point to the boundary point. The behavior of these characteristic vibrations of the lattice, subject to external parameters such as, pressure, temperature, electric and magnetic fields, plays an important role in the interpretation of the stability of the crystal lattice with respect to phase transitions.

6. X-ray Scattering

X-ray photons can, just like light photons in Brillouin scattering, interact with the thermally excited phonons of the crystal lattice. Since, in the case of X-rays, the relative change in frequency between the primary and scattered photons is even smaller than in Brillouin scattering and thus hardly measurable, and furthermore, the anisotropy of the refractive indices for X-rays can be fully ignored, $\boldsymbol{k}_0$ and $\boldsymbol{k}_e$ have, to a sufficient approximation, the same magnitude. Hence, the possibility of directly determining the frequency of the phonons involved in the scattering does not exist. One can gain information on the number of thermally excited phonons in certain frequency ranges from the intensity of the diffracted X-rays close to the Bragg reflections (thermally diffuse scattering). These phonons propagate approximately perpendicular to the Bragg direction. Hence, in the case of simply structured crystals, the application of a model for the frequency spectrum $Z_i(\boldsymbol{g},\nu)$ of the phonons can help to obtain useful estimations of the propagation velocities dependent on the propagation direction $\boldsymbol{g}$ and the frequency. $Z_i(\boldsymbol{g},\nu)d\nu$ is the number of phonons of the ith vibrational type ($i = 1,2,3$) in a frequency interval $d\nu$. The high expectations to receive precise data on elastic properties by such investigations (Wooster, 1962), even on very small crystals, were largely abandoned, because it turned out, that multiple processes make a reliable analysis nearly impossible. This became clear after a comparison of certain elastic constants of two orthorhombic crystals, namely, benzalazine and 1,3,5-triphenyl benzol, determined from thermal diffusion scattering and using ultrasound techniques (Joshi & Kashyap, 1964; Haussühl, 1965; Suresh Chandra & Hemkar, 1973; Haussühl, 1974). Conversely, one can calculate the intensity of thermal diffusion scattering from known elastic constants, neglecting multiple scattering processes. By comparing with experiment one can often draw interesting conclusions concerning anomalies of the phonon spectrum as well as on the existence of certain lattice defects.

The thermal motion of the lattice particles also has a strong influence on the effective scattering power of the atoms, described approximately with the help of the individual temperature factors (Debye–Waller factors). This thermal motion must be taken into account to obtain a firm determination of crystal structure. The evaluation of a large number of measured intensities $I(\boldsymbol{h})$ of reflections $\boldsymbol{h}$ allows, in many cases, statements on temperature factors. These data can contribute to the estimation of elastic properties and to the identification of anomalous dynamic processes, as, for example, observed in connection with phase transitions.

4.5.7
Strategy for the Measurement of Elastic Constants

To complete this section some advices are given for the strategy of determining elastic constants. Whereas simple measurement prescriptions can be provided for an efficient precise determination for crystals with high symmetry, hence, for example, for all crystals containing the PSG 2 or 22 as a subgroup, more extensive measurements are required for triclinic crystals. In any case, it is recommended to firstly concentrate the investigation of elastic waves propagating along the principal directions of the Cartesian reference system and their bisectors. If one succeeds in measuring all three wave types in each of these directions, that is, in a total of nine directions, one obtains for triclinic crystals 27 independent measurements, which should be sufficient for the determination of the 21 components of the elasticity tensor in centrosymmetric crystals. Further measurements in the direction of the four space diagonals of the Cartesian reference system contribute additional 12 independent data. With triclinic crystals, evaluation is conveniently performed using a computer program based on the least-squares method, which calculates a set of elastic constants from the elastodynamic basic equations giving the best agreement with the measured sound velocities.

The dynamic elasticity $\{E_{ij}\}$, a second-rank tensor invariant of the elasticity tensor discussed in Section 4.5 has proven useful for a control of the experimental values. The longitudinal effect E'_{11} represents the sum of squares of the three wave velocities along $\boldsymbol{e}'_1 = u_{1i}\boldsymbol{e}_i$, multiplied by the density, hence, $E'_{11} = u_{1i}u_{1j}E_{ij} = \rho\sum_{s=1}^{3} v_s^2(\boldsymbol{e}'_1)$. Pairs of mutually perpendicular propagation directions $\boldsymbol{e}'_1$ and $\boldsymbol{e}'_2$ then obey

$$\sum_{s=1}^{3} v_s^2(\boldsymbol{e}'_1) + \sum_{s=1}^{3} v_s^2(\boldsymbol{e}'_2) = \sum_{s=1}^{3} v_s^2(\alpha\boldsymbol{e}'_1 + \beta\boldsymbol{e}'_2) + \sum_{s=1}^{3} v_s^2(\beta\boldsymbol{e}'_1 - \alpha\boldsymbol{e}'_2).$$

This relationship is taken directly from the plane principal axes transformation (see Section 4.3.2; $t_{ii} + t_{jj} = t'_{ii} + t'_{jj}$!). Thus, the sum of squares of all such pairs is constant as long as the associated propagation vectors lie in the plane spanned by $\boldsymbol{e}'_1$ and $\boldsymbol{e}'_2$. Accordingly, each of these equations enables a sensitive control for all such wave velocities.

With noncentric crystals, instead of c^E_{ijkl}, one must use the quantities c^D_{ijkl}, which, as we saw in Section 4.5.5, depend on the propagation direction. Here it is appropriate to separately measure the piezoelectric and dielectric tensors, and then with the help of approximate values for the components c^E_{ijkl} arrive at an estimation of the differences $c^D_{ijkl} - c^E_{ijkl} = e_{mij}e_{nkl}g_m g_n / \epsilon^{\varepsilon}_{rs} g_r g_s$, where $e_{mij} = c^E_{ijst} d_{mst}$. d_{mst} are the components of the piezoelectric tensor. Measurements of the coupling coefficients may also be useful. Approximate values for c^E_{ijkl} are obtained from a first evaluation of the measurements of the propagation velocities as in the case of centrosymmetric crystals, i.e., neglecting piezoelectric interactions. Even with large coupling coefficients of about 0.5 one achieves, after taking into account the piezoelectric correction, values of sufficient accuracy after only a few iteration steps (relative error for the principal constants c_{iiii}, c_{iijj} and c_{ijij} below 1%). As an example, we again cite the triclinic crystal lithium hydrogen oxalate-monohydrate. With the aid of the dielectric constants given in Section 4.3.3 and the piezoelectric constants determined by the methods discussed in Section 4.4.1, as well as the data of the measurements of a total of 34 propagation velocities in different directions and with different displacement vectors, the elasticity tensor could be completely determined (Haussühl, 1983). In doing so, however, measurements in other directions were included, which differed from those proposed, because the separate observation of the three possible wave types in each direction did not always succeed due to the strong coupling of the waves.

At this juncture, the necessity of an additional correction in crystals with strong pyroelectric effects must be pointed out (see Section 5).

Table 4.14 presents favorable measurement arrangements for an efficient determination of the components of the elasticity tensor and the associated solutions of the elastodynamic basic equations for nonpiezoelectric crystals. With the exception of triclinic crystals, the respective strategy allows an effortless determination of c_{ijkl} without employing a computer. Of course, other data in nondistinct directions can be used. It is especially important to take care, that the position of the reference system in the crystal is unequivocally fixed. In many cases, as, for example, in the PSG 4/m or $\bar{3}$m it is not sufficient to alone obey the rule, introduced in Section 2.2 for the position of $\boldsymbol{e}_i$ with respect to the crystallographic basic vectors $\boldsymbol{a}_i$; rather an indication is required for which of the possible alternatives one has decided selecting a right-handed system (see exercise 8).

4.5.7.1 General Elastic Properties; Stability

From the knowledge of the complete elasticity tensor one can derive all elastic material properties for arbitrary-shaped samples under any boundary conditions. This includes, for example, the propagation of surface waves or the calculation of elastic properties of pressed powders (see exercise 25) as well as the phenomena of refraction and reflection of sound waves. As mentioned in the introduction to this section, certain invariants, which are more easily accessible to discussion, as, for example, dynamic elasticity, linear compressibility under hydrostatic pressure, volume compressibility or the Debye temperature, deserve special interest. Even more complex properties, as, for example, the transversal contraction coefficient under uniaxial pressure or tension (Poisson's ratio), can be derived from these invariants with an accuracy hardly

Table 4.14 Propagation directions g, displacement vectors ξ and ρv^2-values of distinct wave types in the Laue classes. l pure longitudinal, t pure transvere wave, lt combination wave. For most other propagation directions the possible ρv^2-values and ξ-vectors can only be calculated from the elastodynamic basic equations $(-\rho v^2 \delta_{ik} + c_{ijkl} g_j g_l)\xi_k = 0$ after part of the elastic constants are known.

$g \parallel$	$\xi \parallel$	type	ρv^2
monoclinic (2/m; $e_2 \parallel 2$)			
e_1	e_2	t	c_{66}
e_1	$\xi \cdot e_2 = 0$	lt	$\frac{1}{2}(c_{11}+c_{55}) \pm \frac{1}{2}\sqrt{(c_{11}-c_{55})^2+4c_{15}^2}$
e_2	e_2	l	c_{22}
e_2	$\xi \cdot e_2 = 0$	t	$\frac{1}{2}(c_{44}+c_{66}) \pm \frac{1}{2}\sqrt{(c_{44}-c_{66})^2+4c_{46}^2}$
e_3	e_2	t	c_{44}
e_3	$\xi \cdot e_2 = 0$	lt	$\frac{1}{2}(c_{33}+c_{55}) \pm \frac{1}{2}\sqrt{(c_{33}-c_{55})^2+4c_{35}^2}$
$e_1 \pm e_3$	e_2	t	$\frac{1}{2}(c_{44}+c_{66} \pm 2c_{46})$
$e_1 \pm e_3$	$\xi \cdot e_2 = 0$	lt	$\frac{1}{4}(c_{11}+c_{33}+2c_{55} \pm 2c_{15} \pm 2c_{35})$ $\pm\frac{1}{2}\sqrt{\frac{1}{4}(c_{11}-c_{33} \pm 2c_{15} \mp 2c_{35})^2+(c_{13}+c_{55} \pm c_{15} \pm c_{35})^2}$
orthorhombic (2/mm)			
e_i	e_i	l	c_{ii}
e_i	e_j $(i \neq j)$	t	$c_{9-i-j,9-i-j}$
$e_i \pm e_j$	e_k $(k \neq i,j)$	t	$\frac{1}{2}(c_{9-i-k,9-i-k}+c_{9-j-k,9-j-k})$
$e_i \pm e_j$	$\xi \cdot e_k = 0$ $(k \neq i,j)$	lt	$\frac{1}{4}(c_{ii}+c_{jj}+2c_{9-i-j,9-i-j})$ $\pm\frac{1}{2}\sqrt{\frac{1}{4}(c_{ii}-c_{jj})^2+(c_{ij}+c_{9-i-j,9-i-j})^2}$
trigonal ($\bar{3}$)			
e_3	e_3	l	c_{33}
e_3	$\xi \cdot e_3 = 0$	t	$c_{44} = c_{55}$
trigonal ($\bar{3}2/m$; $e_1 \parallel 2$)			
e_1	e_1	l	c_{11}
e_1	$\xi \cdot e_1 = 0$	t	$\frac{1}{2}(c_{44}+c_{66}) \pm \frac{1}{2}\sqrt{(c_{44}-c_{66})^2+4c_{14}^2}$
e_2	e_1	t	$c_{66} = \frac{1}{2}(c_{11}-c_{12})$
e_2	$\xi \cdot e_1 = 0$	lt	$\frac{1}{2}(c_{11}+c_{44}) \pm \frac{1}{2}\sqrt{(c_{11}-c_{44})^2+4c_{14}^2}$
e_3	e_3	l	c_{33}
e_3	$\xi \cdot e_3 = 0$	t	$c_{44} = c_{55}$
$e_2 \pm e_3$	e_1	t	$\frac{1}{2}(c_{44}+c_{66} \pm 2c_{14})$
$e_2 \pm e_3$	$\xi \cdot e_1 = 0$	lt	$\frac{1}{4}(c_{11}+c_{33}+2c_{44} \mp 2c_{14})$ $\pm\frac{1}{2}\sqrt{\frac{1}{4}(c_{11}-c_{33} \mp 2c_{14})^2+(c_{13}+c_{44} \mp c_{14})^2}$
tetragonal (4/m)			
e_1	e_3	t	$c_{44} = c_{55}$
e_1	$\xi \cdot e_3 = 0$	lt	$\frac{1}{2}(c_{11}+c_{66}) \pm \frac{1}{2}\sqrt{(c_{11}-c_{66})^2+4c_{16}^2}$
e_3	e_3	l	c_{33}
e_3	$\xi \cdot e_3 = 0$	t	c_{44}
$e_1 \pm e_2$	e_3	t	c_{44}
$e_1 \pm e_2$	$\xi \cdot e_3 = 0$	lt	$\frac{1}{2}(c_{11}+c_{66}) \pm \frac{1}{2}\sqrt{(c_{12}+c_{66})^2+4c_{16}^2}$

$g \parallel$	$\xi \parallel$	type	ρv^2
tetragonal (4/mm), like 4/m with $c_{16} = 0$			
$e_1 \pm e_3$	e_2	t	$\frac{1}{2}(c_{44} + c_{66})$
$e_1 \pm e_3$	$\xi \cdot e_2 = 0$	lt	$\frac{1}{4}(c_{11} + c_{33} + 2c_{44})$ $\pm\frac{1}{2}\sqrt{\frac{1}{4}(c_{11} - c_{33})^2 + (c_{13} + c_{44})^2}$
hexagonal (6/m und 6/mm) further ∞/m and ∞/mm			
e_1	e_1	l	c_{11}
e_1	e_2	t	$c_{66} = \frac{1}{2}(c_{11} - c_{12})$
e_1	e_3	t	$c_{44} = c_{55}$
e_3	e_3	l	c_{33}
e_3	$\xi \cdot e_3 = 0$	t	$c_{44} = c_{55}$
$e_1 \pm e_3$	e_2	t	$\frac{1}{2}(c_{44} + c_{66})$
$e_1 \pm e_3$	$\xi \cdot e_2 = 0$	lt	$\frac{1}{4}(c_{11} + c_{33} + 2c_{44})$ $\pm\frac{1}{2}\sqrt{\frac{1}{4}(c_{11} - c_{33})^2 + (c_{13} + c_{44})^2}$
cubic (m3, 4/m3			
e_1	e_1	l	$c_{11} = c_{22} = c_{33}$
e_1	$\xi \cdot e_1 = 0$	t	$c_{44} = c_{55} = c_{66}$
$e_1 \pm e_2$	e_3	t	c_{44}
$e_1 \pm e_2$	$e_1 \mp e_2$	t	$\frac{1}{2}(c_{11} - c_{12})$
$e_1 \pm e_2$	$e_1 \pm e_2$	l	$\frac{1}{2}(c_{11} + c_{12} + 2c_{44})$
$e_1 \pm e_2 \pm e_3$	$e_1 \pm e_2 \pm e_3$	l	$\frac{1}{3}(c_{11} + 2c_{12} + 4c_{44})$
$e_1 \pm e_2 \pm e_3$	$\xi \cdot (e_1 \pm e_2 \pm e_3) = 0$	t	$\frac{1}{3}(c_{11} - c_{12} + c_{44})$

achievable using straightforward static methods. Here, we shall merely discuss the mechanical stability criteria, giving a definite answer to the question, which ratios, at all, can the elastic constants take on in a stable crystal lattice. Firstly, we must demand that all quantities, which can be considered as elastic resistances, take on positive values, while otherwise, for example, even the smallest stress (tension) would suffice to contract or, under pressure, expand the crystal and thus violate the second law of thermodynamics. This means, for example, that the volume compressibility must always take on positive values, thus, for example, with cubic crystals, it must always be true $K = 3/(c_{11} + 2c_{12}) > 0$, i.e., $c_{11} > -2c_{12}$. Furthermore, all longitudinal effects must be positive, i.e., $c'_{11} > 0$ for arbitrary directions $e'_1 = u_{1i}e_i$. The same is true for Young's modulus in each direction $1/s'_{11}$. Moreover, all expressions derived for ρv^2 from the elastodynamic basic equations must be positive. A number of these conditions for distinct directions is found in Table 4.14. For cubic crystals, we have, for example, $c_{11}, c_{44}, c_{11} - c_{12} > 0$, and hence $c_{11} > c_{12}$. Further, the determinant of 6×6 matrix of the elastic constants as well as their main adjuncts must take on positive values. This follows from the requirement that each change of a deformation component must be followed by a change of the corresponding stress component with the same sign. From the trend of the elastic constants as a function of temperature and uniaxial or hydrostatic pressure, one can, with certain phase transition, in particular those of strongly second-order character, often observe a tendency to instability long before the transition is reached.

For example, certain wave velocities, on approaching a phase boundary, can continually decrease, causing the associated elastic resistances to become smaller, and in some cases nearly approach zero (elastic "soft modes"). Hence, to judge the stability, a knowledge of the temperature and pressure behavior of the elastic properties is of major importance. By virtue of the known rules concerning the T and P dependence of elastic constants in stable crystals, one can, in individual cases, recognize hints at possible phase transitions and characteristic changes in bonding coupled to elastic anomalies.

4.5.8
The Dependence of Elastic Properties on Scalar Parameters (Temperature, Pressure)

The quantities dc_{ijkl}/dX can be determined from the measurement of the shift of resonance frequencies under the influence of an external parameter X (temperature T or pressure P) or by employing one of the other highly sensitive methods previously discussed.

If the relationship $\rho v^2 = f(c_{ijkl})$ is valid, the measured frequency shift, on account of $v = 2L\nu_m/m$, yields the following relation:

$$\frac{1}{f(c_{ijkl})}\frac{df(c_{ijkl})}{dX} = \frac{d\rho}{\rho dX} + \frac{2d\nu_m}{\nu_m dX} + \frac{2dL}{LdX}.$$

Instead of the differential quotients, one can also write the difference quotients, as occurring in the measurement. If X is the temperature T, the first term represents the negative thermal volume expansion, and the last, the twofold linear thermal expansion. Similarly, in the case of pressure dependence, the first term represents the negative volume compressibility, and the last, the twofold linear compressibility. If one now carries out the measurements in the arrangements proposed in Table 4.14, one can determine the complete tensors dc_{ijkl}/dX. For the interpretation of these quantities it was found convenient to describe the temperature dependence by the *thermoelastic constants*

$$T_{ij} = \frac{dc_{ij}}{c_{ij}dT} = \frac{d\log c_{ij}}{dT}$$

and the pressure dependence by the *piezoelastic constants*

$$P_{ij} = \frac{dc_{ij}}{dp}.$$

T_{ij} have the dimension (Grad Kelvin^{-1}), P_{ij} are dimensionless. One observes only small variations of these quantities in groups of isotypic crystals with the same particle charge. The coefficients take on characteristic values for different structural types. If one considers other derivatives, this important fact for the crystal-chemical interpretation would not appear. As an example let us mention the temperature dependence of the reciprocal compressibility dK^{-1}/dp, a dimensionless quantity which takes on the value of around 5 in all stable crystals. This universal property deserves a more detailed discussion.

The temperature dependence of the given resonance frequency often plays a decisive role in the application of piezoelectric crystals for frequency generators and stabilizers in electrical engineering and electronics. From the relation given above we obtain the following conditions for a so-called *zero-cut* of the temperature T:

$$\frac{d\nu}{\nu dT} = \frac{df(c_{ijkl})}{2f(c_{ijkl})dT} - \frac{d\rho}{2\rho dT} - \frac{dL}{LdT} = 0.$$

This means, the temperature dependence of the given effective elastic constants $f(c_{ijkl})$ must be compensated for by the effects of thermal expansion contained in the second and third terms. This is satisfied by very few of the presently known crystal types, and only in certain distinct directions. For example, in α-quartz perpendicular to the twofold axis, there exists a direction of a zero-cut of a transverse wave. By combining plates with opposite temperature behavior, it is possible to construct piezoelectric oscillators of α-quartz possessing a frequency stability of $d\nu/\nu dT < 10^{-8}/\mathrm{K}$ over a wide temperature range. This property together with the excellent mechanical properties is the reason for the undisputed unique position of quartz for use in such devices.

In other applications, the temperature dependence of the resonance frequency of an oscillator is used to measure temperature. These devices (e.g., quartz thermometers) are not only highly sensitive, but also exhibit excellent stability in calibration. If the heat capacity of the oscillator is kept sufficiently small by miniaturization of the construction, one can employ these devices as radiation detectors and radiation measuring instruments.

4.5.9
Piezooptical and Elastooptical Tensors

The change in refractive indices, i.e., the velocity of light, under the influence of an external mechanical stress or a deformation is described by *piezooptical* or *elastooptical tensors*. For the tensor representation of the optical properties, one selects, as with the electrooptical effect, the polarization tensor (see Section 4.4.2), whose components are interlinked with the refractive indices according to $a_{ij} = (n_{ij})^{-2}$. To a first and mostly sufficient approximation, both effects are represented by

$$\Delta a_{ij} = q_{ijkl}\sigma_{kl} \quad (\{q_{ijkl}\} \text{ piezooptical tensor})$$
$$\Delta a_{ij} = p_{ijkl}\varepsilon_{kl} \quad (\{p_{ijkl}\} \text{ elastooptical tensor})$$

σ_{kl} and ε_{kl} are the components of the stress- and deformation tensor, respectively. From the symmetry of the stress- and deformation tensor it is found that $q_{ijkl} = q_{jikl} = q_{ijlk} = q_{jilk}$ and correspondingly $p_{ijkl} = p_{jikl} = p_{ijlk} = p_{jilk}$. At the end of this section we will return to a deviation from this symmetry. The number and type of independent components in the individual Laue classes was discussed in Section 4.5. The pair-wise permutation of the index positions is now not allowed. Thus the number of independent components compared to the elastic constants is substantially increased, for example, to 36 in triclinic crystals as against 21 in the elasticity tensor. Both tensors are not independent of one another, rather they carry over into each other according to $p_{ijkl} = q_{ijmn}c_{mnkl}$, as is easily seen by setting $\sigma_{mn} = c_{mnkl}\varepsilon_{kl}$.

4.5.9.1 **Piezooptical Measurements**

The measurement of piezooptical effects can be performed according to the methods discussed in Section 4.4.2 (Electrooptics). The electrical field is now replaced by uniaxial or hydrostatic pressure. With perpendicular incidence, one obtains for the change in path difference of a plane parallel plate of thickness L

$$\Delta G = -\frac{L}{2}(n'^3_{jj}q'_{jjlm}\sigma'_{lm} - n'^3_{kk}q'_{kklm}\sigma'_{lm}) + (n'_{jj} - n'_{kk})Ls'_{iilm}\sigma'_{lm}$$

(summation is only over l and m!).

The direction of radiation runs parallel e'_i, and the direction of vibration of both waves parallel e'_j and e'_k. To specify this relationship, we proceed in the same manner as with the electrooptical effect, i.e., under conditions of strong double refraction $(n'_{jj} - n'_{kk})$ we can neglect the rotation of the sectional ellipse by the piezooptical effect. We then describe the piezooptical effect in the Cartesian coordinate system $\{e'_i, e'_j, e'_k\}$. In the case of degeneracy $n'_{jj} = n'_{kk}$, as, for example, in cubic crystals, the position of the sectional ellipse must be calculated taking into consideration the piezooptical effect. In the case of so called absolute measurements, the value for air (for example, $n'_{kk} \approx 1$, $q'_{kklm} = 0$) must be written for one of both waves. Knowledge of the elasticity coefficients s'_{ijkl} is necessary for the evaluation of the measurements. These are calculated from the elastic constants by matrix inversion of the system $\sigma'_{ij} = c'_{ijkl}\varepsilon'_{kl}$. Furthermore, the aspects set forth in Section 4.4.2 (Electrooptics) are to be noted.

As with all optical measurements, one must set high demands on the optical quality of the specimens. Whereas with hydrostatic loads one can go to very high pressure values, in the case of uniaxial stress one is usually limited to small loads of a few kp/cm^2 to prevent damaging the crystal due to inhomogeneous stress and plastic deformation. Reliable measurements of the change in the path difference can usually be achieved under these conditions, however, absolute measurements, made with the help of interferometers often cause quite a few problems because of the small effects involved. The sensitivity can be substantially improved by employing phase-lock techniques together with dynamic methods in which laud speaker membranes or a motor driven eccenter are used to transfer alternating pressure to the crystal. If one works with purely static or quasistatic methods, then the complete determination of the piezooptical tensor requires a certain number of absolute measurements. However, if elastooptical measurements are included then under favorable circumstances one can forgo absolute measurements and still achieve high accuracy. Because $p_{ijmn} = q_{ijkl}c_{klmn}$ one can feed the measured elastooptical values into the system of equations of q_{ijkl} in order to obtain sufficient overdetermination for the solution. It is advantageous to use those elastooptical measurements combining minimal effort with high accuracy. We will return to this shortly.

For example, with the help of purely static methods employing uniaxial stress, the complete piezooptical tensor of monoclinic aminoethanesulfonic acid (Taurine) with its 20 independent components was determined (Haussühl & Uhl, 1969). Such measurements are more easily performed on optically inactive cubic crystals because of the low number of independent components (4 in PSG m3, 3 in PSG 4/m3) and because $n'_{jj} = n'_{kk} = n$. Table 4.15 presents the relationships for some important arrangements of cubic crystals.

If an optically active crystal type is present, then the interplay of induced double refraction and natural optical activity must be analyzed according to the aspects discussed in Section 4.3.6.8, whereby, to a first approximation it is assumed that the optical activity remains unchanged under the influence of mechanical stress. To what extent this assumption is correct is as yet not clearly resolved. No reliable measurements concerning the tensor $\Delta g_{ijk} = q_{ijklm}\sigma_{lm}$ are available which describe the change of optical activity under the influence of mechanical stress.

4.5.9.2 **Elastooptical Measurements**

The *elastooptical* tensor is amenable to measurements via diffraction experiments as we have come to know from the Schaefer–Bergmann method or from Brillouin scattering. The elastic waves generate a periodic distortion in the crystal. The relation $a_{ij} = n_{ij}^{-2}$ leads to a measurable change in the refractive

Table 4.15 Piezooptical and second-order electrooptical effects for some special directions in cubic crystals upon uniaxial stress along $e_1'(\sigma_{11}')$ or under the influence of an electric field along $e_1'(E_1')$. The light wave impinges perpendicularly onto the crystal surface, sample thickness in the direction $g \parallel e_i'$ of the incident light wave is L. g, $D'(\parallel e_j')$ and $D''(\parallel e_k')$ form a right coordinate system. In the case of the electrooptical effect, the coefficients r_{ij} are assumed inplace of constants q_{ij} (in Voigt notation).

σ_{11}' resp. $E' \parallel e_1'$	$g \parallel e_i'$	$D' \parallel e_j'$	$D'' \parallel e_k'$	$\Delta a_{jj}'/\sigma_{11}'$ resp. $\Delta a_{jj}'/E_1'$ $\Delta a_{kk}'/\sigma_{11}'$ resp. $\Delta a_{kk}'/E_1'$ $\Delta G/\sigma_{11}'$ resp. $\Delta G/E_1'$
[100]	[010]	[001]	[100]	$q_{31} = q_{12}$ q_{11} $-Ln^3(q_{31} - q_{11})/2$
[100]	[001]	[100]	[010]	q_{11} q_{21} $-Ln^3(q_{11} - q_{21})/2$
[100]	[100]	[010]	[001]	q_{21} $q_{31} = q_{12}$ $-Ln^3(q_{21} - q_{12})/2$
[110]	[110]	[001]	[1$\bar{1}$0]	$(q_{12} + q_{21})/2$ $(2q_{11} + q_{12} + q_{21} - 4q_{66})/4$ $-Ln^3(-2q_{11} + q_{12} + q_{21} + 4q_{66})/8$
[110]	[001]	[1$\bar{1}$0]	[110]	$(2q_{11} + q_{12} + q_{21} - 4q_{66})/4$ $(2q_{11} + q_{12} + q_{21} + 4q_{66})/4$ Ln^3q_{66}
[110]	[1$\bar{1}$0]	[110]	[001]	$(2q_{11} + q_{12} + q_{21} + 4q_{66})/4$ $(q_{12} + q_{21})/2$ $-Ln^3(2q_{11} - q_{12} - q_{21} + 4q_{66})/8$
[111]	[111]	[1$\bar{1}$0]	[11$\bar{2}$]	$(q_{11} + q_{12} + q_{21} - 2q_{66})/3$ $(q_{11} + q_{12} + q_{21} - 2q_{66})/3$ 0
[111]	[110]*)	[112]*)	[111]	$(q_{11} + q_{12} + q_{21} - 2q_{66})/3$ $(q_{11} + q_{12} + q_{21} + 4q_{66})/3$ Ln^3q_{66}

*) The same result is obtained for arbitrary directions perpendicular to [111].

indices according to

$$\Delta n_{ij} = -(n_{ij})^3 \Delta a_{ij}/2.$$

The diffraction geometry is determined by the conservation of the quasimomentum (see Section 4.5.6). Energy conservation induces a change in the frequency of the diffracted photons. The intensity of the diffracted light wave can be calculated in a similar manner as the diffraction of X-rays on a grating. It is

$$I = I_0 L^2 (\Delta n_{ij})^2_{\text{eff}} K,$$

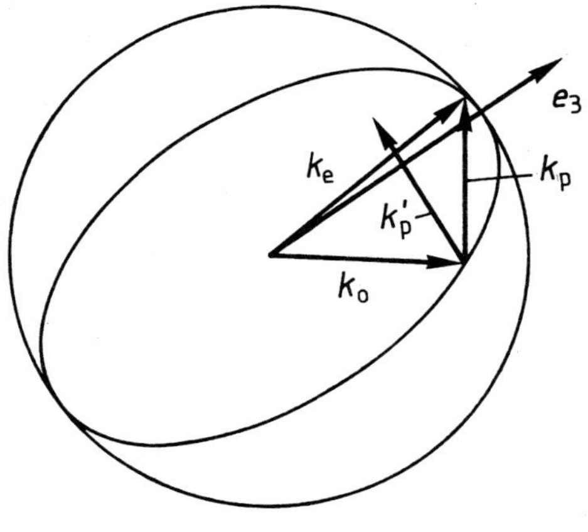

Figure 4.54 Geometrical construction of the diffraction of a light wave with propagation vector $\boldsymbol{k}_0$ using phonons with a propagation vector $\boldsymbol{k}_p$ analogosly to the Ewald construction. Here, the propagation sphere is replaced by an index surface scaled by a factor of $1/\lambda_0$ (where λ_0 denotes the vacuum wavelength) which indicates the allowed propagation vectors for each direction. For quasi-infinite crystals, no diffraction is possible unless the propagation vector $\boldsymbol{k}_p$, starting at the tip of the vector $\boldsymbol{k}_0$, ends on this surface. The interference and dynamical effects observed for crystals with a sufficiently small ratio of transmission path length to acoustical wavelength are completely analogous to those in electron diffraction in thin crystals (see, e.g., *Kristallstrukturbestimmung*). The illustration shows an optically uniaxial crystal with an optical axis $\parallel e_3$, $\boldsymbol{k}_0$ and $\boldsymbol{k}_p$ in the drawing plane.

where $(\Delta n_{ij})_{\text{eff}}$ is the effective Fourier coefficient of the optical phase grating and L is the length of the optical path. K is a constant taking into account the respective unique experimental arrangement.

In the concrete case of an arbitrary plane elastic wave strict conditions must be adhered to for the direction of the primary wave to even make diffraction possible. The propagation vectors $\boldsymbol{k}_0$ of the optic wave and $\boldsymbol{k}_p$ of the acustic wave define the diffraction plane. The intercept of this plane with the indicatrix shows whether a propagation direction exists at all which guarantees momentum conservation. The permissible vectors $\boldsymbol{k}_e$ of the diffracted wave must lie on the given sectional ellipse (Fig. 4.54). This corresponds to the Ewald construction. The calculation of the effective Fourier coefficients is only readily feasible in a few distinct cases. We will return to this experimentally important arrangement later.

First we have to clarify another aspect regarding the true symmetry of the elastooptical tensors. Nelson & Lax (1970) have pointed out that the generally accepted valid relationship $\Delta a_{ij} = p_{ijkl}\varepsilon_{kl}$ must be amended. Namely, in certain cases, the general relation

$$\Delta a_{ij} = p_{ijkl}\frac{\partial \xi_k}{\partial x_l}$$

is written, i.e., the rotational part

$$r_{ij} = \frac{1}{2}\left(\frac{\partial \xi_i}{\partial x_j} - \frac{\partial \xi_j}{\partial x_i}\right)$$

of the distortion vector

$$\frac{\partial \xi_i}{\partial x_j} = \varepsilon_{ij} + r_{ij}$$

may not be neglected. In pure longitudinal waves $r_{ij} = 0$. In contrast, elastic waves with strong transverse components cause a periodic rotational motion of the volume element, which can lead to a periodic change of the effective refractive indices for certain propagation directions. This contribution to the elastooptical effect can be calculated directly from the refractive indices as follows. A rigid rotation of the indicatrix is linked to the rotation of the volume element. The rotation of a vector $\boldsymbol{x}$ is described by $x_i' = x_i + r_{ik}x_k = u_{ik}x_k$ with $u_{ik} = \delta_{ik} + r_{ik}$ (δ_{ik} is the Kronecker symbol). The rotation of the polarization tensor $\{a_{ij}\}$, noted in the crystal-physical coordinate system, is obtained by tensor transformation with the transformation matrix $(u_{ik}) : a_{jj}' = u_{ik}u_{jl}a_{kl}$. The contribution to the elastooptical effect is then

$$\Delta_r a_{ij} = a_{ij}' - a_{ij} = u_{ik}u_{jl}a_{kl} - a_{ij}.$$

With $u_{ik} = \delta_{ik} + r_{ik}$ we get

$$\Delta_r a_{ij} = (\delta_{ik} + r_{ik})(\delta_{jl} + r_{jl})a_{kl} - a_{ij} = r_{ik}a_{kj} + r_{jl}a_{il},$$

whereby the quadratic term with the factor $r_{ik}r_{jl}$ was neglected. Due to the fact that $r_{ik} = -r_{ki}$ and with the help of the index commutation relation $r_{ki} = \delta_{il}r_{kl}$, we come to the following representation:

$$\Delta_r a_{ij} = (a_{il}\delta_{jk} - a_{kj}\delta_{il})r_{kl}.$$

In order to estimate the consequence of this change, we go into the optical principal axis system where $a_{ij} = 0$ for $i \neq j$. This gives $\Delta_r a_{ii} = 0$ and for $i \neq j$:

$$\Delta_r a_{ij} = (a_{jj} - a_{ii})r_{ij} = \left(\frac{1}{n_{jj}^2} - \frac{1}{n_{ii}^2}\right) r_{ij},$$

where n_{jj} and n_{ii} are the principal refractive indices. We then obtain for the general elastooptical effect in this coordinate system

$$\Delta a_{ij} = p_{ijkl}\varepsilon_{kl} + (a_{jj} - a_{ii})r_{ij}.$$

The second term, in the meantime designated as optorotation, delivers only in certain cases a measurable contribution. In cubic crystals and in crystals with small double refraction, optorotation is not or hardly measurable. Pure longitudinal waves do not generate optorotation.

Accordingly, optorotational effects are only expected with waves possessing strong transverse components in crystals with large double refraction. In such crystals, as seen in the examples of rutile and calcite, they can exceed the usual elastooptical effects a multiple of times. In any case the allowance of optorotation with knowledge of the distortion vector is unproblematic. Optorotation then plays a subordinate role for the measurement of elastooptical constants when the light wave is transmitted within a principal plane of the indicatrix and one stays in the range of small diffraction angles of maximum one degree. The orientation of the sectional ellipses for the primary and diffracted light waves then remain unaltered. Because $\Delta_r a_{ii} = 0$ the respective refractive indices do not experience any measurable change due to optorotation.

Another complication appears in piezoelectric crystals when elastic waves are accompanied by an electric field producing an electrooptical effect. As we have seen in Section 4.5.5, the electric field of an elastic wave can be described to sufficient approximation by

$$E_m = -(\operatorname{grad}\phi)_m = -2\pi i k_m \phi \quad \text{mit} \quad \phi = \phi_0 e^{2\pi i(\boldsymbol{k}\cdot\boldsymbol{x}-\nu t)}.$$

$\boldsymbol{E}$ then runs parallel to the propagation direction $\boldsymbol{g} = \boldsymbol{k}/|\boldsymbol{k}|$. From the conditions

$$D_l = e_{lmn}\varepsilon_{mn} + \epsilon^{\varepsilon}_{lm}E_m \quad \text{and} \quad \boldsymbol{D}\cdot\boldsymbol{k} = 0$$

we obtain

$$D_l g_l = e_{lmn}\varepsilon_{mn}g_l - \epsilon^{\varepsilon}_{lm}2\pi i k_m g_l \phi = 0$$

and hence, an expression for f ϕ and

$$E_k = -2\pi i(g_k/\lambda)\phi = -e_{lmn}\varepsilon_{mn}g_l g_k/\epsilon^{\varepsilon}_{rs}g_r g_s.$$

The expression $\epsilon^{\varepsilon}_{rs}g_r g_s$ represents the effective dielectric constant in the propagation direction $\boldsymbol{g}$. With this electric field we expect an electrooptical effect superimposed on the normal elastooptical effect:

$$\Delta a''_{ij} = r_{ijk}E_k = -r_{ijk}e_{lmn}\varepsilon_{mn}g_l g_k/\epsilon^{\varepsilon}_{rs}g_r g_s.$$

This contribution can be calculated from the dielectric, piezoelectric and electrooptical tensors when the propagation direction is given and the distortion vector is known. It is proportional to the amplitude of the elastic wave and

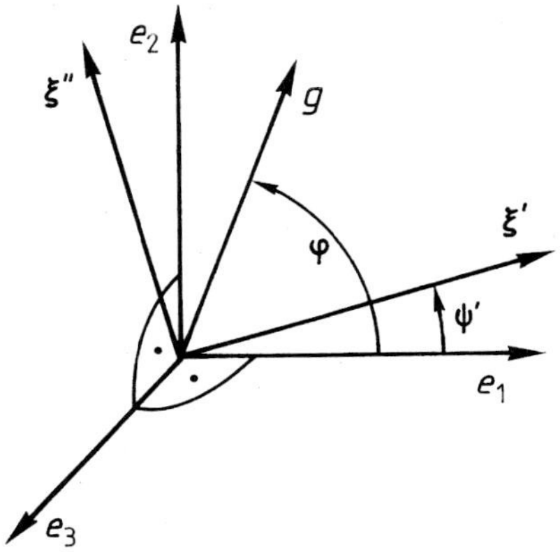

Figure 4.55 Derivation of the change of the indicatrix of cubic crystals under the influence of sound waves. g is the direction of propagation of the sound wave, and ξ' and ξ'' are the corresponding distortion vectors in the plane perpendicular to e_3.

thus superimposes itself with the respective fixed part of the normal elastooptical effect. An estimation for α-quartz with its small piezoelectric and electrooptical constants shows that the contribution in this case remains far under 1% of the normal elastooptical effect for each propagation direction. In crystals with high piezoelectric and electrooptical effects, as, for example, α-iodic acid or α-$LiIO_3$, one can count on substantial contributions of the elastooptical effect. This can be of advantage in practice (see below).

We now want to discuss some proven methods. Mueller (1938) showed that one can derive statements on the elastooptical constants from the state of polarization of the diffracted light. These methods are mainly suited for cubic crystals and for directions of optical axes of anisotropic crystals.

Let us consider the concrete case of an optical inactive cubic crystal cut in the form of a rectangular parallelepiped with light transmitted in approximately perpendicular incidence parallel [001]. The elastic waves generated by the elastooptical effect shall propagate in the plane (001), hence perpendicular to the light ray. The normalized propagation vector of the elastic wave is $g = g_i e_i = \cos\varphi e_1 + \sin\varphi e_2$ (Fig. 4.55). The direction of the distortion vectors of the elastic waves is gained, as discussed, from the electrodynamic basic equations when we insert the associated propagation velocities. From

$$(-\rho v^2 \delta_{ik} + c_{ijkl} g_j g_l)\xi_k = 0$$

we obtain for cubic crystals

$$\begin{vmatrix} -\rho v^2 + c_{11}g_1^2 + c_{44}g_2^2 & (c_{12}+c_{44})g_1g_2 & 0 \\ (c_{12}+c_{44})g_1g_2 & -\rho v^2 + c_{11}g_2^2 + c_{44}g_1^2 & 0 \\ 0 & 0 & -\rho v^2 + c_{44} \end{vmatrix} = 0.$$

As a result we get

$$\rho v'^2 \quad \text{or} \quad \rho v''^2 = \frac{c_{11}+c_{44}}{2} \pm \frac{1}{2}\sqrt{(c_{11}-c_{44})^2(g_1^2-g_2^2)^2+4(c_{12}+c_{44})^2 g_1^2 g_2^2}$$

and $\rho v'''^2 = c_{44}$.

For the associated distortion vectors, one finds

$$\begin{pmatrix}\xi_1\\ \hline \xi_2\end{pmatrix}' \quad \text{resp.} \quad \begin{pmatrix}\xi_1\\ \hline \xi_2\end{pmatrix}'' = \frac{\rho v'^2\ (\text{or } \rho v''^2) \ - c_{11}g_2^2 - c_{44}g_1^2}{(c_{12}+c_{44})g_1 g_2}$$
$$= \cot\psi' \quad \text{resp.} \quad \cot\psi'', \qquad \text{with} \quad \psi = \angle(\boldsymbol{\xi}, \boldsymbol{e}_1),$$

furthermore,

$$\xi_3' = \xi_3'' = 0 \quad \text{as well as} \quad \xi_1''' = \xi_2''' = 0,$$

Since $\boldsymbol{\xi}'$ and $\boldsymbol{\xi}''$ are mutually perpendicular, $\tan\psi'' = -\cot\psi'$. The notation $'$ or $''$ specifies the faster or slower of both waves, respectively. We require the associated components of the deformation tensor to calculate the elastooptical effect. Let

$$\boldsymbol{\xi}' = \boldsymbol{\xi}_0' e^{2\pi i(\boldsymbol{k}\cdot\boldsymbol{x}-\nu t)} = C'\cos\psi'\boldsymbol{e}_1 + C'\sin\psi'\boldsymbol{e}_2$$

the same applies with $\boldsymbol{\xi}''$.

With

$$k_1 = \frac{1}{\lambda}\cos\varphi \quad \text{and} \quad k_2 = \frac{1}{\lambda}\sin\varphi$$

we have

$$\begin{aligned}
\varepsilon_{11}' &= \frac{\partial\xi_1}{\partial x_1} = 2\pi i k_1'\xi_1' = A'\cos\varphi\cos\psi',\\
\varepsilon_{22}' &= A'\sin\varphi\sin\psi',\\
\varepsilon_{12}' &= \frac{1}{2}\left(\frac{\partial\xi_1'}{\partial x_2} + \frac{\partial\xi_2'}{\partial x_1}\right) = \frac{A'}{2}(\cos\psi'\sin\varphi + \sin\psi'\cos\varphi) = \frac{A'}{2}\sin(\varphi+\psi'),\\
\varepsilon_{11}'' &= -A''\cos\varphi\sin\psi',\\
\varepsilon_{22}'' &= A''\sin\varphi\cos\psi',\\
\varepsilon_{12}'' &= \frac{A''}{2}\cos(\varphi+\psi'),\\
\varepsilon_{23}' &= \varepsilon_{13}' = \varepsilon_{23}'' = \varepsilon_{13}'' = 0,\\
\varepsilon_{22}''' &= \varepsilon_{12}''' = 0.
\end{aligned}$$

The sectional ellipse perpendicular [001] experiences the following change:

$$\begin{aligned}
\Delta a'_{11} &= p_{1111} A' \cos\varphi \cos\psi' + p_{1122} A' \sin\varphi \sin\psi',\\
\Delta a'_{22} &= p_{2211} A' \cos\varphi \cos\psi' + p_{1111} A' \sin\varphi \sin\psi',\\
\Delta a'_{12} &= p_{1212} A' \sin(\varphi + \psi'),\\
\Delta a''_{11} &= -p_{1111} A'' \cos\varphi \sin\psi' + p_{1122} A'' \sin\varphi \cos\psi'\\
\Delta a''_{22} &= -p_{2211} A'' \cos\varphi \sin\psi' + p_{1111} A'' \sin\varphi \cos\psi'\\
\Delta a''_{12} &= p_{1212} A'' \cos(\varphi + \psi').
\end{aligned}$$

The wave $\rho v^2 = c_{44}$, distortion vector parallel [001], does not contribute to the elastooptical effect for this sectional ellipse. We now inquire as to the position of the principal axes of this sectional ellipse noted in our crystal–physical coordinate system. The plane principal axis transformation (see Section 4.3.2) delivers

$$\begin{aligned}
\tan 2\vartheta' = \frac{2\Delta a'_{12}}{a'_{11} - a'_{22}} &= 2p_{1212} \sin(\varphi + \psi')/(p_{1111} \cos(\varphi + \psi')\\
&\quad + p_{1122} \sin\varphi \sin\psi' - p_{2211} \cos\varphi \cos\psi')
\end{aligned}$$

and

$$\begin{aligned}
\tan 2\vartheta'' &= -2p_{1212} \cos(\varphi + \psi')/(p_{1111} \sin(\varphi + \psi')\\
&\quad - p_{1122} \sin\varphi \cos\psi' - p_{2211} \cos\varphi \sin\psi').
\end{aligned}$$

In crystals of PSG 43, $\bar{4}3$ and 4/m3 we have $p_{1122} = p_{2211}$, so that

$$\tan 2\vartheta' = 2p_{1212} \tan(\varphi + \psi')/(p_{1111} - p_{1122})$$

and

$$\tan 2\vartheta'' = -2p_{1212} \cot(\varphi + \psi')/(p_{1111} - p_{1122}).$$

If one knows the angles ϑ' and ϑ'' for both elastic waves belonging to an arbitrary angle φ, then the product of both equations delivers

$$\tan 2\vartheta' \tan 2\vartheta'' = -4p_{1212}^2 (p_{1111} - p_{1122})^{-2},$$

an expression independent of the angles φ nd ψ', and hence, amenable to observation without their knowledge.

The position of the sectional ellipse does not depend on the amplitude of the elastic waves. The diffracted light waves possess directions of vibration parallel to the principal axes of the respective sectional ellipse. If one selects the direction of vibration of the incoming light wave parallel to one of the principal axes, then the diffracted light wave also has the same direction of vibration. This allows the measurement of the direction of vibration. If the crystal is placed between a pair of rotatable crossed polarizers the extinction position directly yields the angles ϑ' and ϑ''. If one works with the Schaefer–Bergmann method, then all elastic waves with arbitrary propagation directions perpendicular to the direction of transmission are virtually continuously excited at the same time. It is especially advantageous to furnish the probe with a "gothic arch" grind as mentioned in Section 4.5.6, so as to generate a sufficiently intensive sound field with a minimum of sound power. For each angle φ one then has the two diffraction spots with the angles of the semiaxes ϑ' and ϑ''. The associated angles ψ' and ψ'' of the distortion vectors are found from the angle φ and the elastic constants taken from the formula above. Hence, all the coefficients of the above-mentioned homogeneous linear equations for p_{1212}, p_{1111}, p_{1122} and p_{2211} may be determined.

A second assertion on elastooptical coefficients is easily accessible with the observation of pure longitudinal waves propagating in the directions [100] or [110]. These longitudinal waves generate, as seen in the derivation above, a sectional ellipse whose principal axes run parallel and perpendicular to the propagation directions $\boldsymbol{g}$ of the longitudinal wave. Consequently, we obtain two components of the diffracted light, one parallel to the propagation direction and one vibrating perpendicular to this. Both components combine to form a linear polarized wave, whose direction of vibration arises from the amplitudes of the components. It is

$$\cot\zeta = A_{\mathrm{I}}/A_{\mathrm{II}},$$

whereby both directions of vibration are specified by $\boldsymbol{e}_{\mathrm{I}}$ and $\boldsymbol{e}_{\mathrm{II}}$. If the light wave is incident, as before, perpendicular to the propagation direction of the longitudinal wave and a polarizer is placed in front of the crystal in a position diagonal to $\boldsymbol{e}_{\mathrm{I}}$ and $\boldsymbol{e}_{\mathrm{II}}$ so that the components of the primary wave vibrating in $\boldsymbol{e}_{\mathrm{I}}$ and $\boldsymbol{e}_{\mathrm{II}}$ have the same amplitude, one obtains the following $A_{\mathrm{I}}/A_{\mathrm{II}}$-values for the diffracted waves:

$\boldsymbol{e}_{\mathrm{I}} \parallel$	$\boldsymbol{e}_{\mathrm{II}} \parallel$	$\boldsymbol{g} \parallel$	$A_{\mathrm{I}}/A_{\mathrm{II}} = \cot\zeta$
[100]	[010]	[100]	p_{1111}/p_{2211}
[100]	[001]	[100]	p_{1111}/p_{1122}
[110]	$[\bar{1}10]$	[110]	$\dfrac{(2p_{1111}+p_{1122}+p_{2211}+4p_{1212})}{(2p_{1111}+p_{1122}+p_{2211}-4p_{1212})}$

The extinction position of the light diffracted due to the pure longitudinal wave is measured with the help of a rotatable analyzer. Thus one has ζ and the associated $A_{\mathrm{I}}/A_{\mathrm{II}}$-values when the polarizer is in the diagonal position. Again, one must ensure that no higher order diffraction effects come into play which superimpose on the first order diffraction. The method can also be applied along the directions of optical isotropy in noncubic crystals, whereby the valid electrooptical relationships are to be employed.

A useful method based on the measurement of the intensity of the diffracted light wave was first presented by Bergmann and Fues (1936). The incident light wave, before entering the probe, is decomposed into two mutually perpendicular vibrating components of equal intensity with the aid of a Wollaston prism, whereby both partial waves experience a spatial separation. The directions of vibration are adjusted parallel to the semiaxes of the sectional ellipse. Simultaneous measurements of the intensity of both diffracted waves—using a photographic plate or photo detector—ensure that during the measurement the same sound amplitude is effective in both directions of vibration. Hence, $I_{\mathrm{I}}/I_{\mathrm{II}} = (A_{\mathrm{I}}/A_{\mathrm{II}})^2$. For cubic crystals, the values $A_{\mathrm{I}}/A_{\mathrm{II}}$ given above apply. The method can also be employed for noncubic crystals when the position of the semiaxes of the sectional ellipse is independent of the amplitude of the elastic wave. In this case the intensity ratio is

$$\frac{I_{\mathrm{I}}}{I_{\mathrm{II}}} = \left(\frac{\Delta n_{\mathrm{I}}}{\Delta n_{\mathrm{II}}}\right)^2 = \frac{n_{\mathrm{I}}^6}{n_{\mathrm{II}}^6}\frac{p_{\mathrm{I}}^2}{p_{\mathrm{II}}^2},$$

where n_{I}, n_{II} and p_{I}, p_{II} are the effective refractive indices and effective elastooptical constants respectively, which for the given elastic wave and the effective sectional ellipse are calculated by tensor transformation.

The measurement accuracy can be decisively increased by employing lock-in amplifiers. Instead of the Wollaston prism it is convenient to use a switchable mechanical or electrooptical polarizer, to alternately drive the polarization of the primary light wave with a fixed frequency in both directions of vibration. in this manner, it is possible to measure the intensity ratio of both diffracted light waves to an accuracy of around 1 per mill without further measures. A related method, which even enables the direct determination of individual elastooptical coefficients referred to a standard crystal, was developed by Dixon & Cohen (1966). A standard crystal, for example, a plane parallel rectangular plate of quartz glass is attached to the test object by means of a suitable liquid. Short sound pulses with a fixed repetition rate are first transmitted through the test object then through the standard crystal and after reflection on the free face again back into the test object (Fig. 4.56). Now, if a light wave is sent into the test object and simultaneously another light wave into the standard crystal one obtains, per transit path of the elastic wave train 4 diffracted light flashes: 1. from the test object, 2. from the standard crystal, 3. again from the standard crystal after the reflection and 4. from the test object after the wave train re-enters. The length of the elastic wave train is so adjusted that the incoming wave and the reflected wave do not superimpose. From the ratio of the intensities of the 1. and 4. flashes of light one can determine the attenuation coefficient Z of the amplitude of the wave train passing through the boundary faces test object-standard crystal using $I_4/I_1 = Z^4$. This assumes that $I_2 = I_3$, that is, no reflection losses appear at the free faces of the

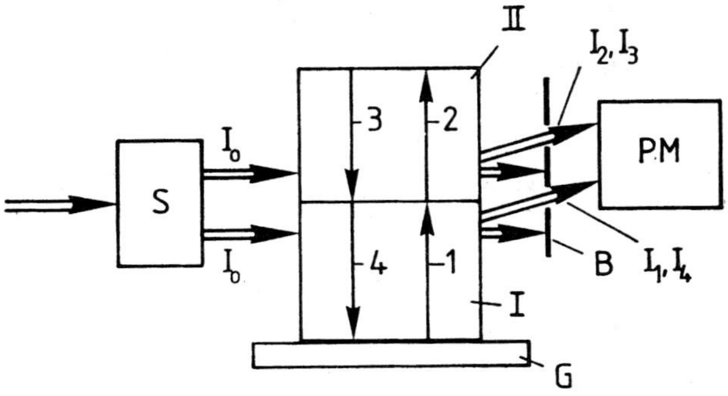

Figure 4.56 Illustration of the Dixon & Cohen method for approximately perpendicular incidence (low acoustic frequency). G Sound generator, I crystal, II standard crystal (e.g., quartz), S beam splitter, B aperture to eliminate primary light rays, PM photomultiplier. 1, 2, 3, and 4 denote the paths of the acoustic pulse through the arrangement of test object and standard crystal.

standard crystal. For test object (I) and standard crystal (II) we then have

$$I_{\mathrm{I}} = K I_{\mathrm{I0}} n_{\mathrm{I}}^6 p_I^2 \xi_{\mathrm{I0}}^2 / \rho_{\mathrm{I}} v_{\mathrm{I}}^3$$

and

$$I_{\mathrm{II}} = K I_{\mathrm{II0}} n_{\mathrm{II}}^6 p_{\mathrm{II}}^2 \xi_{\mathrm{II0}}^2 / \rho_{\mathrm{II}} v_{\mathrm{II}}^3,$$

where I_{I0} and I_{II0} are the intensities of the primary light waves in test object (I) and in standard crystal (II). $|\xi_{\mathrm{II0}}| = Z|\xi_{\mathrm{I0}}|$ is true for amplitudes of the elastic waves in I and II. Hence, one obtains for the effective elastooptical coefficients

$$\frac{p_{\mathrm{I}}^2}{p_{\mathrm{II}}^2} = \frac{I_{\mathrm{I}}(I_1/I_4)^{1/2} I_{\mathrm{II0}} \rho_{\mathrm{I}} v_{\mathrm{I}}^3 n_{\mathrm{II}}^6}{I_{\mathrm{II}} I_{\mathrm{I0}} \rho_{\mathrm{II}} v_{\mathrm{II}}^3 n_{\mathrm{I}}^6}.$$

The assumption is made here that the transmission lengths in I and II are the same and that the wave type does not change during its travel through I and II. This only applies to pure longitudinal- and transverse waves, whose wave normal is perpendicular on the boundary face. At higher diffraction angles, the dependence of the transmission length on the angle of diffraction must be taken into account (Dixon, 1967).

As indicated above, the combination of static piezooptical and dynamic elastooptical methods opens the possibility for the complete determination of tensors, with high accuracy, even in crystals of low symmetry. Thus one can forgo measurements difficult to access or afflicted with large errors, as, for example, "absolute measurements."

To estimate elastooptical and piezooptical constants of cubic crystals, and to a certain extent crystals of low symmetry, we can draw upon the Lorentz–Lorenz relationship

$$\frac{(n^2-1)}{(n^2+2)} = \frac{\rho}{M}R,$$

where R designates the specific refraction (mole refraction) of the substances (see Section 4.3.6.9).

From experiments it is found that the specific refraction is a combination of the quasiadditive contributions of the ions or atoms, which are only weakly dependent on the structural surroundings. Since ρ = mole weight/mole volume, one can crudely estimate the change in the refractive index due to a change of the mole volume brought about by hydrostatic pressure. We will return to this question in Exercise 13.

Finally, we wish to point out the technological importance gained by elastooptical devices as deflectors or modulators of light. The deflection of light can be controlled by diffraction on elastic waves with variable frequency because the diffraction angle is a function of the wavelength and hence the frequency of the elastic wave. One can control a light beam in any arbitrary angle within a certain angular range with an arrangement of two deflectors connected one after another. The areas of application are mainly material processing with laser light, fast light pens and in combination with several colors, the reproduction of large-area color pictures in video technology. A measure of the efficiency of such devices is the elastooptical scattering factor $n^6p^2/\rho v^3$ containing the material properties contributing to the elastooptical effect. Especially important is the strong dependence on the refractive index. The response time of elastooptical devices is about a factor 10^5 longer than with the electrooptical effect (ratio of the velocity of light to the velocity of sound).

Closely related to the piezooptical tensor is the piezodielectric tensor, which reproduces the change of the components of the dielectric tensor due to external mechanical stress

$$\Delta\epsilon_{ij} = Q_{ijkl}\sigma_{kl}.$$

The symmetry properties fully correspond to those of the piezooptical tensor. For the measurements, which so far have only been successful in a few cubic crystals, one uses the method discussed in Section 4.3.3 where the change in the capacity of a crystal plate is measure under the influence of mechanical stress. As we shall see in Section 5, there exists a close relationship between the piezodielectric and the quadratic electrostrictive tensor as well as the dependence of the elastic constants on the electric field.

4.5.10
Second-Order Electrostrictive and Electrooptical Effects

The relationships

$$\varepsilon_{ij} = d_{ijkl} E_k E_l \quad \text{and} \quad \Delta a_{ij} = r_{ijkl} E_k E_l$$

describe, to a first approximation, the *electrostrictive* and the *electrooptical* effects respectively, not covered by the linear effects previously discussed. Both second-order effects appear in all substances since in contrast to the linear effects, there is no total existence constraint. The number and type of independent components of the tensors $\{d_{ijkl}\}$ and $\{r_{ijkl}\}$ was already discussed in Section 4.5.

Up until lately the opinion was that second-order electrostrictive effects were too small to be measured. Such effects were known only in the case of ferroelectric crystals or close to a phase transition. By using dynamic measurement techniques, employing lock-in amplification, the complete second-order electrostrictive tensors of numerous cubic crystals as well as that of rhombic calcium formate were recently determined (Bohatý & Haussühl, 1977). In doing so, the change in the length of the crystal in the longitudinal as well in the transverse arrangement, induced by double the frequency of the electric field was converted into a change in capacitance of a capacitor working as the frequency determining element in a high frequency oscillating circuit. Using an auxiliary crystal exhibiting a linear electrostrictive effect to which a voltage of double the frequency was applied in phase to the voltage at the test object, detuning of the capacitor could be compensated for by the quadratic effect. In this manner it was possible to detect changes in length of around 10^{-13}m on probes with lengths of approximately 1 cm. All independent tensor components can be determined from longitudinal and transverse measurements in different orientations similar to first-order effects. The observed effects on a large collective of crystals were, in fact, found to be extremely small. Nevertheless, some interesting regularities could be uncovered, as, for example, the general phenomena of longitudinal dilatation and also volume expansion in an electric field. We will talk about the connection between second-order electrostriction and the pressure dependence of the dielectric tensor in due course. Second-order electrostriction also plays a decisive role in stimulated Brillouin scattering and in electrooptical effects.

The second-order electrooptical effect, also called the electrooptical Kerr effect, is in all details closely related to the piezooptical effect. The tensors of both effects have the same symmetry properties and the measurement arrangements are similar. In the absence of a linear effect, the change in the optical path difference of a plane parallel plate of thickness L, irradiated in perpendicular incidence obeys

$$\Delta G = -\frac{L}{2}(n'^3_{jj} r'_{jjlm} - n'^3_{kk} r'_{kklm}) E'_l E'_m + (n'_{jj} - n'_{kk}) d'_{iilm} E'_l E'_m.$$

Propagation direction e'_i and the directions of vibration e'_j and e'_k are selected as in the corresponding relations of the piezooptical effect. An evaluation of the measurements of the change of the optical path under the action of an electric field, which are carried out in a manner completely analogous to the methods discussed for piezooptical measurements in Section 4.5.9.1, demand a knowledge of the second-order electrostrictive effect. The belief that this correction may be neglected was refuted by the investigations mentioned above. Rather, one observes that the electrostrictive contribution is of the same order of magnitude as the actual second-order electrooptical effect.

Simple measurement arrangements can be specified for crystals of the subgroup 22 which allow a reliable determination of the complete tensors with a minimum of effort. Table 4.15 presents the corresponding relations for cubic crystals.

Crystals with large coefficients of the second-order electrooptical effect can also be employed for the linear modulation of light. In this regard, one applies a high electrical dc voltage to the given crystal and superimposes this on the control voltage of the modulation. A ΔG then appears corresponding to the linear electrooptical effect, as is directly seen on the basis of the relationship given above for the optical path difference. This means that the second-order electrooptical effect may also be interpreted as a neutralization of the inversion center by the electric field.

4.5.11
Electrogyration

Optical activity can be induced or changed by external fields. In particular, certain crystals can first be optically active under the influence of an external field. We start from the relationship

$$D_i = \epsilon_{ij} E_j + g_{ijk} \frac{\partial E_j}{\partial x_k}$$

for optical activity (see Section 4.3.6.7) and assume a dependence of the components g_{ijk} on the electric field strength given by $\Delta g_{ijk} = g_{ijkl} E_l$.

The *electrogyration tensor* $\{g_{ijkl}\}$ is, just as $\{g_{ijk}\}$, antisymmetric in both first index positions. Thus, it can be carried over to a third-order axial tensor as in $g_{ijkl} \rightarrow \gamma_{mkl}$ or $-\gamma_{mkl}$ with $m \neq i, j$ and i, j, m cyclic or anticyclic in 1, 2, 3. Hence, the formulae derived in Section 4.3.6.7 can be used here when one writes the quantity $\gamma_{ij} + \gamma_{ijk} E_k$ instead of γ_{ij}.

The effects of electrogyration are most easily accessible in the propagation direction of optical isotropy. The situation is especially simple in crystals not possessing natural optical activity. The only cubic PSG in which a pure electrogyration effect appears is m3. In the cubic PSGs with fourfold axes, all components g_{ijkl} vanish as one can directly establish by symmetry reduction. Both independent components g_{1212} and g_{2121} can exist in m3. Since, however, only the scalar product $\boldsymbol{G} \cdot \boldsymbol{g} = \gamma_{ij} g_i g_j$ or $\gamma_{ijk} g_i g_j E_k$ (see Section 4.3.6.7) enters into the equation for the refractive indices of optically active crystals in quasi-isotropy, both components cannot be separately determined by experiments measuring the refractive indices or related phenomena. This means only one independent component is effective. It is convenient to set $g_{1212} = \gamma_{312} = -g_{2121} = +\gamma_{321}$. The electrogyration effects a rotation of the plane of polarization of polarized light. The angle of rotation is given by the relation $\rho = \pi(n' - n'')/\lambda n$, discussed in Section 4.3.6.7, where $n' = n + \boldsymbol{g} \cdot \boldsymbol{G}$ and $n'' = n - \boldsymbol{g} \cdot \boldsymbol{G}$. The gyration vector is given by $G_i = \gamma_{ijk} g_j E_k$ when, as here, natural optical activity is missing. The following table presents some arrangements to be discussed for the measurement of electrogyration in PSG m3 ($q = \boldsymbol{E} \cdot \boldsymbol{a}$):

$\boldsymbol{g} \parallel$	$\boldsymbol{E} \parallel \boldsymbol{a}$	$\boldsymbol{G}$	$\boldsymbol{g} \cdot \boldsymbol{G}$
[110]	[001]	$\frac{\sqrt{2}}{2} E_3 \gamma_{123} (\boldsymbol{e}_1 + \boldsymbol{e}_2)$	$\gamma_{123} E_3$
[001]	$[ab0]$	$\frac{1}{\sqrt{a^2+b^2}} \gamma_{123} \lvert \boldsymbol{E} \rvert \operatorname{sign} q (b\boldsymbol{e}_1 + a\boldsymbol{e}_2)$	0
$[1\bar{1}0]$	[111]	$\frac{\sqrt{2}}{2} E_3 \gamma_{123} (-\boldsymbol{e}_1 + \boldsymbol{e}_2)$	$-\frac{\sqrt{3}}{3} \gamma_{123} \lvert \boldsymbol{E} \rvert \operatorname{sign} q$
[111]	[111]	$\frac{2}{\sqrt{3}} \gamma_{123} E_3 (\boldsymbol{e}_1 + \boldsymbol{e}_2 + \boldsymbol{e}_3)$	$\frac{2}{\sqrt{3}} \gamma_{123} \lvert \boldsymbol{E} \rvert \operatorname{sign} q$
[111]	$[ab.\bar{a} + b]$	*see* Exercise 33	0

If the propagation vector and the electric field lie parallel ($\boldsymbol{g} = \boldsymbol{E}/\lvert \boldsymbol{E} \rvert$), one obtains $\boldsymbol{g} \cdot \boldsymbol{G} = 6\gamma_{123} g_1 g_2 g_3 \lvert \boldsymbol{E} \rvert$. This means electrogyration vanishes in crystals of PSG m3 under longitudinal observation when $\boldsymbol{g}$ is normal to a twofold axis.

The measurement is best performed using a dynamic method in which an electric alternating field is applied to the crystal. The vibration of the plane of polarization oscillating with the same frequency as the field is measured with a lock-in amplifier (Weber & Haussühl, 1974). The effects observed on alums lie in the order of magnitude of 10^{-6} degree rotation of the plane of polarization per cm of irradiated length at a field strength of 1 V/cm. In isotypic crystals one observes marked changes of the coefficients γ_{123} when one replaces symmetrical particles with antisymmetrical one, as, for example, in the β-alums of caesium and methyl ammonium (Weber & Haussühl, 1976). Little is known as yet concerning the electrogyration of other crystals.

4.5.12
Piezoconductivity

The electric conductivity changes under the influence of an external mechanical stress. the relation is specified by

$$\Delta s_{ij} = w_{ijkl}\sigma_{kl}.$$

If one writes

$$\sigma_{mn} = c_{mnkl}\varepsilon_{kl}$$

for the stress tensor then

$$\Delta s_{ij} = w_{ijmn}c_{mnkl}\varepsilon_{kl} = y_{ijkl}\varepsilon_{kl}.$$

The tensors $\{w_{ijkl}\}$ and $\{y_{ijkl}\}$ are of type A and exist in all PSGs.

The corresponding effects also occur in thermal conductivity and in bulk conductivity.

The measurement of piezoconductivity is best performed with the help of longitudinal and transverse effects, whereby the change in the current density is observed as a function of hydrostatic or uniaxial pressure. The measurement scheme, even with triclinic crystals, is so simple that it can be effortlessly educed by the reader. If periodic changes in pressure are employed, one achieves, through lock-in amplification of the resulting change in current density, a high sensitivity.

Up to now, complete tensors have only been found for a few materials.

The effect also appears naturally in polycrystalline materials. The change in resistance of metal wires, as used in strain gauges, is based on piezoconductivity. High pressure sensors, where a change in the electrical resistance provides information on the prevalent pressure are also examples of the application of piezoconductivity.

4.6
Higher Rank Tensors

4.6.1
Electroacoustical Effects

Analogous to the electrooptical effect, an electrically induced change in the propagation direction of elastic waves may appear: *electroacoustical effect*. The tensorial description is

$$\Delta c_{ijkl} = z_{ijklm}E_m + z_{ijklmn}E_mE_n + \cdots$$

The polar tensor $\{z_{ijklm}\}$ only exists in acentric crystals except in PSG 43, whereas the tensor $\{z_{ijklmn}\}$ appears in all PSGs. The change of the elastic constants Δc_{ijkl} leads, via the elastodynamic basic equations, to a change in the propagation velocities of the elastic waves. The measurement of this very small effect is exacerbated by the fact that first- or second-order electrostriction becomes effective, which may cause a change in the dimensions and density of the crystal. These contributions can be calculated with knowledge of the elastic, piezoelectric or electrostrictive coefficients and used for correction of the measured change of the propagation velocities according to the method discussed in Section 4.5.6. So far only isolated measurements of this type have been successful. The connection between the coefficients of the electroacoustical effect and the second-order piezoelectric effect, represented by

$$D_i = d_{ijklm}\sigma_{jk}\sigma_{lm}$$

is referred to in Section 5. σ_{ij} are the components of the mechanical stress tensor.

4.6.2
Acoustical Activity

Analogous to optical activity, there also exists the phenomenon of *acoustical activity*, specified by the fifth-rank tensor $\{c_{ijklm}\}$, the gyration tensor, which can be represented by the following relationship:

$$\sigma_{ij} = c_{ijkl}\varepsilon_{kl} + c_{ijklm}\frac{\partial \varepsilon_{kl}}{\partial x_m}.$$

Just as with optical activity, we also set the condition that wave propagation progresses without additional energy terms. The elastic energy content is then

$$\Delta W = \sum_{i,j}\int \sigma_{ij} d\varepsilon_{kl} = \sum_{i,j}\int c_{ijkl}\varepsilon_{kl} d\varepsilon_{ij} + \sum_{i,j}\int c_{ijklm}\frac{\partial \varepsilon_{kl}}{\partial x_m} d\varepsilon_{ij}.$$

The second term, in the case of plane waves

$$\xi = \xi_0 e^{2\pi i(\mathbf{k}\cdot\mathbf{x}-\nu t)} \quad \text{with} \quad \varepsilon_{ij} = \frac{1}{2}\left(\frac{\partial \xi_i}{\partial x_j} + \frac{\partial \xi_j}{\partial x_i}\right)$$

takes the form

$$\sum_{i,j}\int 2\pi i k_m c_{ijklm}\varepsilon_{kl} d\varepsilon_{ij}.$$

This expression vanishes when $c_{ijklm} = -c_{ijlkm}$, i.e., when the gyration tensor is antisymmetric with respect to the 3. and 4. index position. Since the gyration tensor is an odd-rank polarer tensor, there exists no acoustical activity in centrosymmetric crystals. With pure transverse waves, in certain preferential directions, a rotation of the plane of vibration is observed, namely along the so-called acoustic axes in which two transverse waves of equal velocity can propagate, similar to optical activity. From this rotation one can determine individual components of the gyration tensor. The measurement of rotation is possible with the help of pulsed ultrasound methods, whereby the pulses are recorded with a piezoelectric transverse oscillator, connected as a detector, after traveling through the crystal. The rotation can be determined, after making corrections for reflection and damping losses, from the amplitude of the signals generated by the pulses. The amplitudes are functions of the setting of the direction of vibration of the decor to the direction of vibration of the primary transverse sound generator. In arbitrary directions, the effects of acoustical activity are largely masked by anisotropy and optical activity. In this case the elastic waves take on a rather complicated form so that an analysis of this effect is experimentally laborious.

4.6.3
Nonlinear Elasticity: Piezoacoustical Effects

Hooke's law, i.e., the linear relationship between stress and deformation is a sufficient approximation for most elastic interactions as long as the stress is kept adequately small. Contrary to the long held opinion that deviations from Hooke's law are difficult to measure, the situation in the meantime has completely changed. The methods described in Section 4.5.6 allow observations of deviations from Hooke's law already with mechanical stresses of a few Newton/cm^2. We use the Lagrangian deformation tensor as an inducing quantity to describe these elastic nonlinearities and specify the following relationship:

$$\sigma_{ij} = c_{ijkl}\eta_{kl} + c_{ijklmn}\eta_{kl}\eta_{mn} + c_{ijklmnop}\eta_{kl}\eta_{mn}\eta_{op} + \cdots$$

For small deformation, c_{ijkl} are identical with the components of the elasticity tensor defined in Section 4.5.1.

The components of the sixth-rank tensor are designated as *third-order elastic constants* and those of the eighth-rank tensor as *fourth-order elastic constants*. We recall that

$$\eta_{kl} = \frac{1}{2}\left(\frac{\partial \xi_k}{\partial x_l} + \frac{\partial \xi_l}{\partial x_k}\right) + \frac{1}{2}\frac{\partial \xi_m}{\partial x_k}\frac{\partial \xi_m}{\partial x_l}.$$

The first term corresponds to the components of the ordinary deformation tensor. Here, the pair-wise permutability of the indices follows from the reversibility of the deformation work, just as with the elasticity tensor, hence, for example,

$$c_{ijklmn} = c_{klmnij} = c_{mnklij} = \cdots$$

Table 4.16 Number of independent components of the elasticity tensors of fourth, sixth, and eighth rank.

Laue group	$\bar{1}$	2/m	2/mm	$\bar{3}$	$\bar{3}$m	4/m	4/mm
$Z(c_{ijkl})$	21	13	9	7	6	7	6
$Z(c_{ijklmn})$	56	32	20	20	14	16	12
$Z(c_{ijklmnop})$	126	70	42	42	28	36	25

Laue group	6/m, ∞/m	6/mm, ∞/mm	m3	4/m3	∞/m∞
$Z(c_{ijkl})$	5	5	3	3	2
$Z(c_{ijklmn})$	12	10	8	6	3
$Z(c_{ijklmnop})$	24	19	14	11	4

The indices within the pairs are naturally permutable due the the symmetry of the stress- and deformation tensor. Symmetry reduction in the individual PSGs is performed in a manner similar to the procedures discussed earlier. Table 4.16 presents the number Z of independent components of the tensor $\{c_{ijklmn}\}$ for the eleven Laue groups.

Cubic crystals with fourfold axes are distinguished from those with twofold axes. In abbreviated notation, the independent components are

- m3, 23 (Laue group m3): c_{111}, c_{112}, c_{113}, c_{123}, c_{144}, c_{155}, c_{166}, c_{456}
- 4/m3, 43, $\bar{4}$3 (Laue group 4/m3): c_{111}, c_{112} c_{123}, c_{144}, c_{166}, c_{456}.

Only three independent components exist in isotropic substances c_{111}, c_{112}, c_{123}. Furthermore, $c_{144} = (c_{111} - c_{123})/2$, $c_{166} = (c_{111} - c_{112})/4$ and $c_{456} = (c_{111} - 3c_{112} + 2c_{123})/8$, as is established from the condition of invariance of an arbitrary rotation about an arbitrary axis, for example, $\mathbf{e}_3$ (see Exercise 35).

One important aspect of nonlinear elastic behavior is exhibited by the dependence of the elastic constants on hydrostatic pressure, described by the coefficients dc_{ij}/dp (see Section 4.5.8). The connection of these coefficients with the third-order elastic constants is found from the relationships to be discussed below.

The complete determination of third-order elasticity tensors requires a carefully prepared strategy. A method, since found to be successful, was proposed by Thurston & Brugger (1964). An exposition of this method is found in an overview article by Wallace (1970). One measures changes in delay times or shifts in the resonance frequencies of plane parallel plates under the influence of external hydrostatic and uniaxial stress. These effects correspond to those of piezooptics, whereby here, however, only "absolute measurements," i.e., measurements of the absolute changes in velocity are taken into consideration, because a quantitative analysis of the interference of elastic waves would require a much too higher effort.

We now want to outline the considerations establishing a connection between piezoacoustical effects and nonlinear elastic properties. In an experiment we compare the delay time of a sound pulse or the resonance frequencies of plane parallel plates with and without a state of external stress characterized by $\{\sigma^0_{ij}\}$. In both situations, the elastodynamic basic equations in the form

$$\frac{\partial \sigma_{ij}}{\partial x_j} = \rho \frac{\partial^2 \xi_i}{\partial t^2}$$

are valid. The familiar equations $(-\rho^0 v^{02}\delta_{ik} + c^0_{ijkl} g_j g_l)\xi_k = 0$ $(i = 1,2,3)$ result from the prestress-free state. Furthermore, the coordinates of a mass point in the stress-free state are specified by x^0_i and in the prestressed crystal by x_i. The same applies to all other quantities. The prestresses σ^0_{ij} induce the deformations $\eta_{ij} = s^0_{ijkl}\sigma^0_{kl}$. The deformations carry over the fixed material position vector $\boldsymbol{x}^0$ into a position vector $\boldsymbol{x}$, referred to the same Cartesian coordinate system, according to

$$x_i = x^0_i + \frac{\partial x_i}{\partial x^0_j} x^0_j = (\delta_{ij} + \eta_{ij})x^0_j = u_{ij} x^0_j.$$

We assume that no rotational components appear. Formally, we can interpret this relationship as a coordinate transformation from a Cartesian base system to a nonlinear coordinate system among others. The components of the elasticity tensor are now to be transformed in accordance with this transformation. However, when we wish to keep the original Cartesian system as the coordinate system, we must guarantee the invariance of the density of the elastic deformation work. This is accomplished by furnishing all transformed quantities with the factor $V^0/V = |u_{ij}|^{-1}$. $|u_{ij}|$ are also called the Jacobian determinants. Accordingly, the components of the elasticity tensor are then

$$c'_{ijkl} = |u_{ij}|^{-1} u_{ir} u_{js} u_{kt} u_{lv} c^0_{rstv}.$$

Writing $u_{ij} = (\delta_{ij} + \eta_{ij})$ and expanding the deformation tensor η_{ij} to first order, taking into account $|u_{ij}|^{-1} = (1 + \eta_{11} + \eta_{22} + \eta_{33} + \cdots)^{-1} = (1 - \eta_{11} - \eta_{22} - \eta_{33} \cdots)$, gives

$$c'_{ijkl} = (1 - \delta_{rs}\eta_{rs})c^0_{ijkl} + c^0_{ijkr}\eta_{lr} + c^0_{ijrl}\eta_{kr} + c^0_{irkl}\eta_{jr} + c^0_{rjkl}\eta_{ir}.$$

The corresponding transformation of the components c_{ijklmn} delivers, for the effective components c_{ijkl}, the contribution $c_{ijklmn}\eta_{mn}$ as the lowest order term in η_{ij}. Hence, we have

$$c_{ijkl} = c'_{ijkl} + c_{ijklmn}\eta_{mn}.$$

Thus the deformations generated by an elastic wave $\boldsymbol{\xi} = \boldsymbol{\xi}_0 e^{2\pi i(\boldsymbol{k}\cdot\boldsymbol{x}-\nu t)}$ superimpose on the deformations generated by the prestresses. We require the resulting total stress σ_{ij}. This is calculated in an analogous fashion to c_{ijkl}, employing, however, that transformation describing the change of the coordinates of the position vector of the prestressed state to the effective state described by the deformation resulting from the elastic waves. It is

$$x_i' = x_i + \Delta\xi_i = x_i + \frac{\partial\xi_i}{\partial x_j}x_j = \left(\delta_{ij} + \frac{\partial\xi_i}{\partial x_j}\right)x_j = u_{ij}'x_j.$$

Hence, $\sigma_{ij}' = |u_{ij}'|^{-1}u_{ir}'u_{js}'\sigma_{rs}^0$, where σ_{ij}^0 is the prestress. Expanding again, to a first approximation with respect to the deformation quantities $\partial\xi_i/\partial x_j$, one finds

$$\sigma_{ij}' = \left(1 - \frac{\partial\xi_k}{\partial x_k}\right)\sigma_{ij}^0 + \frac{\partial\xi_j}{\partial x_k}\sigma_{ik}^0 + \frac{\partial\xi_i}{\partial x_k}\sigma_{kj}^0.$$

Together with the deformation of the elastic waves gives

$$\sigma_{ij} = \sigma_{ij}' + c_{ijkl}\varepsilon_{kl}.$$

The elastodynamic basic equations then take on the following form:

$$\frac{\partial\sigma_{ij}}{\partial x_j} = (\sigma_{lj}^0\delta_{ik} + c_{ijkl})\frac{\partial^2\xi_k}{\partial x_j\partial x_l} = \rho\frac{\partial^2\xi_i}{\partial t^2}.$$

Using the plane wave approach results in

$$-\rho v^2\xi_i + (\sigma_{lj}^0\delta_{ik} + c_{ijkl})g_j g_l \xi_k = 0.$$

To an approximation, one may assume that the position of the propagation vector and the deformation vector depend on the prestress σ_{ij}^0. Multiplying the above equation with ξ_i and summing over i gives

$$\rho v^2 = (\sigma_{lj}^0\delta_{ik} + c_{ijkl})g_j g_l \xi_i \xi_k/\boldsymbol{\xi}^2.$$

The values derived above are inserted for c_{ijkl}. Substituting the deformation η_{ij} by $s_{ijkl}^0\sigma_{kl}^0$, which is approximately correct, explicitly gives ρv^2 as a function of the prestress σ_{ij}^0 and hence

$$\begin{aligned}\frac{\partial\rho v^2}{\partial\sigma_{pq}^0} &= g_p g_q + \frac{\partial c_{ijkl}}{\partial\sigma_{pq}^0}g_j g_l \xi_i \xi_k/\boldsymbol{\xi}^2 \\ &= g_p g_q + (-c_{ijkl}^0 s_{rspq}^0\delta_{rs} + c_{ijkr}^0 s_{lrpq}^0 + c_{ijrl}^0 s_{krpq}^0 + c_{irkl}s_{jrpq}^0 \\ &\quad + c_{rjkl}^0 s_{irpq}^0 + c_{ijklmn}s_{mnpq})g_j g_l \xi_i \xi_k/\boldsymbol{\xi}^2.\end{aligned}$$

With the help of these relationships, the components of the third-order elasticity tensor can be determined in a given propagation direction and position of the deformation tensor and with known elastic constants. If one works with static prestresses, which can be of hydrostatic or uniaxial nature, the isothermal constants must be inserted for s^0_{ijkl}. For sound propagation, one employs the adiabatic constants c^0_{ijkl}. The connection between isothermal and adiabatic quantities is discussed in Section 5.2. The third-order constants are a mixture of isothermal–adiabatic types. However, a closer inspection has shown that the difference between isothermal and adiabatic or third-order mixed constants is so small that they are hardly measurable (Guinan & Ritchie, 1970). Conveniently one employs computer programs to evaluate measurements performed under hydrostatic pressure and uniaxial stress. A direct evaluation is possible with cubic crystals. From the equation above, one obtains the following relationship for the pressure dependence of the elastic constants $(-p = \sigma^0_{11} = \sigma^0_{22} = \sigma^0_{33})$ of cubic crystals

$$\frac{dc_{11}}{dp} = -1 - (K^0/3)(c^0_{11} + c_{111} + c_{112} + c_{113}) \quad \text{with} \quad g_1 = \xi_1/|\xi| = 1,$$

$$\frac{dc_{44}}{dp} = -1 - (K^0/3)(c^0_{44} + c_{144} + c_{155} + c_{166}) \quad \text{with} \quad g_1 = \xi_2/|\xi| = 1,$$

$$\frac{d(c_{11} - c_{12})/2}{dp} = -1 - (K^0/3)(c^0_{11} - c^0_{12} + c_{111} - c_{123})/2 \quad \text{with}$$
$$g_1 = g_2 = \xi_1/|\xi| = -\xi_2/|\xi| = \sqrt{2}/2,$$

and hence,

$$\frac{dc_{12}}{dp} = 1 - (K^0/3)(c^0_{12} + c_{112} + c_{113} + c_{123}),$$

where K^0 is the volume compressibility

$$K^0 = 3(s^0_{11} + 2s^0_{12}) = 3(c^0_{11} + 2c^0_{12})^{-1}.$$

Similar relationships result for the dependence of the constants c_{11} and c_{44} on uniaxial pressure parallel or normal to the propagation direction [100] (see Exercise 36). It follows that the pressure dependence of the reciprocal volume compressibility, an important scalar invariant, is given by

$$\frac{dK^{-1}}{dp} = \frac{d(c_{11} + 2c_{12})}{3dp} = -(K^0/9)(c_{111} + 3c_{112} + 3c_{113} + 2c_{123})$$

whereby the difference between isothermal and adiabatic compressibility was neglected. The above expression is identical, except for the factor $-K^0/27$, to the invariants $c_{ijklmn}\delta_{ij}\delta_{kl}\delta_{mn}$.

Elastic nonlinearity, analogous to dielectric nonlinearity, generates, from an incident wave a frequency doubled secondary wave which can be used to a certain extent to measure nonlinear elastic properties. However, acoustic frequency doubling does not come into consideration as a precise measurement of routine nonlinear elastic properties.

If one takes into account higher order deformations in the relationships derived above, then fourth and higher order elasticity tensors come into play which are only partially and with low accuracy accessible to measurement.

5
Thermodynamic Relationships

5.1
Equations of State

The behavior of a crystal under the influence of different external (inducing) quantities can be calculated with knowledge of the given property. We call the existing relationships equations of state. If, for example, the mechanical stress in the form of the stress tensor $\{\sigma_{ij}\}$, an electric field $\boldsymbol{E}$, a magnetic field $\boldsymbol{H}$, and the temperature difference ΔT are accepted as independent (inducing) variables, then for the corresponding (dependent) induced quantities such as mechanical deformation $\{\varepsilon_{ij}\}$, electric displacement $\boldsymbol{D}$, magnetic induction $\boldsymbol{B}$ and entropy difference ΔS per unit volume, one must specify, to a first approximation, the following linear equations:

$$
\begin{aligned}
\varepsilon_{ij} &= s_{ijkl}^{E,H,T}\sigma_{kl} + c_{ijk}^{H,T}E_k + m_{ijk}^{E,T}H_k + \alpha_{ij}^{E,H}\Delta T \\
\Delta D_i &= d_{ijk}^{H,T}\sigma_{jk} + \epsilon_{ij}^{\sigma,H,T}E_j + q_{ij}^{\sigma,T}H_j + p_i^{\sigma,H}\Delta T \\
\Delta B_i &= n_{ijk}^{E,T}\sigma_{jk} + b_{ij}^{\sigma,T}E_j + \mu_{ij}^{\sigma,E,T}H_j + m_i^{\sigma,E}\Delta T \\
\Delta S &= \beta_{ij}^{E,H}\sigma_{ij} + q_i^{\sigma,H}E_i + n_i^{\sigma,E}H_i + \frac{C^{\sigma,E,H}}{T}\Delta T
\end{aligned}
$$

The attached symbols σ, E, H, and T mean that each respective quantity is fixed. The relationships are mostly familiar and otherwise easy to interpret. $C^{\sigma,E,H}$ is the specific heat per unit volume. In a following second approximation one has to include the tensor relationships describing the quadratic dependence of the quantities ε_{ij}, D_i, B_i and ΔS on the inducing quantities, hence on the pure quadratic and mixed products, as, for example, $\sigma_{ij}E_k$, $E_i\Delta T$ and so on. The same applies to higher order approximations. Accordingly, the first, second and higher approximation represent nothing else as a Taylor expansion of the dependent quantities with respect to the independent quantities in first, second, and higher orders.

Under suitable experimental arrangements we can also introduce, as independent variables, other combinations, as, for example, mechanical deforma-

Physical Properties of Crystals. Siegfried Haussühl.

ISBN: 978-3-527-40543-5

tion, the electric and magnetic field and temperature. One then obtains analogous equations of state:

$$\begin{aligned}
\sigma_{ij} &= c_{ijkl}^{E,H,T}\varepsilon_{kl} + f_{ijk}^{H,T}E_k + p_{ijk}^{E,T}H_k + \gamma_{ij}^{E,H}\Delta T \\
\Delta D_i &= e_{ijk}^{H,T}\varepsilon_{jk} + \epsilon_{ij}^{\varepsilon,H,T}E_j + q_{ij}^{\varepsilon,T}H_j + p_i^{\varepsilon,H}\Delta T \\
\Delta B_i &= q_{ijk}^{E,T}\varepsilon_{jk} + b_{ij}^{\varepsilon,T}E_j + \mu_{ij}^{\varepsilon,E,T}H_j + m_i^{\varepsilon,E}\Delta T \\
\Delta S &= \delta_{ij}^{E,H}\varepsilon_{ij} + q_i^{\varepsilon,H}E_i + n_i^{\varepsilon,E}H_i + \frac{C^{\varepsilon,E,H}}{T}\Delta T
\end{aligned}$$

We inquire as to the relationships existing under the different coefficients. In this case it is convenient to draw upon the so called thermodynamic potential. The density of the internal energy U of a body, measured, for example, in Jm^{-3}, represents the sum of the total energy content. Accordingly, the change ΔU is given by the following expression:

$$\Delta U = \Delta Q + \sigma_{ij}\Delta\varepsilon_{ij} + E_i\Delta D_i + H_i\Delta B_i = \Delta Q + \Delta W,$$

where ΔQ signifies the change of the caloric energy content. In the case of reversible processes $\Delta Q = T\Delta S$. ΔW specifies all noncaloric energy content. A second important function of state is the free enthalpy G, also called the *Gibbs potential* arising from U through the following Legendre transformation:

$$G = U - \sigma_{ij}\varepsilon_{ij} - E_iD_i - H_iB_i - TS.$$

Its differential change is given by

$$\begin{aligned}
\Delta G &= \Delta U - \sigma_{ij}\Delta\varepsilon_{ij} - \varepsilon_{ij}\Delta\sigma_{ij} - E_i\Delta D_i - D_i\Delta E_i \\
&\qquad - H_i\Delta B_i - B_i\Delta H_i - T\Delta S - S\Delta T \\
&= -\varepsilon_{ij}\Delta\sigma_{ij} - D_i\Delta E_i - B_i\Delta H_i - S\Delta T.
\end{aligned}$$

In thermodynamic equilibrium, i.e., here, in a state of constant mechanical stress, constant electric and magnetic field as well as constant temperature, $\Delta G = 0$. This statement plays a fundamental role in the stability of a crystal type under isobaric ($\Delta\sigma_{ij} = 0$), isagrischen ($\Delta E_i = \Delta H_i = 0$) and isothermal ($\Delta T = 0$) conditions. From several arrangements (modifications) of the same chemical constituents existing under the same conditions, the most stable is the one possessing the smallest Gibbs free energy.

We now assume that G is an arbitrarily differentiable function of the variables $(\sigma_{ij}, E_i, H_i, T)$. We can then expand G as a Taylor series:

$$\Delta G = \left(\frac{\partial G}{\partial \sigma_{ij}}\right)_{E,H,T} \Delta\sigma_{ij} + \left(\frac{\partial G}{\partial E_i}\right)_{\sigma,H,T} \Delta E_i + \left(\frac{\partial G}{\partial H_i}\right)_{\sigma,E,T} \Delta H_i + \left(\frac{\partial G}{\partial T}\right)_{\sigma,E,H} \Delta T + \frac{1}{2}\left(\frac{\partial^2 G}{\partial \sigma_{ij}\partial\sigma_{kl}}\right)_{E,H,T} \Delta\sigma_{ij}\Delta\sigma_{kl} + \frac{1}{2}\left(\frac{\partial^2 G}{\partial E_i \partial E_j}\right)_{\sigma,H,T} \Delta E_i \Delta E_j + \ldots$$

Comparison with the expression derived from the definition of G gives the following relations:

$$\left(\frac{\partial G}{\partial \sigma_{ij}}\right)_{E,H,T} = -\varepsilon_{ij}, \qquad \left(\frac{\partial G}{\partial E_i}\right)_{\sigma,H,T} = -D_i,$$

$$\left(\frac{\partial G}{\partial H_i}\right)_{\sigma,E,T} = -B_i, \qquad \left(\frac{\partial G}{\partial T}\right)_{\sigma,E,H} = -S.$$

Differentiating these expressions again and making use of the permutability of the sequence of differentiation, gives, for example,

$$\left(\frac{\partial^2 G}{\partial \sigma_{ij}\partial E_k}\right)_{H,T} = -\left(\frac{\partial \varepsilon_{ij}}{\partial E_k}\right)_{H,T} = -\left(\frac{\partial D_k}{\partial \sigma_{ij}}\right)_{H,T},$$

hence,

$$c_{ijk}^{H,T} = d_{kij}^{H,T}.$$

In a similar manner we obtain

$$m_{ijk}^{E,T} = n_{kij}^{E,T}, \quad \alpha_{ij}^{E,H} = \beta_{ij}^{E,H}, \quad q_{ij}^{\sigma,T} = b_{ij}^{\sigma,T}, \quad p_i^{\sigma,H} = q_i^{\sigma,H}, \quad m_i^{\sigma,E} = n_i^{\sigma,E}.$$

If one takes into consideration isochore ($\Delta\varepsilon_{ij} = 0$), isagric and isothermal conditions, then analogous relationships can be derived from the *electric–magnetic Gibbs potential* $F = U - E_i D_i - H_i B_i - TS$.

In the corresponding equilibrium state $\Delta F = 0$ because

$$\Delta F = \sigma_{ij}\Delta\varepsilon_{ij} - D_i \Delta E_i - B_i \Delta H_i - S\Delta T.$$

To begin with

$$\left(\frac{\partial F}{\partial \varepsilon_{ij}}\right)_{E,H,T} = \sigma_{ij}, \qquad \left(\frac{\partial F}{\partial E_i}\right)_{\varepsilon,H,T} = -D_i,$$

$$\left(\frac{\partial F}{\partial H_i}\right)_{\varepsilon,E,T} = -B_i, \qquad \left(\frac{\partial F}{\partial T}\right)_{\varepsilon,E,H} = -S.$$

From this, we obtain by differentiation, taking into account the permutability of the sequence of differentiation

$$e_{ijk}^{H,T} = -f_{kij}^{H,T}, \quad p_{ijk}^{E,T} = -q_{kij}^{E,T}, \quad \gamma_{ij}^{E,H} = -\delta_{ij}^{E,H},$$
$$q_{ij}^{\varepsilon,T} = b_{ij}^{\varepsilon,T}, \quad p_i^{\varepsilon,H} = q_i^{\varepsilon,H}, \quad m_i^{\varepsilon,H} = n_i^{\varepsilon,H}.$$

We have used some of these relationships, for example, $c_{ijk}^{H,T} = d_{kij}^{H,T}$ and $e_{ijk}^{H,T} = -f_{kij}^{H,T}$, in preceding sections.

Further such relations (*first-order Maxwell relations*) can be calculated in an analogous fashion for all other arbitrary combinations of auxiliary conditions from the associated functions of state F_X, derived from the internal energy U with the help of a Legendre transformation according to $F_X = U - X$. X can be one or even several of the energy terms of type $\sigma_{ij}\varepsilon_{ij}$, E_iD_i, H_iB_i or TS. These terms represent products of two conjugate quantities A_k and $\tilde{A}_k$, also designated as intensive and extensive quantities, respectively (A_k: σ_{ij}, E_i, H_i, ΔT; $\tilde{A}_k$: ε_{ij}, D_i, B_i, ΔS).

It is then true that $\partial F_X/\partial A_k = -\tilde{A}_k$ or $\partial F_X/\partial \tilde{A}_k = A_k$, when X contains the product $A_k\tilde{A}_k$ or not respectively. The first-order Maxwell relations then take the form

$$\frac{\partial^2 F_X}{\partial A_k \partial A_l} = -\frac{\partial \tilde{A}_k}{\partial A_l} = -\frac{\partial \tilde{A}_l}{\partial A_k}.$$

Similar results are obtained for $\frac{\partial^2 F_X}{\partial \tilde{A}_k \partial \tilde{A}_l}$.

If one again differentiates these relations with respect to the variables A_i and $\tilde{A}_i$ noting the permutability of the sequence of differentiation, we get the second-order Maxwell relations

$$\frac{\partial^3 F_X}{\partial A_i \partial A_j \partial A_k} = -\frac{\partial^2 \tilde{A}_k}{\partial A_i \partial A_j} = -\frac{\partial^2 \tilde{A}_j}{\partial A_i \partial A_k} = -\frac{\partial^2 \tilde{A}_i}{\partial A_j \partial A_k}$$

and correspondingly

$$\frac{\partial^3 F_X}{\partial \tilde{A}_i \partial \tilde{A}_j \partial \tilde{A}_k} = \frac{\partial^2 A_k}{\partial \tilde{A}_i \partial \tilde{A}_j} = \cdots,$$

whereby certain auxiliary conditions must be kept to with respect to the other variables.

From the many relations, we point out here some examples of importance:

Third-rank elasticity tensor $\sigma_{ij} = c_{ijkl}\eta_{kl} + \underline{c_{ijklmn}}\eta_{kl}\eta_{mn}$:

$$\frac{\partial^2 \sigma_{ij}}{\partial \eta_{kl} \partial_{mn}} = \frac{\partial^2 \sigma_{kl}}{\partial \eta_{ij} \partial \eta_{mn}} = \frac{\partial^2 \sigma_{mn}}{\eta_{ij}\eta_{kl}} \rightarrow c_{ijklmn} = c_{klijmn} = c_{mnijkl} = \dots$$

Second-order dielectric tensor $D_i = \epsilon_{ij}E_j + \underline{\epsilon_{ijk}}E_jE_k$:

$$\frac{\partial^2 D_i}{\partial E_j \partial E_k} = \frac{\partial^2 D_j}{\partial E_i \partial E_k} = \frac{\partial^2 D_k}{\partial E_i \partial E_j} \rightarrow \epsilon_{ijk} = \epsilon_{jik} = \epsilon_{kij} = \dots \quad \text{(totally symmetric)}.$$

This corresponds to "Kleinmans rule" for nonlinear dielectric susceptibility (see Section 4.4.4).

Second-order piezoelectric effect $D_i = d_{ijk}\sigma_{jk} + \overline{d_{ijklm}}\sigma_{jk}\sigma_{lm}$ and the mechanoelectric deformation tensor $\varepsilon_{jk} = s_{jklm}\sigma_{lm} + \underline{q_{jklmi}}\sigma_{lm}E_i$:

$$\frac{\partial^2 D_i}{\partial \sigma_{jk}\partial \sigma_{lm}} = \frac{\partial^2 \varepsilon_{jk}}{\partial \sigma_{lm}\partial E_i} \rightarrow d_{ijklm} = q_{jklmi}.$$

Second-order electrostriction $\varepsilon_{ij} = d_{ijk}E_k + \overline{d_{ijkl}}E_kE_l$ and pressure dependence of the dielectric tensor $D_k = \epsilon_{kl}E_l + \underline{e_{klij}}\sigma_{ij}E_l$:

$$\frac{\partial^2 \epsilon_{ij}}{\partial E_k \partial E_l} = \frac{\partial^2 D_k}{\partial \sigma_{ij}\partial E_l} = \frac{\partial^2 D_l}{\partial \sigma_{ij}\partial E_k} \rightarrow d_{ijkl} = e_{klij}.$$

Accordingly, the components of the tensor of quadratic electrostriction are numerically equal to the corresponding components of the tensor describing the dependence of the dielectric constants on external mechanical stress. Hence, we have a simple independent check of these tensors (Preu u. Haussühl, 1983). Substituting the electric quantities by the corresponding magnetic ones results in completely analogous relations.

5.2 Tensor Components Under Different Auxiliary Conditions

The question often arises, whether tensor components differ when measured under different auxiliary conditions, as, for example, at constant electric field and constant electric displacement or at constant temperature (isothermal) and constant entropy (adiabatic). The expected differences are usually very small, often of the order of the measurement accuracy. However, cases exist where substantial differences appear, as we shall now see.

Firstly, we again start from a system of equations of state, for example, from the system of variables σ_{ij}, E_i, H_i and T. Then, for example, $S =$ const. means that

$$\Delta S = \alpha_{ij}^{E,H}\sigma_{ij} + q_i^{\sigma,H}E_i + m_i^{\sigma,E}H_i + \frac{C^{\sigma,E,H}}{T}\Delta T = 0.$$

Hence, with the absence, for example, of an electric and a magnetic field ($E_i = H_i = 0$) $\Delta T = -\frac{\alpha_{ij}^{E,H}\sigma_{ij}}{C^{\sigma,E,H}}T$. This is the so-called adiabatic temperature increase, appearing as an effect of an external mechanical stress. We want to calculate the order of magnitude of this effect on a simple example. In a pressure cell, we apply a pressure of 1000 bar (10^8 Pa) on a NaCl crystal, when, for example, investigating some property as a function of pressure. With $\sigma_{ii} = -p = 10^8$ Pa, we expect, with a linear thermal expansion $\alpha_{11}^{E,H} = \alpha_{ii}^{E,H} = 40 \cdot 10^{-6}\,\mathrm{K}^{-1}$ and specific heat $C^{\sigma,E,H} = 1,6 \cdot 10^6\,\mathrm{Jm}^{-3}\mathrm{K}^{-1}$, a temperature increase of

$$\Delta T = (\alpha_{11}^{E,H} + \alpha_{22}^{E,H} + \alpha_{33}^{E,H})pT/C^{\sigma,E,H} \approx 2,5\,\mathrm{K}$$

at around 300 K.

From the equation of state $\varepsilon_{ij} = s_{ijkl}^{E,H,T}\sigma_{kl} + \alpha_{ij}^{E,H}\Delta T$ and with the above expression for ΔT by $E_i = H_i = 0$, we obtain

$$\begin{aligned}\varepsilon_{ij} &= s_{ijkl}^{E,H,T}\sigma_{kl} - \alpha_{ij}^{E,H}\alpha_{kl}^{E,H}\sigma_{kl}T/C^{\sigma,E,H},\\ &= (s_{ijkl}^{E,H,T} - \alpha_{ij}^{E,H}\alpha_{kl}^{E,T}T/C^{\sigma,E,H})\sigma_{kl};\end{aligned}$$

hence,

$$s_{ijkl}^{E,H,S} = s_{ijkl}^{E,H,T} - \alpha_{ij}^{E,H}\alpha_{kl}^{E,H}T/C^{\sigma,E,H}.$$

We obtain a generally valid expression for such differences as follows. Let Y be a function dependent on the variables X_1 and X_2 where X_1 is a function of X_2 and X_3. Then

$$\Delta Y = \left(\frac{\partial Y}{\partial X_1}\right)_{X_2}\Delta X_1 + \left(\frac{\partial Y}{\partial X_2}\right)_{X_1}\Delta X_2$$

and

$$\Delta X1 = \left(\frac{\partial X_1}{\partial X_2}\right)_{X_3}\Delta X_2 + \left(\frac{\partial X_1}{\partial X_3}\right)_{X_2}\Delta X_3.$$

Writing the second condition in the first gives

$$\Delta Y = \left(\frac{\partial Y}{\partial X_1}\right)_{X_2}\left(\frac{\partial X_1}{\partial X_2}\right)_{X_3}\Delta X_2 + \left(\frac{\partial Y}{\partial X_1}\right)_{X_2}\left(\frac{\partial X_1}{\partial X_3}\right)_{X_2}\Delta X_3 + \left(\frac{\partial Y}{\partial X_2}\right)_{X_1}\Delta X_2.$$

It follows that

$$\left(\frac{\partial Y}{\partial X_2}\right)_{X_3} - \left(\frac{\partial Y}{\partial X_2}\right)_{X_1} = \left(\frac{\partial Y}{\partial X_1}\right)_{X_2}\left(\frac{\partial X_1}{\partial X_2}\right)_{X_3} \quad \text{with} \quad \Delta X_3 = 0.$$

This formula is valid for arbitrary conditions in all the other variable. These we will only specify below in case of ambiguity.

For example, let $Y = \varepsilon_{ij}$, $X_2 = \sigma_{kl}$, $X_1 = T$ and $X_3 = S$. Then

$$\left(\frac{\partial \varepsilon_{ij}}{\partial \sigma_{kl}}\right)_S - \left(\frac{\partial \varepsilon_{ij}}{\partial \sigma_{kl}}\right)_T = \left(\frac{\partial \varepsilon_{ij}}{\partial T}\right)_{\sigma_{kl}} \left(\frac{\partial T}{\partial \sigma_{kl}}\right)_S = -\alpha_{ij}\alpha_{kl}T/C^{\sigma}$$

in agreement with the above result for $(s^S_{ijkl} - s^T_{ijkl})$.

If Y is dependent on several variables X_i ($i = 1, \ldots n$) and if the quantities X may depend on one another as well as on other variables X'_j ($j = 1, \ldots, n'$), we have

$$\Delta Y = \sum_{i=1}^{n-1} \left(\frac{\partial Y}{\partial X_i}\right)_{X \neq X_i} \Delta X_i + \left(\frac{\partial Y}{\partial X_n}\right)_{X \neq X_n} \Delta X_n$$

and

$$\Delta X_i = \sum_{\substack{j=1 \\ (j \neq i)}}^{n} \left(\frac{\partial X_i}{\partial X_j}\right)_{X \neq X_i} \Delta X_j + \sum_{j=1}^{n'} \left(\frac{\partial X_i}{\partial X'_j}\right)_{X \neq X'_j} \Delta X'_j.$$

We now inquire as to the partial differentiation of Y with respect to the variables X_n under the auxiliary condition that all variables X' are kept constant. It is

$$\Delta Y = \sum_{i=1}^{n-1} \left(\frac{\partial Y}{\partial X_i}\right)_{X \neq X_i} \sum_{\substack{j=1 \\ (j \neq i)}}^{n} \left(\frac{\partial X_i}{\partial X_j}\right)_{X \neq X_j} \Delta X_j + \left(\frac{\partial Y}{\partial X_n}\right)_{X \neq X_n} \Delta X_n$$

and hence,

$$\left(\frac{\partial Y}{\partial X_n}\right)_{X' = \text{ const.}} - \left(\frac{\partial Y}{\partial X_n}\right)_{\substack{X = \text{ const.} \\ (\text{except } X_n)}} = \sum_{i=1}^{n-1} \left(\frac{\partial Y}{\partial X_i}\right)_{X \neq X_i} \left(\frac{\partial X_i}{\partial X_n}\right)_{X \neq X_n}.$$

Contrary to the notation for tensor components, here we have set the indices, characterizing the auxiliary conditions, as subscripts corresponding to the usage in differential calculus. Finally, it should be pointed out that with a first selection of the independent variables already all material constants of the equation of state are fixed. This means that the material constants appearing with another arbitrary selection of independent variables can be calculated from the constants of the first selection of variables. Already familiar examples are the elastic c and s tensors or the dielectric ϵ and a tensors. As a further example let us look at the tensor γ_{ij} appearing in both equations of state

$$\sigma_{ij} = c^T_{ijkl}\varepsilon_{kl} + \cdots + \gamma_{ij}\Delta T$$

$$\Delta S = -\gamma_{ij}\varepsilon_{ij} + \cdots + \frac{C^{\varepsilon}}{T}\Delta T.$$

Under adiabatic conditions we obtain $\Delta T = \gamma_{ij}\varepsilon_{ij}T/C^{\varepsilon}$, whereby $\boldsymbol{E}$ and $\boldsymbol{H}$ are constant.

Table 5.1 Relations between material constants under various conditions.

	Y	X_i	X_n	X'_j
Specific heat				
$C^{\sigma} - C^{\varepsilon} = -\alpha_{ij}\gamma_{ij}T$ with $\gamma_{ij} = -\alpha_{kl}c_{klij}\frac{C^{\varepsilon}}{C^{\sigma}}$	S	ε_{ij}	T	σ_{ij}
Pyroelectric effect				
$p_i^{\sigma} - p_i^{\varepsilon} = e_{ijk}^{T}\alpha_{jk}^{\sigma}$	D_i	ε_{jk}	T	σ_{jk}
Thermal expansion				
$\alpha_{ij}^{D} - \alpha_{ij}^{E} = d_{kij}^{T}\left(\frac{\partial E_k}{\partial T}\right)_D$	ε_{ij}	E_k	T	D_k
Dielectricity tensor				
$\epsilon_{ij}^{S} - \epsilon_{ij}^{T} = -p_i p_j T/C^{\sigma}$	D_i	T	E_j	S
$\epsilon_{ij}^{\sigma} - \epsilon_{ij}^{\varepsilon} = e_{ikl}d_{jkl}$	D_i	ε_{kl}	E_j	σ_{kl}
Piezoelectricity tensor				
$d_{ijk}^{S} - d_{ijk}^{T} = -\alpha_{jk}p_i^{\sigma}T/C^{\sigma}$	D_i	T	σ_{jk}	S
Elasticity tensor				
$s_{ijkl}^{S} - s_{ijkl}^{T} = -\alpha_{ij}\alpha_{kl}T/C^{\sigma}$	ε_{ij}	T	σ_{kl}	S
$c_{ijkl}^{S} - c_{ijkl}^{T} = \alpha_{mn}\alpha_{pq}c_{ijmn}c_{klpq}T\frac{C^{\varepsilon}}{C^{\sigma 2}}$	σ_{ij}	T	ε_{kl}	S

With the value for the adiabatic increase in temperature discussed previously

$$\Delta T = -\alpha_{kl}\sigma_{kl}\frac{T}{C^{\sigma}}$$

one finds, using $\sigma_{kl} = c_{klij}\varepsilon_{ij}$, the following relation:

$$\gamma_{ij} = -\alpha_{kl}c_{klij}\frac{C^{\varepsilon}}{C^{\sigma}}.$$

Further relations are derived in a similar manner.

Table 5.1 presents some important differences which can be derived from the differentiation relation just discussed.

Similar expressions are found for the corresponding case of magnetic auxiliary conditions. Especially large differences, sometimes even a change of sign, can be expected in the case of strong pyroelectric and piezoelectric effects, as, for example, in ferroelectric crystals, when thermal expansion is also large. An influence of mechanical boundary conditions on dielectric constants is only present in piezoelectric crystals.

Phase transitions are often accompanied by a substantial change in certain properties, for example, thermal expansion and pyroelectric effect, which may result in unusually large differences of other coefficients under different auxiliary conditions.

5.3 Time Reversal

Time reversal, that is, a change in the sense of direction of time, acts to change the sense of direction of all quantities possessing a linear time dependence, in particular, velocities, current densities and certain quantities derived from these. The most important types of quantities are the electric current density vector, the magnetic field strength, the magnetization and the magnetic moment which are all coupled to the motion charges. The same applies to all mass current densities. The operation of time reversal, designated by T, effects here a change in sign. Other quantities, conserved in the static case under time reversal are, for example the electric field, the electric moment, the mechanical state of stress, the mechanical deformation and the temperature. We now want to take a closer look at the effects of time reversal for magnetic interactions.

In the linear equations of state with the magnetic field strength $\boldsymbol{H}$ as the independent variable, we have the following magnetic terms:

$$\begin{aligned} \varepsilon_{ij} &= \ldots m_{ijk} H_k \ldots \\ \Delta D_i &= \ldots q_{ij} H_j \ldots \\ \Delta B_i &= \ldots \mu_{ij} H_j \ldots \\ \Delta S &= \ldots m_i H_i \ldots \end{aligned}$$

Under time reversal, the induced quantities on the left-hand side do not change except B_i. On the right-hand side $\boldsymbol{H}$ changes its sign. The question arises whether the quantities m_{ijk}, q_{ij}, μ_{ij} and m_i are invariant or not under time reversal. In crystals, where the magnetic moments of the lattice particles are statistically distributed, thus producing no total moment, as in diamagnetic and paramagnetic materials, time reversal cannot effect a change in the properties. This means that the tensors $\{m_{ijk}\}$, $\{q_{ij}\}$ and $\{m_i\}$ must vanish. If higher powers of H_i are included in the equations of state, then the following is true: The magnetomechanical, magnetoelectrical, and magnetothermal tensors vanish for all odd powers or products of the components of the magnetic field strength, respectively. On the other hand, terms with even powers are not forbidden.

Other relationships prevail in crystals in which the magnetic moments are aligned (ferromagnetic: all magnetic moments are aligned; ferrimagnetic: different types of magnetic moments exist which can assume various ordered orientations, however, one nonvanishing total moment remains; antiferromagnetic: ordered magnetic moments compensate to zero). In these, time reversal effects a change in sign of the magnetic moment and hence in the material properties. The combination of spacial symmetry and time reversal leads to a classification of the "magnetic" crystals in the form of so-called *point sym-*

metry groups, magnetic, identical to the black–white point symmetry groups (see, for example, Taschentext *Kristallgeometrie*). There exists a total of 58 real black–white groups aside from the familiar 32 crystallographic point symmetry groups. An analysis of the tensor properties of the "magnetic" crystals requires a consideration of the combined space-time symmetry.

As an example, we consider the magnetoelectrical effect in a rhombic crystal of the magnetic point symmetry group m′m′2. The symbol m′ or n′ specifies a mirror plane or n-fold rotation axis respectively, combined with a change in sign under time reversal. Hence, we examine the effect described by $D_i = q_{ij}H_j$. $\{q_{ij}\}$ is an axial tensor, because $\boldsymbol{H}$ is axial and $\boldsymbol{D}$ not. The negative sign of time reversal then compensates the negative sign factor when the axial tensor is transformed by a rotation-inversion. Symmetry reduction thus leads to the same result as with a second-rank polar tensor. It is $q_{12} = q_{13} = q_{21} = q_{31} = q_{23} = q_{32} = 0$. In contrast, q_{11}, q_{22} and q_{33} remain without restrictions. Consequently, the magnetoelectrical tensor here possesses the three components q_{11}, q_{22} and q_{33}.

As a further example for the effect of time reversal we consider the magnetostriction $\varepsilon_{ij} = m_{ijk}H_k$ in cubic crystals of the magnetic point symmetry group m3m′ (complete 4′/m$\bar{3}$2′/m′). It is enough to draw on the generating symmetry operations. Since 23 is a subgroup of the PSG at hand, it is first of all true, as with all third-rank polar tensors that

$$m_{123} = m_{231} = m_{312},$$

and furthermore, due to the symmetry of the deformation tensor $m_{ijk} = m_{jik}$. With

$$\boldsymbol{R}_{\bar{2}\|(e_1+e_2)} = \begin{pmatrix} 0 & \bar{1} & 0 \\ \bar{1} & 0 & 0 \\ 0 & 0 & 1 \end{pmatrix}$$

we obtain, taking into account time reversal and axiality $m'_{123} = m_{123}$. The same results from the symmetry plane

$$\boldsymbol{R}_{\bar{2}\|e_1} = \begin{pmatrix} \bar{1} & 0 & 0 \\ 0 & 1 & 0 \\ 0 & 0 & 1 \end{pmatrix}.$$

An inspection of the symmetry elements 4′, $\bar{3}$ and 2′ leads to no further condition. This means, the first-order magnetostriction tensor exists in the PSG m3m′. The only independent component is m_{123}. One can see straight away (subgroup relationships to m3m′!), that the following PSGs 23, m3, $\bar{4}$′3m and 4′32′can also exhibit such an effect. All other cubic PSGs exhibit no first-order magnetostriction.

The effect of time reversal in transport processes requires special consideration. In the case of magnetic interactions we must again distinguish between "nonmagnetic" crystals (*diamagnetic* and *paramagnetic*) and "magnetic" crystals (*ferromagnetic* and so on). For example, let a second-rank tensor, the tensor of electric or thermal conductivity, be dependent on the magnetic field, thus $s_{ij}(\boldsymbol{H})$. A time reversal effects a reversal of the direction of $\boldsymbol{H}$. This also means that an interaction, initially described in a right-handed system, given, for example, by $\boldsymbol{e}_i$, $\boldsymbol{e}_j$ and $\boldsymbol{H}$, is to be represented, after time reversal, in a left-handed system $\boldsymbol{e}_i$, $\boldsymbol{e}_j$, $\boldsymbol{H}'$ $(= -\boldsymbol{H})$ with the same tensor. The Onsager relation $s_{ij} = s_{ji}$ valid for transport processes considered in the magnetic interaction above for "nonmagnetic" crystals takes on the form $s_{ij}(\boldsymbol{H}) = s_{ji}(-\boldsymbol{H})$.

A dependence of the magnetic field, expanded according to powers of the components of the magnetic field strength, then demands the following:

$$\begin{aligned} s_{ij}(\boldsymbol{H}) &= s_{ij}^0 + s_{ijk}H_k + s_{ijkl}H_kH_l + \cdots \\ s_{ji}(-\boldsymbol{H}) &= s_{ji}^0 - s_{jik}H_k + s_{jikl}H_kH_l + \cdots \quad . \end{aligned}$$

This means that $s_{ijk\ldots} = -s_{jik\ldots}$ is true for the odd-rank s-tensors and $s_{ijk\ldots} = s_{jik\ldots}$ is true for those of even rank. We encounter such conditions with the Hall effect, with the magnetic resistance tensor, with the Righi–Leduc effect as well as with the thermomagnetic resistance tensor.

In "magnetic" crystals one must also take into consideration the change in direction of ordered magnetic moments under time reversal. Hence, if a transport property is also dependent on the position of the magnetic moment then the general Onsager relation takes the form $s_{ij}(\mathcal{M}, \boldsymbol{H}) = s_{ji}(-\mathcal{M}, -\boldsymbol{H})$, where $\mathcal{M}$ designates the magnetic moment. This can lead to a modification of magnetic interactions in transport processes. For example, a quasi Hall effect, i.e., the build-up of an electric field perpendicular to the current, is conceivable in "magnetic" crystals, even without an external magnetic field, when the corresponding symmetry properties are present.

5.4 Thermoelectrical Effect

The driving force for electrical charge transport or caloric energy transport are the corresponding gradients of the electric potential and temperature respectively. The Joule heat generated in each volume element traversed by an electric current influences the temperature distribution and hence the heat current. This suggests the use of a general approach of the type

$$\begin{aligned} I_i &= -s_{ij}(\operatorname{grad}\varphi)_j - k_{ij}(\operatorname{grad}T)_j \\ Q_i &= -t_{ij}(\operatorname{grad}\varphi)_j - l_{ij}(\operatorname{grad}T)_j \end{aligned}$$

for the description of the simultaneous existence of transport processes of electric charge and caloric energy.

Consequently, the tensors $\{s_{ij}\}$, $\{k_{ij}\}$, $\{t_{ij}\}$ and $\{l_{ij}\}$ represent generalized conductivities. If one interchanges the inducing quantities with the currents, which, at least is formally conceivable, then the corresponding basic equations are

$$-(\text{grad}\,\phi)_i = \sigma_{ij} I_j + \kappa_{ij} Q_j$$
$$-(\text{grad}\,T)_i = \tau_{ij} I_j + \lambda_{ij} Q_j.$$

The tensors emerging here have the property of general resistors. In all these processes we must note that ϕ, T, $\boldsymbol{I}$ and $\boldsymbol{Q}$ are functions of position. This means, the basic equations are only valid for correspondingly small volume elements. Thus, the equations here have the character of differential equations. Moreover, one must keep in mind that the material constants must be specified as functions dependent on position. This is especially true for the temperature dependence of the electrical conductivity. In this sense, the above equations can be drawn upon as basic equations for transport processes under inhomogeneous conditions. The tensors $\{\sigma_{ij}\}$, $\{\kappa_{ij}\}$, $\{\tau_{ij}\}$, and $\{\lambda_{ij}\}$ appearing in the first-order approximation of the interaction are not only linked to electrical and thermal conductivity discussed in Sections 4.3.7 and 4.3.8 but, moreover, describe the appearance of certain additional phenomena. For example, they describe the onset of an electric field in a heat current or a temperature gradient with the passage of an electric current even in directions normal to the current vector. Such phenomena are designated as thermoelectrical effects.

If one introduces an additional magnetic field, whereby the material tensors are to be represented as a Taylor series of the components of the magnetic field, in a manner similar to that discussed in the preceding section, one expects further interesting effects. At present, however, little is known concerning the experimental clarification of these effects.

6
Non-Tensorial Properties

6.1
Strength Properties

If solid bodies are exposed to ever increasing mechanical stress, one normally observes, after a range of stress proportional deformation, an increase in deviations from Hooke's Law, whereby, all processes remain reversible. After further deformation one finds an increasing resistance (strain hardening) associated with irreversible deformation (plastic deformation) and finally breaking processes. Figure 6.1 presents such a behavior in tensile testing.

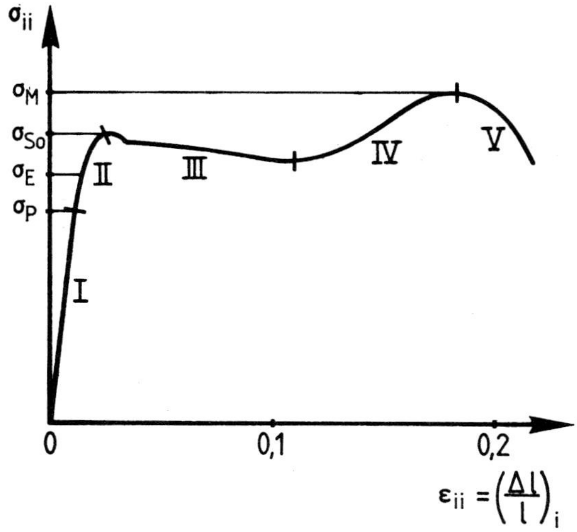

Figure 6.1 Typical stress–strain curve of a cylinder during uniaxial tensile testing along e_i. I: Region of proportionality between stress and strain (Hooke's law), II: Deviation from Hooke's law (quasi-reversible deformation), III: Plastic flow, IV: Strain hardening due to plastic deformation, V: Yield. σ_p denotes the region of quasi-proportionality, σ_E the region of reversible elasticity, σ_{So} the upper elastic bound, at which flow processes start to occur, and σ_M the yield strength. σ_{ii} relates to the cross-section of the cylinder without strain.

Physical Properties of Crystals. Siegfried Haussühl.

ISBN: 978-3-527-40543-5

We want to discuss the application of some important strength properties in more detail. We divide the strength properties roughly in the following domains:

1. Resistance against elastic deformation (elastic strength),
2. Resistance against plastic deformation (hardness),
3. Breaking strength.

The definition of elastic properties and their measurements was discussed in detail in Section 4.5.2. The question, to what extent elastic properties co-determine other strength properties, effective under totally different auxiliary conditions, will be answered in the following.

6.1.1 Hardness (Resistance Against Plastic Deformation)

Eminent technological importance is attached to plastic deformation processes not only in the processes of forming of workpieces, but also in those processes resulting, for the most part, in undesired changes of form during loading. It is therefore understandable that for a long time a focal point of research activities in industrial laboratories was dedicated to the clarification of these properties, resulting in the meantime, in the extensive literature on the subject. Here, we can only discuss some fundamental aspects which should make it easier to understand the phenomena. A first access to plastic deformation is provided by experience with highly viscous fluids, which can take on given forms under the influence of external mechanical forces. Here, the important thing is the time progression of the deformation and expenditure of deformation work. The latter can be assumed unmeasurably small in ideal fluids. As a measure of the deformation resistance of the viscous fluid, however, not the only reasonable one, one can use the dynamic shear resistance which must vanish in ideal fluids. The measurement of the propagation velocity v_T of a transverse wave of sufficiently high frequency (for example, higher than 10 KHz) yields the frequency dependent shear resistance of the viscous fluid $c_{44} = \rho v_T^2$. where ρ is the density. In any case, with fast loading, we expect a certain correlation between c_{44} and the viscosity and with resistance to plastic deformation. With quasistatic, that is, very slow loading we observe, even in highly viscous fluids, a vanishing c_{44}. These properties are impressively demonstrated by elastomeric materials, which behave like rubber under fast loading and like plastic clay under slow loading.

When we now carry over these ideas formed from practical experience to crystals, we expect that under low loading conditions plastic deformations only then appear on an easily measurable scale, when loading occurs slowly and the crystal in question, possesses a very small shear resistance. This is at least qualitatively observed. However, if one compares different materials with approximately the same shear resistance, as, for example, silver chloride ($AgCl$) and potassium alum ($KAl(SO_4)_2 \cdot 12H_2O$), just to mention two familiar types of crystal, one finds extreme differences in resistance to plastic deformation. Crystals of silver chloride are almost "soft as butter." They can be easily deformed and bent. In contrast, an alum crystal offers a power of ten higher resistance to plastic deformation. It breaks like glass before appreciable deformation sets in. The reasons are of a structural nature.

It has been long known that morphological changes on the surface of crystals accompany deformation. This involves parallel grooves- or step-like recesses given the name *glide lines*. It was found that these lines could be understood as intersections of certain preferred lattice planes with the crystal surface. In such instances plastic deformation progresses via a parallel displacement of complete crystal layers parallel to the given planes as with the deformation of a stack of paper sheets or a book parallel to the pages. Furthermore, it was found that the deformation within these planes takes place in distinct directions, the so-called *glide directions*. Hence, such processes were characterized by two crystal-geometric quantities: *glide planes* and glide directions, together referred to as a *glide system*. If external mechanical stress is so applied that strong shear components appear in a glide direction, then plastic shear is easily effected. If several glide systems exist, as, for example, in cubic crystals, then several glide processes can simultaneously lead to a complicated deformation. Table 6.1 presents the glide systems for a few structural types.

Twin formation can also participate in an essential way in the deformation process. This involves the twinning of individual domains of a single crystal, which sometimes can take on macroscopic proportions under the influence of external stress. A long known example is twinning in calcite by uniaxial pressure approximately along one of the three long space diagonals of the rhombohedral elementary cell (Fig. 6.2). The achievable deformation is considerable. It results from the angle the individual twins seem tilted compared to the initial position. These kinds of twin formation can be produced in many crystals under appropriate stress conditions. The geometric details of twin formation are best investigated using X-ray methods (for example, the Laue method).

There exists a range of crystals which form twins even under extremely small mechanical stresses. Belonging to this class are, in particular, the ferroelastic crystals, which in a certain temperature range below the ferroelastic phase transformation often assume different twinning orientations even under weak finger pressure. This phenomenon does not appear in the temperature range above the transformation (prototype). Especially well suited for demonstration purposes are Sb_5O_7I (Nitsche et al., 1977) and $Rb_2Hg(CN)_4$ (Haussühl, 1978). A classification of the ferroelastics on the basis of the possi-

Table 6.1 Glide systems in some selected crystal types. The structure types are denoted according to the structure report (Ewald & Hermann, 1930). Glide planes are usually densely-packed lattice planes which appear morphologically as major growth planes. The glide directions are almost always along densely-packed lattice directions.

Structure	Examples	Glide plane	Glide direction
Face-centered cubic (A1)	Cu, Ag, Au, Pb	{111}	⟨1$\bar{1}$0⟩
Face-centered cubic (A1)	Al, Pt	{001}	⟨110⟩
Body-centered cubic (A2)	α-Fe, Mo, Nb, W	{011}	⟨1$\bar{1}$1⟩
Diamond (A4)	C, Si, Ge	{111}	⟨1$\bar{1}$0⟩
Zinc blende (B3)	ZnS, GaAs, InSb	{111}	⟨1$\bar{1}$0⟩
NaCl (B1)	Alkali halides of NaCl type; MgO, PbS	{110}, {001}	⟨1$\bar{1}$0⟩
CsCl (B2)	Alkali halides of CsCl type (CsCl, CsBr, CsI); TlCl, TlBr, NH_4Cl, NH_4Br	{110}	⟨001⟩
Fluorite (C1)	CaF_2, BaF_2	{001}	⟨110⟩
Calcite	$CaCO_3$, $NaNO_3$	{1$\bar{1}$1}*	⟨110⟩
α-Quartz (C8)	SiO_2	(001)*	⟨110⟩
		{100}	[001]
Baryte	$BaSO_4$, $KClO_4$	(001)	[100]
Gypsum	$CaSO_4 \cdot H_2O$	(010)	[001]

* Trigonal-hexagonal arrangement.

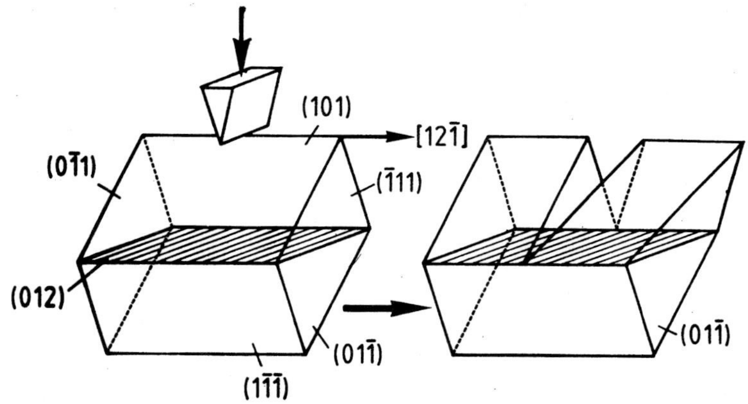

Figure 6.2 Example of twin formation in a calcite rhombohedron induced by uniaxial pressure perpendicularly to [12$\bar{1}$]. All indices refer to a trigonal-hexagonal setting. [12$\bar{1}$] is the glide direction, (012) the glide plane which becomes also the twin plane. Therefore, only a discrete amount of gliding ist possible.

ble combinations of point symmetry of prototype and ferroelastics was given by Aizu (1969). Similar to the usual twin formation caused by pressure, a further deformation after twinning requires overcoming a new much higher resistance threshold.

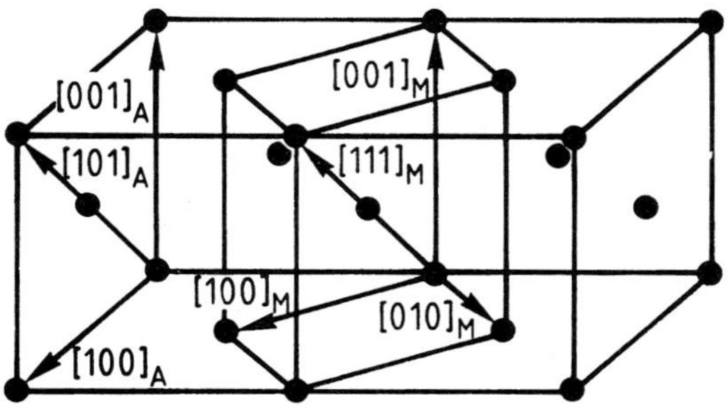

Figure 6.3 Position of the martensitic elementary cell in the austenitic lattice. The transition ist accompanied by the following changes in the interatomic distances: $a_{1,M} = 1.12a_A\sqrt{2}/2$, $a_{3,M} = 0.80a_A$. This results in an increase in volume of about 4%. In the $[111]_M$ directions, the interatomic distances remain virtually unchanged as compared to austenite, and $[111]_M$ is approximately parallel to $[101]_A$. (M and A denote martensite and austenite, respectively).

In some crystals, one observes plastic deformation coupled with a *phase transformation*. Such processes are favored when the transformation temperature is strongly dependent on the external mechanical stress; especially uniaxial pressure. At the working temperature of the deformation the crystals are then already in the region of instability or, the mechanical stress facilitates the start of the transition kinetics or even initiates them. Many ferroelastic crystals in the nonferroelastic high temperature phase belong to this category. Another long known and technologically very important type in this category is represented by the martensite transformation in the cooling of austenite (iron with a carbon content of over 0.1 % by weight; the iron atoms form a face-centered Bravais lattice) in martensite (the iron atoms form a tetragonal body-centered Bravais lattice). The transformation occurs only under a slight displacement of the lattice particles (Fig. 6.3). The phase transformations NaCl type $\rightarrow$ CsCl type or monoclinic ZrO_2 $\rightarrow$ tetragonal ZrO_2 (rutile type) and numerous transformations of alloys also belong to this category. Such a deformation leaves strong mechanical stress inhomongeneities in the deformed region which are coupled to an increase in mechanical strength. In many cases, the deformed regions spring back to the initial phase after a temperature increase, whereby the original shape of the crystal is restored (shape memory). The required threshold values for mechanical stresses and temperatures can, in most cases, be varied within a wide range by the addition of slight amounts of impurities. Hence, one can, for example, adapt the strength properties of steel to specific requirements.

We now want to dwell briefly on the mechanism of plastic deformation in the micro-range. The process of plastic shear along a glide plane demands a roll-off of atoms of neighboring lattice planes on top of one another. The critical resistance is found to be about a third of the corresponding shear resistance from simple model calculations using spherical particles. The actual observed critical values for the onset of plastic deformation, the so-called critical flow stress, lies, however, several orders of magnitude lower. A qualitative understanding for this enormous discrepancy is found first in the close study of lattice defects, in particular, dislocations. These defects mostly originate in crystal growth as a consequence of unavoidable inhomogeneities (chemical composition, temperature).

Dislocations play a decisive role in plastic deformation. As shown by simple model calculations, dislocations can be made to move with far smaller mechanical shear stresses than the critical shear stress of the defect-free crystal, hence enabling a glide process. The resistance (Peierls potential) inhibiting the movement of the dislocations is overcome because only small pieces of the respective dislocation line stay in the domain of higher potential (kink formation), while the remaining dislocation line ends in potential wells (Peierls valley). Thus, the required threshold stress for plastic deformation is extremely reduced. A further condition is natural that a sufficient number of dislocations are available which do not interfere with one another in their ability to move. In the meantime one knows several mechanisms to generate dislocations with low energy expenditure as, for example, the Frank–Read source. The required activation energy depends essentially on the binding properties of the lattice particles. This explains the large variation of plastic properties in crystals with similar elastic properties. If, in the course of the formation of new dislocations under strong plastic deformation, a build-up of dislocations of various orientations takes place, then this can lead to an increase in resistance against further deformation due to the mutual interference of movement (*hardening*). These fundamental results were backed up by observations of dislocation structures before and after deformation with the aid of optical etching experiments, electron microscopy and X-ray topography. In few cases one could even make visible the movement of dislocations directly in the electron microscope.

The time progression of plastic deformation can be described to a first approximation by the change in the components of the deformation tensor with respect to time:

$$\frac{d\varepsilon_{ij}}{dt}.$$

If one assumes an approximately linear relationship between the components of the stress tensor and this deformation velocity, one would expect the following:

$$\frac{d\varepsilon_{ij}}{dt} = H_{ijkl}\sigma_{kl}(t).$$

Because of the complicated directional dependence of the plastic deformation process, which is not coupled in a simple way to the elastic anisotropy, H_{ijkl} cannot be freely interpreted as tensor components. Since plastic deformation of a probe in the ideal case (ideal plastic deformation), evolves under constant volume, we then have $\sum_i \varepsilon_{ii} = 0$ and hence, $\sum_i d\varepsilon_{ii}/dt = 0$.

However, due to the experimental difficulties involved to even measure good reproducible values with respect to this mathematical statement, it was only possible so far to collect data on cubic crystals, in particular, metals and alloys. According to practical experience, one expects, for the initial process of small plastic deformations, at most a certain possibility of a meaningful application of these formal tensor relations.

6.1.2 Indentation Hardness

The measurement of plastic properties on large single crystals in a macroscopic deformation experiment, for example, a tensile test, is comparatively expensive. Hence, for such a test it is advisable to employ a microscopic method which has been developed for perfection over a long time. It concerns the measurement of the so-called impression hardness. This involves pressing a diamond pyramid into the surface of the object. The resulting impression (so-called indentation), due to irreversible deformation, normally corresponds to a negative form of the diamond pyramid. From the dimensions, in particular, from the diameter of the impression, one can derive a quantity which, over a wide scale, turns out to be independent of the load of the diamond. The quantitative connection is as follows: In equilibrium, after the action of the diamond for a specific time, the resistance of the crystal against deformation absorbs just that force K loading the diamond. The resistance produced per unit area σ_0 is equivalent to the yield point. This means that the deformation would first advance after a further increase in load. The total resistance is equal to the area of the impression multiplied by σ_0; hence, $K = F\sigma_0$ with $F = Qd_{\text{eff}}^2$, where Q is the form factor of the pyramid and d_{eff} is a specific diameter. From this one obtains a measure for σ_0, the indentation hardness H,

$$H = \frac{K}{Q(d + d_0)^2},$$

where d_0 takes into account the elastic part of the deformation in equilibrium. Hence, $d = d_{\text{eff}} - d_0$ is the diameter of the impression after releasing the diamond. H as well as d_0 can be determined to amazingly good reproducibility after graphical evaluation of the measured values for different loads K and the associated diameters d. This involves plotting d as abscissa against $\sqrt{K}$. The gradient of the line of best fit yields $\sqrt{HQ}$ and hence H. The intersection of the line with the $\sqrt{K}$-axis ($d = 0$!) gives d_0 (Fig. 6.4).

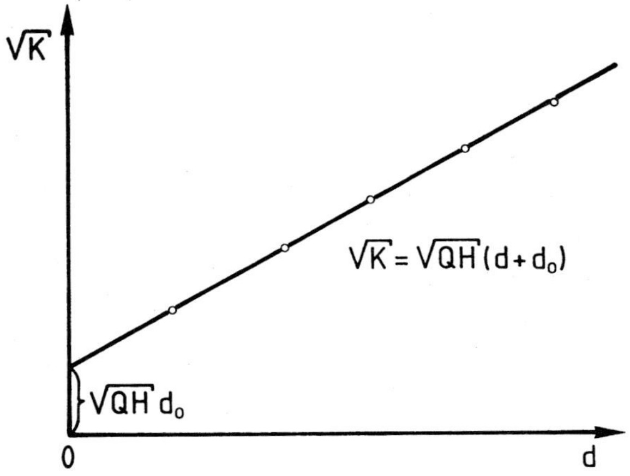

Figure 6.4 Graphical analysis of Vickers impression measurements.

In practice, the standard form of the diamond as a tetragonal pyramid with a height to base diagonal ratio of 1/7 (Vickers pyramid) has become widely accepted. If one measures the diagonal in micrometers, the applied force in Pond (1 kp = 9.807 N) and specifies H, the *Vickers hardness*, in kp/mm^2, the constant Q acquires the value $0.5392 \cdot 10^{-3}$. The Vickers hardness is then

$$H = 1854.4\, K/(d + d_0)^2 \quad (\mathrm{kp/mm^2}).$$

If the mechanical stress comes close to the flow stress as in the measurement of Vickers hardness, the plastic deformation does not come to a standstill, even after long observation times, due to continuous thermal activation. For this reason the loading time in these types of measurements must be limited, for example, to 20 s when measuring Vickers hardness. The method also allows hardness measurements on very small crystals with little time expenditure. Thus the method has proven itself as an analytical aid in qualitative phase analysis of coarse crystalline rocks, in particular, ores. A great advantage of the method is that only small areas on the surface of the crystals are damaged. The anisotropy of Vickers hardness in crystals is usually not very pronounced. In low-symmetry crystals, for example, gypsum or rhombic potassium nitrate, one observes not only different hardness values on different surfaces, but also surprisingly large deviations from the quadratic ideal form of the impression. Often, the size of the impression is dependent on the orientation of the diagonals within the surface of the test specimen. For example, most alkali halides display a smaller impression on (100) when the diagonal runs parallel to [011], and a larger one when it runs parallel to [001] (Fig. 6.5). From this, one can conclude that the glide system (110)/[001] in these crystals has a larger share in the plastic deformation processes than the glide system (100)/[011]. In many

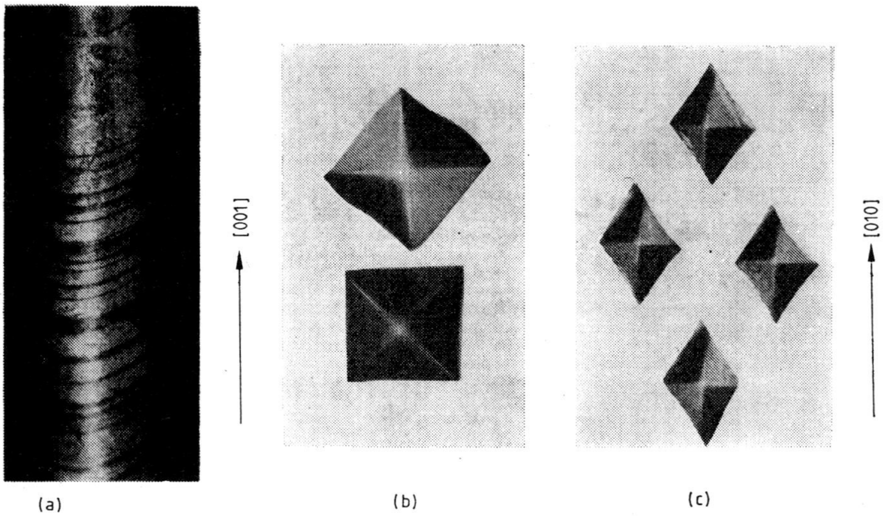

Figure 6.5 (a) Glide directions in a melt-grown Zn crystal (Bridgman–Stockbarger method). (b) Vickers impressions on a (100) surface in KCl. The minimum impression hardness is observed if the base edge of the pyramid is parallel $[011]$. (c) Anisotropic impressions on an (100) surface of KNO_3. The cross-sections of the impressions deviate markedly from a square.

cases a careful analysis of such details permits an insight into the kinetics of the deformation, which otherwise can only be obtained by rather tedious experiments. The use of diamonds with pyramid rhombic cross-section (Knoop pyramid) or other special forms can lead to further predictive statements in this field.

The investigation of Vickers hardness is especially easy to conduct with the well and comfortably equipped commercial units available today (micro hardness tester).

6.1.3 Strength

In contrast to plastic deformations, a breaking process is coupled with a substantial increase in the boundary surface. If one neglects the plastic deformation preceding the breakage, one expects that the energy expanded in the breaking process corresponds approximately to the energy required to create a new boundary surface. This also applies to a certain extent to glass fibres and to whisker-like crystals. Breaking processes can be initiated in macroscopic crystals with far lower energy. The reason is the existence of cracks and other inhomogeneities, which, in particular, considerably facilitate the initial process of breakage, namely first crack formation. However, even under conditions of existing micro-cracks, the creation of the new boundary surface is the dominant principle. Consequently, a decrease in fracture work is expected when the boundary surface energy can be reduced, for example, by the absorption of surface-active molecules on the surface. Here we come across the phenomena, empirically investigated for hundreds of years, of the improvement of the effectiveness of tools by employing suitable liquids or sprays when drilling, milling, cutting, and grinding materials.

The experimental investigation of breaking properties in large single crystals is, just like the investigation of plastic deformation, an expensive exercise, which moreover, in single tests yields rather unreliable values. This is because normally the distribution of primary cracks and other so-called "notches" is not sufficiently known, except in those cases, where artificial crack systems are introduced. Fortunately, there exists a material property linked to breaking properties, namely abrasive strength, which is easily accessible to experiments and yields excellent, reproducible data. We will take a closer look at the two most important methods, which can also be used for the investigations on crystals provided they are of adequate size.

6.1.4 Abrasive Hardness

In the grinding process, the simultaneous action of sharp edges and corners of a great number of abrasive grains initiate numerous breaking processes on the surface. As a result, the surface is furrowed under a sometimes considerable fraction of plastic deformation, which can lead to hardening phenomena and finally to cracks and breaks with the formation of small particles. Their form and size depends on the grain size distribution of the abrasive as well as on the grinding liquid and of course decisively on the object itself. If one measures the loss in weight or volume of the object after a grinding process conducted under defined conditions, one finds a surprisingly good reproducibility of the abrasion. This mirrors the statistic character of abrasive hardness. Furthermore, it emerges that under otherwise similar conditions, the abrasion test over a wide range, independent of cross-section, is directly proportional to the force, with which the grinding tool presses against the specimen, as long as the abrasive does not possess too fine or a too coarse grains. An interpretation of this behavior turned out to be not quite simple. However, it is the prerequisite for the two most important methods to measure abrasive hardness . Kusnezow (1961) described in detail a method in which two crystals A and B are ground against each other and the ratio of the volume loss $V(A,B)$ is considered as a quantitative measure for the ratio of the abrasive hardnesses $F(A)/F(B)$. A closer study on calcium formate (Haussühl, 1963) showed, however, that the relation necessary to measure relative abrasive strengths

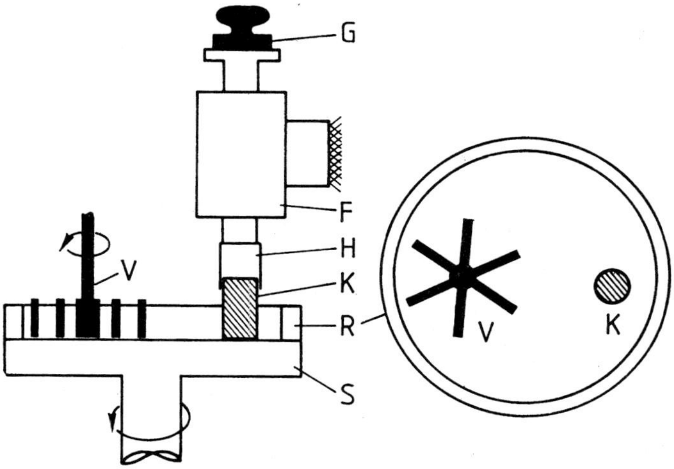

Figure 6.6 Illustration of the grinding apparatus according to v. Engelhardt und Haussühl. S Brass grinding plate with frame R, V abrasive distributor, K crystal with holder H, F bearing of crystal holder axis, G load.

using the *Kusnezow method*, $V(A,B)/V(A,C) = V(C,B)$, where A, B, C are three arbitrary crystal faces, is only approximately fulfilled (deviation under 1%), when the abrasive used possesses coarse granulation (about 100 μ grain diameter) or surface-active substances are added to the grinding liquid. The special advantage of this method consists of the low experimental expenditure and the fact that it is largely independent on secondary condition. The crystals are weighed before and after the grinding process; grinding pressure, grinding time and grinding motion have practically no influence on the measured result $V(A,B)$. NaCl (cubic faces) has proven quite useful as a reference crystal for soft crystals, while quartz (prism faces (100)) has proven as a reference for hard materials.

Another method used by Engelhardt and Haussühl (1960, 1965) to determine the abrasive hardness of many crystals allows the absolute measurement of the abrasive hardness under standardized conditions. Figure 6.6 shows a scheme of the equipment. The crystal, roughly machined to a cylindrical shape, is cemented in a holder to move freely in the vertical direction and pressed with a distinct weight G against the circular grinding plate. The distance of the axis of the crystal holder to the center of the grinding plate is 4 cm, so that the grinding path per revolution amounts to about 25 cm. The recommended standard conditions enabling a good reproducibility and comparability of the measured values are: rotational speed of the grinding plate: 24 revolutions/min; abrasive: high-grade corundum with a mean grain diameter of 140 μ ("Bikorit 100"); per test 10 g abrasive and 15 cm^3 grinding liquid;

100 revolutions at 100 g load on the crystal. More details are given in the cited work. A measure for the abrasive strength F is then the reciprocal of the abraded volume, hence, $F = V^{-1}$, where V is determined by the weight loss of the crystal after the grinding process.

By means of the considerations given below, we recognize a connection between the measured abrasive hardness F and the specific free boundary surface energy γ_i. This is the energy required to generate a boundary surface of the size of a unit area in the orientation denoted by the index i. The total mechanical energy E expended in the test is distributed over three parts: boundary surface work E_g, plastic deformation work E_p and frictional heat E_w. Hence, $E = E_g + E_p + E_w$.

We can formally specify the boundary surface work E_g as the product of a mean boundary surface energy γ with the total surface generated O: $E_g = \gamma O$. Using the simplification that the abraded particles possess the same form, we obtain $O = \alpha N^{1/3} V^{2/3}$, where α is a form factor and N is the number of abraded particles. Hence,

$$F^{-1} = V = (E - E_p - E_w)^{3/2} \alpha^{-3/2} \gamma^{-3/2} N^{-1/2}.$$

As an approximation, we further assume that the total applied energy E is constant in the standard grinding process. Furthermore, we expect that structurally related crystals do not exhibit too large difference with respect to E_p, E_w, α, and N. Thus, for such crystals, it is at least possible to make a qualitative estimation of the specific free boundary surface energy from the measurement of the abrasive strength. For two crystal types A and B it should be approximately $F_A/F_B = (\gamma_A/\gamma_B)^{3/2}$.

In particular, this allows an interpretation of the influence of boundary surface active liquids. For example, with alkali halides of the NaCl-type, one observes in pure xylol in a dry nitrogen atmosphere, an abrasive strength almost double as high as in a solution of 0.1 M stearic acid per liter xylol. Stearic acid turns out to be especially effective in reducing boundary surface energy just like, for example, certain amines as dodecylamine or other polar molecules. In order to obtain good reproducible values for abrasive strength it is advisable to use a sufficiently concentrated solution of polar molecules instead of a pure nonpolar liquid as a grinding liquid. Even the daily fluctuating values of air humidity can lead to changes in the measured values of up to several percent. These effects, however, can also be drawn to determine the relative boundary surface energy $\gamma_c' = \gamma_c/\gamma_0$. γ_0 and γ_c are the boundary surface energies opposing the pure liquid and opposing a solution of the polar molecules of concentration c (in M/l), respectively. With the aid of Gibbs adsorption equation

$$c_g = -\frac{c}{RT}\frac{d\gamma}{dc}$$

and the Langmuir adsorption isotherm

$$c_g = \frac{c_\infty c}{1/b + c}$$

one obtains the Szyskowski relation

$$\gamma_0 - \gamma_c = c_\infty RT \log(1 + cb).$$

where R is the universal gas constant, T is the absolute temperature, c_g and c_∞ are the concentrations and saturation concentration, respectively, of the molecules in the boundary surface and $b = ve^{L/RT}$, where v is the mole volume of adsorbed molecules and L is the adsorption energy of the molecules (per mole). If now one plots, instead of γ_c/γ_0, the ratio $(F_c/F_o)^{2/3}$ against the concentration c, the quantities c_∞/γ_0 and L by a best fit procedure and hence, gain information concerning γ_0 as well as the adsorption energy of the molecules on the given crystal. Full particulars are found in the work cited above (Engelhardt & Haussühl, 1960, 1965).

An overview of the general strength behavior of crystals is found in the diagram presented by Engelhardt and Haussühl (1965) in which the values for indentation hardness H (Vickers hardness) are plotted against the values of abrasive strength F, measured under comparable conditions (Fig. 6.7). This shows the expected relationship between the mechanical strength properties. The values of crystals of the respective isotypic series with fixed cation lie approximately on straight lines. The arrangement on these lines corresponds almost throughout to the sequence of the mean elastic resistance, expressed by the reciprocal volume compressibility K^{-1}. Crystals with very small shear resistances, for example, AgCl or KCN possess especially small hardness values in agreement with the models for plastic deformation discussed above. In contrast, crystals with low abrasive strength exhibit relatively large shear resistances. Interestingly, crystals of very low hardness, as AgCl or KCN exhibit comparatively high values of abrasive strength. This distinctive feature, also observed in the alkali halides, is certainly not solely due to the delayed effect of the plasticity on the breakage, but also due to the ability of such substances to adsorb grains on the surface during abrasion and hold these for a while, whereby, naturally an apparent increase in strength occurs. This is especially noticeable when using abrasives with fine granulation. Moreover, one finds that in the collective of crystals of approximately the same hardness, those with pronounced cleavage, i.e., with a distinct direction of especially low boundary surface energy, exhibit the smallest abrasive strength. In Fig. 6.7, the lines of approximately the same reciprocal compressibility run perpendicular to the main diagonals of the field. The distance of these lines from the zero point increases with increasing elastic resistance. The crystals located in the left and right outer regions of these lines are plastic or highly brittle substances, respectively.

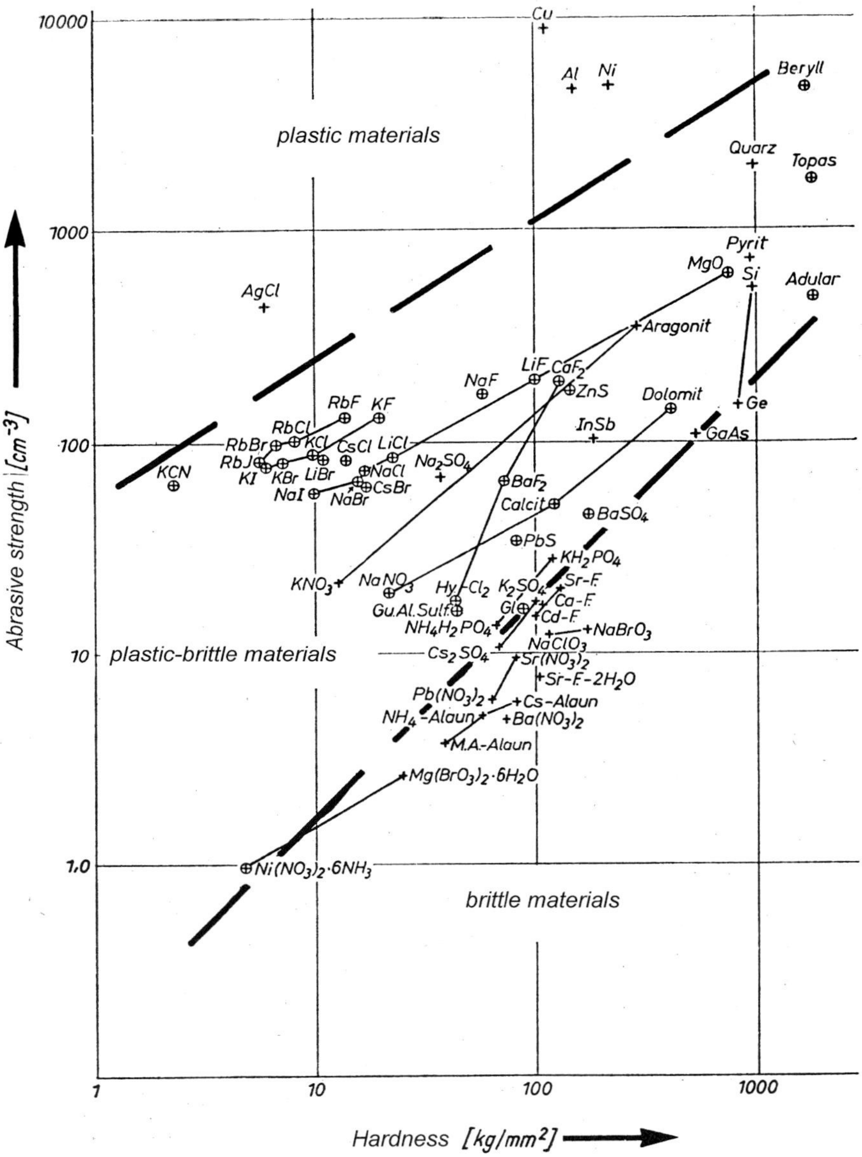

Figure 6.7 Abrasive strength vs. hardness diagram for different types of crystals in xylene after v. Engelhardt and Haussühl. Crystals with particularly low shear resistivity are denoted by ⊕. Abbreviations: $HyCl_2$ = hydrazinium dichloride, Gu.Al.Sulf. = guanidine aluminium sulfate hexahydrate, Ca-F., Sr-F., Cd.-F. = Ca, Sr, Cd formiate, M.A.-alum = methylammonium alum, Gl = triglycinium sulfate.

The numerous further strength properties, which play a role in materials science are naturally closely correlated with both the properties H and F discussed in this section. This also applies to the *scratch hardness*, used in qualitative mineral diagnosis. The *Mohs Hardness Scale* (actually scratch hardness scale) is a sequence of 10 minerals, whereby the one able to scratch the other is the harder of the minerals. There is no special order in which the scratch test is conducted. The hardness is measured on a scale from 1 to 10.

Talc	1	Apatite	5	Topaz	8
Gypsum	2	Orthoclase feldspar	6	Corundum	9
Calcite	3	Quartz	7	Diamond	10
Fluorite	4				

If, for example, a crystal scratched by apatite can scratch fluorite, it is given a Mohs hardness of 4.5. In practice, assigning such a number often causes difficulties because the scratch hardness on different faces also depends on the direction of the scratch in the face. Even scratching in one end in the opposite direction of a certain face can give widely different values, as, for example, with the mineral Disthen (Al_2SiO_5). A careful discussion of these properties is found in Tertsch (1949).

6.2 Dissolution Speed

The dissolution speed characterizes the ablation or etching behavior in a certain solvent. Whether the respective process is described by a physical solubility or a chemical dissolution is not important. The practical measurement is carried out as follows: A sufficiently large plane area of a crystal face is polished perpendicular to the direction $\boldsymbol{e}$. For all measurements, the diameter of the area should not be below a minimum value, for example 10 mm. A certain amount of solvent, say 100 g, is sprayed on the face of the crystal under fixed conditions, i.e., at a definite rate, fixed nozzle and equivalent flow conditions, fixed temperature and so on. An arrangement shown in Fig. 6.8 with the nozzle positioned close to the crystal face has been found to be satisfactory. Directly after the dissolution process the crystal is dried and the loss in weight is measured. This is a direct measure of the dissolution speed. The amount of solvent is so determined that not all-too deep dissolution furrows develop on the surface of the crystal. A proven method of evaluation is to plot the weight loss as a function of the amount of solvent in order to eliminate the expected influences due to the character of the surface as well as the wipe effect occurring independently of the amount of solvent used. The face

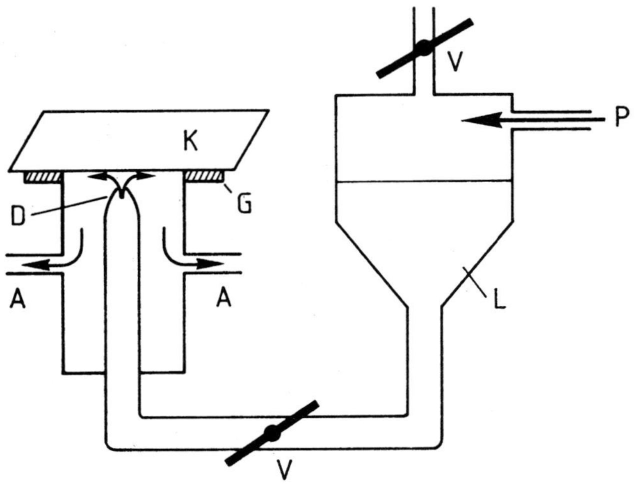

Figure 6.8 Arrangement for measuring of the dissolution speeds of crystals. K crystal plate facing the stream of solvent from the nozzle D with a circular area (diameter 14 mm). A drainage, L solvent, G rubber fitting, P pressurized air, V valves, distance between nozzle and crystal 4 mm, stream velocity 0.7 m s^{-1}.

is to be polished anew for each individual measurement so that the same initial state is present. The gradient of the resulting straight line is a measure of the dissolution speed. A more detailed description of the measurement procedure and results is found in an investigation by Haussühl & Müller (1972). It is worth mentioning that the dissolution speed in directions e and $-e$ can strongly differ when these directions, as in polar crystals, are not symmetry equivalent. Furthermore, one observes that already small additives in the solvent, of the order of a millionth, can effect extreme changes in the dissolution speed. These are often the same substances which in the growth of crystals influence crystal habit and produce a change in growth rate. In this respect, dissolution studies, which are simple and quick to carry out, can impart useful information on growth properties.

6.3 Sawing Velocity

Crystal sawing velocity has as yet received little attention and has hardly been investigated. Its measurement, as with dissolution speed, requires clear agreement as to the conditions to be kept. The respective object is in the form of a cylinder with the cylinder axis e. It is advisable, however, not essential, to select a fixed diameter, for example, 10 mm for all measurements. The saw

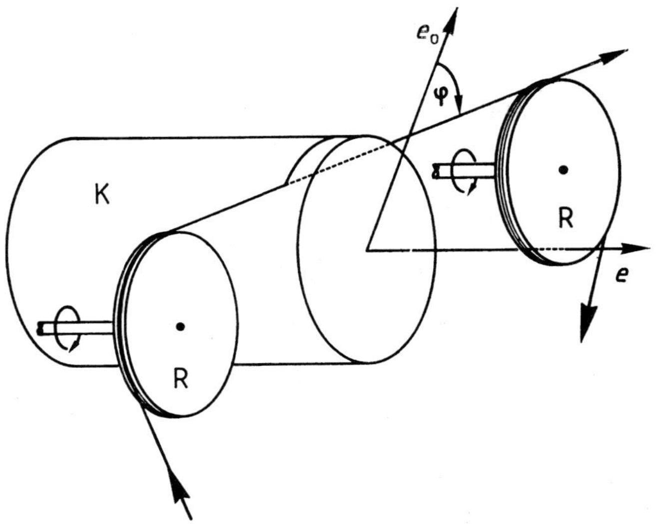

Figure 6.9 Measurement of sawing speed using a thread saw. The thread, uniformly wetted with a suitable liquid, runs with constant speed over the two wheels R. The cylindrical crystal K is pressed against the thread with a constant force. $\boldsymbol{e}$ is the normal to the cutting surface, $\boldsymbol{e}_0$ a reference direction within the cutting surface.

thread, usually a nylon thread or a wolfram wire with a diameter of about 0.2 mm is guided with a constant velocity perpendicular to the cylinder axis, whereby the crystal is pressed against the thread with a given force. During the slicing process, the thread is evenly tensioned and wetted with a suitable liquid. The thread in advance of the crystal should be freed of excessive liquid with the help of a sponge. One measures the time required to cut through the cylinder at a fixed angle φ between the cutting direction and a reference direction $\boldsymbol{e}_0$ (Fig. 6.9). The reciprocal value of the time multiplied by the surface correction factor πR^2 (R is the radius of the cylinder) gives the sawing velocity in mm^2/s. With soluble crystals one preferably uses suitable solvents as the lubricant and with metals, hardly soluble carbonates and silicates, one uses diluted acids or a suspension of abrasives in a viscous oil. For each cylinder orientation $\boldsymbol{e}$ one determines the dependence of the sawing velocity on the angle φ. The values for each $\boldsymbol{e}$ are plotted in the form of an even curve, whereby the position of $\boldsymbol{e}_0$ is marked. A large number of such reference curves is required for the complete representation of the directional dependence. Here, it is also interesting that by reversing the sawing direction in certain low symmetry crystals, a change in the sawing velocity is observed.

6.4 Spectroscopic Properties

Most of the macroscopic properties discussed so far describe the behavior of homogeneous regions of a volume under the action of inducing quantities. A further group of properties, which are not directly representable by tensors, can be collected under the term *spectroscopic properties*. These are based on the interaction of photons or other particles or quasiparticles with the particles of the crystal lattice and the possible excitations resulting from these. We speak of spectroscopic properties in the narrow sense when the interaction is strongly dependent on the frequency or energy of the given radiation. For example, the propagation velocity of electromagnetic waves in crystals is weakly frequency dependent, the absorption in certain frequency ranges, however, is strongly frequency dependent. For a rough overview we distinguish between

(a) localized interactions or excitations of individual particles (electron states, nuclear states, vibrational, and rotational states)

(b) collective interactions tied to large undisturbed regions of the crystal (phonons, magnons, etc.).

In case (a) there appears characteristic differences to the interactions observed in free atoms or molecules. These are influenced by the binding of individual particles to neighboring particles and hence reflect the nature of the crystal field in the vicinity of the particles. Depending on the type of interaction, we understand under the notion "crystal field" the electric or magnetic field or the force field in the neighborhood of the given particle. The symmetry of the local crystal field, the so-called site symmetry, referred to the center of mass of the particle, is determined by a point symmetry group derived from the geometric space group symmetry at the given site. This shapes the finer details of the spectroscopic phenomena. Conversely, from such spectroscopic observations, one can make inferences on the crystal field and the interactions of the individual particles with their neighbors and hence on the binding resulting from such interactions.

As examples, we mention electron and nuclear spin resonance, nuclear resonance fluorescence (Mößbauer effect), high resolution X-ray fluorescence spectroscopy and optical spectroscopy, in particular infrared spectroscopy for the analysis of intramolecular vibrational states as well as Raman spectroscopy.

The effective crystal field is represented with the help of symmetry matched spacial functions. For example, the spacial change of the electric field is described to a first approximation by the tensor of the electric field gradient

$$Q_{ij} = \frac{\partial E_i}{\partial x_j} = -\frac{\partial^2 U}{\partial x_i \partial x_j},$$

where U is the electrostatic potential. More precise information on the tensor of the electric field gradient can be gained with the help of the Mößbauer effect when investigating certain types of atoms in crystals whose resonance absorption exhibits a particularly high sensitivity with respect to the action of the crystal field, as for example, ^{57}Fe. An analogous situation applies for electron and nuclear spin resonances and for the fine structure of the X-ray fluorescence radiation.

The collective vibrational states mentioned in case (b), that is phonons, magnons, and so on and their frequency distribution, i.e., energy distribution, are principally acquired through scattering and diffraction experiments using X-ray and neutron beams. An important aim in these investigations is the complete determination of the dispersion curves, i.e., the relationship $\lambda(\nu)$ between wavelength and frequency for different propagation directions and vibrational forms (modes) in the crystal. Complete knowledge concerning these properties has as yet only been achieved in crystals of high symmetry.

The interpretation of the spectroscopic observations mentioned in this section is based on a very essential manner on the symmetry properties of the crystal under investigation. Group theoretical methods have proven themselves especially for the classification and allocation of the states.

We refer the interested reader to the rich literature available not only on experimental spectroscopic methods, but also on crystal–chemical discussions of the spectra.

7
Structure and Properties

7.1
The Interpretation of Properties from Model Calculations, Correlations Between Properties

Among the most important tasks of material sciences today, aside from the experimental characterization via structural determination and the measurement of physical properties is, in particular, the interpretation and prediction of crystal properties from the chemical constituents, a field residing in the domain of crystal chemistry. In this section we present a brief overview on the current situation.

The first step is the determination of the structure, i.e., the atomic arrangement of the lattice particles which we denote by the motif and metric of the lattice. The furthermost goal, namely, with the help of *ab-initio* calculations based on quantum theoretical and thermodynamic considerations to make assured statements on possible stable crystal structures was successfully realized for a few simple chemical constituents only in the last years. This was due not only to the immense increase in computational power of the computer systems but also to the further development of mathematical tools such as the Hartree–Fock theory or density functional methods (see, for example, Winkler, 1999). Although, in principle there seems to be no insurmountable barriers in the prediction of structures of complex substances, one may not anticipate that all problems of structure formation will be satisfactorily solved in the near future. This applies, in particular, to substances containing large structural units of low inherent stability, for example, large organic molecules or complex adducts of inorganic and organic groups of substances.

The same applies to the second step of the prediction, namely the determination of chemical and physical properties of arbitrary substances of known crystal structure (including chemical constituents). This has only been achieved to a satisfactory degree for a fraction far under one per mill of the approximately 0.5 million structurally known substances to date (listed in the crystal structure data bank ICSD and CSD). Since the experimental evaluation

Physical Properties of Crystals. Siegfried Haussühl.

ISBN: 978-3-527-40543-5

of most of the crystal-physical properties discussed in this book are still considerably time consuming, in particular, even the manufacture of sufficiently large single crystals of high quality (low density of imperfections), the automatic determination of the properties remains an urgent goal should a rich collection of data be made available not only for a broad crystal-chemical interpretation but also for the selection of application specific materials. And, because the precision of the experimental data is often far higher than that achieved through calculation, one will not be able to abandon the experimental method at the present stage, i.e., for the time being, one cannot do away with single crystal growth and practical crystal physics in many areas of work. This is especially true, for example, for the determination of the temperature dependence of the properties.

If one is content with a few quantitative or only qualitative predictions, then one can fall back on the models already developed at the beginning of the last century. The largest advance at that time was provided by the Born lattice theory (see, for example Geiger–Scheel: Handbuch der Physik XXIV, 1933, and Born & Huang 1954). This theory allowed the approximate calculation of a series of properties from simple models of the interaction between lattice particles. The attractive forces, in particular, the Coulomb interaction in ionic crystals and the dipole–dipole or multipole–multipole interactions in the so-called molecular crystals are in equilibrium with the repulsive forces of the negatively charged electron shells. The latter also result from Coulomb interactions as well as from Pauli's exclusion principle. One of the simplest statements for the interaction potential of two ions is represented by the sum $\Phi_{ij} = Kq_iq_j/r_{ij} + b/r_{ij}^n - c/r_{ij}^6$. The first term is the Coulomb potential. K is the constant of the given system of measurement, q_i and q_j are the charges of the ith and jth point-shaped ions and r_{ij} is the distance between the ions. The second term takes into account the repulsion, whereby the exponent n, for example, in alkali halides takes on a value between 10 and 12. Finally, the multipole–multipole interactions are approximated by the third term, the van der Waals term. These quantities fix the given interaction. The summation over all ion pairs ij of the crystal then yields the lattice energy. This is equivalent to the energy released when the ionized, however, not isolated particles come together to form the lattice from a virtual infinite distance. This procedure is also suitable for the calculation of simple properties in types of cubic crystals such as the alkali halides.

In even simpler models one assigns point masses to the particles coupled via springs (spring model, harmonic approximation). The unique advantage of these models lies in their direct visual quality in contrast to the complex operations of *ab-initio* calculations. That is, correlations are more likely to be recognized and qualitatively understood.

The effects of particle thermal motion are first considered in a second step by allocating the spring constants a certain dependence on the amplitude of the vibrations. In this manner the so-called anharmonic effects, for example, thermal expansion and the temperature and pressure dependence of many properties can be calculated. Since, except for symmetry properties, the description of the interaction, at least in simple crystals, requires very little data, namely only atomic distance and the spring constant associated with the respective binding, the different types of properties calculated with this method must exhibit close correlation to these parameters and among themselves. In particular, a strong correlation is expected to the elastic properties because of their direct coupling to the spring constants thus playing a key role in these types of considerations. In Table 7.1 we present an overview of some important relationships, which we, however, are unable to discuss here in any detail. Corresponding correlations must also exist with respect to the anisotropy of the properties. Table 7.2 lists some interesting examples.

Deviations from such correlations always provide an indication of special structural details and binding properties not taken into consideration in simple models. But just because of this, the models can be improved by the addition of further criteria.

For a crude review of the relationship between structure and properties we make a broad division of the properties into two groups which, however, is not always unequivocal:

1. Additive or quasiadditive properties,
2. Nonadditive properties.

7.1.1 Quasiadditive Properties

Additivity refers to the corresponding properties of the particles (constituents: atoms, ions, molecules or particle complexes such as chains, bands, layers, polyhedric structures). That is, an additive property of the given substance should be able to be calculated from the sum of the parts of the constituent properties. Here we are mainly dealing with scalar properties, hence independent of direction and thus independent of a property of a reference system such as spacial mean values. Scalar invariants come into consideration with tensor properties as they are easy to calculate from the complete tensor. This applies especially to cubic crystals.

For our purposes we describe the chemical constituents of a substance A by the sum formula $A = (A_1)n_1(A_2)n_2 \ldots (A_q)n_q$, where each A_j stands for a certain element. Unfortunately, the sequence is not fixed in general except

Table 7.1 Correlation of physical properties and elastic properties in groups of isotypic and chemically related crytals. ↑: positively correlated with c_{ij}, ↓: negatively correlated with c_{ij}.

Property	Correlation	Explanation
$\frac{\text{lattice energy}}{\text{volume}} = E_0$	↑	lattice-theoretical models yield $c_{ij} \sim E_0$.
average surface energy $\bar{\sigma}$	↑	$\bar{\sigma} \sim E_0$.
melting temperature T_0	↑	melting occurs when thermal activation results in a certain losening of bonds (Lindemann formula).
impression hardness H (resistance against plastic deformation)	↑	critical flow strain $\sim$ elastical shear resistance.
abrasive strength F (yield strength)	↑	critical yield strain $\sim$ c_{ij}.
Debye temperature $\Theta_D = \frac{h\nu_D}{k}$ h Planck constant, k Boltzmann constant, ν_D Debye frequency	↑	Debye theory yields $\Theta_D^{-3} = \left(\frac{k}{h}\right)^3 \frac{V_0}{9} \int \sum_{i=1,2,3} \frac{1}{v_i^3(\boldsymbol{g})} \mathrm{d}\Omega$ V_0 average volume of particles, $v_i(\boldsymbol{g})$ propagation velocities of waves in direction $\boldsymbol{g}$. Integration over all spatial directions is implied.
specific heat C_V at low temperatures	↓	At low temperatures, the Debye model yields $C_V = 12\pi^4 Nk/5(T/\Theta_D)^3$ für $T < \Theta_D$, N number of particles per mole.

Table 7.1 *(continued)*

Property	Correlation	Explanation
thermal conductivity λ	$\uparrow$	The Debye formula for thermal conductivities of dielectrics is $\lambda \sim C_V \bar{v} \bar{l}$, $\bar{v}$ average speed of sound in the respective direction, $\bar{l}$ mean free path of phonons.
thermal expansion α	$\downarrow$	The Grüneisen relation $\alpha = \frac{\gamma C_V}{3VK^{-1}}$ holds, where K is the volume compressibility. The Grüneisen constant $\gamma = -\mathrm{d}\log\Theta_D/\mathrm{d}\log V$ varies only slightly in isotypical series (e.g. $\gamma \approx 2$ for NaCl).
infrared limiting frequencies	$\uparrow$	The force constants (spring constants) of the oscillators are proportional to c_{ij}.
molar polarization and dielectricity constants of non-ferroelectric crystals, refractivities	$\downarrow$	The deformability of electron clouds correlates inversely with bond energy.
piezoelectric effects in polar crystals	$\downarrow$	Elastically more "soft" crystals suffer more pronounced deformation upon mechanical stress.
piezooptic effects	$\downarrow$	Elastically more "soft" crystals suffer more pronounced deformation upon mechanical stress.

Table 7.2 Examples of correlations of the anisotropic character of physical properties in strongly anisotropic crystals. As a measure of the anisotropy, the ratio of the maximum and minimum values of the respective property is given. $E = \rho \sum_{i=1,2,3} v_i^2$ dynamic elasticity, K linear compressibility under hydrostatic pressure, F abrasive strength (in xylene using corundum M35), α linear thermal expansion, λ thermal conductivity, n refractivity (at 600 nm). All values for 293 K. $\oplus$ and $\ominus$ indicate that the maximum and minimum values are observed in the direction $\boldsymbol{u}$ from the maximum of E, respectively.*)

Crystal type	$\frac{E_{max}}{E_{min}}$	$\boldsymbol{u}$	$\frac{K_{max}}{K_{min}}$	$\frac{F_{max}}{F_{min}}$	$\frac{\alpha_{max}}{\alpha_{min}}$	$\frac{\lambda_{max}}{\lambda_{min}}$	$\frac{n_{max}}{n_{min}}$
$NaNO_3$ ($\bar{3}$m)	1.52	[100]**)	4.08 $\ominus$	3.1 $\oplus$	12.2 $\ominus$	1.3 $\oplus$	1.19 $\oplus$
turmaline (3m)	1.53	[100]**)	1.86 $\ominus$	– $\oplus$	2.38 $\ominus$	1.35 $\oplus$	1.015 $\oplus$
$C(NH_2)_3(SO_4)_2 \cdot 6H_2O$ (3m)	1.90	[100]**)	49.6 $\ominus$	1.9 $\oplus$	9.2 $\ominus$	3.0 $\oplus$	1.06 $\oplus$
rutile (TiO_2, 4/mm)	1.23	[001]	1.97 $\ominus$	– $\oplus$	1.29 $\oplus$	1.43 $\oplus$	1.11 $\oplus$
KH_2PO_4 ($\bar{4}$m)	1.32	[100]	1.06 $\ominus$	1.5 $\ominus$	1.59 $\ominus$	1.3 $\oplus$	1.03 $\oplus$
α-KNO_3 (mmm)	1.56	[100]	−2.99 $\ominus$	6.1 $\oplus$	8.3 $\ominus$	1.2 $\oplus$	1.13 $\oplus$
calcium formiate (mmm)	1.54	[100]	−4.7 $\ominus$	1.7 $\oplus$	−4.1 $\ominus$	2.16 $\oplus$	1.45 $\oplus$
gypsum ($CaSO_4 \cdot 2H_2O$, 2/m)	1.7	≈ [$\bar{1}$01]	3.64 $\ominus$	8.5 $\oplus$	26.5 $\ominus$	2.4 $\oplus$	1.005 $\oplus$

*) In monoclinic and triclinic crystals, these directions can only approximately be identical.

**) In trigonal-hexagonal arrangement.

for organic substances. In another representation the constituents are decomposed in thermodynamically stable components B_j according to $A = (m_1 B_1) \cdot (m_2 B_2) \cdot \ldots \cdot (m_q B_q)$. Finally, it is common to designate chemical constituents according to cations and anions as well as molecular components as with the hydrated salts for instance. As an example we mention potassium alum $A = H_{24}O_{12}AlS_2K = \frac{1}{2}K_2O \cdot \frac{1}{2}Al_2O_3 \cdot 2SO_3 \cdot 12H_2O = KAl(SO_4)_2 \cdot 12H_2O$.

To calculate the mole weight M per formula unit of substance A we set $M = \sum n_j A_j$, where for A_j we write the average atomic weight of the given element. For a direct comparison of a substance A with its constituents it is often convenient to express n_j, the relative fraction of the jth type of atom of the mole weight by the mole fraction $n'_j = n_j / \sum n_j$. The modified sum formula $(A_1)n'_1 (A_2)n'_2 \ldots (A_q)n'_q$ then contains L atoms ($L = 6,022 \cdot 10^{23}\text{mol}^{-1}$ Avogadro's number).

Examples of the 1. group are:

(a) The average magnetic susceptibility κ of nonferromagnetic ionic crystals is $\kappa(A) \approx \sum n_j \kappa_j$ with $\kappa_j = \kappa(A_j)$ per mole.

(b) Faraday effect, expressed by the Verdet constant Ver. $Ver(A) \approx \sum n_j Ver_j$ with $Ver_j = Ver(A_j)$ for cubic crystals and is valid approximately for noncubic ionic crystals (Haussühl, Effgen 1988; Kaminsky, Haussühl, 1993).

(c) Mole polarization of ionic crystals. The Clausius–Mossotti formula links the average polarizability α per unit volume in the low frequency range with the average relative dielectric constant ϵ_{rel} according to $(\epsilon_{\text{rel}} - 1)(\epsilon_{\text{rel}} + 2) = L\rho\alpha/3\epsilon_0 M \approx \sum N_j \alpha_j / 3\epsilon_0$ (see Section 4.3.3). ρ is the density, M is the mole weight, N_j is the number of jth types of ions per unit volume and α_j is their polarizability. Then $\alpha \approx \sum n_j \alpha_j$ with $n_j = N_j M / L\rho$. α and α_j are then functions of pressure, temperature, frequency and so on.
In the range of optical frequencies ϵ_{rel} is replaced by the average refractive index $n = (n_1 + n_2 + n_3)/3$ (n_i principal refractive index). One then obtains the Lorenz–Lorenz formula $(n^2 - 1)(n^2 + 2) = L\rho R/3\epsilon_0 M \approx \sum N_j R_j / 3\epsilon_0$. R is the mole refraction and R_j is the refraction of the jth type of atom. Hence $R \approx \sum n_j R_j$. The quasiadditivity allows the summation over the R-values of the stable components of the given substance. For example, $R(Mg_2SiO4) \approx 2R(MgO) + R(SiO_2)$.
Also, the average optical activity α_{optact} can be estimated according to $\alpha_{optact} \approx \sum n_j \alpha_{optact}(A_j)$ (per mole) provided it is generated by primarily chiral particles possessing an average chirality of $\alpha_{optact}(A_j)$.

(d) Constituents with similar types of binding possess a mole volume V_{mol} that according to $V_{\text{mol}} \approx \sum m_j V_j$ with $V_j = V_{\text{mol}}(A_j)$ can be estimated from the mole volumina of the components.

(e) The lattice energy Φ in ionic crystals (per formula unit) can, according to Fersman and Kapustinsky, be estimated quasiadditive with surprising accuracy from the "energy coefficient" Φ_j of the given substance according to $\Phi \approx \sum m_j \Phi_j$ (see for example Saukow 1953). The quasiadditivity is also valid for components. For example, one obtains for the spinel $MgAl_2O_4 = MgO \cdot Al_2O_3$ the value $\Phi(MgO \cdot Al_2O_3) \approx \Phi(MgO) + \Phi(Al_2O_3)$.

(f) The X-ray mass attenuation coefficient μ^* which exhibits a hardly measurable anisotropy in the wavelength range $0.03\,\text{nm} < \lambda < 0.3\,\text{nm}$ and is thus quasiisotropic, can be calculated from the mass attenuation coefficients μ_j^* of the atomic constituents ($\mu* = \mu/\rho$; μ is the linear absorption coefficient defined by $I = I_0 e^{-\mu x}$, where I_0 is the primary intensity and I is the observed intensity after passing through a plane parallel plate of thickness x): $\mu^*(A) = \sum n_j A_j \mu_j^* / \sum n_j A_j$. The substance A is noted in the form of the sum formula. The associated atomic weights are written for A_j. The wavelength dependence of μ_j^* varies considerably under the elements. Hence, from the measurement of $\mu^*(A)$ at different wavelengths one obtains a statement on the chemical constituents of A, in particular with very small and thin probes (X-ray absorption analysis; see for example Taschentext Kristallstrukturbestimmung). Incidentally, this relationship is also quite well fulfilled for nonionic bound crystals.

(g) The volume compressibility $K = -d \log V/dp$ is expressed by $K \approx \sum m_j (V_j/V) K_j$, where V_j/V is the estimated fractional volume of the jth component of a substance and K_j is its volume compressibility. The prerequisite is that the components only show a small mutual interaction and that $V \approx \sum V_j$.

(h) The product S of the average elastic strength C and molecular volume MV, $S = C \cdot MV$, where $C = (c_{11} + c_{22} + c_{33} + c_{44} + c_{55} + c_{66} + c_{12} + c_{13} + c_{23})/9$ represents a scalar invariant of the elasticity tensor also shows a quasiadditive behavior. In ionic crystals with similar types of binding we have $S(A) \approx \sum m_j S(B_j)$, i.e., the S-values of the compound A can be calculated from the S-values of the components (Haussühl 1993). Thus one can approximately determine the average elastic strength C of a crystal type from the elastic properties of its components at known molecular volume. As an example let us consider garnet $A = Y_3Al_5O_{12}$. We expect $S(A) \approx 1{,}5S(Y_2O_3) + 2{,}5S(Al_2O_3)$. Experimentally one finds $S(A) = 4031$, $S(Y_2O_3) = 1077$ and $S(Al_2O_3) =$

1107 (S in $[10^{-20}\text{Nm}]$). The estimated value of $4383 \cdot 10^{-20}$Nm exceeds the expected value by about 8%. The deviations in many substances of similar types of binding lie in this order of magnitude. However, it turns out that the deviations, in part, result from a different coordination of ions. A higher coordination in the substance A compared to the average coordination of the components effects an increase of $S(A)$ compared to the average S-value of the components and vice versa. This is seen, for example, with the halides of lithium, sodium, potassium and rubidium with a NaCl type of structure (6-coordination) possessing S-values of on average about $110 \cdot 10^{-20}$Nm and the corresponding value of $123 \cdot 10^{-20}$Nm with the halides of cesium with a CsCl type of structure (8-coordination). If large deviations appear, as for example, in the high pressure modification of SiO_2 (coesite: 400; stichovite: 801) compared to α-quartz (195), then this is an indicator for a changed binding state which can also be tied to a considerable change of coordination. Hence, deviations of the additivity of the S-values can provide concrete pointers concerning structural details (Haussühl 1993).

Finally, it should be mentioned that the product of reciprocal volume compressibility and molecular volume tend to possess similar properties as the S-values.

We should not forget that persistent properties also appear in certain spectroscopic experiments. Among these are, for example, the emission and absorption lines of transitions in the deeper lying electronic shells that are little or hardly influenced in a measurable way by the binding states of the valency shells. This also applies to intramolecular excitation in the optical and infrared spectral region and can be called upon for qualitative and, in part, quantitative chemical analysis.

A special case of quasiadditivity is presented by mixed crystals with isotype components. We consider the simple case of a substance consisting of two components according to $A = (1-x)B_1 \cdot xB_2$. To a first approximation, most properties E can be expressed by the linear relationship $E(A) = (1-x)E(B_1) + xE(B_2)$ if x is sufficiently small (for example $x < 0.1$) and all particles of component B_2 assume the position of the corresponding particle B_1 in the crystal lattice (diadoch substitution): Vegard's rule. Naturally, this rule can be extended to mixed crystals with several isotype components. With the arbitrary substitution of particles in nondiadoch positions, which often occurs in the doping of electrically conducting substances, one observes, as expected, considerable deviations from Vegard's rule. Such properties were discussed by Smekal (1933) and designated as "structure sensitive". However, this nomenclature is no longer compatible with the concept of structure we know today.

7.1.2
Nonadditive Properties

The examples listed in Tables 7.1 and 7.2 as well as the constraints mentioned in the previous section indicate that there exists many properties closely correlated with the properties of the given component, however, not quasiadditive. We cannot discuss these individually in this book. Rather we will address those examples which on the one hand show structural features responsible for the appearance of certain properties and on the other demonstrate the existence of complicated interactions of the particles.

In the first case we think of crystals with noncentric structures. If we are dealing with tensor properties as, for example, pyroelectric, piezoelectric or nonlinear electrical and optical effects, then the associated odd-rank polar tensors must exist as argued in the respective sections. The given structures may not possess inversion centers, i.e. noncentric units of particle with preferred orientations must exist so that, in total, a structural polarity results. This can happen through stable, primary noncentric particles as, for example, the ions of CN, NO_2, BO_3, CO_3, NO_3, SO_3, ClO_3, BrO_3, IO_3, BO_4, NH_4, SiO_4,PO_4, SO_4, ClO_4, AsO_4, many amines and carbonic acids or the molecules NH_3, H_2O and alcohol. The prerequisite for the appearance of polar properties is that the polarity of the particles is not neutralized by the symmetric arrangement. The chiral stable particles make an exception. These include many amines and most amino acids whose intrinsic screw always prevents the occurrence of an inversion center unless the structure contains exactly the same number of enantiomeric particles of the same type (100% racemic mixture). This means that the chiral organic particles represent an almost inexhaustible arsenal for the directed synthesis of new crystals with polar effects.

Some crystals build a polar structure from primarily centrosymmetric atoms and ions, as, for example, low quartz α-SiO_2, silicon carbide SiC or zinc blende ZnS and their polytypes, in which the tetrahedric polarity first develops during the growth phase. There also exist numerous examples of structural chirality, resulting from the chiral arrangement of nonchiral particles (α-quartz, $NaClO_3$, $NaBrO_3$, $Na_3SbS_4 \cdot 9H_2O$).

For over twenty years many laboratories have been searching for new materials for optical data storage and data processing. This involves the application of electric or magnetic fields which result in changes to the optical properties of crystals i.e., refractive indices or absorption, which remain fixed in a metastable state for a definite time. Among these are photochromatic or holographic effects. In crystals such as lithium niobate $LiNbO_3$ or barium titanate $BaTiO_3$, doped with impurities such as iron or hydrogen, charge displacements by the electric field can generate an intense coherent light wave (laser radiation) which first vanishes after the application of heat (photo-galvanic effect). Information superimposed on the light wave in the form of amplitude modulation is then stored in the crystal as a holographic lattice. The information can be retrieved and processed by suitable means so enabling the development of new and highly efficient devices for information technology. The same applies, however, with less favorable prospects, to magneto-optical storage where the light wave modulates the magnetization of thin layers of, for example, $Y_3Fe_5O_{12}$ (yttrium iron garnet) grown epitaxially on $Y_3Ga_5O_{12}$.

Many groups of substances exist in which a light wave can induce electronic transitions in the lattice particles which cause large metastable changes in optical absorption in certain temperature ranges. These changes can also be used to store the information content of the light wave. Special mention should be given to $Na_2[Fe(CN)_5NO] \cdot 2H_2O$ and related substances, which when irradiated with blue light below 200 K trigger strong photochromatic effects coupled to a change in the electronic state of the NO group. These excited states can be quenched by increasing the temperature or also by irradiating with red light (Hauser et al. 1977,1978; Woike et al. 1993). Furthermore, it could be shown that the photochromatic properties of these groups of substances are also suitable for the generation of holographic effects thereby exhibiting the largest as yet known efficiencies (Haussühl et al. 1994; Imlau et al. 1999). The reason is that in these substances doping is not necessary, because all NO groups of the crystal are available for excitation. Apparently, these effects are not linked to a symmetry condition of the given crystal. Especially favorable is the fact that the holograms can be quickly quenched using red light just as with photochromatic effects. One disadvantage which must be still overcome is the low working temperature which, however, could be brought in the range of 300 K by varying the chemical constituents (see, for example, Schaniel et al. 2004).

A further property that can only be realized by a very special combination of the chemical composition, is superconductivity. This has been known for a long time. However, the discovery of the unexpected electrical conductivity of complicated mixed oxides with copper content by Bednorz and Müller (1986) ignited a lively and in the meantime successful search for substances with still higher transition temperatures. These efforts resulted in the achievement of a transition temperature of 125 K in $Tl_2Ba_2Ca_2Cu_3O_{10}$ a benchmark apparently not to be surpassed (Sheng, Hermann 1988). Their technological application in the generation of high magnetic fields is well under way. Apparently, not only is the participation of a certain component, namely CuO necessary, but also an unusual cooperation of several other oxides whose special participation is, however, difficult to recognize.

7.1.2.1 **Thermal Expansion**

We now come to discuss the the temperature and pressure dependence of the physical properties. In the concrete case we consider thermal expansion as well as elastic properties. For simply built molecular crystals as for example, naphthalene, Kitaigorodskii (1973) was able to calculate, with the help of a potential approach that assigned characteristic parameters to the attractive and repulsive forces between two respective atoms of neighboring molecules and simultaneously took into consideration the thermal energy of the molecules, a series of properties such as lattice constants, thermal expansion, elastic constants and also thermoelastic constants in good agreement with experimental values. Other attempts, also with the alkali halides and other simply built types of crystals, to predict the more complicated properties such as thermoelastic constants have been less satisfactory.

The thermally activated motions of lattice particles of a crystal can be studied by optical methods such as Brillouin scattering (see Section 4.5.6) or in the course of crystal structure analysis. The effects of a large change in temperature influencing the lattice vibrations and excitation of certain atomic and molecular states is often made noticeable through phase transformations, changes in optical absorption (thermochromatic effects) and a few especially easily accessible macroscopic effects. Among these are thermal expansion and the temperature dependence of elastic properties. The relation derived by Grüneisen for cubic crystals $\alpha = \gamma K C_V / 3V$, where α is the linear thermal expansion coefficient, $\gamma = -d \log \Theta / d \log V$ is the Grüneisen constant with a value between 1.5 and 3, K is the volume compressibility, C_V is the specific heat per mole volume, V is the mole volume and Θ is the Debye temperature, couples at least qualitatively the thermal expansion with the specific heat and a measure for the elastic strength K^{-1}. Hence, large thermal expansions are correlated with large specific heats and small elastic strengths. For example, diamond at $300K$ possesses with its large elastic strength one of the smallest thermal expansion coefficients of about $0.87 \cdot 10^{-6}\mathrm{K}^{-1}$. In contrast, very soft organic crystals posses α values of over 100-fold higher. The temperature dependence $\alpha(T)$ close to absolute zero is determined by the T^3 law of specific heat and at very high temperatures by the quasilinear rise in compressibility with increasing temperature. Changes in the mean amplitudes of the thermal vibrations of atoms and ions as a function of temperature, determined from crystal structure analysis often agree quite well with the experimental values. In a first approximation the center of mass motion of each atomic particle is assigned a vibration ellipsoid whose semi-axis is set proportional to the amplitude of the thermal displacement vector in the respective directions of the principal axes (Debye–Waller factors; see, for example, Taschentext Kristallstrukturbestimmung). Simple isotype series such as the alkali halides or the oxides of bivalent cations of NaCl-type possess almost the same α-values. In the course of phase transformations, even second-order ones, drastic changes in thermal expansion often appear. Furthermore, it has been found that many noncubic crystals exhibit thermal expansion accompanied with considerable anisotropic effects. From this we can infer that in general thermal expansion is hardly amenable to a quantitative structural interpretation.

7.1.2.2 **Elastic Properties, Empirical Rules**

From the beginnings of the lattice theory of ionic crystals, based on a Coulomb term and a term for the repulsive potential with the exponent n, hence, without the van der Waals term, one obtained the relationship $c_{ij} = F_{ij}r_{ij}^4 = Q_{ij}\cdot$mole lattice energy/mole volume. In isotype series the F_{ij}-values increase weakly with increasing mole weight due to a rise in the van der Waals potential and the exponent n of the repulsive term in heavy ions. For example, in the alkali halides of NaCl-type one observes approximately a doubling of F_{11} in passing from LiF to RbI. Also, the factors Q_{ij} do not depend on separation of the ions but are proportional to n. A similar formula exists for molecular crystals with predominantly van der Waals binding. That is, the repulsive potential determines in a very decisive way the elastic behavior. To a certain extent this statement applies qualitatively to each crystal.

A simple path for the interpretation and prediction of elastic properties is presented by the quasiadditivity of S-values already mentioned above. These allow a rough estimate of the average elastic strength from the properties of the components. A further fundamental property of the S-values consists in the fact that in isotypes and structurally or chemically closely related crystals the S-values fluctuate very little (Haussühl 1993). However, the deformation work and the stability of the electronic shells of the particles is seen to influence the S-values to a certain extent. This can be taken into account by a correction when one assigns a qualitative "hardness" to the particles which is correlated to the exponent of the repulsive term and, for example, in ions such as H^-, the electron e^- or CN^- turns out to be especially low. The experimental data compiled so far allows the formulation of a few simple insightful rules which otherwise would not be directly discernable from the complicated processes of *ab-initio* calculations.

In a next step, one can try to interpret the elastic anisotropy and the relationship to elastic shear resistances and longitudinal resistances. This must take into consideration density, strength and direction of the principal binding chains in the given structures. Directions in which many strong principal binding chains run exhibit strong elastic longitudinal effects. As examples, we mention graphite and β-succinic acid $(CH_2)_2(COOH)_2$. In graphite, strong $sp^2 - C - C$ binding chains, comparable to the sp^3-hybride bindings in diamond, lie in the plane perpendicular to the hexagonal axis, whereas along this axis far weaker interactions appear. This results in a ratio c_{11}/c_{33} of about 29, the largest known longitudinal-elastic anisotropy effect. In β-succinic acid, a monoclinic crystal of the point symmetry group $2/m$, one finds the molecules along $\boldsymbol{a}_3$ due to the H-bridges connecting the chains while in all other directions only weak van der Waals bonding occur. This results in a ratio of maximum to minimum longitudinal resistance of 12.7—the longest observed value in ionic- or molecular crystals to date.

Information concerning the ratio c_{kj}/c_{ii} of certain elastic constants can be gained from the association of the elementary cell to the Bravais lattices in which the interaction of nearest neighbors is best approximated. For example, for the cubic Bravais lattice one finds the following relationships which are easy to calculate using spring models: P-lattice: $c_{44}/c_{11} = 0$, F-lattice: $c_{44}/c_{11} = 0.5$ and I-lattice: $c_{12}/c_{11} = 1$. An instructive example is presented by the alkali halides of NaCl-type. If one goes over from LiBr or LiI with the weakest overlap of electronic shells of cations and anions to rubidium iodide with the largest corresponding overlap, then the face centered lattice of the Br^- or I^--ions must be the physically effective form of both lithium compounds, while for RbI it must be the primitive lattice formed commonly by the center of mass of the Rb and I ions. As a matter of fact one finds for c_{44}/c_{11} the values 0.49 for LiBr and 0.11 for LiI and RbI, respectively.

A further aspect is offered by the deviations from the Cauchy relations which we described by the tensor invariants $\{g_{mn}\} = \{\frac{1}{2}e_{mik}e_{njl}c_{ijkl}\}$. $\{e_{mik}\}$ is the Levi-Civitá tensor (see Section 4.5.1). These deviations show a macroscopic observable effect of the atomic properties of the particles with respect to elastic behavior. Among these are asphericity, that is the deviation of the electronic shells from the spherical form, polarizability, covalence of the binding and a liquid-type interaction of the particles (Haussühl 1967). As is well known liquids possess nonmeasurable, small shear resistances in the low frequency region. Since isotropic bodies such as liquids and glasses exhibit at least the symmetry of cubic crystals, the pure shear resistances c_{66} and $(c_{11} - c_{12})/2$ corresponding to those of the cubic crystals must vanish. This means $c_{66} = 0$ and $c_{12} = c_{11}$. This gives the largest allowed deviations from the Cauchy relations, namely $g_{33} = c_{1122} - c_{1212} = c_{12} - c_{66} = c_{11}$. Particles, which because of their deformability (polarizability) contribute to conserving the volume, similar to the particles of a liquid, can be expected to show an increase in the deviations from the Cauchy relations. The same applies to strongly aspherical particles which act to magnify the transverse contraction effect, i.e., the constants $c_{ij} (i \neq j)$. In contrast, directed covalent bonding tends to increase the shear resistance, i.e., lessens the deviations from the Cauchy relations.

For practical comparison it is meaningful to refer the g_{mn} to the invariant C. The so defined quantities $\Delta_{mn} = g_{mn}/C$ are then dimensionless. As examples we consider the alkali halides, alkali cyanides, AgCl, and several oxides with cubic symmetry and the following values for the invariants $\Delta_{11} = 3(c_{12} - c_{66})/(c_{12} + 2c_{12})$:

	NaCl	KCl	NaCN	KCN	LiH	Li_2O	LiF
Δ_{11}	0.010	0.031	0.78	0.72	−0.95	−0.45	−0.26

	BeO	MgO	BaO	AgCl
Δ_{11}	−0.24	−0.39	0.114	0.81

The largest effects appear in the cyanides and in AgCl, in which the strongly aspheric cyanide ions and the highly polarizable silver ion respectively, bring Δ_{11} close to the liquid value $\Delta_{11} = 1$. This is also directly seen in the highly plastic behavior of these crystals. In contrast, one finds negative Δ_{11} in crystals with cations and anions of low atomic number. These are to be interpreted as the effects of a strong covalent bonding content. The smallest value is observed in LiH.

As an example of the just discussed application of empirical rules, we will attempt to estimate the elastic constants of Na_2O, a substance crystallizing in a flourspar lattice and not easily produced in the form of a large single crystal. The necessary data come from the Landolt–Börnstein compilation and from the data banks of the lattice constants.

At first we assume that Na_2O possesses a similar S-value as Li_2O, namely $233.1 \cdot 10^{-20}$Nm. The molecular volume of Na_2O is $42.97 \cdot 10^{-30}\text{m}^3$. Hence, $C(Na_2O) \approx 5{,}43 \cdot 10^{10}\text{Nm}^{-2} = (c_{11} + c_{12} + c_{66})/3$. The deviation from the Cauchy relations Δ_{11} is taken from the relationship $\Delta_{11}(Na_2O)/\Delta_{11}(NaF) \approx \Delta_{11}(Li_2O)/\Delta_{11}(LiF)$. The experimental values are $\Delta_{11}(NaF) = -0.08$, $\Delta_{11}(Li_2O) = -0.45$ and $\Delta_{11}(LiF) = -0.26$, furthermore, $\Delta_{11}(Na_2O) \approx -0.14$. For c_{66}/c_{11} we assume the value 0.29 measured in Li_2O as well as in NaF. From the three values for C, Δ_{11} and c_{66}/c_{11} we then obtain $c_{11}(Na_2O) \approx 10.80$, $c_{12} \approx 2.33$ and $c_{66} = c_{44} \approx 3.13 \cdot 10^{10}\text{Nm}^{-2}$. From experiences gathered so far, the actual values can deviate by about 10% from these estimates.

Another path uses the rule that the S-values of chemically similar constituents differ only by a small amount. We consider, for example, a hypothetical substance $A = Na_2CaF_2O$ and expect $S(A) \approx 2S(NaF) + S(CaO) \approx S(CaF_2) + S(Na_2O)$. With $S(NaF) = 123$, $S(CaO) = 339$ and $S(CaF_2) = 328$ one finds $S(Na_2O) \approx 257 \cdot 10^{-20}$Nm, a value exceeding the one assumed above by about 10%.

More exact predictions are achieved with little effort when one investigates specimens of pressed fine grained powder rather than single crystals. For example, one can easily determine the two possible sound velocities of such isotropic substances, namely the pure longitudinal wave and the pure transverse wave which are coupled to the constants c_{11} and $c_{44} = c_{66}$ respectively, from the resonance frequencies of plane parallel plates or even measure the volume compressibility in a simple arrangement. With the aid of such experimental average values in combination with the estimates discussed above, one can, at least with cubic crystals, almost always come to rather useful predictions (see also Section 11, Exercise 25).

We now want to briefly address the testing of bonding properties with the help of the S-values (units: $[10^{-20}\text{Nm}]$). Let us again inspect the S-values of Na_2O. Since the properties of Na_2GeO_3 and GeO_2 are known, one can check the relation $S(Na_2GeO_3) \approx S(GeO_2) + S(Na_2O)$. Inserting the experimental values gives $466 \approx 739$ (for GeO_2 in the high pressure modification of the rutile-type) or about 200 (for GeO_2 in the α-quartz-type)$+S(Na_2O)$. It is obvious that the Ge-O bond in Na_2GeO_3 is in no way equivalent to the bond in the high pressure modification of GeO_2.

7.1.2.3 Thermoelastic and Piezoelastic Properties

The thermal energy content of a solid expresses itself mainly in the spectrum of the lattice vibrations and in the excitations of atomic and molecular states. The rise in amplitude of the lattice vibrations with increasing temperature not only causes a change in the mean separation between particles and hence the thermal expansion, but also a change in the interaction potential. This means, that apart from the consequences of excitation, it may be expected that the elastic properties are affected by the temperature dependence of the separation r_{ij} of the particles and also by the effective coefficients of the potential terms, in particular the repulsive potential and the van der Waals potential. The change of the elasticity tensor with temperature is described, at least in a limited temperature interval, by the tensor $\{T^*_{ijkl}\} = \{dc_{ijkl}/dT\}$ which possesses the symmetry properties of the elasticity tensor. However, for the discussion of thermoelastic properties, the logarithmic derivatives of the elastic constants $T_{ij} = d\log c_{ij}/dT$ have gained acceptance, which just like the elastic constants c_{ij} do not represent tensor components. We will use these quantities, which designate the relative change of the elastic constants per Kelvin in what is to follow. With a suitable temperature factor as, for example, the Debye temperature they become dimensionless and hence assume the tensor character. We will return to this point latter.

From the relationship $c_{ij} = F_{ij} \cdot r_{ij}^4$ given above, one obtains for ionic crystals $T_{ij} = d\log F_{ij}/dT - 4d\log r_{ij}/dT$. In cubic crystals such as the alkali halides, the second term corresponds to the negative four-fold linear thermal expansion α_{11}, for which experimental values of around $40 \cdot 10^{-6}\text{K}^{-1}$ exist. The constant T_{11} is about $-0.80 \cdot 10^{-3}\,\text{K}^{-1}$. This means that the change of the longitudinal resistance c_{11} with temperature is here more strongly determined by the influence of the temperature on F_{11}, hence, in particular on the effective exponent of repulsion and on the fraction of van der Waals bonds than on the separation of the ions. The change of F_{44} is made far less noticeable by the constants T_{44} describing the influence of the temperature on the shear resistance c_{44}. Thus the alkali halides demonstrate that thermoelastic properties are closely correlated with thermal expansion, but by no means can they alone be attributed to this. Further details are described in the literature (for example, Haussühl 1960).

In the following we present the more important empirically found rules for thermoelastic behavior.

1. The T_{ij} are, with a few exceptions, always negative, except close to a higher order phase transformation. Its order of magnitude in stable crystals ranges from $-0.01 \cdot 10^{-3}\mathrm{K}^{-1}$ (diamond) to about $-5 \cdot 10^{-3}\mathrm{K}^{-1}$ (organic molecular crystals as, for example, benzene). The anisotropy of T_{ij} can be considerable.
2. The T_{ij} begin at absolute zero with vanishingly small values and with rising temperature take on increasingly negative values ($dT_{ij}/dT < 0$).
3. When approaching a strong 1. order phase transformation as, for example, the melting point, no particularly noticeable anomalies are observed until shortly before the transformation.
4. In isotype crystal series T_{ij} take on characteristic values. For example, crystals of NaCl-type and of CsCl-type possess strongly different T_{ij}. If symmetric ions are substituted by less symmetric ones, then significant deviations appear (for example, the substitution of K or Rb by NH_4 or CH_3NH_3; substitution of A subgroup elements by B subgroup elements).
5. The value of T_{ij} in isotype ionic crystals decreases with increasing charge (for example, $|T_{ij}(\mathrm{MgO})| < |T_{ij}(\mathrm{LiF})|$).
6. In isotypes or structurally and chemically related ionic crystals, those species with heavy particles exhibit normally a larger value of T_{ij} (for example, alkali halides, alums, sulfates, nitrates, phthalates, among others). Exceptions: with increasing solidification, for example, in organic crystals or in the sequence chlorate-bromate-iodate.
7. A general elastic isotropization or a reduction in the deviations from the Cauchy relations with increasing temperature is not observed.
8. The type of bonding is mirrored in the thermoelastic behavior: $|T_{ij}(\text{covalent})| < |T_{ij}(\text{ionic})| < |T_{ij}(\text{van der Waals})|$ at comparable average elastic resistance (C-value).
9. Characteristic thermoelastic anomalies appear close to second and higher order phase transformations, which can be attributed to certain transformation mechanisms (for example, rotative-librative; order-disorder; ferroelastic, ferroelectric and ferromagnetic phenomena).
10. In a large collective of crystals one finds a distinct correlation between thermoelastic properties and thermal expansion as, for example, between T_{ij} and α_{ij} ($T_{ij} \sim -\alpha_{ij}$) in rhombic crystals. The same sign for

T_{ij} and α_{ij} appears seldom which points to partly different causes of the anharmonicity.

11. The anisotropy of T_{ij} in cubic crystals, expressed by the ratio of longitudinal effects along [011] and [100], thus T'/T_{11} with $c' = (c_{11} + c_{12} + 2c_{44})/2$, takes on structure typical values. For example, NaCl-type including oxides: $T'/T_{11} \approx 0.60 \pm 0.05$; CsCl-type: $T'/T_{11} \approx 1.70 \pm 0.05$.
12. The invariants $T^* = \log C/dT$ with $C = (c_{11} + c_{22} + c_{33} + c_{44} + c_{55} + c_{66} + c_{12} + c_{13} + c_{23})/9$ and $z = d(\Delta/C)/dT$ with $\Delta = (c_{12} + c_{13} + c_{23} - c_{44} - c_{55} - c_{66})/3$ = average deviation from the Cauchy relations, take on similar values in different isotype series with equivalent ions. T^* drops to about one quarter when the ionic change is doubled. z also exhibits lower values.

As hinted above, when one now multiplies T_{ij} with the Debye temperature, which can be estimated from the elastic data, the range of variability of T_{ij} is reduced considerably, because elastically harder substances exhibit lower T_{ij}-values and larger C-values than softer substances. To what extent the general discussion can thereby be simplified is not yet decided, because the Debye temperatures must be calculated from the elastic data in the low temperature range and often these cannot be extrapolated with sufficient accuracy from the values measured at substantially higher temperatures. Whether one succeeds, with these dimensionless data, in setting up rules for thermoelastic properties similar to those for the S-values requires more detailed inspection.

Furthermore, we note that the mentioned rules also apply qualitatively in collectives of chemically related crystal types. Groups of substances with H-bridge bonding, however, often show a strong scattering in their thermoelastic constants due to the large variability of the binding strength (examples: carbonic acids and their salts; α and β alums). Tables 12.17 to 12.19 present experimental values of elastic properties of some of the substances discussed here.

Similar rules also exist for the piezoelastic constants $P_{ij} = dc_{ij}/dp$, where p is the hydrostatic pressure. These quantities are dimensionless. They exhibit similar regular traits as the thermoelastic constants, however, with opposite sign. In general, one can say that an increase of hydrostatic pressure is equivalent to a decrease in temperature. For example, often a phase transformation initiated at normal pressure by dropping the temperature can also be set into motion by increasing the pressure at normal temperature. An especially impressive property is shown by the pressure dependence of the reciprocal volume compressibility dK^{-1}/dp, which in almost all substances, virtually independent of structure and chemical constituents, assumes a value between approximately 4 and 6. Large deviations appear only in the region of phase transformations. This comparatively easy to measure quantity is thus suited to the detection of a phase transformation, all the more, because a deviation from normal behavior is recognizably far removed from the transition temperature and from the transition pressure, and finally the measurements can also be carried out on very small probes.

With continuing advances in automation and miniaturization of measurement methods one can anticipate that in the near future the pool of data on physical properties will grow substantially so that the rules already formulated will become firmly cemented and new relationships in compliance with the laws will be found.

7.2 Phase Transformations

All crystal types investigated so far exhibit the phenomena of phase transformation during a transition through a certain temperature- or pressure range. If during this process a homogeneous crystal transforms into another homogeneous substance in the physical-chemical sense, as for example, in congruent melting or in the transformation of metallic tin in grey tin, then we talk of a homogeneous phase transformation. If, on the other hand a crystal breaks down, for example, during heating, into different components associated with many homogeneous substances, as, for example, in the dehydration of gypsum, then we are dealing with a heterogeneous phase transformation. At this point we will only concern ourselves with homogeneous phase transformations of the type solid–solid.

The phase transformations (PT) that interest us are firstly specified by the structure of the initial phase (mother phase) and the resulting phase (daughter phase). The processes involved in the rearrangement of the particles to form the daughter phase (transformation kinetics) can be of a diverse nature. It will be the aim of further research to elucidate just these processes occurring in the atomic range to gain a deeper understanding of the phenomena.

At first we recognize the phase transformation by certain erratic changes in macroscopic properties. In many cases a PT is observed visually or microscopically as a result of changes in the index and the formation of cracks which cloud the appearance of the crystal. In other cases, gross changes in the thermal, electrical and elastic properties can occur. On the other hand, a number of PTs can only be detected with sensitive measurement techniques. This applies, for example to most ferromagnetic, ferroelectric and the so-called order–disorder transformations as well as the just recently discovered incommensurable phases. In the latter, the periodic lattice structure motif is superimposed with a variation of periodic structure not compatible with the lattice periodicity in distinct directions. Table 7.3 presents some typical examples of phase transformations. The PTs can be classified according to various aspects.

Table 7.3 Examples of phase transitions at standard pressure. T_0 equilibrium temperature of the phases I and II. RG: space group.

Phase I ($< T_0$) Phase II ($> T_0$)	T_0 [K] $\lvert\Delta V/V_{II}\rvert$ ΔQ [Jg^{-1}]	Differences in physical behavior	RG I II	Transition kinetics Structural type of transition Thermodynamic type of transition
Graphite (C) Diamond (C) (metastable)	1200*) 0,54 158	large	$P6_3/mmc$ Fd3m	slow reconstruction 1st order
β-Fe γ-Fe**)	1183 $2 \cdot 10^{-2}$ 16,5	small	Im3m Fm3m	fast dilative–martensitic 1st order
α-Fe β-Fe	1033 – 90	α phase ferromagnetic	Im3m Im3m	fast order–disorder strongly 2nd order
α-CsCl β-CsCl	743 0,16 22,4	large	Pm3m Fm3m	slow dilative–martensitic 1st order
α-Quartz β-Quartz	847 $< 10^{-6}$ 12,1	α phase longitudinal piezoelectric effect	$P3_12$ $P6_22$	fast displacive strongly 2nd order
β-NaCN γ-NaCN	284 $1,2 \cdot 10^{-2}$ 48	γ phase anomal elasticity	Immm Fm3m	fast rotatory order–disorder strongly 2nd order, weakly 1st order

Table 7.3 *(continued)*

γ-$BaTiO_3$	393	γ phase	P4mm	fast
δ-$BaTiO_3$	$< 10^{-5}$	ferroelectric	Pm3m	displacive order–disorder
	0,84			strongly 2nd order
α-Nb_3Sn	18,05	α phase	Pm3m	fast
β-Nb_3Sn	–	supraconducting	Pm3m	order–disorder
	> 0			2nd order
α-$Rb_2Hg(CN)_4$	398	α phase	$R\bar{3}c$	fast
β-$Rb_2Hg(CN)_4$	$1,5 \cdot 10^{-2}$	ferroelastic	Fd3m	displacive
	1,03			strongly 2nd order
α-$TlAl(SO_4)_2 \cdot 12H_2O$	≈ 360	all	Pa3	upon activation by touching
β-$TlAl(SO_4)_2 \cdot 12H_2O$	$0,5 \cdot 10^{-2}$	properties	Pa3	reconstruction, $\beta \to \alpha$ irreversible
(metastabil)	0,8			1st order
α-$CaCO_3$ (Aragonite)	–	all	Pnam	upon heavy mechanical stress, slow
β-$CaCO_3$ (Calcite)	$7,5 \cdot 10^{-2}$	properties	$R\bar{3}c$	reconstruction, $\beta \to \alpha$ irreversible
	2			1st order
α-$Li_2Ge_7O_{15}$	284	α phase	Pmm2	fast
β-$Li_2Ge_7O_{15}$	$< 10^{-4}$	pyro- and	Pmmm	displacive order–disorder
	0,1	ferroelectric		strongly 2nd order
α-TGS***)	322	α phase	$P2_1$	fast
β-TGS	$< 10^{-4}$	pyro- and	$P2_1/m$	displacive order–disorder
	1,94	ferroelectric		strongly 2nd order

*) The transition diamond → graphite can easily be observed above this temperature.

**) The lowest-temperature stable phase is usually denoted α or I, while the higher-temperature phases are subsequently denoted β, γ, δ or II, III, IV.

***) TGS: Triglycine sulfate, $CH_2NH_2COOH)_3 \cdot H_2SO_4$.

Thermodynamic considerations stood in the forefront for a long time. If two phases are in thermodynamic equilibrium then they must possess the same values for the free enthalpy (pro mole) under the given auxiliary conditions. As an example, we consider the case of constant mechanical stresses, hence, also constant hydrostatic pressure, constant electric and magnetic fields and constant temperature. As we saw in Section 5, the Gibbs free enthalpy G assumes a minimum under isobaric, isagric, and isothermal conditions ($\Delta G = 0$; minimum from the point of view of stability). If both phases I and II are in equilibrium, then $G_{\mathrm{I}} = G_{\mathrm{II}}$. From this condition one obtains, for example, for the simple special case of hydrostatic pressure and no electric and magnetic field strengths, the Clausius–Clapeyron relation

$$\frac{dp}{dT} = \frac{\Delta S}{\Delta V} = \frac{\Delta Q}{T \Delta V},$$

where ΔQ is the transformation enthalpy, T is the equilibrium- or transformation temperature and ΔV is the change of volume in the transition from I to II. As proof one makes use of the expansion

$$G = G_0 + \frac{\partial G}{\partial p} \Delta p + \frac{\partial G}{\partial T} \Delta T$$

and the relations

$$\frac{\partial G}{\partial p} = V \qquad \text{and} \qquad \frac{\partial G}{\partial T} = -S.$$

The curve dp/dT describes the progress of the transformation pressure as a function of temperature and vice versa (vapor–pressure curve). If we cross from phase I to phase II by increasing the temperature, for example, then this is only possible by supplying energy. In equilibrium, ΔQ is the amount of energy required for the transformation. It is always positive, i.e., energy must always be supplied in a transition from a low temperature- to a high temperature phase, provided one stays close to equilibrium. The deeper background for this is the second law of thermodynamics, which in this case demands an increase in entropy ($S_{\mathrm{II}} > S_{\mathrm{I}}$). This means that the internal energy of the high temperature phase is higher than that of the low temperature phase.

If we work with uniaxial mechanical stresses instead of hydrostatic pressure and at the same time allow the action of electrical and magnetic fields then instead of the Clausius–Clapeyron relation one obtains a more complicated relationship, which we will not pursue any further at this point.

Crystals can be roughly divided into two groups on the basis of their different transformation enthalpies. The first group has easily measurable ΔQ-values of the order of a few percent of the lattice energy and more (lattice energy = formation enthalpy from the isolated lattice particles); in contrast, the ΔQ-values of the second group are hardly or virtually not measurable. Intermediate substances are comparatively seldom. The transformations of the first group are termed *first-order transformations*, and those of the second as *second-order transformations*. A more exact definition states that in first-order transformations the first derivatives of a thermodynamic potential with respect to temperature or pressure (or another inducing quantity) exhibit a discontinuity. In second-order transformations the first derivatives show a continuous behavior, while the second and higher derivatives exhibit discontinuities. This formal classification appears quite elegant although it only captures a part on the transformations. In particular, the theory developed by Landau (1937) allows a number of general formulations concerning certain types of transformations, primarily those of second order. In this connection, an important role is played by the loss of symmetry elements of the mother phase during the phase transition leading to a characteristic change of certain tensor properties. The general discussion involves the variation ΔG of the Gibbs free energy on approaching the transformation point as a function of the inducing quantities. In the forefront is the behavior of so called *order parameters*, which are coupled to the interactions driving the transformations.

A crystallographic classification of PTs on the basis of structural aspects was proposed by Buerger (1948, 1951). Buerger distinguishes between four groups:

I. Transformations of the first coordination sphere,

II. Transformations of the second coordination sphere,

III. Order–disorder transformations,

IV. Transformations of the bonding type.

If the kinetics of transformation proceed hesitantly, as in all cases requiring an activation energy, then we are mainly dealing with crude structural changes. These phase transformations were called *reconstructive* by Buerger. Among these, in particular, most first-order transformations. If the structural difference of phases I and II is caused by small displacements of the particles, then the transformation is called *displacive*. In the case of homogeneous deformations, as for example, in martensitic PT, one speaks of *dilative transformations*, in transformations coupled with the activation of the rotational motion of groups of particles one speaks of *rotative transformations*. This classification has the great advantage that structural occurrences stand in the forefront in the treatment of transformations. From experience, one can set up common rules for transformation properties of the individual groups such as speed of the transformation process, reversibility and magnitude of the transformation enthalpy, which, for example, in the case of displacive PTs of the second coordination sphere, agree with some of the Landau rules for the second-order PTs.

The most important rules are cited as follows:

1. Reconstructive first-order PTs require a nucleation of the new phase. The work of nucleation prevents the prompt occurrence of the PT when exceeding the equilibrium conditions of temperature, pressure and electric or magnetic field strengths. This is due to the phenomena of under-cooling or over-heating of the mother phase in the region of the stable daughter phase. This delay of transformation can often extend over a long time, as is seen from the existence of certain minerals, for example, calcite, which were formed millions of years ago and in the mean time have not transformed into the more stable phase (here aragonite). The possibility to synthesis a variety of crystals, that cannot be produced under normal conditions is based on this behavior. The crystals are synthesized under high temperature and pressure conditions and then brought into a region outside their stability range by fast cooling or pressure relief without them breaking up (quenching). This applies, for example, to diamond synthesis.

In principle, some is true for first-order transformations induced by electric or magnetic fields.

As already alluded above, the transformation enthalpies are very large. Due to the often considerable change in volume, the homogeneity of single crystals seldom remains conserved. The new phase is created mostly in the form of small crystalline domains (crystallites) with irregular orientation. Quite often the crystals sustain many cracks and break into smaller pieces. One of the causes for this phenomenon is also the inhomogeneous temperature distribution. Likewise, the reverse transformation proceeds almost always strongly delayed (hysteresis).

2. Displacive, rotative, dilative (for example, also some martensitic) and order–disorder transformations occur without substantial delay when exceeding the equilibrium conditions. The reverse transformation commences very quickly. The observed small delays (hysteresis) are partly due to the defect structures of the crystal which can hinder transformation. The transformation enthalpies are extremely small. Single crystals of the mother phase pass over into single crystals of the daughter phase, whereby the formation of new defects hardly occurs, because virtually no change in volume takes place. The reverse transformation brings the crystal back to its initial situation, including the original defects. Finally, there exists almost always a close structural relationship between both phases. In particular, almost without exception, the space group of the low temperature phase is a subgroup of the space group of the high temperature phase. The classification of ferroelectric and ferroelastic PTs is based on this property (Aizu, 1969; Shuvalov, 1970).

A detailed analysis of the phenomena has shown that first-order reconstructive PTs are unequivocally identified by the criteria given here. Numerous phase transformations possess the characteristics mentioned under 2. with the exception of the transformation enthalpy, which although small, is clearly observed, as well as a certain weak hysteresis of a few degrees. Often these transformations are also accompanied by a minute change in volume and an optical clouding arising from microcracks. In these cases we conveniently talk of a strong second-order or a weak first-order transformation, respectively. The crucial difference as opposed to the first type of PTs consists of the fact that these crystals possess an anomalous temperature- or pressure dependence with respect to certain properties over a wide temperature- or pressure range (or a corresponding range of the electric or magnetic field strength). Under anomalous we understand a basically different behavior as we would observe in a collective of crystals which over a large range of the state variables does not suffer such phase transformations. Especially strong are the anomalies appearing in the temperature- and pressure coefficients of the elastic constants (see, for example, Fig. 7.1). When approaching the phase boundary the anomaly often amplifies itself to a singular behavior to then slowly normalize

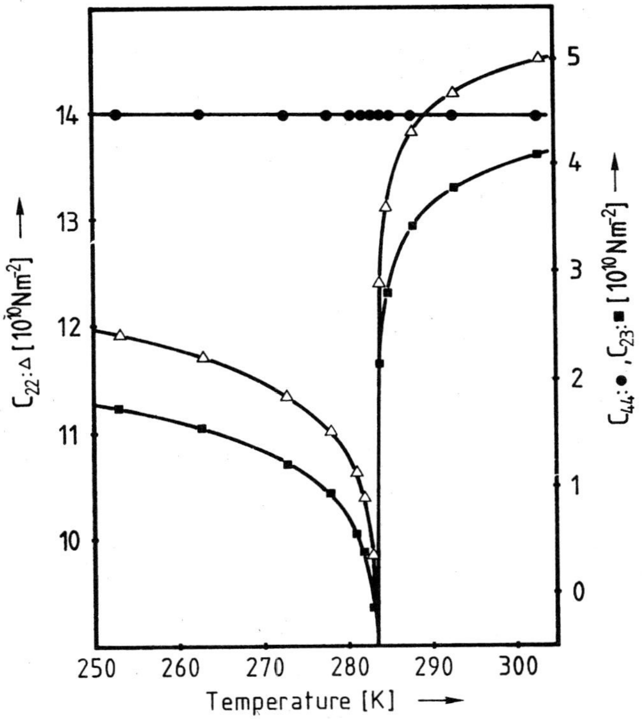

Figure 7.1 Anomalous temperature dependence of the elastic behavior of $Li_2Ge_7O_{15}$ (c_{22} and c_{23} indicated) in a broad range of temperatures around the transition temperature at 284 K (see also Fig. 4.5(a) and (b)).

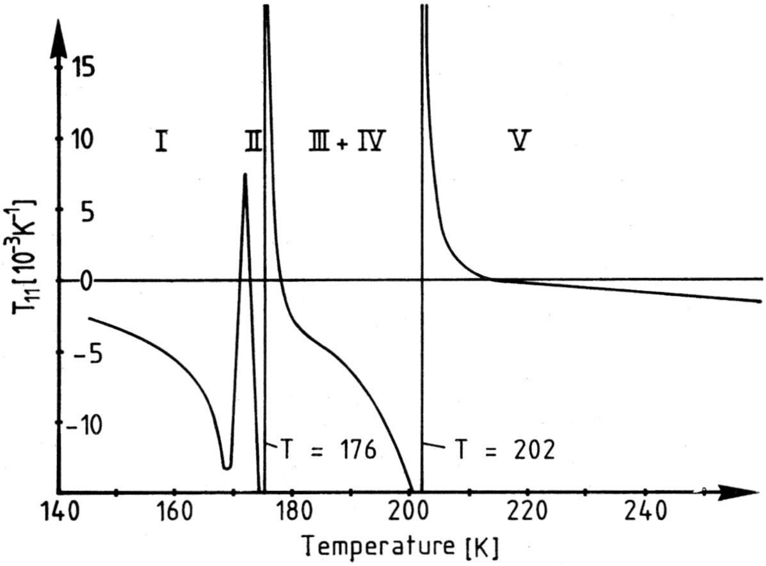

Figure 7.2 Temperature dependence of the thermoelastic constant T_{11} of thiourea. Three regions of rapid sign inversion can be observed, corresponding to the transitions at $T_1 = 202$ K, $T_3 = 176$ K, and $T_4 = 173$ K. The phase transition ($III \leftrightarrow IV$) expected around 180 K from other experiments is barely visible in the behavior of T_{11}.

in the daughter phase farther removed from the phase boundary. The term λ-transformation was derived from this behavior. As far as the corresponding investigations on single crystals are concerned, there does not exist one single PT of this type which is not distinguished by such anomalies in elastic behavior already far from the transformation boundary. In contrast, in distinctive first-order PTs one does not observe such elastic anomalies even in the immediate vicinity of the stability limit. First-order PTs occur, in a manner of speaking, as a catastrophy without warning and lead to abrupt changes in properties.

Finally, there exists a series of crystals in which both types of transformation appear in certain ranges of the state variables. Within a specific temperature- or pressure range such crystals strive to attain a second-order transformation, however, before reaching the new phase may be destroyed by a first-order transformation. Furthermore, there also exists examples of several strong second-order PTs especially in organic crystals. Figure 7.2 presents the interesting case of thiourea ($CS(NH_2)_2$) which shows five PTs in the range between 168 and 300 K.

Furthermore, there also exists the possibility of carrying crystals over into another state by irradiation with light, whereby the external thermodynamic variables do not change. This observed change in the electronic states of the atoms or molecules in crystals causes a change in the optical absorption behavior. In some cases these effects are directly seen as a change in color of the crystal (photochromatic effects). An interesting case of this type is illustrated by $Na_2FeNO(CN)_5 \cdot 2H_2O$, a rhombic crystal, which when irradiated with intense shortwave laser light at liquid nitrogen temperatures experiences a drastic change in the Mößbauer spectrum of the iron atoms over a period of many hours, and at the same time one observes a strong change in the optical absorption properties (Hauser et al., 1977). All such excitation states return to the initial state at higher temperatures due to thermal activation.

Related in a way to these phenomena is the desired effect of the formation of color centers, i.e., of excited states and specific defects, which in part rely on the existence of impurity atoms, (for example, color centers in the alkali halides, fluorite (CaF_2), smoky quartz, amethyst) by irradiation with high energy particles (γ-quantums, α-particles and so on). After intensive irradiation it may come to a partial destruction of the regular lattice structure, as for example, in the case of the isotropization of zircon ($ZrSiO_4$) and related types of crystals after radioactive irradiation. The energy stored in the damaged regions can be released by thermal activation, whereby the crystal, to a large part, again assumes its ordered structure.

A phase transition is interpreted by specifying a definite driving force resulting from a special interaction of the particles with one another. One anticipates that PTs based on analogous mechanisms and driving forces display similar traits of macroscopic transformation behavior and the same or a related functional dependence of the kinetics of transformation on the corresponding state variables (scaling).

As mentioned many times, the elastic properties play a fundamental role in all considerations on the stability of crystal phases. Let explain this aspect a little more. A crystal can only exist in a stable state at all, when its thermal energy is insufficient to deform large volume elements simultaneously into another, at least metastable arrangement, similar, for instance, to plastic deformation, which requires overcoming a certain threshold resistance. The elastic stability conditions are (see also Section 4.5.7.1):

All static and dynamic elastic resistances must exceed a certain minimum value given by the structural details and the thermal energy content of the crystal per unit volume. In particular, all the values of ρv^2 (ρ density, v velocity of sound) calculated from the elastic constants out of the elastodynamic basic equations for any arbitrary direction of propagation must lie above a certain limit. In the case of cubic crystals, these relations are:

$$c_{11},\ c_{44},\ c_{11} - c_{12},\ c_{11} + 2c_{12} > 0.$$

For c_{12} this means

$$-c_{11}/2 < c_{12} < c_{11}.$$

In fact, one normally observes in crystals with very small shear resistances (for example, in cubic crystals c_{44} or $(c_{11} - c_{12})/2$) a tendency to transformation, especially then, when one approaches a temperature- or pressure range, in which the given coefficients dc_{ij}/dT and dc_{ij}/dp, effect a further reduction in the respective shear resistances (example, alkali cyanides). For similar reasons one can make rather confident predictions and also obtain clear details on the processes driving the transformation from a careful study of the temperature- and pressure dependence of the elastic constants wholly within the field of stability of strong second-order phase transitions.

8
Group Theoretical Methods

The methods of group theory are of valuable service for the analysis of the mathematical structure of tensors, in particular for the investigation and the determination of the number and type of independent tensor components. We mention especially the general importance of the application of group theoretical methods for the description of electronic states and their transitions, for the analysis of the vibrational states of molecular and crystalline systems as well as for the classification of phase transitions based on symmetry properties. In this book, however, we shall only discuss a few basic aspects.

In the following we will recapitulate some important definitions and theorems to provide a helpful introduction for those unfamiliar with group theory. In particular, this chapter should be read in conjunction with introductory books on group theory and the rules practiced by working through concrete examples.

8.1
Basics of Group Theory

A set of elements $g_1, g_2, \ldots, g_3$ forms a *group* G, when the following conditions are fulfilled:

1. Between any two elements g_i and g_j of the group an operation is defined which also leads to an element belonging to the group, hence $g_i g_j = g_k$. The commutative law for "multiplication" need not be fulfilled.

2. The operation is associative, i.e., for arbitrary g_i, g_j, g_k then it is always true that $g_i(g_j g_k) = (g_i g_j) g_k$.

3. There is an identity element e of the group, also called the neutral element, such that $e g_i = g_i e = g_i$ for all elements g_i of the group.

4. Each element g_i possesses an inverse element g_i^{-1} also belonging to the group such that $g_i g_i^{-1} = e$.

The number of elements of a group is called the order h of the group. We denote a *subgroup* as a subset of the group satisfying all the conditions of a group.

Physical Properties of Crystals. Siegfried Haussühl.

ISBN: 978-3-527-40543-5

Table 8.1 Some symmetry operations for transformations of cartesian frames in matrix notation.

n-fold rotation (n) parallel to $\boldsymbol{e}_3$; angle $\varphi = 2\pi/n$:

$$A_{n\|e_3} = \begin{pmatrix} \cos\varphi & \sin\varphi & 0 \\ -\sin\varphi & \cos\varphi & 0 \\ 0 & 0 & 1 \end{pmatrix};$$

n-fold rotation–inversion ($\bar{n}$) parallel to $\boldsymbol{e}_3$:

$$A_{\bar{n}\|e_3} = \begin{pmatrix} -\cos\varphi & -\sin\varphi & 0 \\ \sin\varphi & -\cos\varphi & 0 \\ 0 & 0 & \bar{1} \end{pmatrix};$$

threefold rotation parallel to the space diagonal of the cartesian frame:

$$A_{3\|[111]} = \begin{pmatrix} 0 & 1 & 0 \\ 0 & 0 & 1 \\ 1 & 0 & 0 \end{pmatrix};$$

twofold rotation parallel to the axis dissecting $\boldsymbol{e}_1$ and $\boldsymbol{e}_2$:

$$A_{2\|[110]} = \begin{pmatrix} 0 & 1 & 0 \\ 1 & 0 & 0 \\ 0 & 0 & \bar{1} \end{pmatrix};$$

reflection at a mirror plane perpendicular to the axis dissecting $\boldsymbol{e}_1$ and $\boldsymbol{e}_2$:

$$A_{\bar{2}\|[110]} = \begin{pmatrix} 0 & \bar{1} & 0 \\ \bar{1} & 0 & 0 \\ 0 & 0 & 1 \end{pmatrix};$$

All other symmetry operations with different orientations with respect to the crystallographic reference frame can be deduced from those indicated by a similarity transformation $\boldsymbol{A} = \boldsymbol{U}^{-1}\boldsymbol{A}'\boldsymbol{U}$. The transformation matrix $\boldsymbol{U}$ creates from the crystallographic reference frame $\{\boldsymbol{e}_i\}$ the new reference frame $\{\boldsymbol{e}'\}$ according to $\boldsymbol{e}' = \boldsymbol{U}\boldsymbol{e}$, in which the respective symmetry operator assumes the form $\boldsymbol{A}'$ of one of the matrices given above.

Those elements of a group, which through their operation build a complete group are called generating elements. If the elements are concrete things such as geometric figures, permutations of numbers or letters, functions of the position vector or matrices then we speak of concrete *representations* of the group. In the broader sense, we will be mainly dealing with groups whose elements consist of quadratic matrices (number of rows = number of columns). Among these, in particular, are all transformation matrices describing symmetry operations. Table 8.1 lists these matrices for distinct directions.

The properties of a group are taken from the associated *group table*. This contains all products $g_i g_j$. In the case of point symmetry groups the simplest way to generate the table is by means of a stereographic projection in which all the symmetry operations occurring are written. We select a point P inside the respective elementary triangle and first let the symmetry element g_j act on P to generate $g_jP = P'$. Then we let g_i act on P' and obtain $g_ig_jP = g_iP' = P''$. From the stereographic projection we can see how P'' is to be directly generated with a single operation g_k of the group according to $P'' = g_kP$. Hence, we find the product $g_ig_j = g_k$, which we then write in the ith row and jth column of the group table.

Two elements g_i and g_j are *conjugated* when there exists one element x in G such that $g_i = x^{-1}g_jx$. All elements conjugated to an element g_i form a *class*. If g_i and g_j as well as g_j and g_k are conjugated, then g_i and g_k are also conjugated. Since two classes of a group do not possess a common element, one can decompose each finite group in a finite number of classes $K_1, K_2 \ldots$. The number of elements n_i of a class K_i is a divisor of the order of G, hence, $h = n_ih'$, h' whole. As an example we select PSG 3m. The generating symmetry elements are the threefold axis and a symmetry plane containing the threefold axis. The group consists of six elements $g_1 = e$, $g_2 = R_3$, $g_3 = R_3^2$, $g_4 = m_1 = R_{\bar{2}\|\boldsymbol{a}_1}$, $g_5 = m_2 = R_{\bar{2}\|\boldsymbol{a}_2}$ and $g_6 = m_3 = R_{\bar{2}\|\boldsymbol{a}_1'}$ with $\boldsymbol{a}_1' = -\boldsymbol{a}_1 - \boldsymbol{a}_2$. In order to carry out the decomposition into classes, we multiply each element g_i with all elements x according to $x^{-1}g_ix$. This procedure can be carried out comfortably by means of the given group table. Groups with the same group table are called *isomorphic groups* (example, m3 and 23).

Group table of PSG 3m

$g_1 = e$	$g_2 = R_3$	$g_3 = R_3^2$	$g_4 = m_1$	$g_5 = m_2$	$g_6 = m_3$
g_2	g_3	g_1	g_6	g_4	g_5
g_3	g_1	g_2	g_5	g_6	g_4
g_4	g_5	g_6	g_1	g_2	g_3
g_5	g_6	g_4	g_3	g_1	g_2
g_6	g_4	g_5	g_2	g_3	g_1

The following classes result K_1: $g_1 = e$, K_2: g_2, g_3; K_3: g_4, g_5, g_6. The class K_1 with the identity element exists in each group. The three elements of class K_3 belong to the three symmetry equivalent symmetry planes of PSG 3m. In case the group is represented by matrices $g_k = (a_{ij})_k$ then each class can be characterized by an invariant quantity, namely the *trace* $S = \sum_i a_{ii}$ of one of the matrices of the class since all elements of a class possess the same trace; hence, $S(g_i) = S(g_j)$ for the case $g_i = x^{-1}g_jx$. The proof for

$$S(\boldsymbol{A}^{-1}\boldsymbol{B}\boldsymbol{A}) = S(\boldsymbol{B}) = S(\boldsymbol{A}\boldsymbol{B}\boldsymbol{A}^{-1})$$

results from the relation

$$S(\boldsymbol{A}\boldsymbol{B}\boldsymbol{C}) = S(\boldsymbol{B}\boldsymbol{C}\boldsymbol{A}) = S(\boldsymbol{C}\boldsymbol{A}\boldsymbol{B}) = A_{ij}B_{jk}C_{ki}$$

for the case $\boldsymbol{C} = \boldsymbol{A}^{-1}$. The quadratic matrices $\boldsymbol{A}$, $\boldsymbol{B}$ and $\boldsymbol{C}$ are of the same order.

The trace assigned to a class, which we will denote by χ in what is to follow, is called the *character* of the class. The invariance of the character consists in the fact that a transformation matrix retains its trace even after a change of the reference system. From a reference system $\{e'_i\}$ let new basic vectors a'_i arise according to $a'_i = a'_{ij}e'_j$, or in the matrix notation $a' = A'e'$. The basic vectors e'_i and a'_i with the transformation matrix U shall result from the unprimed basic vectors according to

$$e'_i = u_{ij}e_j \quad \text{or} \quad a'_i = u_{ij}a_j \quad (e' = Ue, a' = Ua).$$

Hence, $a = U^{-1}A'Ue$ when we solve the relation $a' = Ua$ with the help of the inverse matrix in $a = U^{-1}a'$ and substitute a' by $A'e'$ and e' by Ue. We denote the matrix $A = U^{-1}A'U$ as a to A' equivalent matrix or symmetry operation. The same relation applies to the coordinates when U represents a unitary matrix, i.e., when $U = (\bar{U}^T)^{-1}$. T means transposed matrix and $\bar{}$ means conjugate complex for the case that the components of the matrix are complex. Consequently, the elements of a class represent the set of symmetry equivalent operations of a certain type within the group. The elements of an arbitrary finite group can always be represented in the form of matrices. This is seen from the *regular representation*, directly constructed from the group table. This is done by rearranging the columns of the group table such that the identity element is present in each principal diagonal. The matrix of the ith element of the regular representation is obtained by writing the number 1 for the element g_i and 0 for all remaining positions in the rearranged group table. The order of these matrices is equal to the order of the group.

The element $g_i g_j^{-1}$ is in the ith row and jth column of the rearranged group table and the element $g_j g_i^{-1} = (g_i g_j^{-1})^{-1}$ is in the jth row and ith column. This means that the transposed matrices are equivalent to the inverse matrices, hence, unitary. That the regular representation actually obeys the group table is recognized as follows. To each three elements g_i, g_j, g_k there always exists two elements g_m and g_n with the property

$$g_i g_k = g_m, \quad g_j g_n = g_k, \quad \text{hence} \quad g_i = g_m g_k^{-1} \quad \text{and} \quad g_j = g_k g_n^{-1}.$$

The associated matrices of the regular representation have the components $(A_i)_{mk} = 1$ and $(A_j)_{kn} = 1$. All the components with other values for mk or nk than those resulting from the above conditions vanish.

We now inquire as to the elements $g_l = g_i g_j$ with the matrix A_l. Because $g_l = g_i g_j = g_m g_k^{-1} g_k g_n^{-1}$, we have $(A_l)_{mn} = 1$. If the regular representation were to obey the group table, then we would have $A_l = A_i A_j$. In point of fact the components

$$(A_i A_j)_{mn} = \sum_k (A_i)_{mk}(A_j)_{kn} = 1$$

agree with $(A_l)_{mn}$ for each pair mn with the above value of k. All other mn yield the component 0 in both the cases. This proves our assertion.

The matrices g_i can be simultaneously transformed into the so-called block matrices with the aid of the unitary transformation $\boldsymbol{U}^{-1}g_i\boldsymbol{U}$ (by applying a suitable unitary matrix). Representations whose matrices cannot be reduced into still smaller block matrices are called *irreducible representations*. These are of fundamental importance in group theory applications. In all further unitary transformations, the individual blocks (see the irreducible representation for the PSG 3m in Table 8.2 arising from the regular representation) are transformed among one another without involving other block matrices and their coefficients. This means that the corresponding block matrices among themselves are also irreducible representations of the given group fulfilling the conditions of the group table. These irreducible representations in the form of block matrices possess a number of important properties which we cite without proof:

1. The number of different nonequivalent irreducible representations of a finite group is equal to the number S of the class.

2. The number of equivalent matrices of an irreducible representation contained in the regular representation is equal to the order of the given matrices. If one divides the identity element of the regular representation, i.e., the unit matric of order h, into unit matrices of the block matrices of the irreducible representation, one sees that the sum of the squares of the matrices of all nonequivalent irreducible representations is equal to the order of the group; hence, $\sum_{s=1}^{S} l_s^2 = h$, where l_s is the order of the block matrix of the sth irreducible representation.

3. If $\chi_k(g_i)$ is the trace of the matrix of the ith element in the kth irreducible representation then the following orthogonality relations are valid:

$$\sum_{s=1}^{S} n_s \chi_k^{\text{irr}}(s)\bar{\chi}_{k'}^{\text{irr}}(s) = \sum_{i=1}^{h} \chi_k^{\text{irr}}(i)\bar{\chi}_{k'}^{\text{irr}}(i) = h\delta_{kk'}.$$

   As an abbreviation, we write $\chi(i)$ instead of $\chi(g_i)$. One sums over the classes in the first summation and over all elements of the group in the second summation. n_s is the number of elements in the sth class. A similar relation for the products of the characters of the different classes is given by

$$\sum_{k=1}^{S} \chi_k^{\text{irr}}(s)\bar{\chi}_k^{\text{irr}}(s') = \delta_{ss'}h/n_s.$$

Table 8.2 Representations of PSG 3m.

Rearranged group table:

g_1	g_3	g_2	g_4	g_5	g_6
g_2	g_1	g_3	g_6	g_4	g_5
g_3	g_2	g_1	g_5	g_6	g_4
g_4	g_6	g_5	g_1	g_2	g_3
g_5	g_4	g_6	g_3	g_1	g_2
g_6	g_5	g_4	g_2	g_3	g_1

Regular representation:

$$g_1 = \begin{pmatrix} 1 & 0 & 0 & 0 & 0 & 0 \\ 0 & 1 & 0 & 0 & 0 & 0 \\ 0 & 0 & 1 & 0 & 0 & 0 \\ 0 & 0 & 0 & 1 & 0 & 0 \\ 0 & 0 & 0 & 0 & 1 & 0 \\ 0 & 0 & 0 & 0 & 0 & 1 \end{pmatrix}, \quad g_2 = \begin{pmatrix} 0 & 0 & 1 & 0 & 0 & 0 \\ 1 & 0 & 0 & 0 & 0 & 0 \\ 0 & 1 & 0 & 0 & 0 & 0 \\ 0 & 0 & 0 & 0 & 1 & 0 \\ 0 & 0 & 0 & 0 & 0 & 1 \\ 0 & 0 & 0 & 1 & 0 & 0 \end{pmatrix}, \quad g_3 = \begin{pmatrix} 0 & 1 & 0 & 0 & 0 & 0 \\ 0 & 0 & 1 & 0 & 0 & 0 \\ 1 & 0 & 0 & 0 & 0 & 0 \\ 0 & 0 & 0 & 0 & 0 & 1 \\ 0 & 0 & 0 & 1 & 0 & 0 \\ 0 & 0 & 0 & 0 & 1 & 0 \end{pmatrix},$$

$$g_4 = \begin{pmatrix} 0 & 0 & 0 & 1 & 0 & 0 \\ 0 & 0 & 0 & 0 & 1 & 0 \\ 0 & 0 & 0 & 0 & 0 & 1 \\ 1 & 0 & 0 & 0 & 0 & 0 \\ 0 & 1 & 0 & 0 & 0 & 0 \\ 0 & 0 & 1 & 0 & 0 & 0 \end{pmatrix}, \quad g_5 = \begin{pmatrix} 0 & 0 & 0 & 0 & 1 & 0 \\ 0 & 0 & 0 & 0 & 0 & 1 \\ 0 & 0 & 0 & 1 & 0 & 0 \\ 0 & 0 & 1 & 0 & 0 & 0 \\ 1 & 0 & 0 & 0 & 0 & 0 \\ 0 & 1 & 0 & 0 & 0 & 0 \end{pmatrix}, \quad g_6 = \begin{pmatrix} 0 & 0 & 0 & 0 & 0 & 1 \\ 0 & 0 & 0 & 1 & 0 & 0 \\ 0 & 0 & 0 & 0 & 1 & 0 \\ 0 & 1 & 0 & 0 & 0 & 0 \\ 0 & 0 & 1 & 0 & 0 & 0 \\ 1 & 0 & 0 & 0 & 0 & 0 \end{pmatrix},$$

Table 8.2 *(continued)*

Equivalent representation constructed from irreducible block-diagonal matrices:

$$g_1 = \begin{pmatrix} 1 & 0 & 0 & 0 & 0 & 0 \\ 0 & 1 & 0 & 0 & 0 & 0 \\ 0 & 0 & 1 & 0 & 0 & 0 \\ 0 & 0 & 0 & 1 & 0 & 0 \\ 0 & 0 & 0 & 0 & 1 & 0 \\ 0 & 0 & 0 & 0 & 0 & 1 \end{pmatrix}, \quad g_2 = \begin{pmatrix} 1 & 0 & 0 & 0 & 0 & 0 \\ 0 & 1 & 0 & 0 & 0 & 0 \\ 0 & 0 & -1/2 & \sqrt{3}/2 & 0 & 0 \\ 0 & 0 & -\sqrt{3}/2 & -1/2 & 0 & 0 \\ 0 & 0 & 0 & 0 & -1/2 & \sqrt{3}/2 \\ 0 & 0 & 0 & 0 & -\sqrt{3}/2 & -1/2 \end{pmatrix},$$

$$g_3 = \begin{pmatrix} 1 & 0 & 0 & 0 & 0 & 0 \\ 0 & 1 & 0 & 0 & 0 & 0 \\ 0 & 0 & -1/2 & -\sqrt{3}/2 & 0 & 0 \\ 0 & 0 & \sqrt{3}/2 & -1/2 & 0 & 0 \\ 0 & 0 & 0 & 0 & -1/2 & -\sqrt{3}/2 \\ 0 & 0 & 0 & 0 & \sqrt{3}/2 & -1/2 \end{pmatrix}, \quad g_4 = \begin{pmatrix} 1 & 0 & 0 & 0 & 0 & 0 \\ 0 & \bar{1} & 0 & 0 & 0 & 0 \\ 0 & 0 & \bar{1} & 0 & 0 & 0 \\ 0 & 0 & 0 & 1 & 0 & 0 \\ 0 & 0 & 0 & 0 & \bar{1} & 0 \\ 0 & 0 & 0 & 0 & 0 & 1 \end{pmatrix},$$

$$g_5 = \begin{pmatrix} 1 & 0 & 0 & 0 & 0 & 0 \\ 0 & \bar{1} & 0 & 0 & 0 & 0 \\ 0 & 0 & 1/2 & -\sqrt{3}/2 & 0 & 0 \\ 0 & 0 & -\sqrt{3}/2 & -1/2 & 0 & 0 \\ 0 & 0 & 0 & 0 & 1/2 & -\sqrt{3}/2 \\ 0 & 0 & 0 & 0 & -\sqrt{3}/2 & -1/2 \end{pmatrix}, \quad g_6 = \begin{pmatrix} 1 & 0 & 0 & 0 & 0 & 0 \\ 0 & \bar{1} & 0 & 0 & 0 & 0 \\ 0 & 0 & 1/2 & \sqrt{3}/2 & 0 & 0 \\ 0 & 0 & \sqrt{3}/2 & -1/2 & 0 & 0 \\ 0 & 0 & 0 & 0 & 1/2 & \sqrt{3}/2 \\ 0 & 0 & 0 & 0 & \sqrt{3}/2 & -1/2 \end{pmatrix},$$

The character of an element of an arbitrary representation $\Gamma(g_i)$ can be resolved according to

$$\chi(g_i) = \sum_{k=1}^{S} m_k \chi_k^{\text{irr}}(i).$$

Hence, m_k specifies how often the kth irreducible representation is contained in the representation $\Gamma(g_i)$. From the orthogonality relation it follows that

$$m_k = \frac{1}{h} \sum_{s=1}^{S} n_s \chi(s) \bar{\chi}_k^{\text{irr}}(s) = \frac{1}{h} \sum_{i=1}^{h} \chi_k(g_i) \bar{\chi}_k^{\text{irr}}(g_i),$$

whereby in the first summation one again sums over the classes and in the second over the elements of the group. For each arbitrary representation Γ it is true that

$$\Gamma = \sum_{k=1}^{S} m_k \Gamma_k^{\text{irr}},$$

where Γ_k^{irr} is the kth irreducible representation.

For many problems it is sufficient to consider the character table of a group instead of the group table. Each column contains the characters of a certain class for the different irreducible representations and the rows contain the characters of the elements of the different classes for a fixed irreducible representation.

In point symmetry groups, the order h of the group, i.e., the number of symmetry operations, is always equal to the number of surface elements of a general form; hence, one can specify h from the known morphological rules.

8.2 Construction of Irreducible Representations

The irreducible representations of the crystallographic point symmetry groups can always be derived without difficulty with the help of the above relations, in particular, with the orthogonality relations. Thus we can forgo a description of a general procedure for the construction of irreducible representations here.

In some cases it is useful to apply an additional rule concerning the *direct matrix product* (*inner Kronecker product*), which also plays an important role in other relationships. The direct matrix product of two matrices $\boldsymbol{A} = (a_{ij})$ and $\boldsymbol{B} = (b_{kl})$, denoted by $\boldsymbol{A} \times \boldsymbol{B}$, is a matrix with the elements

$$(\boldsymbol{A} \times \boldsymbol{B})_{ij,kl} = a_{ik} b_{jl}.$$

The direct product of two representations

$$\Gamma_a(g) \times \Gamma_b(g) = \Gamma_c(g)$$

consists of matrices formed by the direct product of the matrices belonging to each element g of both representations. The direct product is also a representation of the group when the matrices Γ_a and Γ_b are permutable for all elements g. This results because

$$\begin{aligned}(\Gamma_c(f)\Gamma_c(g))_{ij,kl} &= (\Gamma_a(f) \times \Gamma_b(f))_{ij,mn}(\Gamma_a(g) \times \Gamma_b(g))_{mn,kl}\\ &= (\Gamma_a(f)_{im}\Gamma_b(f)_{jn})(\Gamma_a(g)_{mk}\Gamma_b(g)_{nl})\\ &= (\Gamma_a(f)\Gamma_a(g))_{ik}(\Gamma_b(f)\Gamma_b(g))_{jl}\\ &= (\Gamma_a(fg) \times \Gamma_b(fg))_{ij,kl}\\ &= (\Gamma_c(fg))_{ij,kl}.\end{aligned}$$

Since an arbitrary representation can be resolved according to $\Gamma_s(g) = \sum_{k=1}^{S} m_k\Gamma_k^{\mathrm{irr}}$, and for m_k the relation

$$m_k = \frac{1}{h}\sum_{s=1}^{S} n_s\chi(s)\bar{\chi}_k^{\mathrm{irr}}(s)$$

holds as well as the relation

$$\chi(\Gamma_a(g) \times \Gamma_b(g)) = \chi(\Gamma_a(g))\chi(\Gamma_b(g)),$$

proved directly from the Kronecker relation, one has

$$\Gamma_a(g) \times \Gamma_b(g) = \frac{1}{h}\sum_{k=1}^{S}\sum_{g}\chi_a(g)\chi_b(g)\bar{\chi}_k^{\mathrm{irr}}(g)\Gamma_k.$$

If $\Gamma_a(g)$ and $\Gamma_b(g)$ are two irreducible representations, then one can obtain other irreducible representations via the direct product as long as $\Gamma_a(g)$ and $\Gamma_b(g)$ are different from the identity representation.

The construction of the irreducible representation is now made as follows:

1. The group is decomposed into classes by calculating the product $x^{-1}gx$, whereby with given g for x all elements of the group are written down. As already mentioned, this operation can be comfortably carried out using the group table.

2. From the relation $\sum_{s=1}^{S} l_s^2 = h$, one obtains in almost all important applications an unambiguous statement concerning the dimensions of the individual representations. From 1. the number of irreducible representations is known (S = number of classes!).

3. $\Gamma_1 = (1;1;1;\ldots;1)$ is always an irreducible representation (total symmetrical representation, trivial representation). We use the following scheme

for the construction of nontrivial representations for $l = 1, 2$ and 3 (angle of rotation $\varphi = 2\pi/n$):

	$g = 1$	$\bar{1}$	$\mathrm{m} = \bar{2} \parallel e_1$	$\mathrm{n} \parallel e_3$	$\bar{\mathrm{n}} \parallel e_3$
$l = 1$	(1)	$(\bar{1})$	$(\bar{1})$	(1)	$(\bar{1})$
$l = 2$	$\begin{pmatrix} 1 & 0 \\ 0 & 1 \end{pmatrix}$	$\begin{pmatrix} \bar{1} & 0 \\ 0 & \bar{1} \end{pmatrix}$	$\begin{pmatrix} \bar{1} & 0 \\ 0 & 1 \end{pmatrix}$	$\begin{pmatrix} \cos\varphi & \sin\varphi \\ -\sin\varphi & \cos\varphi \end{pmatrix}$	$\begin{pmatrix} -\cos\varphi & -\sin\varphi \\ \sin\varphi & -\cos\varphi \end{pmatrix}$
$l = 3$	$\begin{pmatrix} 1 & 0 & 0 \\ 0 & 1 & 0 \\ 0 & 0 & 1 \end{pmatrix}$	$\begin{pmatrix} \bar{1} & 0 & 0 \\ 0 & \bar{1} & 0 \\ 0 & 0 & \bar{1} \end{pmatrix}$	$\begin{pmatrix} \bar{1} & 0 & 0 \\ 0 & 1 & 0 \\ 0 & 0 & 1 \end{pmatrix}$	$\begin{pmatrix} \cos\varphi & \sin\varphi & 0 \\ -\sin\varphi & \cos\varphi & 0 \\ 0 & 0 & 1 \end{pmatrix}$	$\begin{pmatrix} -\cos\varphi & -\sin\varphi & 0 \\ \sin\varphi & -\cos\varphi & 0 \\ 0 & 0 & \bar{1} \end{pmatrix}$

From the similarity transformation $\boldsymbol{U}\boldsymbol{A}\boldsymbol{U}^{-1}$ one obtains the corresponding matrices $\boldsymbol{A}'$ for the different orientations of the symmetry operators. For example, one has for $g_5 = \boldsymbol{R}_{\bar{2}\parallel \boldsymbol{a}_2}$ in the PSG 3m, in two-dimensional representation:

$$g_5 = \boldsymbol{U} g_4 \boldsymbol{U}^{-1} = \begin{pmatrix} \cos\varphi & \sin\varphi \\ -\sin\varphi & \cos\varphi \end{pmatrix} \begin{pmatrix} \bar{1} & 0 \\ 0 & 1 \end{pmatrix} \begin{pmatrix} \cos\varphi & -\sin\varphi \\ \sin\varphi & \cos\varphi \end{pmatrix}$$
$$= \begin{pmatrix} 1/2 & -\sqrt{3}/2 \\ -\sqrt{3}/2 & -1/2 \end{pmatrix}$$

with $\varphi = 120°$ (rotation of the symmetry plane about the threefold axis).

In any case, one must check whether the orthogonality conditions are adhered to. For cyclic groups, in which all group elements are powers of the generating elements, hence, with all rotation axes and rotation–inversion axes, each element forms a class on its own, i.e., only one-dimensional representations exist. In these cases we can immediately specify the complete irreducible representations. For an n-fold rotation axis, one always has $g^n = 1$, hence, $g = e^{2\pi i k/n}$ (unit roots of nth degree) for the one-dimensional representations. The irreducible representations are then the n powers $\Gamma_k = (e^{2\pi i k m/n})$ with $m = 0, 1, 2, \ldots, n-1$. Thus, for example, for a sixfold rotation axis $\Gamma_2 = (1; e^{2\pi i 2/6}; e^{2\pi i 4/6}; e^{2\pi i 6/6}; e^{2\pi i 8/6}; e^{2\pi i 10/6})$. We recognize the validity of the orthogonality relation when we form the "scalar product" of two such representations

$$\Gamma_k \cdot \bar{\Gamma}_{k'} = \sum_{m=0}^{n-1} e^{2\pi i k m/n} e^{-2\pi i k' m/n} = \sum_{m=0}^{n-1} e^{2\pi i (k-k')m/n}$$
$$= \frac{e^{2\pi i(k-k')} - 1}{e^{2\pi i (k-k')/n} - 1} = \left\{ \begin{array}{ll} 0 & \text{for} \quad k \neq k' \\ 1 & \text{for} \quad k = k' \end{array} \right\}.$$

For rotation inversions with even orders of symmetry one obtains the same irreducible representations (the PSGs n and $\bar{n}$ are isomorphic). In the case of odd orders of symmetry, one makes use of the fact that the group $\bar{n} = n \times \bar{1}$. This results in $2n$ irreducible representations when one writes the identity element

(1) and the element ($\bar{1}$) for the group $\bar{1}$. The irreducible representations then follow directly from the product of the representations of n and $\bar{1}$, because the one-dimensional representations are identical with the corresponding character values.

We will illustrate the procedure using the PSGs 3m, 23, and m3 as examples. As already shown, 3m has six elements divided into three classes[1]:

$$K_1 : g_1 = e;$$
$$K_2 : g_2 = 3^1, g_3 = 3^2;$$
$$K_3 : g_4 = m_1, g_5 = m_2, g_6 = m_3.$$

The dimensions of the irreducible representations result from $\sum_{s=1}^{3} l_s^2 = 6 = 1+1+4$; thus there exists two one-dimensional representations and one two-dimensional irreducible representation. The table below gives the values $\chi(\Gamma_2) = (1,1,\bar{1})$ for the character values of Γ_2 and $\chi(\Gamma_3) = (2,\bar{1},0)$ for the two-dimensional representation. Consequently, the complete character table for the PSG 3m looks like

	e	$(3^1, 3^2)$	(m_1, m_2, m_3)
$\chi(\Gamma_1)$	1	1	1
$\chi(\Gamma_2)$	1	1	$\bar{1}$
$\chi(\Gamma_3)$	2	$\bar{1}$	0

The associated irreducible representations can be calculated with the help of the above table when one generates the elements g_5 and g_6 from g_4, as demonstrated above:

	$g_1 = e$	$g_2 = 3^1$	$g_3 = 3^2$	$g_4 = m_1$	$g_5 = m_2$	$g_6 = m_3$
Γ_1	1	1	1	1	1	1
Γ_2	1	1	1	$\bar{1}$	$\bar{1}$	$\bar{1}$
Γ_3	$\begin{pmatrix} 1 & 0 \\ 0 & 1 \end{pmatrix}$	$\begin{pmatrix} -1/2 & \sqrt{3}/2 \\ -\sqrt{3}/2 & -1/2 \end{pmatrix}$	$\begin{pmatrix} -1/2 & -\sqrt{3}/2 \\ \sqrt{3}/2 & -1/2 \end{pmatrix}$	$\begin{pmatrix} \bar{1} & 0 \\ 0 & 1 \end{pmatrix}$	$\begin{pmatrix} 1/2 & -\sqrt{3}/2 \\ -\sqrt{3}/2 & -1/2 \end{pmatrix}$	$\begin{pmatrix} 1/2 & \sqrt{3}/2 \\ \sqrt{3}/2 & -1/2 \end{pmatrix}$

At this point we note that all groups $n2$ or nm with arbitrary n-fold axis possess easily manageable character tables and irreducible representations. If n is even we have the classes K_1: e; K_2: n^1, n^{n-1}; K_3: n^2, n^{n-2}; ... $K_{n/2+1}$: $n^{n/2}$; $K_{n/2+2}$: $2'$ ($n/2$ times) or m$'$ ($n/2$ times); $K_{n/2+3}$: $2''$ ($n/2$ times) or m$''$ ($n/2$ times). $2'$ and m$'$ are the generating elements of the group $n2$ and nm, respectively; $2''$ and m$''$ are further symmetry elements bisecting the angles between the axes $2'$ and the symmetry planes m$'$, respectively. Thus a total of $n/2+3$ classes exist. From $\sum_{s=1}^{S} l_s^2 = h = 2n$ follows the unique decomposition $2n = 4 + 4(n-2)/2$, i.e., there exists four one-dimensional irreducible representations and $(n-2)/2$ two-dimensional representations.

1) In the following we use the symbol n^q as an abbreviation for $\boldsymbol{R}_n^q$.

If n is odd, then the groups $n2$ and nm, respectively, possess the following classes: K_1: e, K_2: n^1, n^{n-1}; K_3: n^2, n^{n-2}; ...; $K_{(n+1)/2}$: $n^{(n-1)/2}, n^{(n+1)/2}$; $K_{(n+1)/2+1}$: 2 (n-times) and m (n-times) respectively. Thus there exists $(n+1)/2+1 = (n+3)/2$ classes. With $\sum_{s=1}^{S} l_s^2 = h = 2n$, one obtains uniquely $2n = 2 + 4(n-1)/2$ and hence, two one-dimensional irreducible representations and $(n-1)/2$ two-dimensional irreducible representations. The construction of the character table for these groups and the complete determination of the associated irreducible representations for arbitrary n is unproblematic.

We now come to the PSG 23, possessing two threefold axes running in the direction of the space diagonals of a cube, as the generating symmetry operations. The elements of the group, aside from the identity element, are the three twofold axes along the edges of the cube, the four threefold axes 3^1 along the space diagonals and the four threefold axes with an angle of rotation of 240°, which we denote as 3^2. The group contains a total of 12 elements. From the relation $\sum_{s=1}^{S} l_s^2 = 12$, we recognize the following:

If a three-dimensional representation exists, then there must also exist three one-dimensional representations. In the case of a two-dimensional representation we would have eight additional one-dimensional representations and in the case of two two-dimensional representations, a further four one-dimensional representations. From the equivalent symmetry operations mentioned above it clearly emerges that only four classes exist. Thus only the first alternative comes into question: three one-dimensional representations and one three-dimensional representation.

The characters of the nontrivial one-dimensional representations of the classes $K_3 = 3^1$ (4-times) and $K_4 = 3^2$ (4-times) are found by setting the values $e^{2\pi i m/3} = 1$.

We then obtain

$$\chi(\Gamma_2) = (1, 1, e^{2\pi i/3}, e^{2\pi i 2/3})$$

and

$$\chi(\Gamma_3) = (1, 1, e^{2\pi i 2/3}, e^{2\pi i 4/3}).$$

The character values of the three-dimensional representation are obtained by using the values from our table and we find

$$(\Gamma_4) = (3, -1, 0, 0).$$

Checking the orthogonality relation confirms the correctness.

The associated irreducible representations are found directly from the matrix representation of the symmetry elements of the group when one performs the respective transformations. For Γ_4 the result is

	e	$2' \parallel [100]$	$2'' \parallel [010]$	$2''' \parallel [001]$
Γ_4	$\begin{pmatrix} 1 & 0 & 0 \\ 0 & 1 & 0 \\ 0 & 0 & 1 \end{pmatrix}$	$\begin{pmatrix} 1 & 0 & 0 \\ 0 & \bar{1} & 0 \\ 0 & 0 & \bar{1} \end{pmatrix}$	$\begin{pmatrix} \bar{1} & 0 & 0 \\ 0 & 1 & 0 \\ 0 & 0 & \bar{1} \end{pmatrix}$	$\begin{pmatrix} \bar{1} & 0 & 0 \\ 0 & \bar{1} & 0 \\ 0 & 0 & 1 \end{pmatrix}$
	$3^{1'} \parallel [111]$	$3^{1''} \parallel [\bar{1}\bar{1}1]$	$3^{1'''} \parallel [1\bar{1}\bar{1}]$	$3^{1''''} \parallel [\bar{1}1\bar{1}]$
	$\begin{pmatrix} 0 & 1 & 0 \\ 0 & 0 & 1 \\ 1 & 0 & 0 \end{pmatrix}$	$\begin{pmatrix} 0 & 1 & 0 \\ 0 & 0 & \bar{1} \\ \bar{1} & 0 & 0 \end{pmatrix}$	$\begin{pmatrix} 0 & \bar{1} & 0 \\ 0 & 0 & 1 \\ \bar{1} & 0 & 0 \end{pmatrix}$	$\begin{pmatrix} 0 & \bar{1} & 0 \\ 0 & 0 & \bar{1} \\ 1 & 0 & 0 \end{pmatrix}$
	$3^{2'} \parallel [111]$	$3^{2''} \parallel [\bar{1}\bar{1}1]$	$3^{2'''} \parallel [1\bar{1}\bar{1}]$	$3^{2''''} \parallel [\bar{1}1\bar{1}]$
	$\begin{pmatrix} 0 & 0 & 1 \\ 1 & 0 & 0 \\ 0 & 1 & 0 \end{pmatrix}$	$\begin{pmatrix} 0 & 0 & \bar{1} \\ 1 & 0 & 0 \\ 0 & \bar{1} & 0 \end{pmatrix}$	$\begin{pmatrix} 0 & 0 & \bar{1} \\ \bar{1} & 0 & 0 \\ 0 & 1 & 0 \end{pmatrix}$	$\begin{pmatrix} 0 & 0 & 1 \\ \bar{1} & 0 & 0 \\ 0 & \bar{1} & 0 \end{pmatrix}$

The irreducible representations and the respective character vectors for the PSG m3 are obtained from the Kronecker product according to m3 $= 23 \times \bar{1}$. The 24 elements are distributed over eight classes, of which the first four are identical with those of the PSG 23, and the second four result simply from the first four by multiplication with the element $g_{13} = \bar{1}$. Formally, we can represent a group G, formed from a group G' by a Kronecker product with the group $\bar{1} = (g_1 = 1, g_2 = -1)$ as $G = G' \times \bar{1} = G'g_1 + G'g_2$. The character table then has the form

	$K(G')$	$K(G'g_2)$
$\chi(\Gamma(G'))$	$\chi(G')$	$\chi(G')$
$\chi(\Gamma(G'g_2))$	$\chi(G')$	$-\chi(G')$

$K(G')$ means the classes of G' and $K(G'g_2)$ means the classes of $G'g_2$. Correspondingly, $\Gamma(G')$ and $\Gamma(G'g_2)$ are the irreducible representations of G' and $G'g_2$, respectively.

Exercises 37 to 40 provide further practice.

As we have seen from the concrete examples, each individual irreducible representation includes certain partial aspects of the group law. The irreducible representations are therefore associated with the concept of symmetry types referred to the respective group. For example, the identity representation Γ_1 contains absolutely no specific group properties. It is therefore called as totally symmetric. The decomposition of an arbitrary representation into irreducible representations sheds light on the analysis of different symmetry types.

8.3 Tensor Representations

If one carries out an arbitrary symmetry operation, one finds that certain tensor components transform among themselves, independent of the position and the order of symmetry of the rotation axis or the rotation–inversion axis. This is illustrated by taking a general second-rank tensor as an example. We already know the scalar invariant

$$I = t_{ij}\delta_{ij} = t_{11} + t_{22} + t_{33} = t'_{11} + t'_{22} + t'_{33}.$$

A vector invariant is the vector

$$\begin{aligned} \boldsymbol{t} &= t_{ij}e_{ijk}\boldsymbol{e}_k \\ &= (t_{23} - t_{32})\boldsymbol{e}_1 + (t_{31} - t_{13})\boldsymbol{e}_2 + (t_{12} - t_{21})\boldsymbol{e}_3 \\ &= \boldsymbol{t}', \end{aligned}$$

where e_{ijk} are the components of the Levi-Cività tensor. The components of $\boldsymbol{t}$ transform like the coordinates, i.e., only the components of $\boldsymbol{t}$ appear in $\boldsymbol{t}'$.

If we consider a general mth-rank tensor in three-dimensional space, not subject to secondary conditions, we then have 3^m independent components, which we can conceive as the coordinates of a 3^m-dimensional space. The second-rank tensor is then represented as a vector of the form $\boldsymbol{T} = t_i\boldsymbol{e}_i$, where, for example $t_{11} = t_1$, $t_{22} = t_2$, $t_{33} = t_3$, $t_{23} = t_4$, $t_{31} = t_5$, $t_{12} = t_6$, $t_{32} = t_7$, $t_{13} = t_8$ and $t_{21} = t_9$. We now go over to a new reference system with the basic vectors $\boldsymbol{e}_i^0$, which are orthogonal, hence, $\boldsymbol{e}_i^0 \cdot \boldsymbol{e}_j^0 = \delta_{ij}$, and so selected that the invariants I and $\boldsymbol{t}$ fix the first four of these basic vectors, namely

$$\begin{aligned} \boldsymbol{e}_1^0 &= (\boldsymbol{e}_1 + \boldsymbol{e}_2 + \boldsymbol{e}_3)/\sqrt{3} \\ \boldsymbol{e}_2^0 &= (\boldsymbol{e}_4 - \boldsymbol{e}_7)/\sqrt{2} \\ \boldsymbol{e}_3^0 &= (\boldsymbol{e}_5 - \boldsymbol{e}_8)/\sqrt{2} \\ \boldsymbol{e}_4^0 &= (\boldsymbol{e}_6 - \boldsymbol{e}_9)/\sqrt{2}. \end{aligned}$$

For the remaining five new basic vectors $\boldsymbol{e}_i^0 = a_{ij}\boldsymbol{e}_j$, the orthogonality to the first four demands $\boldsymbol{e}_i^0 \cdot \boldsymbol{e}_k^0 = 0$ for $i \neq k$, hence, $a_{i1} + a_{i2} + a_{i3} = 0$ as well as

$$a_{i4} - a_{i7} = a_{i5} - a_{i8} = a_{i6} - a_{i9} = 0.$$

These relationships are fulfilled, for example, by the following mutually orthogonal vectors:

$$\begin{aligned}
e_5^0 &= (e_1 + e_2 - 2e_3)/\sqrt{6} \\
e_6^0 &= (e_1 - e_2)/\sqrt{2} \\
e_7^0 &= (e_4 + e_7)/\sqrt{2} \\
e_8^0 &= (e_5 + e_8)/\sqrt{2} \\
e_9^0 &= (e_6 + e_9)/\sqrt{2}.
\end{aligned}$$

In the abbreviated notation we have $\boldsymbol{e}_i^0 = D_{ij}\boldsymbol{e}_j$ and correspondingly $t_i^0 = D_{ij}t_j$. In this representation I, the vector $\boldsymbol{t}$ as well as each vector composed of the basic vectors $\boldsymbol{e}_5^0$ to $\boldsymbol{e}_9^0$ remain within the space spanned by $\boldsymbol{e}_1^0, \boldsymbol{e}_2^0, \boldsymbol{e}_3^0, \boldsymbol{e}_4^0$ or $\boldsymbol{e}_5^0, \boldsymbol{e}_6^0, \boldsymbol{e}_7^0, \boldsymbol{e}_8^0, \boldsymbol{e}_9^0$ during the transformation (rotation or rotation inversion). This means that the space spanned by the new basic vectors is resolved into three subspaces of dimensions 1, 3, and 5. We call these subspaces the invariant linear subspaces of the corresponding 3^m-dimensional space formed by an mth-rank tensor of that three-dimensional space. The search for these subspaces is carried out in a manner analogous to the above example, even for tensors of higher rank, whereby a knowledge of the tensor invariants is of valuable help. Conversely, resolving a tensor into its invariant linear subspaces allows an overview of the tensor invariants, which in many cases are amenable to a direct geometric interpretation.

A transformation of the basic system, which we introduced with the definition of the tensor concept, that is

$$t'_{ij} = u_{ii^*}u_{jj^*}t_{i^*j^*},$$

takes the form $t'_i = R_{ij}t_j$ or abbreviated $\boldsymbol{t}' = \boldsymbol{R}\boldsymbol{t}$ in the respective linear vector space. The matrix $\boldsymbol{R}$ is derived from the three-dimensional transformation matrix $\boldsymbol{U}$, which carries over the Cartesian basic vectors $\boldsymbol{e}_i$ to the new basic vectors $\boldsymbol{e}'_i$ according to $\boldsymbol{e}'_i = u_{ii^*}\boldsymbol{e}_{i^*}$. From the definition given above for the inner Kronecker product, we recognize straight away that the matrix $\boldsymbol{R}$ is to be understood as the Kronecker product of $\boldsymbol{U}$ with itself. Accordingly, with a mth-rank tensor, we have to use the m-fold Kronecker product of the matrix $\boldsymbol{U}$ for the transformation $\boldsymbol{R}$.

We now have the possibility of calculating the number of independent tensor components for the case that the tensor belongs to a certain symmetry group. The tensor, even in the linear vector space representation, is carried over by the respective symmetry operations into itself. The number of independent components must be equal to the dimension of the linear vector space which the tensor takes up because of symmetry operations or other conditions. For example, the three-dimensional subspace spanned by $\boldsymbol{e}_2^0$, $\boldsymbol{e}_3^0$,

e_4^0 remains empty for the case of a symmetric second-rank tensor, i.e., a symmetric second-rank tensor only possesses the one-dimensional and the five-dimensional invariant linear subspace.

The symmetry operations of the given symmetry group carry the tensor over into itself, i.e., the individual tensor components experience an identical transformation, which in the representation of the matrices $\boldsymbol{R}$ must be conserved as identity representations. Consequently, the number of independent tensor components is equal to the number of identity representations conserved in the representation by the matrices $\boldsymbol{R}$. This number, according to the rules discussed above, is

$$m_1 = \frac{1}{h}\sum_{s=1}^{S} n_s \chi(s)\chi_1(s) \quad \text{and with } \chi_1(s) = 1:$$

$$m_1 = \frac{1}{h}\sum_{s=1}^{S} n_s \chi(s) \quad (n_s \text{ number of elements in the } s\text{-th class}).$$

As an example, consider the third-rank tensor t_{ijk} (without permutability of the indices) in the PSG 3m. The character of the $\boldsymbol{R}$-representation is obtained as the third power of the character value of the three-dimensional representation Γ_3' of the PSG 3m. It is $\Gamma_3' = \Gamma_1 + \Gamma_3$ (Γ_3 is two-dimensional!), hence,

$$\chi(\Gamma_3') = \chi(\Gamma_1) + \chi(\Gamma_3) = (3, 0, 1).$$

Thus

$$\chi(\Gamma_3' \times \Gamma_3' \times \Gamma_3') = (27, 0, 1) \quad \text{and} \quad m_1 = \frac{1}{6}(27 + 3) = 5.$$

This result is obtained far more laboriously as with the method of symmetry reduction discussed in Section 4.4. However, the other way of determining the independent tensor components with the methods of the linear vector space representation of the tensor in linear vector space and the explicit calculation of the invariant subspaces is not essentially different from those discussed in Section 3.8. Nevertheless, there does not exist a simpler method as that just discussed to calculate the number of independent tensor components, particularly with tensors of higher rank. In this regard, let us consider a further example of a ninth-rank tensor (without permutability of the indices) in the PSG 3m. The character values of the representation of the $\boldsymbol{R}$-matrices are the ninth powers of the character of the representation Γ_3'. Hence, $\chi(\boldsymbol{R}) = (3^9, 0, 1^9)$ and thus

$$m_1 = \frac{1}{6}(19683 + 3) = 3281.$$

We now have to investigate the effects of secondary conditions, such as the permutability of indices, which can arise, for example, from physical reasons in almost all tensors describing physical properties (most second-rank tensors, piezoelectric tensor, elasticity tensor etc.).

In many cases one can obtain the transformation matrix $\boldsymbol{R}$ of the tensors through the Kronecker products $\boldsymbol{R} = \boldsymbol{A} \times \boldsymbol{B}$. The associated character values can be calculated with the help of the product rule $\chi(\boldsymbol{A} \times \boldsymbol{B}) = \chi(\boldsymbol{A})\chi(\boldsymbol{B})$, where $\boldsymbol{A}$ and $\boldsymbol{B}$ may themselves be Kronecker products.

The situation in the case of the permutability of indices is less clear. A general derivation of the valid relations is found, for example, in Ljubarski (1962). For our purposes, it suffices to discuss practical applications for some important cases. Firstly, we consider the permutability of indices of second-rank tensors ($t_{ij} = t_{ji}$, second-rank symmetric tensors). To calculate the character of $\boldsymbol{R}$ we need only sum the factors in the principal diagonal of the transformation table:

$$
\begin{aligned}
t_{11} &= u_{11}u_{11}t_{11} + \cdots \\
t_{22} &= \cdots + u_{22}u_{22}t_{22} + \cdots \\
t_{33} &= \cdots + u_{33}u_{33}t_{33} + \cdots \\
t_{12} = t_{21} &= \cdots + (u_{11}u_{22} + u_{12}u_{21})t_{12} + \cdots \\
t_{23} = t_{32} &= \cdots + (u_{22}u_{33} + u_{23}u_{32})t_{23} + \cdots \\
t_{31} = t_{13} &= \cdots + (u_{33}u_{11} + u_{31}u_{13})t_{31} + \cdots
\end{aligned}
$$

hence,

$$
\begin{aligned}
\chi(\boldsymbol{R}) = \chi_{(3\times3)_s}(\boldsymbol{R}) &= \sum_{i=1} u_{ii}^2 + \frac{1}{2}\sum_{i\neq j}(u_{ii}u_{jj} + u_{ij}u_{ji}) \\
&= \frac{1}{2}\sum_{i,j} u_{ij}u_{ji} + \frac{1}{2}\left(\sum_i u_{ii}\right)^2 .
\end{aligned}
$$

As one can immediately verify with

$$
\boldsymbol{U} = \begin{pmatrix} \cos\varphi & \sin\varphi & 0 \\ -\sin\varphi & \cos\varphi & 0 \\ 0 & 0 & 1 \end{pmatrix},
$$

the first sum is just the half trace of $\boldsymbol{U}^2$. The second sum is half the square of the trace of $\boldsymbol{U}$. Hence,

$$
\chi_{(3\times3)_s}(\boldsymbol{R}) = \frac{1}{2}\chi_3(\boldsymbol{U}^2) + \frac{1}{2}\chi_3^2(\boldsymbol{U}),
$$

where the symbols $(3\times3)_s$ denote the symmetric second-rank tensor, 3×3 denote the general second-rank tensor and corresponding higher tensors. For example, $(3 \times 3)_s \times (3 \times 3)$ means a fourth rank tensor in which two index positions are permutable.

Since in this connection we must always use the three-dimensional representation for $\boldsymbol{U}$, it is convenient to note the characters for the simple products: $\chi(\boldsymbol{U}) = \pm(1 + 2\cos\varphi)$ with $\varphi = 2\pi/n$. $+$ specifies rotation operations, $-$ specifies rotation-inversion. Furthermore, $\chi_3(\boldsymbol{U}^2) = (1 + 2\cos 2\varphi)$ and thus

$$\chi_{(3\times3)_s}(\boldsymbol{R}) = 2\cos\varphi + 4\cos^2\varphi \quad \text{for } n \text{ and } \bar{n}.$$

Again, we take the PSG 3m as an example. In this PSG, a symmetric second-rank tensor has

$$m_1 = \frac{1}{6}(6 + 2\cdot 0 + 3\cdot(-2+4)) = 2 \quad \text{independent components.}$$

For $\boldsymbol{U}$ we have written the three-dimension representation Γ_3' already given above.

As the next example, we consider the piezoelectric tensor and other third-rank tensors in which two index positions are permutable, hence, $t_{ijk} = t_{ikj}$. We use the product formula to calculate the character values

$$\begin{aligned}\chi_{3\times(3\times3)_s} &= \chi_3(\boldsymbol{U})\chi_{(3\times3)_s}(\boldsymbol{U})\\ &= \pm(1+2\cos\varphi)(2\cos\varphi + 4\cos^2\varphi)\\ &= \pm 2(\cos\varphi + 4\cos^2\varphi + 4\cos^3\varphi),\end{aligned}$$

where we write $+$ for n and $-$ for $\bar{n}$.

In the PSG 3m, we then have $\chi(\boldsymbol{R}) = (18, 0, 2)$ and thus

$$m_1 = \frac{1}{6}(18 + 6) = 4 \quad \text{independent components}$$

in agreement with the result in Section 4.4.1.

To calculate $\chi(\boldsymbol{R})$ in the case of three and more mutually permutable index positions we proceed as in the case of second-rank tensors and construct the respective transformation formulae. The result is then, for example,

$$\chi_{(3\times3\times3)_s} = \pm 2\cos\varphi(-1 + 2\cos\varphi + 4\cos^2\varphi)$$

with $+$ for n and $-$ for $\bar{n}$ as well as

$$\chi_{(3\times3\times3\times3)_s} = 1 - 2\cos\varphi - 8\cos^2\varphi + 8\cos^3\varphi + 16\cos^4\varphi$$

for n and $\bar{n}$.

Table 8.3 presents a compilation of the more important formulae for the calculation of the characters of tensor representations (polar tensors).

Table 8.3 Characters of tensor representations.

Tensor	Character $\chi(\boldsymbol{R})$ (+ for n, − for $\bar{n}$)
t_i	$\chi_3 = \pm(1+2\cos\varphi)$
t_{ij}	$\chi_{(3\times3)} = (1+2\cos\varphi)^2$
$t_{ij\ldots s}$ (mth rank)	$\chi_{(3\times3\times\ldots3)} = (\pm1)^m(1+2\cos\varphi)^m$
$t_{ij} = t_{ji}$ (totally symmetric)	$\chi_{(3\times3)_s} = \frac{1}{2}\chi_3(\boldsymbol{U}^2) + \frac{1}{2}\chi_3^2(\boldsymbol{U}) = 2\cos\varphi(1+2\cos\varphi)$
$t_{ijk} = t_{ikj}$ (interchangeable within one pair)	$\chi_{3\times(3\times3)_s} = \chi_3\chi_{(3\times3)_s} = \pm2\cos\varphi(1+2\cos\varphi)^2$
t_{ijk} (totally symmetric)	$\chi_{(3\times3\times3)_s} = \frac{1}{3}\chi_3(\boldsymbol{U}^3) + \frac{1}{2}\chi_3(\boldsymbol{U}^2)\chi_3(\boldsymbol{U}) + \frac{1}{6}\chi_3^3(\boldsymbol{U}) = \pm2\cos\varphi(-1+2\cos\varphi+4\cos^2\varphi)$
$t_{ijkl} = t_{ijlk}$	$\chi_{(3\times3)\times(3\times3)_s} = \chi_{3\times3}\chi_{(3\times3)_s} = 2\cos\varphi(1+2\cos\varphi)^3$
$t_{ijkl} = t_{klij}$ (pairwise interchangeable)	$\chi_{((3\times3)\times(3\times3))_s} = \frac{1}{2}\chi_{(3\times3)}(\boldsymbol{U}^2) + \frac{1}{2}\chi^2_{(3\times3)}(\boldsymbol{U}) = 1+4\cos\varphi(1+2\cos\varphi+4\cos^2\varphi+4\cos^3\varphi)$
$t_{ijkl} = t_{jikl} = t_{ijlk} = t_{jilk}$ (interchangeable within pairs)	$\chi_{(3\times3)_s\times(3\times3)_s} = \chi^2_{(3\times3)_s} = 4\cos^2\varphi(1+2\cos\varphi)^2$
$t_{ijkl} = t_{iklj} = t_{ilkj} = \ldots$ (interchangeable within a triple)	$\chi_{(3)\times(3\times3\times3)_s} = \chi_3\chi_{(3\times3\times3)_s} = 2\cos\varphi(1+2\cos\varphi)(-1+2\cos\varphi+4\cos^2\varphi)$
$t_{ijkl} = t_{ijlk} = t_{lkij} = \ldots$ (pairwise and within pairs interchangeable)	$\chi_{((3\times3)_s\times(3\times3)_s)_s} = \frac{1}{2}\chi_{(3\times3)_s}(\boldsymbol{U}^2) + \frac{1}{2}\chi^2_{(3\times3)_s}(\boldsymbol{U})$ $= 1+4\cos^2\varphi(-1+2\cos\varphi+4\cos^2\varphi)$
t_{ijkl} (totally symmetric)	$\chi_{(3\times3\times3\times3)_s} = 1+2\cos\varphi(-1-4\cos\varphi+4\cos^2\varphi+8\cos^3\varphi)$
t_{ijklm} (totally symmetric)	$\chi_{(3\times3\times3\times3\times3)_s} = \pm1\pm4\cos\varphi(1-2\cos\varphi-6\cos^2\varphi+4\cos^3\varphi+8\cos^4\varphi)$
$t_{ijklmn} = t_{ijklnm} = t_{ijlknm} = \ldots$ (interchangeable within pairs)	$\chi_{(3\times3)_s\times(3\times3)_s\times(3\times3)_s} = \chi^3_{(3\times3)_s}(\boldsymbol{U}) = 8\cos^3\varphi(1+2\cos\varphi)^3$
$t_{ijklmn} = t_{klijmn} = t_{klijnm} = \ldots$ (pairwise and within pairs interchangeable)	$\chi_{((3\times3)_s\times(3\times3)_s\times(3\times3)_s)_s} = \frac{1}{3}\chi_{(3\times3)_2}(\boldsymbol{U}^3) + \frac{1}{2}\chi_{(3\times3)_s}(\boldsymbol{U}^2)\chi_{(3\times3)_s}(\boldsymbol{U}) + \frac{1}{6}\chi^3_{(3\times3)_s}(\boldsymbol{U})$ $= 8\cos^2\varphi(2-\cos\varphi-6\cos^2\varphi+4\cos^3\varphi+8\cos^4\varphi$

With axial tensors (pseudo tensors), $\chi(\boldsymbol{R})$ must be furnished with an additional factor (-1) when applying a symmetry operation $\bar{n}$.

All further cases are calculated in an analogous fashion. The formulae for the totally symmetric sixth and higher rank tensors can be easily derived by the reader by summing the coefficients in the principal diagonals of the system of the transformation formulae (see also Exercise 7)

$$t'_{ij\ldots s} = t_{ij\ldots s} = u_{ii}u_{jj}\ldots u_{ss}t_{ij\ldots s},$$

where

$$\boldsymbol{U} = \begin{pmatrix} \cos\varphi & \sin\varphi & 0 \\ -\sin\varphi & \cos\varphi & 0 \\ 0 & 0 & 1 \end{pmatrix}$$

as previously.

8.4
Decomposition of the Linear Vector Space into Invariant Subspaces

We now return to the decomposition of a tensor into invariant linear subspaces. Since, in a unitary transformation, the components spanning a certain invariant subspace again join to an invariant subspace of the same dimension, the transformation matrices can be split into irreducible blocks. In doing so, the basic vectors are selected corresponding to the invariant subspaces, as we explained on hand of the example of the second-rank tensor (see section 8.3). Accordingly, the characters (traces) of the transformation matrices are equal to the sum of the characters of the transformations in the individual subspaces. If one knows the character values for a tensor representation $\chi(\boldsymbol{R})$, then one can extract the nature of the subspace formed by the given tensor. It is $\chi(\boldsymbol{R}) = \sum_j m_j\chi_j(\boldsymbol{R})$. $\chi_j(\boldsymbol{R})$ is the character of the transformation matrix for the j-dimensional subspace for a certain symmetry operation $\boldsymbol{U}$. Accordingly, the tensor representation is then $\Gamma(\boldsymbol{R}) = \sum_j m_j\Gamma_j(\boldsymbol{R})$.

The following matrices furnish a $(2l+1)$-dimensional representation of the group of n-fold rotation axis:

$$\Gamma_{q;2l+1}(\boldsymbol{R}) = \begin{pmatrix} e^{ilq\varphi} & 0 & & \vdots & \\ 0 & e^{i(l-1)q\varphi} & & \cdots 0 \cdots & \\ & \vdots & \ddots & \vdots & \\ \cdots & 0 & \cdots & e^{-i(l-1)q\varphi} & 0 \\ & \vdots & & 0 & e^{-ilq\varphi} \end{pmatrix},$$

where $\varphi = 2\pi/n$. Allowing q to run through the values $0, 1, \ldots, n-1$, results in n matrices, which fulfil the group properties of the existent rotation group.

This representation has the character

$$\chi_{2l+1}(\boldsymbol{R}^q) = \sum_{m=-l}^{m=+l} e^{imq\varphi} = 1 + 2\sum_{m=+1}^{m=+l} \cos mq\varphi = \frac{\sin(2l+1)q\varphi/2}{\sin q\varphi/2}.$$

The simplest way of deriving this relation is with the help of a proof by induction or through the summation formula for a geometric series

$$\sum_{m=-l}^{m=+l} e^{imq\varphi} = \frac{e^{i(l+1)q\varphi} - e^{-ilq\varphi}}{e^{iq\varphi} - 1} = \frac{(e^{i(l+1)q\varphi} - e^{-ilq\varphi})(e^{-iq\varphi} + 1)}{(e^{iq\varphi} - 1)(e^{-iq\varphi} + 1)}$$

$$= \frac{\sin(l+1)q\varphi + \sin lq\varphi}{\sin q\varphi} = \frac{\sin(2l+1)q\varphi/2}{\sin q\varphi/2}$$

(with $\sin\alpha + \sin\beta = 2\sin(\alpha+\beta)/2 \cdot \cos(\alpha-\beta)/2$). In practice, it is convenient to expand the cos terms in powers of $\cos\varphi$. Then one obtains the following values for $q = 1$:

$$\begin{aligned}
\chi_1(\boldsymbol{R}) &= 1 \\
\chi_3(\boldsymbol{R}) &= 1 + 2\cos\varphi \\
\chi_5(\boldsymbol{R}) &= -1 + 2\cos\varphi + 4\cos^2\varphi \\
\chi_7(\boldsymbol{R}) &= -1 - 4\cos\varphi + 4\cos^2\varphi + 8\cos^3\varphi \\
\chi_9(\boldsymbol{R}) &= 1 - 4\cos\varphi - 12\cos^2\varphi + 8\cos^3\varphi + 16\cos^4\varphi \\
\chi_{11}(\boldsymbol{R}) &= 1 + 6\cos\varphi - 12\cos^2\varphi - 32\cos^3\varphi + 16\cos^4\varphi + 32\cos^5\varphi \\
\chi_{13}(\boldsymbol{R}) &= -1 + 6\cos\varphi + 24\cos^2\varphi - 32\cos^3\varphi - 80\cos^4\varphi \\
&\quad + 32\cos^5\varphi + 64\cos^6\varphi \\
\chi_{15}(\boldsymbol{R}) &= -1 - 8\cos\varphi + 24\cos^2\varphi + 80\cos^3\varphi - 80\cos^4\varphi - 192\cos^5\varphi \\
&\quad + 64\cos^6\varphi + 128\cos^7\varphi
\end{aligned}$$

and so on.

As an example, we consider the tensor $t_{ijk} = t_{ikj}$ (for example, the piezoelectric tensor) and the tensor $t_{ijkl} = t_{klij} = t_{jikl} = \ldots$ (for example, the elasticity tensor). Because

$$\chi_{3\times(3\times3)_s} = 2\cos\varphi + 8\cos^2\varphi + 8\cos^3\varphi$$

(see Table 8.3) one obtains the unique decomposition

$$\chi_{3\times(3\times3)_s}(\boldsymbol{R}) = 2\chi_3(\boldsymbol{R}) + \chi_5(\boldsymbol{R}) + \chi_7(\boldsymbol{R}).$$

One begins by assigning the respective highest power to the corresponding $\chi_{2l+1}(\boldsymbol{R})$. The result is that there does not exist a scalar invariant (one-dimensional subspace), rather, two three-dimensional and a five- and seven-dimensional invariant subspace. In the other example we have

$$\chi_{((3\times3)_s\times(3\times3)_s)_s}(\boldsymbol{R}) = 2\chi_1(\boldsymbol{R}) + 2\chi_5(\boldsymbol{R}) + \chi_9(\boldsymbol{R}).$$

Here we have two scalar (one-dimensional) invariants as well as two five-dimensional subspaces and one nine-dimensional subspace. For the elasticity tensor this means, for example, that except for the dynamic elasticity $\bar{E} = \sum_{i,j} c_{ijij} = c_{ijkl}\delta_{ik}\delta_{jl}$ and the trace $\sum_{m=1}^{3} g_{mm}$ of the second-rank tensor describing the deviation from the Cauchy relations, where $g_{mm} = 1/2 e_{mik}e_{njl}c_{ijkl}$, that no further scalar invariants exist that cannot be formed from these two. The same applies accordingly to both second-rank tensor invariants, the dynamic elasticity, and the deviations from the Cauchy relations (see Sections 4.5.1 and 4.5.5).

8.5
Symmetry Matched Functions

For the description of the directional dependence of properties and even of functions of the position coordinates or a wave function of a quantum mechanical system, one can employ upon a system of functions which in themselves obey a certain point symmetry. Of particular interest are homogeneous polynomials in the coordinates x_1, x_2, x_3 of the Cartesian reference system. These polynomials can also be represented as functions of the angles ξ and η with the aid of polar coordinates. ξ is the angle between the position vector $\boldsymbol{x} = x_i\boldsymbol{e}_i$ and the vector $\boldsymbol{e}_3$ and η is the angle between the projection of the position vector on the plane spanned by $\boldsymbol{e}_1$ and $\boldsymbol{e}_2$ and the vector $\boldsymbol{e}_1$. Accordingly,

$$x_1 = |\boldsymbol{x}|\cos\eta\sin\xi, \quad x_2 = |\boldsymbol{x}|\sin\eta\sin\xi$$

and

$$x_3 = |\boldsymbol{x}|\cos\xi \quad \text{with} \quad |\boldsymbol{x}| = \sqrt{x_1^2 + x_2^2 + x_3^2}.$$

If a symmetry operation n or $\bar{n}$ exists, then each polynomial of the lth degree $\Theta = A_{ijk\ldots s}x_i x_j \ldots x_s$ is carried over in an identical form by the respective transformation, i.e., one has $A'_{ijk\ldots s} = A_{ijk\ldots s}$. Hence, these polynomials transform as the quadric of degree l.

The scheme of independent coefficients for each polynomial of the lth degree and the relationships of the coefficients among one another corresponds to the conditions for the components of a totally symmetric lth-rank tensor in the respective PSG. For example, a third-degree polynomial has the following form in the PSG 3m

$$\Theta_3 = A_{113}x_1^2x_3 - 3A_{222}x_1^2x_2 + A_{222}x_2^3 + A_{113}x_2^2x_3 + A_{333}x_3^3$$

in accordance with the result of symmetry reduction in the PSG 3m under the condition of total symmetry (see Section 4.4).

We now consider functions dependent only on the direction of the position vector and not on its magnitude and those containing the radial dependence in a separate factor respectively. We then have

$$\Theta = \Theta(r)\Theta(\xi,\eta) \quad \text{with} \quad r = |\boldsymbol{x}|.$$

Analogous to the expansion of a periodic function in a Fourier series, the angular dependent function $\Theta(\xi,\eta)$ can be expanded in a series with respect to the terms of a suitable system of functions $\{F_l\}$ according to

$$\Theta(\xi,\eta) = \sum_{l=0}^{q} A_l F_l.$$

If we demand that F_l is a polynomial of the lth degree in x_i/r and further, that the function $\Theta(\xi,\eta)$ is optimally approximated by the functions $S_q = \sum_{l=0}^{q} A_l F_l$ in each stage q according to the least squares method, i.e., that

$$\Delta q = \int_{\text{sphere}} [\Theta(\xi,\eta) - S_q(\xi,\eta)]^2 d\Omega$$

becomes a minimum for each q, then we come to the system of spherical harmonic functions, well known from potential theory. These functions are solutions of the Laplace differential equation

$$\frac{\partial^2 Y}{\partial x_1^2} + \frac{\partial^2 Y}{\partial x_2^2} + \frac{\partial^2 Y}{x_3^2} = 0.$$

Since this quantity is a scalar invariant of the differential operator $\{\partial^2/\partial x_i \partial x_j\}$, the solutions always have the same form independent of the respective reference system. The spherical harmonic functions are expressed as follows:

$$Y_l^m = N_l^m P_l^m(\zeta) e^{im\eta}.$$

The index m is an integer number and can only take on the values $|m| \leq l$. They obey the orthogonality- and normalization condition

$$\int_{\text{sphere}} Y_l^m \bar{Y}_{l'}^{m'} d\Omega = \int_{\eta=0}^{2\pi} \int_{\xi=0}^{\pi} Y_l^m \bar{Y}_{l'}^{m'} \sin\xi d\xi d\eta = \delta_{ll'} \delta_{mm'}$$

with

$$d\Omega = \sin\xi d\xi d\eta.$$

For P_l^m there exist a simple recursion formula

$$P_l^m = \frac{(1-\zeta^2)^{m/2} d^{l+m} (\zeta^2 - 1)^l}{2^l l! d\zeta^{l+m}},$$

with

$$\zeta = \cos\xi \quad \text{and} \quad (1 - \zeta^2) = \sin^2\xi.$$

The normalization factor N_l^m has the form

$$N_l^m = \left(\frac{(2l+1)(l-m)!}{4\pi(l+m)!} \right)^{1/2}.$$

Hence, we can expand any arbitrary function $\Theta(\xi, \eta)$, which is numerically or analytically known, in a series with respect to the functions Y_l^m, according to

$$\Theta(\xi, \eta) = \sum_{l=0}^{\infty} \sum_{m=-l}^{m=+l} A_{lm} Y_l^m.$$

We obtain the coefficients A_{lm} directly from the orthogonality relation by inserting the series

$$A_{lm} = \int_{\text{sphere}} \Theta(\xi, \eta) \bar{Y}_l^m d\Omega.$$

Some low indexed Y_l^m are presented in Table 8.4.

We now come to discuss two further important properties of the functions Y_l^m. Let us carry out a transformation of the Cartesian reference system according to $e_i' = u_{ij} e_j$, we then have $Y_l^{m'} = R_{mn} Y_l^n$. This means that $Y_l^{m'}$, a polynomial of the lth degree, can be constructed from a linear combination of Y_l^n. This results from the fact that $Y_l^{m'}$ must also be a polynomial of the lth degree, as well as from the expansion formula. A rotation about e_3 with the angle $2\pi/n$ (n-fold rotation axis) carries Y_l^m over in $Y_l^m e^{2\pi i m/n}$, as is directly seen from the definition of Y_l^m. In a rotation-inversion $\bar{n} \parallel e_3$, one must distinguish between the cases "l even" and "l odd." These are

$$Y_l^{m'} = Y_l^m e^{2\pi i m/n} \quad \text{for } l \text{ even and}$$
$$Y_l^{m'} = Y_l^m e^{2\pi i m/n}(-1) \quad \text{for } l \text{ odd.}$$

Thus the spherical harmonics are subject to the following conditions under the existence of symmetry properties (aside from the condition $|m| \leq l$):

n: $e^{2\pi i m/n} = 1$ hence, $m = nq$, q integer,
$\bar{n}$: $e^{2\pi i m/n} = -1$ for l odd, hence, $m = n(2q+1)/2$, q integer
$\bar{n}$: $e^{2\pi i m/n} = 1$ for l even, hence, $m = nq$, q integer, as in the case of n.

From these we read the following rules, which we have come to know in part already in Section 3 with regards to tensor transformations:

1. All homogeneous polynomials of odd order and thus all odd-rank polar tensors vanish for the case that the respective symmetry groups contain a rotation-inversion $\bar{n}$ with odd n ($\bar{1}, \bar{3}, \bar{5}$, and so on).

Table 8.4 Spherical harmonics Y_l^m for $l = 0$ to 3 ($r = \sqrt{x_1^2 + x_2^2 + x_3^2}$; $x_1 = r \sin\xi \cos\eta$; $x_2 = r \sin\xi \sin\eta$; $x_3 = r\cos\xi$).

$$Y_0^0 = \frac{1}{\sqrt{4\pi}}$$

$$Y_1^{-1} = -\sqrt{\frac{3}{8\pi}}\frac{(x_1 - ix_2)}{r}; \quad Y_1^0 = \sqrt{\frac{3}{4\pi}}\frac{x_3}{r}; \quad Y_1^1 = \sqrt{\frac{3}{8\pi}}\frac{(x_1 + ix_2)}{r}$$

$$Y_2^{-2} = \sqrt{\frac{15}{32\pi}}\frac{(x_1 - ix_2)^2}{r^2}; \quad Y_2^{-1} = \sqrt{\frac{15}{8\pi}}\frac{x_3(x_1 - ix_2)}{r^2}; \quad Y_2^0 = \sqrt{\frac{5}{16\pi}}\frac{(3x_3^2 - r^2)}{r^2};$$

$$Y_2^1 = \sqrt{\frac{15}{8\pi}}\frac{x_3(x_1 + ix_2)}{r^2}; \quad Y_2^2 = \sqrt{\frac{15}{32\pi}}\frac{(x_1 + ix_2)^2}{r^2}$$

$$Y_3^{-3} = -\sqrt{\frac{35}{64\pi}}\frac{(x_1 - ix_2)^3}{r^3}; \quad Y_3^{-2} = \sqrt{\frac{105}{32\pi}}\frac{x_3(x_1 - ix_2)^2}{r^3};$$

$$Y_3^{-1} = -\sqrt{\frac{21}{64\pi}}\frac{(5x_3^2 - r^2)(x_1 - ix_2)}{r^3}; \quad Y_3^0 = \sqrt{\frac{7}{16\pi}}\frac{x_3(5x_3^2 - 3r^2)}{r^3};$$

$$Y_3^1 = \sqrt{\frac{21}{64\pi}}\frac{(5x_3^2 - r^2)(x_1 + ix_2)}{r^3}; \quad Y_3^2 = \sqrt{\frac{105}{32\pi}}\frac{x_3(x_1 + ix_2)^2}{r^3}; \quad Y_3^3 = \sqrt{\frac{35}{64\pi}}\frac{(x_1 + ix_2)^3}{r^3}.$$

Real functions can be constructed from simple linear combinations of the complex Y_l^m, e.g.

$$Y_2^2 + Y_2^{-2} = \sqrt{\frac{15}{8\pi}}\frac{(x_1^2 - x_2^2)}{r^2}; \quad i(Y_2^2 - Y_2^{-2}) = -2\sqrt{\frac{15}{8\pi}}\frac{x_1x_2}{r^2};$$

$$i(Y_2^1 + Y_2^{-1}) = -2\sqrt{\frac{15}{8\pi}}\frac{x_3x_2}{r^2}; \quad (Y_2^1 - Y_2^{-1}) = 2\sqrt{\frac{15}{8\pi}}\frac{x_3x_1}{r^2}.$$

In general, for each l a total of $(2l + 1)$ real orthogonal and normalized functions can be constructed according to

$$\frac{1}{\sqrt{2}}(Y_l^m \pm Y_l^{-m})i^{(m-1/2\pm 1/2)} \quad \text{mit} \quad 0 < m \le l \quad \text{und} \quad Y_l^0.$$

This statement can easily be proven from the relation $Y_l^{-m} = (-1)^m \overline{Y_l^m}$ that derives directly from the definition of the Y_l^m.

2. All homogeneous polynomials of even order and thus all even-rank polar tensors have the same form when a rotation axis n or a rotation-inversion $\bar{n}$ of the same order of symmetry exists. This means that polynomials and tensors of even order and rank respectively are "centrosymmetric"!

Furthermore, due of the limitations of m, we can take in at a glance which spherical harmonics, at all, can be found with the existence of an n-fold axis or a rotation-inversion $\bar{n}$. Thus, we can now derive all symmetry matched spherical harmonics for the PSG of type n, $\bar{n}$ and n/m. For the PSG $n2$, $n\mathrm{m}$ and the cubic PSG, the respective conditions for a second distinct direction must also be fulfilled. In this manner, the homogeneous polynomials, allowed for each l (degree of the polynomial), are now accessible for each arbitrary point symmetry group.

From the transformation behavior of Y_l^m we obtain a $(2l+1)$-dimensional representation, which in the case of a rotation axis n assumes the form of the representation $\Gamma_{m;2l+1}(\boldsymbol{R})$ discussed in the previous section. The individual Y_l^m for a respective fixed l represent, to a certain extent, the coordinates of a $(2l+1)$-dimensional vector. Thus, we also recognize the character of the given transformation matrices and are then in the position to apply the known rules of group theory to analyze symmetry matched functions.

As an example, consider the PSG 3m and m3, for which we want to calculate the $(2l+1)$-dimensional representation. We use the formula for the number m_j of the jth irreducible representation Γ_j of the group contained in an arbitrary representation Γ

$$m_j = \frac{1}{h} \sum_{s=1}^{S} n_s \chi(s) \bar{\chi}_j(s)^{\mathrm{irr}}.$$

It is

$$\chi(\boldsymbol{R}) = \pm \frac{\sin(2l+1)\varphi/2}{\sin \varphi/2},$$

where $\varphi = 2\pi/n$; $+$ for n and $\bar{n}$ when l is even and $-$ for $\bar{n}$ when l is odd. The procedure runs as follows:

1. Calculate the characters of the classes for each l with the help of the above formula,

2. Calculate the values m_j and hence, the decomposition

$$\Gamma_{(2l+1)} = \sum_{j=1}^{S} m_j \Gamma_j.$$

The index "$2l+1$" was placed in brackets to distinguish irreducible representations.

Tables 8.5 and 8.6 present the results for the PSGs m3 and 3m.

Table 8.5 Decomposition of the $(2l+1)$-dimensional representations in irreducible representations of the PSG m3 = 23×$\bar{1}$. Classes: $K_1 = e$; $K_2 = 2$ (3×); $K_3 = 3^1$ (4×); $K_4 = 3^2$ (4×); $K_5 = K_1' = \bar{1}$; $K_6 = K_2' = m$ (3×); $K_7 = K_3' : \bar{3}^1$ (4×); $K_8 = K_4' = \bar{3}^2$ (4×).

Characters of irreducible representations

	e	2 (3×)	3^1 (4×)	3^2 (4×)	$\bar{1}$	m (3×)	$\overline{3^1}$ (4×)	$\overline{3^2}$ (4×)
Γ_1	1	1	1	1	1	1	1	1
Γ_2	1	1	$e^{2\pi i/3}$	$e^{2\pi i 2/3}$	1	1	$e^{2\pi i/3}$	$e^{2\pi i 2/3}$
Γ_3	1	1	$e^{2\pi i 2/3}$	$e^{2\pi i 4/3}$	1	1	$e^{2\pi i 2/3}$	$e^{2\pi i 4/3}$
Γ_4	3	$\bar{1}$	0	0	3	$\bar{1}$	0	0
$\Gamma_5 = \Gamma_1'$	1	1	1	1	$\bar{1}$	$\bar{1}$	$\bar{1}$	$\bar{1}$
$\Gamma_6 = \Gamma_2'$	1	1	$e^{2\pi i/3}$	$e^{2\pi i 2/3}$	$\bar{1}$	$\bar{1}$	$-e^{2\pi i/3}$	$-e^{2\pi i 2/3}$
$\Gamma_7 = \Gamma_3'$	1	1	$e^{2\pi i 2/3}$	$e^{2\pi i 4/3}$	$\bar{1}$	$\bar{1}$	$-e^{2\pi i 2/3}$	$-e^{2\pi i 4/3}$
$\Gamma_8 = \Gamma_4'$	3	$\bar{1}$	0	0	$\bar{3}$	1	0	0

Characters $\chi_{(2l+1)}$ of the $(2l+1)$-dimensional representations

	e	2 (3×)	3^1 (4×)	3^2 (4×)	$\bar{1}$	m (3×)	$\overline{3^1}$ (4×)	$\overline{3^2}$ (4×)	$\Gamma_{(2l+1)}$
$\Gamma_{(1)}$	1	1	1	1	1	1	1	1	Γ_1
$\Gamma_{(3)}$	3	$\bar{1}$	0	0	$\bar{3}$	1	0	0	Γ_4'
$\Gamma_{(5)}$	5	1	$\bar{1}$	$\bar{1}$	$\bar{5}$	$\bar{1}$	1	1	$\Gamma_2' + \Gamma_3' + \Gamma_4'$
$\Gamma_{(7)}$	7	$\bar{1}$	1	1	$\bar{7}$	1	$\bar{1}$	$\bar{1}$	$\Gamma_1' + 2\Gamma_4'$
$\Gamma_{(9)}$	9	1	0	0	$\bar{9}$	$\bar{1}$	0	0	$\Gamma_1' + \Gamma_2' + \Gamma_3' + 2\Gamma_4'$
$\Gamma_{(11)}$	11	$\bar{1}$	$\bar{1}$	$\bar{1}$	$\overline{11}$	1	1	1	$\Gamma_2' + \Gamma_3' + 3\Gamma_4'$

Table 8.6 Decomposition of the $(2l+1)$-dimensional representations in irreducible representations of the PSG 3m. Classes: $K_1 = e$; $K_2 = 3^1, 3^2$; $K_3 =$ m (3×).

Characters of irreducible representations

	e	$3^1, 3^2$	m (3×)
Γ_1	1	1	1
Γ_2	1	1	$\bar{1}$
Γ_3	2	$\bar{1}$	0

Characters $\chi_{(2l+1)}$ of the $(2l+1)$-dimensional representations

	e	$3^1, 3^2$	m (3×)	$\Gamma_{(2l+1)}$
$\Gamma_{(1)}$	1	1	1	Γ_1
$\Gamma_{(3)}$	3	0	1	$\Gamma_1 + \Gamma_3$
$\Gamma_{(5)}$	5	$\bar{1}$	$\bar{1}$	$\Gamma_2 + 2\Gamma_3$
$\Gamma_{(7)}$	7	1	1	$2\Gamma_1 + \Gamma_2 + 2\Gamma_3$
$\Gamma_{(9)}$	9	0	$\bar{1}$	$\Gamma_1 + 2\Gamma_2 + 3\Gamma_3$
$\Gamma_{(11)}$	11	$\bar{1}$	1	$2\Gamma_1 + \Gamma_2 + 4\Gamma_3$
$\Gamma_{(13)}$	13	1	$\bar{1}$	$2\Gamma_1 + 3\Gamma_2 + 4\Gamma_3$
$\Gamma_{(15)}$	15	0	1	$3\Gamma_1 + 2\Gamma_2 + 5\Gamma_3$
$\Gamma_{(25)}$	25	1	$\bar{1}$	$4\Gamma_1 + 5\Gamma_2 + 8\Gamma_3$

9
Group Algebra; Projection Operators

In this section we will become familiar with a method allowing us to carry out symmetry reduction of tensors, discussed in Section 3.8.2, by means of group theory, whereby the existence of the individual tensor components can be directly checked by a certain computational procedure. First, we introduce the concept of the linear operator $\boldsymbol{A}$, which generates from each vector $\boldsymbol{x}$ of an n-dimensional linear vector space V exactly one vector $\boldsymbol{y}$ in V according to $\boldsymbol{Ax} = \boldsymbol{y}$, whereby the following linearity relations hold:

$$\boldsymbol{A}(\boldsymbol{u}+\boldsymbol{v}) = \boldsymbol{Au}+\boldsymbol{Av} \quad \text{and} \quad \boldsymbol{A}(q\boldsymbol{u}) = q(\boldsymbol{Au}). \quad q \text{ is an arbitrary number.}$$

The symmetry operations are linear operators. In a system of n-multiple differentiable functions $f_i(\boldsymbol{x})$, all differential operators $\partial/\partial x_j$ and correspondingly higher differential operators are linear operators.

For two linear operators $\boldsymbol{A}$ and $\boldsymbol{B}$ one can define a sum and a product, namely $(\boldsymbol{A}+\boldsymbol{B})\boldsymbol{x} = \boldsymbol{Ax}+\boldsymbol{Bx}$ and $(\boldsymbol{AB})\boldsymbol{x} = \boldsymbol{A}(\boldsymbol{Bx})$. Similar to matrix multiplication $\boldsymbol{A}(\boldsymbol{Bx})$ means that the operator $\boldsymbol{B}$ operates on $\boldsymbol{x}$ and then the operator $\boldsymbol{A}$ operates on s$(\boldsymbol{Bx})$. It is clear that the operators so combined are also linear operators.

If one now constructs arbitrary linear combinations of the type

$$g = x_i g_i,$$

with the elements g_i of a group, taken as linear operators, then all such g represent a linear vector space of dimension h of the group. In this vector space, multiplication is defined in the way introduced above, namely

$$(x_i g_i)(y_j g_j) = x_i y_j (g_i g_j),$$

since $(g_i g_j)$ also represent group elements. A vector space in which multiplication is defined is called an algebra. The vector space $\bar{G}$ constructed from the elements G of a group is called *group algebra*.

We also require the notion of the *center* Z of a group G: All elements of a group, which commute with all other elements with respect to operations, form the center of the group. The center of a group possess the property of a

Physical Properties of Crystals. Siegfried Haussühl.

ISBN: 978-3-527-40543-5

subgroup, as is easily checked. Accordingly, we can introduce the center $\bar{Z}$ of the group algebra $\bar{G}$, whose elements commute with all other elements of $\bar{G}$. $\bar{Z}$ forms a subgroup of $\bar{G}$, as one sees from the subgroup properties of Z. We now come to the important theorem: The sum of the elements of each class represent a complete basis of the center **Z** of the group algebra. Let the sum of the classes be specified by ${}^iK$, hence, ${}^iK = {}^ig_1 + {}^ig_2 + \ldots$, where ${}^ig_j$ is the jth element of the ith class. The proof is as follows: if q is an arbitrary element taken of G, then the element $q^{-1i}g_jq$ also belongs to the ith class. If one lets ${}^ig_j$ run through all the elements of the class then this generates different elements in each case. If $q^{-1i}g_jq = q^{-1i}g_kq$, then we would have ${}^ig_j = {}^ig_k$, contrary to the assumption. This means, however, that the ith class sum can also be written as

$${}^iK = q^{-1i}g_1q + q^{-1i}g_2q + \cdots .$$

Therefore,

$${}^iK = q^{-1i}Kq, \qquad \text{hence} \quad q^iK = {}^iKq,$$

thus ${}^iK$ lies in $\bar{Z}$. Since the class sums are different, we obtain a basis for the linear vector space of the center of the group algebra, whose dimension is equal to the class number S. In particular, each arbitrary element taken from the center can be represented as a linear combination of the class sums.

Among these linear combinations there exists several especially interesting elements p given the name *Idempotent*, which have the property $p^2 = p$. Furthermore, if for two idempotents p_i and p_j one has $p_ip_j = 0$ for $i \neq j$, then the idempotents are orthogonal. A system of orthogonal idempotents is always linearly independent, i.e., the equation $\sum_i a_ip_i = 0$ exists only when all $a_i = 0$. In fact multiplying with p_i gives for each i: $a_ip_i^2 = a_ip_i = 0$, thus $a_i = 0$. Moreover, one finds that the sum of orthogonal idempotents is always an indempotent:

$$p = \sum_i p_i, \quad p^2 = \sum_{i,j} p_ip_j = \sum_i p_i^2 = \sum_i p_i = p.$$

There always exists a maximal set of orthogonal idempotents p_i from the center of the group algebra possessing the following properties:

1. $\sum_i^S p_i = e$ (identity element of the group)

2. Each p_i is indivisible, i.e., no orthogonal idempotents exist in $\bar{Z}$ with $p_i = p_i' + p_i''$.

3. All other idempotents can be represented as the sum of a few idempotents.

4. The maximal set is uniquely determined.

To prove 1. we form $(e - p')$ with $p' = \sum_{i=1}^{s} p_i$. This quantity is a new idempotent if it is different from zero, as one can easily check. If p' is complete, however, then a further idempotent cannot exist, i.e., $(e - p') = 0$.

Furthermore, from $p_i = p_i' + p_i''$ we would get by multiplication $p_i' = p_i' p_i$ and $p_i'' = p_i'' p_i$, and consequently $p_i' p_j = p_i' p_i p_j = 0$ and $p_i'' p_j = p_i'' p_i p_j = 0$. This means, p_i' and p_i'' would be two other orthogonal idempotents, which could take the place of p_i, contrary to the assumption that the system is maximal. The other two assertions can be easily proved by the reader. The orthogonal idempotents can now be constructed from the rules just discussed. It is seen, however, that they can also be found directly with the help of the character table, then

$$p_i = \frac{l_i}{h} \sum_{j=1}^{h} \chi_i^{\text{irr}}(g_j^{-1}) g_j,$$

where l_i is the dimension of the ith representation and h is the order of the group. For unitary matrices $\chi(g^{-1}) = \chi(g)$! The most important idempotent for our purposes is $p_1 = \frac{1}{h} \sum_{j=1}^{h} g_j$. The proof that we are dealing with an idempotent results simply from $p_1^2 = \frac{1}{h^2} \sum_{j,k}^{h} g_j g_k$.

The sum represents nothing else as the sum of all elements of the group table, hence, $\sum_{j,k}^{h} g_j g_k = h \sum_j g_j$, because each element is found once in each row. Therefore, $p_1^2 = p_1$. According to the above formula p_1 corresponds to the totally symmetric representation Γ_1. The general proof for the validity of this relationship is found, for example, with the help of the orthogonality relations of the characters.

As an example, let us again consider the PSG 3m. According to the character table (see Table 8.3)

$$p_1 = \frac{1}{6}(e + 3^1 + 3^2 + m_1 + m_2 + m_3)$$

$$p_2 = \frac{1}{6}(e + 3^1 + 3^2 - m_1 - m_2 - m_3)$$

$$p_3 = \frac{2}{6}(2e - 3^1 - 3^2).$$

The same result is obtained by using the relations

$$p_i p_j = \delta_{ij} p_i \quad \text{with} \quad p_i = k_{il} g_l \quad \text{and} \quad p_j = k_{jl} g_l.$$

The operators p_i decompose the linear vector space V in linear subspaces V_i, which in turn carry over into themselves through the respective elements of the group (G-invariant subspaces). The vectors ${}^i\boldsymbol{x}$ are called of the ith symmetry type when $p_i {}^i\boldsymbol{x} = {}^i\boldsymbol{x}$. The set of all these ith symmetry type vectors from V span the subspace V_i. Consequently, V_i is assigned the ith irreducible representation.

We now consider an important theorem: Each vector $\boldsymbol{x}$ of the vector space V is uniquely decomposed in a sum of vectors ${}^i\boldsymbol{x}$ from the subspaces V_i according to

$$\boldsymbol{x} = {}^1\boldsymbol{x} + {}^2\boldsymbol{x} + {}^3\boldsymbol{x} + \ldots {}^s\boldsymbol{x}.$$

This is recognized as follows: With $\sum_i p_i = e$ one obtains

$$\boldsymbol{x} = e\boldsymbol{x} = \left(\sum_i^s p_i\right) x = p_1 x + p_2 x + \ldots p_s x.$$

Due to this property, these idempotents are also referred to as *projection operators*. According to Neumann's principle, the following is true for the transformation in a symmetry equivalent system

$$t'_{ij\ldots s} = u_{ii^*} u_{jj^*} \cdots u_{ss^*} t_{i^* j^* \ldots s^*} = t_{ij\ldots s}.$$

The transformation matrices in the respective linear vector space are therefore equal to (1) and the associated irreducible representation is $\Gamma_1 = (1, 1, \ldots, 1)$. Thus, one obtains with the help of p_1, operating on arbitrary vectors $\boldsymbol{t}$ of the linear vector space spanned by the tensor components, those components lying in V_1 which construct the tensor.

We can directly employ this property for the symmetry reduction of tensors. For this purpose p_1 is applied to the vectors $\boldsymbol{t}$ of the linear vector space spanned by the tensor components. Thus

$$\boldsymbol{t}' = p_1(\boldsymbol{t}) = \frac{1}{h} \sum_{l=1}^{h} g_l(\boldsymbol{t}) = \boldsymbol{t}.$$

This is explained using the first- to third-rank tensors of the PSG 3m as examples. For each g_l we write the Kronecker product of the respective transformation matrices. In the case of the first-rank tensor we use the three-dimensional vector representation $\Gamma'_3 = \Gamma_3 + \Gamma_1$. Thus with Γ'_3:

$$g_1 = e = \begin{pmatrix} 1 & 0 & 0 \\ 0 & 1 & 0 \\ 0 & 0 & 1 \end{pmatrix}, \quad g_2 = 3^1 = \begin{pmatrix} -1/2 & \sqrt{3}/2 & 0 \\ -\sqrt{3}/2 & -1/2 & 0 \\ 0 & 0 & 1 \end{pmatrix},$$

$$g_3 = 3^2 = \begin{pmatrix} -1/2 & -\sqrt{3}/2 & 0 \\ \sqrt{3}/2 & -1/2 & 0 \\ 0 & 0 & 1 \end{pmatrix}, \quad g_4 = \mathrm{m}_1 = \begin{pmatrix} \bar{1} & 0 & 0 \\ 0 & 1 & 0 \\ 0 & 0 & 1 \end{pmatrix}, \tag{9.1}$$

$$g_5 = \mathrm{m}_2 = \begin{pmatrix} 1/2 & -\sqrt{3}/2 & 0 \\ -\sqrt{3}/2 & -1/2 & 0 \\ 0 & 0 & 1 \end{pmatrix}, \quad g_6 = \mathrm{m}_3 = \begin{pmatrix} 1/2 & \sqrt{3}/2 & 0 \\ \sqrt{3}/2 & -1/2 & 0 \\ 0 & 0 & 1 \end{pmatrix}$$

(see 8.2)

$$p_1(\boldsymbol{t}) = \frac{1}{6}\Big(g_1(\boldsymbol{t}) + g_2(\boldsymbol{t}) + g_3(\boldsymbol{t}) + g_4(\boldsymbol{t}) + g_5(\boldsymbol{t}) + g_6(\boldsymbol{t})\Big)$$

$$= \frac{1}{6}\left\{ (t_1, t_2, t_3) + \left(\frac{1}{2}t_1 + \frac{\sqrt{3}}{2}t_2, -\frac{\sqrt{3}}{2}t_1 - \frac{1}{2}t_2, t_3\right)\right.$$

$$+ \left(-\frac{1}{2}t_1 - \frac{\sqrt{3}}{2}t_2, \frac{\sqrt{3}}{2}t_1 - \frac{1}{2}t_2, t_3\right)$$

$$+ (-t_1, t_2, t_3) + \left(\frac{1}{2}t_1 - \frac{\sqrt{3}}{2}t_2, -\frac{\sqrt{3}}{2}t_1 - \frac{1}{2}t_2, 0\right)$$

$$\left. + \left(\frac{1}{2}t_1 + \frac{\sqrt{3}}{2}t_2, \frac{\sqrt{3}}{2}t_1 - \frac{1}{2}t_2, t_3\right)\right\}$$

$$= (0, 0, t_3) = t.$$

This means that only the component t_3 exists.

If the components $t_{ij\ldots s}$ are independent of one another, then p_1 can be applied to vectors only exhibiting one component, respectively. However, if relations exist among the components as in the case of degeneracy (in trigonal crystals, for example, $t_{11} = t_{22}$), then it is convenient to apply the projection on the complete vector $\boldsymbol{t}$ of the given linear vector space. This situation is recognized by the fact that the projection p_1, applied to certain vectors with only one component leads to a vector with several components. This is illustrated by the second-rank tensor of the PSG 3m. We have, for example,

$$p_1(t_{11}, 0, 0 \ldots 0) = \left(\frac{1}{2}t_{11}, \frac{1}{2}t_{11}, 0, 0, 0, 0, 0, 0, 0\right),$$

when the vector $\boldsymbol{t}$ has the form

$$\boldsymbol{t} = (t_{11}, t_{22}, t_{33}, t_{23}, t_{31}, t_{12}, t_{32}, t_{13}, t_{21}).$$

From this, it would follow that $t_{11} = t_{22} = 0$, which would be correct in the case of $t_{22} = 0$. If we now construct $p_1(\boldsymbol{t})$, we find

$$p_1(\boldsymbol{t}) = \left(\frac{1}{2}(t_{11} + t_{22}), \frac{1}{2}(t_{11} + t_{22}), t_{33}, 0, 0, 0, 0, 0, 0\right).$$

The existence of the independent components $t_{11} = t_{22}$ and t_{33} follows from the fact that $p_1(\boldsymbol{t}) = \boldsymbol{t}$. In practice, in has proved useful, as a first step, to calculate projections with only one component, respectively. We can then determine which tensor components vanish and which components are coupled with others. The projections carried out in a second step with the complete vector $\boldsymbol{t}$ are then easier to calculate, because the vanished components need no longer be taken into consideration.

This is again illustrated by the third-rank tensor of the PSG 3m.

1. Step: Calculating the projection of vectors $\boldsymbol{t}$ with only one component respectively. For this one requires the factors $u_{ii}u_{jj}u_{kk}$ (for $i = i^*, j = j^*, k = k^*$) appearing in the transformation $t'_{ijk} = u_{ii^*}u_{jj^*}u_{kk^*}t_{i^*j^*k^*}$, in other words, the products of the diagonal coefficients of g_i in the three-dimensional representation Γ'_3 (see above).

$$
\begin{aligned}
p_1(t_{111}, 0, 0\ldots) &= 0;\\
p_1(0, t_{222}, 0, 0\ldots) &= (0, \frac{1}{4}t_{222}, \ldots),\\
p_1(\ldots 0\ldots, t_{112}, \ldots 0\ldots) &= (\ldots, \frac{1}{4}t_{112}, \ldots),\\
p_1(\ldots 0\ldots, t_{113}, \ldots 0\ldots) &= (\ldots, \frac{1}{2}t_{113}, \ldots),\\
p_1(\ldots 0\ldots, t_{123}, \ldots 0\ldots) &= 0,\\
p_1(\ldots 0\ldots, t_{223}, \ldots 0\ldots) &= (\ldots, \frac{1}{2}t_{223}, \ldots),\\
p_1(\ldots 0\ldots, t_{133}, \ldots 0\ldots) &= 0,\\
p_1(\ldots 0\ldots, t_{233}, \ldots 0\ldots) &= 0,\\
p_1(\ldots 0\ldots, t_{333}) &= (\ldots 0\ldots, t_{333}).
\end{aligned}
$$

Similar projections apply to t_{121}, t_{211}, t_{131}, t_{311}, t_{232}, t_{322}. Consequently, the following tensor components vanish: t_{111}, t_{123}, t_{231}, t_{312}, t_{132}, t_{321}, t_{213}, t_{133}, t_{313}, t_{331}, t_{233}, t_{323}, t_{332}. Those remaining are: t_{222}, t_{112}, t_{121}, t_{211}, t_{113}, t_{131}, t_{311}, t_{223}, t_{232}, t_{322} and t_{333}.

2. Step: One lets p_1 act on a vector containing only the nonvanishing components. Because $p_1(\boldsymbol{t}) = \boldsymbol{t}$ we obtain the following relations:

$$
\begin{aligned}
&t_{112} = t_{121} = t_{211} = -t_{222} = \frac{1}{4}(t_{112} + t_{121} + t_{211} - t_{222}),\\
&t_{113} = t_{223} = \frac{1}{2}(t_{113} + t_{223}),\\
&t_{131} = t_{232} = \frac{1}{2}(t_{131} + t_{232}),\\
&t_{311} = t_{322} = \frac{1}{2}(t_{311} + t_{322}),\\
&t_{333} = t_{333}.
\end{aligned}
$$

Accordingly, the components of this tensor span a five-dimensional space. This result is identical with that derived in Section 4.4.

The question now arises, when is the projection procedure, discussed here, preferred to that discussed in Section 4.2.1. With respect to the projection method, the calculation procedure shows that the projection method deals with a stringent prescription, which in the form presented can be easily transferred to a computer program. Thus, in many cases, in particular, with higher rank tensors, one can attain complete symmetry reduction more quickly than with the method of the transformation of the components in symmetry equivalent reference systems. Special emphasis is placed on the possibility of checking the existence of individual tensor components with the help of the projections. A substantial contraction of the calculation procedure is achieved when one uses the following conditions taken from symmetry operations as, for example, in the case of a two-fold axis parallel $\boldsymbol{e}_i$ or a symmetry plane perpendicular $\boldsymbol{e}_i$ (index i an odd number of times or an even number of times with odd-rank tensors and an even number of times or odd number of times with even-rank tensors). In particular, the cyclic permutability of the indices in cubic crystals should be noted. The projection method represents an extremely useful instrument for many other applications including the analysis of the vibrational states in molecules and crystals as well as in the classification of phase transformations and the interactions resulting from these (normal coordinate analysis, symmetry types of the interactions).

10
Concluding Remarks

Building upon the main features of crystals physics presented in this book, the interested reader should have little problem in analyzsing and handling most of the other phenomena which were only touched upon or completely left out in the text. In such cases one first has to clarify the question of inducing and induced quantities as well as their mathematical relationship. One must check whether a tensorial or non-tensorial property is present, and which characteristic symmetry results from the relationship. The influence of the symmetry properties of the crystal then follows from the rules of symmetry reduction when working with tensors, or with the help of symmetry matched functions. When looking for simple observation- and measurement arrangements, the aspects of ""longitudinal effect" and " "transverse effect" should be in the foreground. However, this only addresses one area of crystal physics today. The fundamental question as to the structural interpretation of the properties, especially the explanation of the observed anisotropy effects falls into the area of crystal chemistry and could not be covered here in any detail. This also includes the problem of which extreme values of the properties are attainable at all. Here, for example, we ask for the largest attainable heat conductivity, the lowest velocity of light, the largest pyroelectric or piezioelectric effect in crystals. Such considerations come to the fore in " "chemical engineering"," i.e., the constructive search for crystals with novel or improved properties. Therein exists one of the most important challenges of modern materials research. The acquisition of crystal–physical data and often also the resulting application depends decisively on the size and quality of the available crystals. The collective of crystals available for measurements and applications could be substantially increased, if it were possible to further miniaturize the already existing methods, i.e., to achieve a reduction in the required dimensions of the crystal specimens. This aspect plays, for example, an important role in the construction of micro-miniaturized electronic and optical devices. Furthermore, by no means have the possibilities been exhausted of investigating novel material properties with the help of inhomogeneous crystalline devices, as has been utilized for a long time in semiconductor technology (inhomogeneous doped mixed crystals, heterogeneous mixed crystals, crystals

Physical Properties of Crystals. Siegfried Haussühl.

ISBN: 978-3-527-40543-5

with inhomogeneous defect distribution). The same applies to the configuration of the geometric boundary of the devices and the application of inhomogeneous boundary conditions, as well as to the effects of inhomogeneous inducing quantities, such as, for example, from torsional stresses, which lead to finite gradients of the deformation tensor and thus to certain higher- order effects.

Finally, more efficient data collection in the future requires a more vigorous development of automatic measurement methods. There can be no doubt that the prospective material scientist awaits diverse and highly interesting tasks in all these fields.

11
Exercises

1. Prepare a set of glass or brass grinding plates of good evenness according to the following method: From a stack of 12 approximately quadratic or circular disks with a thickness of ca. 8 mm and an edge length of ca. 100 mm select pairs of plates with about the same camber using a straightedge. One then grinds the plates pairwise against each other so that the respective cambered or concave surfaces are abraded away. A suitable abrasive is water with corundum powder of about 30 μm grain size. Thereafter, the plates are again sorted with a straightedge according to about the same deviation from flatness and repeat the pairwise mutual abrasion of plates of the same curvature. After repeating the procedure again one normally finds that over half the plates are suitable as a grinding base for making plane grinds with a deviation of at most 1 μm per 10,000 μm. Use a finer grain size (for example, 15 μm) in the last steps.

2. Practice grinding plane faces on crystals of various hardness using the grinding plates prepared in Exercise 1. In doing so take special care that grinding is performed under uniform motion in long strokes with little rotation and not too much pressure on the specimen. Otherwise the danger exists that the edges are more strongly abraded. Sometimes it is convenient to use a weakly convex grinding plate to work against this effect. The flatness can be checked with the help of optical interference methods or with a precision straightedge. The next step is to prepare a plane parallel plate. Firstly, plane grind a face and then grind the opposite face under continuous control using a micrometer screw or better using a thickness gauge so that the distances between the faces is constant within the given tolerances over the complete surface. In the last steps, one should use a very fine corundum powder as an abrasive together with a suitable liquid (water, propanol, ethylene glycol, ethylether), whereby it is essential to ensure that no coarse grained abrasives from previous grinding processes are introduced into the fine grinding step. In order to prevent corners and edges from breaking out it is recommended to level the edges by careful grinding.

Physical Properties of Crystals. Siegfried Haussühl.

ISBN: 978-3-527-40543-5

The preparation of plane parallel plates and other geometrically defined specimens is made considerably easier by employing ring-shaped holders, for example, machined from brass. The raw specimens are cemented in the holders such that the area to be abraded away protrudes over the rim of the given holder thus allowing targeted removal during the grinding process.

3. Prove by coordinate transformation applying Neumann's principle that the fundamental polynomials (in the coordinates x_i, homogeneous polynomials of the nth degree) have the following form in the PSG 22:

$$P_0 = a_0, \quad P_1 = 0, \quad P_2 = a_1 x_1^2 + a_2 x_2^2 + a_3 x_3^2, \quad P_3 = a_4 x_1 x_2 x_3,$$
$$P_4 = a_5 x_1^4 + a_6 x_2^4 + a_7 x_3^4 + a_8 x_1^2 x_2^2 + a_9 x_2^2 x_3^2 + a_{10} x_3^2 x_1^2.$$

Which form has P_5 (polynomial of the 5. degree in x_1, x_2, x_3)?

From these polynomials what can one infer concerning the existence of the corresponding nth rank tensors?

4. In the cylinder groups ∞, ∞m, ∞2 and so on, the tensors up to the fourth rank take on the same form as in the corresponding hexagonal groups PSG 6, 6m, 62 and so on. Prove this by tensor transformation. The condition is that a symmetry equivalent position results from a rotation of an arbitrary angle about the cylinder axis.

5. What is the change in a third- and fourth-rank tensor after a small rotation of the reference system described by the antisymmetric tensor $\{r_{ij}\}$? Calculate

$$\Delta t_{ijk} = t'_{ijk} - t_{ijk} \quad \text{and} \quad \Delta t_{ijkl} = t'_{ijkl} - t_{ijkl} \quad \text{respectively.}$$

6. Prove that symmetric functions $F(x_i)$ under symmetric secondary conditions $G(x_1, x_2, x_3)$ always take on an extreme value along the space diagonals of the Cartesian reference system.

For this purpose, one constructs the auxiliary function

$$H = F(x_1, x_2, x_3) - \lambda G(x_1, x_2, x_3)$$

and substitutes x_i by ru_i, where u_i are the directional cosines of the position vector. According to the method applied in Section 4.3.6, calculate the condition for extreme values of H.

7. The number of independent components of a total symmetric mth-rank tensor in n-dimensional space is

$$Z(m, n) = \frac{(n + m - 1)!}{(n - 1)!m!}.$$

Confirm this formula by a proof of induction. In a triclinic crystal, how large is the number of independent components of an eighth-rank tensor in which the eight positions are divided in pairs of four that are mutually permutable?

8. With the convention introduced in Section 2.2, the position of the basic vectors of the crystal-physical reference system is not uniquely fixed in each case when different equivalent crystallographic arrangements exist. For example, in the PSGs 4/m and 3 both crystallographic reference systems $\{\boldsymbol{a}_1, \boldsymbol{a}_2, \boldsymbol{a}_3\}$ and $\{\boldsymbol{a}'_1 = \boldsymbol{a}_1, \boldsymbol{a}'_2 = -\boldsymbol{a}_2, \boldsymbol{a}'_3 = -\boldsymbol{a}_3\}$ are equivalent (both are right-handed systems and possess the same metric). How do the components c_{1112} and c_{1113}, respectively, of the elasticity tensor transform in a transition from $\boldsymbol{a}_i$ to $\boldsymbol{a}'_i$? Accordingly, one is to check how the components c_{1123} in the PSG $\bar{3}$m transform when an equivalent crystallographic reference system is selected, which results from a 180° rotation about the three-fold axis based initially on the system $\{\boldsymbol{a}_i\}$. In such cases it is necessary to specify the position of the selected reference system via morphological information (indexing certain forms, generated, for example, by spherical growth- or etching methods) or X-ray information (structure factors of certain reflexes) in order to uniquely determine the tensor properties.

9. In which direction appears the largest deviation from a pure longitudinal effect in second-rank tensors (maximum of the angle between inducing and induced vector quantities, for example, in the case of electrical conductivity, between field strength $\boldsymbol{E}$ and current density vector $\boldsymbol{I}$)? The simplest way to carry out the calculation is in the principal axis system. The searched for direction lies in the plane of the smallest and largest principal value.

10. Show that enantiomorphous individuals possess opposite polar effects (polar tensors of odd rank) and that they do not differ in properties described by even-rank tensors when based on the same reference system, respectively. Note that the tensors for right and left individuals in the corresponding right- and left-handed systems have the same tensor components.

11. In double refracting crystals, the maximum ray double refraction appears, to a first approximation, in the direction of the bisector of the largest and smallest semi-axis of the indicatrix when the double refraction is sufficiently small. Prove this assertion.

12. Minimal deflection: when a prism, formed by an optically isotropic medium II or an optically uniaxial crystal, transmits light such that the

deflection of the refracted ray is a minimum with respect to the primary ray, then one has for the deflection angle $\overline{\alpha}$ and prism angle φ (only for ordinary rays in optically uniaxial crystals):

$$\frac{\sin(\overline{\alpha}+\varphi)/2}{\sin\varphi/2} = \frac{n_{\mathrm{II}}}{n_{\mathrm{I}}}.$$

Prove this relationship by calculating $d\overline{\alpha}/d\alpha_{\mathrm{I}} = 0$ using the law of refraction. α_{I} is the angle of incidence of the primary ray on the face of a prism. The ray passes symmetrically through the prism in minimal deflection.

13. Calculate from the Lorentz–Lorenz formula

$$\frac{n^2-1}{n^2+2}\cdot\frac{M}{\rho} = R$$

the dependence of the refractive index on hydrostatic pressure under the assumption that the mole refraction R is pressure independent.

14. A fine parallel beam striking a plane parallel plate at perpendicular incidence with normals parallel to an optical axis, experiences a double refraction. Show that the partial rays of all directions of vibration of the incident beam emerge from the crystal on a cylinder with circular cross-section (inner conical refraction). The radius of the cylinder is $r = L\tan\mu$. μ is given by

$$\tan 2\mu = \frac{\sqrt{(n_3^2-n_2^2)(n_2^2-n_1^2)}}{n_1 n_3},$$

where L is the thickness of the crystal plate. Hint: place a Cartesian reference system in the crystal such that the direction of the center principal axis of the indicatrix is parallel $\boldsymbol{e}_2'$ and the direction of the optical axis is parallel $\boldsymbol{e}_1'$. The propagation vector then runs parallel $\boldsymbol{e}_1'$ and an arbitrary $\boldsymbol{D}$-vector of the incident wave has the form

$$\boldsymbol{D} = \boldsymbol{D}_0(\cos\varphi\boldsymbol{e}_2' + \sin\varphi\boldsymbol{e}_3').$$

Now calculate the relation $D_i' = \epsilon_{ij}'E_j'$ by tensor transformation and from this the reversal $E_i' = a_{ij}'D_j'$. This gives the direction of the ray vector $\boldsymbol{s}' \parallel \boldsymbol{E}'\times\boldsymbol{H}'$ with $\boldsymbol{H}' \parallel \boldsymbol{D}'\times\boldsymbol{e}_1'$ as a function of the angle φ and thus a way to calculate the geometrical details.

15. In which direction has a rhombic crystal (calcium formate) a vanishing thermal expansion ($\alpha_{11}' = 0$) when the components of the tensor of thermal expansion have the values $\alpha_{11} = -16.6$, $\alpha_{22} = 68.6$ and $\alpha_{33} = 29.8\cdot 10^{-6}\mathrm{K}^{-1}$?

16. Derive the conditions existing between the components of the elasticity tensor c_{ijklmn} in isotropic substances. One sets out from the independent components of cubic crystals of the PSG 4/m3 and calculates the effect of an arbitrary n-fold axis, for example, parallel e_3, on these components.

17. In the cubic PSG there does not exist a pure converse piezoelectric longitudinal effect ($\hat{d}'_{221}, \hat{d}'_{331} \neq 0$ for any arbitrary electric field $\boldsymbol{E} = E_i \boldsymbol{e}_i \parallel \boldsymbol{e}'_1$). Why?

18. In all crystals possessing the subgroup 22, no change in volume is generated by the first-order electrostrictive effect. Why?

19. The converse piezoelectric effect can be used to construct an electronic position transducer, which experiences a defined change in length proportional to the applied electric voltage. Which voltage must be applied to a 5 mm thick quartz plate, cut perpendicular to a two-fold axis, in order to achieve a change in length of 100 Å(=0.01 μm)? $d_{111} = 2.31 \cdot 10^{-12}$m/V.

20. Why is the deviation from Ohm's law in bismuth (PSG $\bar{3}$m) first observed in the third power of the electric field strength?

21. Under which angle to the six-fold axis in $LiIO_3$ (PSG 6) does phase matching of the fundamental wave and the frequency doubled harmonic appear for the vacuum wavelength $\lambda = 1.06\mu$m? The refractive indices are

$$n_0^{\nu} = 1.860, \quad n_e^{\nu} = 1.719, \quad n_0^{2\nu} = 1.901, \quad n_e^{2\nu} = 1.750.$$

22. With uniaxial tension along $\boldsymbol{e}_i$ one observes, aside from longitudinal dilatation ε_{ii} in the direction of tension, a lateral contraction ε_{jj} perpendicular to the direction of tension. The Poisson relation

$$\nu_{ij} = \frac{\varepsilon_{jj}}{\varepsilon_{ii}} = \frac{s_{iijj}}{s_{iiii}}$$

specifies the magnitude of lateral contraction. Prove that this relation in cubic crystals, for tension along a cube edge, is isotropic and assumes the value $-c_{12}/(c_{11} + c_{12})$ for all directions perpendicular to the edge. Furthermore, derive the lateral contraction relation in rhombic crystals for tension parallel to a rhombic principal direction $\boldsymbol{a}_i$ and for a lateral contraction in the direction $\boldsymbol{a}_j$ perpendicular to this principal direction. It is

$$\nu_{ij} = \frac{-c_{ik}c_{jk} + c_{ij}c_{kk}}{c_{jj}c_{kk} - c_{jk}^2},$$

where $i, j \neq k$.

23. Which form has the elasticity tensor of a cubic crystal when the Cartesian reference system is so selected that e_3 is parallel to a three-fold axis, hence

$$e_3 = (a_1 + a_2 + a_3)/a\sqrt{3},$$

and e_1 is parallel to a bisector of two cubic principal axes ($e_1 = (a_1 - a_2)/a\sqrt{2}$)?

24. For crystals with subgroup 22, prove, on the basis of the expressions given in Table 4.14 for both ρv^2-values c' and c'' of the waves vibrating in the plane spanned by e_i and e_j, that the relationships, $c' - c'' = c_{iijj} + c_{ijij}$ are approximately true, under weak anisotropy conditions, when $g = (\sqrt{2}/2)(e_i \pm e_j)$.

25. In many cases, the physical properties of polycrystalline aggregates can be easily calculated from the properties of single crystals when simple assumptions are made concerning the grain distribution of the aggregate and reasonable boundary conditions are introduced for the transfer of inducing quantities from grain to grain. As an example, we consider the elastic constants of a polycrystalline aggregate. In a first model case, we assume that the deformations are homogeneously distributed over all grains, in a second, that the stress propagates homogeneously through all grains. If the crystals possess anisotropic elastic properties, then both assumptions are only approximately correct. In the first case (Voigt case) we can directly use the relation $\sigma_{ij} = c_{ijkl}\varepsilon_{kl}$ to obtain an average value, in the second case (Reuss case) we must apply the relation $\varepsilon_{ij} = s_{ijkl}\sigma_{kl}$. We calculate the properties of an isotropic medium conveniently by first assuming cubic crystal symmetry and then setting the condition that the anisotropy of the cubic medium is canceled by, for example, letting the material quantities along [100], [110] or [111] and in other directions respectively, take on the same values. Also, the condition of simultaneously fulfilling hexagonal and cubic symmetry leads to isotropy in crystals. The elasticity tensor in cubic crystals has only the three constants $c_{11} = c_{1111}$, $c_{12} = c_{1122}$ and $c_{66} = c_{44} = c_{1212}$.

The condition of isotropy is given when, for example, the longitudinal resistances along [100] and [110] are equal, hence,

$$c_{1111} = (c_{1111} + c_{1122} + 2c_{1212})/2$$

and thus $2c_{44} = c_{11} - c_{12}$. We now demand that the single crystal and the polycrystalline aggregate exhibit the same scalar invariants. In the case of the c-tensor we are dealing with $I_1 = c_{ijkl}\delta_{ij}\delta_{kl}$ and $I_2 = c_{ijkl}\delta_{ik}\delta_{jl}$.

One finds for the single crystal

$$I_1 = c_{1111} + c_{2222} + c_{3333} + 2(c_{1122} + c_{2233} + c_{3311}),$$
$$I_2 = c_{1111} + c_{2222} + c_{3333} + 2(c_{1212} + c_{2323} + c_{3131}).$$

The polycrystalline aggregate gives: $\bar{I}_1 = 3\bar{c}_{1111} + 6\bar{c}_{1122}$ and $\bar{I}_2 = 3\bar{c}_{1111} + 6\bar{c}_{1212} = 6\bar{c}_{1111} - 3\bar{c}_{1122}$ (with $2c_{1212} = c_{1111} - c_{1122}$). This allows the calculation of the elastic constants of the aggregate:

$$\bar{c}_{1111} = \bar{c}_{11} = (I_1 + 2I_2)/15 \quad \text{and} \quad \bar{c}_{1122} = \bar{c}_{12} = (2I_1 - I_2)/15.$$

Completely analogous expressions are valid for the coefficients $\bar{s}_{ijkl}$ and and the corresponding invariants. For cubic crystals, calculate the difference in the values $\bar{c}_{11}$ and $\bar{c}_{12}$ for both cases (homogeneous deformation and homogeneous stress, respectively). The latter is obtained by matrix inversion. In practice, it turns out that the arithmetic mean of both pairs of values comes close to those observed experimentally (see also Kröner, 1958).

26. Show that the number of independent components of the elasticity tensors, in the case of triclinic crystals, is reduced from 21 to 18 when one selects Cartesian axes parallel to the deformation vectors of the three elastic waves propagating in the direction of a dynamic longitudinal effect c'_{1111} (here, a pure longitudinal wave and two pure transverse waves exist; c'_{1111} takes on an extreme value, as shown in Section 4.5.5).

Furthermore, prove that in monoclinic crystals the elastic constant c'_{15} ($= c'_{1113}$) vanishes when one rotates the Cartesian reference system by an angle φ about the axis e_2 ($\|2$ and m, respectively) so that the longitudinal effect c'_{1111} takes on an extreme value ($dc'_{1111}/d\varphi = 0$!).

27. Show that in all PSGs with the subgroup 22 no piezoelectric coupling occurs with elastic waves in the propagation direction e_i ($c^E = c^D$).

28. In cubic crystals the energy current in the propagation of elastic transverse waves along the three-fold axis forms a finite angle ζ with the wave normals (elastic internal refraction). Calculate this angle with the help of the relationship discussed at the end of Section 4.5.5. For this purpose, the components s_j of the ray vector, which are proportional to c'_{1j13}, are to be calculated for $j = 1, 2, 3$ in a reference system $\{e'_i\}$, whose basic vector e'_3 runs parallel to a three-fold axis. For example, let

$$e'_1 = \frac{1}{\sqrt{2}}(e_1 - e_2), \quad e'_2 = \frac{1}{\sqrt{6}}(e_1 + e_2 - 2e_3), \quad e'_3 = \frac{1}{\sqrt{3}}(e_1 + e_2 + e_3).$$

In the case of cubic crystals, transformation results in the following components:

$$c'_{1113} = 0, \quad c'_{1213} = (c_{11} - c_{12} - 2c_{44})/3$$

and

$$c'_{1313} = (c_{11} - c_{12} + c_{44})/3.$$

If the deformation vector runs parallel e'_1, then one obtains

$$\cos\zeta = \frac{c'_{1313}}{\sqrt{c'^2_{1213} + c'^2_{1313}}}.$$

29. The Grüneisen tensor (after Kitaigorodskii)

$$\gamma_{ij} = \frac{V}{C_V} c^T_{ijkl} \alpha_{kl}$$

is a generalization of the ordinary Grüneisen relation

$$\gamma = \frac{3V}{C_V} \frac{\alpha}{K},$$

where α_{ij} are the components of the tensor of thermal expansion, V is the mole volume, C_V is the specific heat at constant volume (per mole volume) and K is the volume compressibility. Prove the identity of both relations for cubic crystals.

30. Which differences appear between adiabatic and isothermal piezoelectric constants in the following PSGs: 23, 3, mm2?

31. Calculate the form of the magnetoelastic tensor

$$\Delta c_{ijkl} = c_{ijklmn} H_m H_n$$

for the PSG m3. (There exists a total of 13 independent components, as one can immediately calculate from the formula

$$n = \frac{1}{h} \sum_{j=1}^{h} \chi(g_j)!$$

32. Which form has the tensor of first-order magnetostriction in the magnetic PSGs $\underline{\text{m}}3\underline{\text{m}}$, $\underline{\text{m}}3$ and mm$\underline{\text{m}}$ ($\varepsilon_{ij} = m_{ijk} H_k$)?

33. Why does the important scalar product $\boldsymbol{g} \cdot \boldsymbol{G}$ for electrogyration vanish in crystals of the PSG m3 when we have $\boldsymbol{g} \parallel [111]$ and $\boldsymbol{E} \parallel [a\, b . \bar{a} + \bar{b}]$?

34. In the cubic PSGs, which form has the fifth-rank tensor describing the second-order piezoelectric effect according to $D_i = d_{ijklm}\sigma_{jk}\sigma_{lm}$?

35. Calculate the conditions for the direction of the vanishing and maximal longitudinal piezoelectric effect for crystals of the PSG m from the components of the piezoelectric tensor.

36. From the general equation for $\partial v^2/\partial\sigma_{pq}$ derived in Section 4.6.3, derive $\partial c_{1111}/\partial\sigma_{11}$ and $\partial c_{2323}/\partial\sigma_{11}$ for a longitudinal wave and a transverse wave, respectively, which propagate in the direction [100] under uniaxial pressure along [100].

37. The 60 elements of the icosaheder group 235 are distributed over the following five classes: K_1: e, K_2: 2 (15×), K_3: 3^1, 3^2 (each 10×), K_4: 5^1, 5^4 (each 6×) and K_5: 5^2, 5^3 (each 6×). Prove the correctness of the following character table of irreducible representations with the help of the orthogonality relations.

	K_1	K_2	K_3	K_4	K_5
Γ_1	1	1	1	1	1
Γ_2	3	-1	0	$\frac{1}{2}(1+\sqrt{5})$	$\frac{1}{2}(1-\sqrt{5})$
Γ_3	3	-1	0	$\frac{1}{2}(1-\sqrt{5})$	$\frac{1}{2}(1+\sqrt{5})$
Γ_4	4	0	1	-1	-1
Γ_5	5	1	-1	0	0

38. In the cylindrical symmetry, the number of independent tensor components can be calculated from the formula valid for finite groups

$$n = \frac{1}{h}\sum_{j=1}^{h} \chi(g_j),$$

when one sets correspondingly high orders of symmetry of the axes of rotation. Since, however, cylindrical symmetry allows all arbitrary angles of rotation, one can, through a boundary transition, apply the integral

$$n = \frac{1}{2\pi}\int_0^{2\pi} \chi(\varphi)d\varphi$$

instead of the sum.

With the help of this formula calculate the number of independent components for tensors of rank from 1 to 6 for the case of a cylindrical symmetry ∞.

39. As seen in Section 8.3, a second-rank tensor spans a nine-dimensional linear vector space, which can be decomposed into a one-, three- and five-dimensional invariant subspace. In the case of second-rank symmetrical tensors the three-dimensional subspace is empty, and the basic vectors of the remaining subspaces are:

$$V_1 : e_1^0 = (e_1 + e_2 + e_3)/\sqrt{3}; \quad V_5 : e_2^0 = (e_1 + e_2 - 2e_3)/\sqrt{6},$$
$$e_3^0 = (e_1 - e_2)/\sqrt{2}, \quad e_4^0 = e_4, \quad e_5^0 = e_5, \quad e_6^0 = e_6.$$

In highly symmetric crystals, it is often convenient to describe the relationship between two second-rank tensors in the coordinates of these subspaces, the so-called symmetry coordinates. For example, Hooke's law in cubic crystals has the following form:

$$\begin{aligned}
\sigma_1^0 &= \sigma_{11} + \sigma_{22} + \sigma_{33} = (c_{1111} + 2c_{1122})(\varepsilon_{11} + \varepsilon_{22} + \varepsilon_{33} \\
&= (c_{11} + 2c_{12})\varepsilon_1^0, \\
\sigma_2^0 &= \sigma_{11} + \sigma_{22} - 2\sigma_{33} = (c_{1111} - c_{1122})(\varepsilon_{11} + \varepsilon_{22} - 2\varepsilon_{33}) \\
&= (c_{11} - c_{12})\varepsilon_2^0, \\
\sigma_3^0 &= \sigma_{11} - \sigma_{22} = (c_{1111} - c_{1122})(\varepsilon_{11} - \varepsilon_{22}) = (c_{11} - c_{12})\varepsilon_3^0, \\
\sigma_4^0 &= \sigma_{23} = 2c_{2323}\varepsilon_{23} = c_{44}\varepsilon_4^0, \\
\sigma_5^0 &= \sigma_{13} = 2c_{3131}\varepsilon_{31} = c_{44}\varepsilon_5^0, \\
\sigma_6^0 &= \sigma_{12} = 2c_{1212}\varepsilon_{12} = c_{44}\varepsilon_6^0,
\end{aligned}$$

Verify these relationships and establish that the quantities ε_i^0 and σ_i^0 connected with the elastic constants

$$c_1^0 = c_{11} + 2c_{12}, \quad c_2^0 = c_3^0 = (c_{11} - c_{12})$$

and

$$c_4^0 = c_5^0 = c_6^0 = c_{44}$$

transform as the spherical harmonic functions

$$Y_0, \quad Y_2^0, \quad Y_2^2 + Y_2^{-2}, \quad Y_2^1 + Y_2^{-1}, \quad Y_2^1 - Y_2^{-1}, \quad Y_2^2 - Y_2^{-2}$$

$$\left(Y_0 \sim \frac{x_1^2 + x_2^2 + x_3^2}{r^2} = 1, \quad Y_2^0 \sim \frac{1}{r^2}(x_1^2 + x_2^2 - 2x_3^2), \right.$$
$$Y_2^2 + Y_2^{-2} \sim \frac{1}{r^2}(x_1^2 - x_2^2), \quad Y_2^1 + Y_2^{-1} \sim \frac{1}{r^2}x_2 x_3,$$
$$\left. Y_2^1 - Y_2^{-1} \sim \frac{1}{r^2}x_1 x_3, \quad Y_2^2 - Y_2^{-2} \sim \frac{1}{r^2}x_1 x_2 \right).$$

That means, that all quantities of a relationship $\sigma_i^0 = c_i^0 \varepsilon_i^0$ transform to the same type of symmetry.

40. How many scalar invariants has a fifth-rank tensor of type $t_{3\times 3\times 3\times (3\times 3)_S}$, a sixth-rank tensor of type $t_{(3\times 3)_S\times (3\times 3)_S\times (3\times 3)_S}$ and a third-rank pseudotensor of type $t_{3\times (3\times 3)_S}$?

12 Appendix

12.1 List of Common Symbols

(multiple meanings possible!)

α_i	Angle of the crystallographic metric
a_i	Length of crystallographic basis vectors
a_{ij}	Components of the polarization tensor
α_{ij}	Components of the tensor of thermal expansion
B_i	Components of the magnetic induction vector
C_V, C_p	Specific heats at constant volume and constant hydrostatic pressure
c	Speed of light in vacuum
c_{ijkl}, c_{ij}	Components of the elastic tensor or elastic constants
δ_{ij}	Kronecker symbol
d_{ijk}	Components of the piezoelectric tensor
$\hat{d}_{ijk}$	Components of the electrostrictive tensor
D_i	Components of the dielectric displacement vector
ϵ_{ij}	Components of the dielectricity tensor
ϵ_0	Vacuum permittivity ($8.8542 \cdot 10^{-12}\,\mathrm{C\,V^{-1}\,m^{-1}}$)
ε_{ij}	Components of the deformation tensor
E_i	Components of the electric field vector
e_{ijk}	Components of the piezoelectric e tensor or the Levi-Civita tensor
F	Abrasive strength or Helmholtz energy
γ_{ij}	Components of the Grüneisen tensor
G	Gibbs energy
g_i	Components of the normalized propagation vector
g_{ijk}	Components of the gyration tensor
g_{ijkl}	Components of the electrogyration tensor
H_i	Components of the magnetic field vector

Physical Properties of Crystals. Siegfried Haussühl.

ISBN: 978-3-527-40543-5

h	Planck constant ($6.6262 \cdot 10^{-34}$ J s)
h_i	Miller indices
I_i	Components of the electric current density vector
k	Boltzmann constant ($1.380662 \cdot 10^{-23}$ J K^{-1})
K	Volume compressibility
k_i	Components of the propagation vector
K_i	Components of a force vector
κ_{ij}	Components of the magnetic susceptibility vector
k_t	electromecanical coupling constant for a longitudinal thickness vibration
λ	Wavelength
λ_{ij}	Components of the thermal conductivity tensor
L	Transmission path length
μ_{ij}	Components of the magnetic permeability tensor
μ_0	Vacuum permeability ($4\pi \cdot 10^{-7}$V s A^{-1}m^{-1})
ν	Frequency
n_i	Principal refractivities
π_i	Components of the pyroelectric tensor
p	Hydrostatic pressure
$P_{ij} = dc_{ij}/dp$	Piezoelastic constants
p_{ijkl}	Components of the elastooptic tensor
Q_i	Components of the heat flow density vector
q_{ijkl}	Components of the piezooptic tensor
ρ	Density
r_{ijk}, r_{ijkl}	Components of the first-and second-order electrooptic tensor
σ_{ij}	Components of the mechanical strain tensor
S	Entropy
s_{ij}	Elastic coefficients or components of the electrical conductivity tensor
s_{ijkl}	Components of the elastic s tensor
t	Time
T	Temperature
$T_{ij} = \frac{d \log c_{ij}}{dT}$	thermoelastic constants
V	Volume
U	Internal Energy
v	Velocity of light or sound
$\omega = 2\pi\nu$	Angular frequency

12.2 Systems of Units, Units, Symbols and Conversion Factors

The continuing use of several systems of units has created substantial confusion particularly among beginners and laymen. The introduction of the SI sytem (Système International d'Unités) in 1960 has reduced this confusion to some extent. Today, the use of SI-compliant units, of which some are collected in Table 12.1, is strongly recommended. The definition of the base units is assumed to be known and will not be repeated here. Prefixes that may be used with a unit to indicate decimal multiples and submultiples are (symbol in parantheses):

10^{12}:	tera	(T)	10^{-1} :	deci	(d)
10^{9} :	giga	(G)	10^{-2} :	centi	(c)
10^{6} :	mega	(M)	10^{-3} :	milli	(m)
10^{3} :	kilo	(k)	10^{-6} :	micro	(μ)
10^{2} :	hecto	(h)	10^{-9} :	nano	(n)
10^{1} :	deka	(d)	10^{-12}:	pico	(p)

A number of outdated units are still encountered; their use is, however, not recommended. Some examples are:

Force	:	$1\ \text{dyn} = 10^{-5}\ \text{N}$
Pressure	:	$1\ \text{bar} = 10^{5}\ \text{Pa} = 10^{5}\ \text{N}\,\text{m}^{-2}$
Density	:	$1\ \text{g}\,\text{cm}^{-3} = 10^{3}\ \text{kg}\,\text{m}^{-3}$
Energy	:	$1\ \text{erg} = 10^{-7}\ \text{J}$,
	:	$1\ \text{cal} = 4.187\ \text{J}$,
	:	$1\ \text{kW}\,\text{h} = 3.6 \cdot 10^{6}\ \text{J}$,
	:	$1\ \text{eV} = 1.602 \cdot 10^{-19}\ \text{J}$
Charge	:	$1\ \text{esu} = \frac{1}{3} \cdot 10^{-9}\ \text{C}$

In some disciplines, the cgs system of units is still in use. For mechanical units, the conversion factors to SI units are always simple powers of ten. This is different for electric and magnetic quantities, where more complex factors arise because of the respective definitions of electrical charge in the different measurement systems. The conversion into other units thus requires particular care. To avoid these difficulties, the exclusive use of SI units is strongly recommended especially for electric and magnetic quantities. For theoretical derivations, however, the choice of a different measurement system may be appropriate if this allows for a more concise mathematical description; we have also taken this liberty in a few cases in this text.

The derived units for elastic, piezoelectric, electrooptical or magnetooptical quantities can be determined without difficulty. Some interesting remarks regarding units are given in *Pure and Applied Chemistry*, Vol. 21, pp. 1–113 (1970).

Table 12.1 SI units.

Quantity	Unit	Symbol
Length	meter	m
Area	square meter	m^2
Volume	cubic meter	m^3
Plane angle	degree or radian	$^\circ$ or rad ($360^\circ = 2\pi$ rad)
Solid angle	steradian	sr
Mass	kilogram	kg
Density	kilogram per cubic meter	$kg\,m^{-3}$
Time	second	s
Frequency	hertz (one cycle per second)	Hz
Speed, velocity	meter per second	$m\,s^{-1}$
Acceleration	meter per second squared	$m\,s^{-2}$
Force	newton	N
Pressure, mech. stress	pascal	Pa (1 Pa $= 1\,N\,m^{-2}$)
Energy	joule	J (1 J $= 1$ N m)
Power	watt	W ($1\,W = 1\,J\,s^{-1} = 1\,N\,m\,s^{-1}$)
Electric potential	volt	V
Resistance	ohm	Ω
Charge	coulomb	C
Electric displacement	coulomb per square meter	$C\,m^{-2}$
Electric current	ampére	A
Capacitance	farad	F
Electric field strength	volt per meter	$V\,m^{-1}$
Magnetic flux	weber	Wb
Magnetic induction	tesla	T
Magnetic field strength	ampére per meter	$A\,m^{-1}$
Inductance	henry	H
Temperature	kelvin (degree centigrade)	K (°C) ($T/K = 273.15 + T/^\circ C$)

12.3
Determination of the Point Space Group of a Crystal From Its Physical Properties

Group A: acentric PSG (1, 2, m, 22, mm, 3, 32, 3m, 4, 42, 4m, $\bar{4}$, $\bar{4}2$, 6, 62, 6m, $\bar{6}$, $\bar{6}2$, 23, 43, $\bar{4}3$);

Group B: centrosymmetric PSG ($\bar{1}$, 2/m, mmm, $\bar{3}$, $\bar{3}$m, 4/m, 4/mm, 6/m, 6/mm, m3, 4/m3).

A preliminary investigation of a crystalline powder using optical frequency doubling (SHG test, see Section 4.4.4) and the piezoelectric effect (Giebe–Scheibe method, see Section 4.4.1) or the longitudinal piezoelectric effect (micro-miniaturized for small crystals) yields the classification of a crystal structure into group A or B except for the PSG 42, 62, and 43, provided the observed effects are large enough.

Investigations using polarization microscopy enable the classification intothe following groups:

	A:	B:
optically isotropic	23, 43, $\bar{4}3$	m3, 4/m3
optically uniaxial	3, 32, 3m, 4, 42, 4m	$\bar{3}$, $\bar{3}$m, 4/m, 4/mm,
	$\bar{4}$, $\bar{4}$2, 6, 62, 6m, $\bar{6}$, $\bar{6}$2	6/m, 6/mm
optically biaxial	1, 2, m, 22, mm	$\bar{1}$, 2/m, mmm

Crystals from group A can be further classified according to their optical activity. Cubic crystals display optical activity only in PSG 23 and 43. In 1, 2, 22, 3, 32, 4, 42, 6, and 62, a rotation of the plane of polarization along the optical axis is observed.

Other properties that can be described by polar tensors of rank two, such as thermal expansion, thermal conductivity, dielectric peoperties, exhibit the same symmetry as the optical properties.

A pyroelectric effect can occur in the following PSG of group A: 1, 2, m, mm, 3, 3m, 4, 4m, 6, 6m.

The form of the piezoelectric (or electrooptic) tensor allows a unique determination of the PSG of acentric crystals, except for PSG 43 (no effect) and the pairs 42 and 62, 4m and 6m, 23 and $\bar{4}$3m, which cannot be distinguished. If one analyzes, in addition, the elastic properties, the pairs 42 and 62 as well as 4m and 6m can be distinguished, but not 23 and $\bar{4}$3. Crystals from group B (Laue groups) can be distinguished from their elastic properties, with the exception of 6/m and 6/mm, and m3 and 4/m3. The pairs 23 and $\bar{4}$3, m3 and 4/m3, 6/m and 6/mm can be distinguished macroscopically only with the aid of certain 4th-rank tensors (such as the piezooptic effect, electrogyration, or 2nd-order electrostriction). The PSG 43 and 4/m3 can best be distinguished from an analysis of their optical activity. If this provides no clear distinction, higher-order effects must be employed. Note, however, that all properties that are described by tensors of even rank can at most indicate to which of the eleven Laue groups the crystal belongs.

This procedure can in most cases be avoided, however, by analyzing the morphological properties (particularly spherical growth and etching tests) and diffractograms (particularly Laue diagrams). Also, a complete structure determination provides almost always a clear indication of the space group and the PSG of the investigated crystal.

12.4 Electric and Magnetic Effects and Properties up to 4th-Rank Tensors

See Tables 12.2 and 12.3.

Table 12.2 Electric and magnetic effects. $\boldsymbol{D}$ electric displacement, $\boldsymbol{E}$ electric field strength, $\boldsymbol{B}$ magnetic induction, $\boldsymbol{H}$ magnetic field strength, $\boldsymbol{I}$ electric current density, T temperature, $\{\varepsilon_{ij}\}$ tensor of mechanical deformation, $\{\sigma_{ij}\}$ tensor of mechanical stress. In place of the relations used here with independent quantities $\boldsymbol{E}, \boldsymbol{H}, \sigma$, and T, corresponding relations between $\boldsymbol{D}, \boldsymbol{B}$, and ε as inducing quantities can also be used in many cases. p: polar tensor, a: axial tensor.

Quantity	Electric phenomena	Type
Pyroelectricity	$\Delta P_i = \pi_i \Delta T$	p
Dielectric susceptibility	$D_i = \epsilon_{ij} E_j = \epsilon_0(\delta_{ij} + \kappa_{ij}) E_j$	p
Magnetoelectric susceptibility	$D_i = \alpha_{ij} H_j$	a
Electric conductivity	$I_i = s_{ij} E_j$	p
Piezoelectricity	$P_i = d_{ijk} \sigma_{jk}$	p
1st-order electrostriction	$\varepsilon_{ij} = \hat{d}_{ijk} E_k$	p
1st-order nonlinear dielectric susceptibility	$D_i = \epsilon_{ijk} E_j E_k$	p
1st-order electrooptic effect (Pockels)	$\Delta a_{ij} = r_{ijk} E_k$	p
Optical activity	$D_i = \epsilon_{ij} E_j + g_{ijk} \frac{\partial E_j}{\partial x_k}$ $(g_{ijk} = -g_{jik})$	p
1st-order magnetoelectric effect	$D_i = \gamma_{ijk} H_j E_k$	a
1st-order electric resistance change	$\Delta s_{ij} = s_{ijk} E_k$	p
2nd-order magnetoelectric susceptibility	$D_i = \alpha_{ijk} H_j H_k$	p
Hall effect	$E_i = k_{ijk} I_j H_k$	a
2nd-order electrostriction	$\varepsilon_{ij} = \hat{d}_{ijkl} E_k E_l$	p
2nd-order electric resistance change	$\Delta s_{ij} = s_{ijkl} E_k E_l$	p
Magnetoelectrostriction	$\varepsilon_{ij} = v_{ijkl} H_k E_l$	a
2nd-order nonlinear dielectric susceptibility	$D_i = \epsilon_{ijkl} E_j E_k E_l$	p
higher-order magnetoelectric effects	$D_i = u_{ijkl} H_j E_k E_l$	a
higher-order magnetoelectric effects	$D_i = \gamma_{ijkl} H_j H_k E_l$	p
higher-order magnetoelectric effects	$D_i = \alpha_{ijkl} H_j H_k H_l$	a
Electrically induced piezoelectricity	$P_i =^e d_{ijkl} E_j \sigma_{kl}$	p
Electrically induced optical activity (electrogyration)	$D_i = g_{ijkl} E_j \frac{\partial E_k}{\partial x_l}$	p
Magnetically induced piezoelectricity	$P_i =^m d_{ijkl} H_j \sigma_{kl}$	a
2nd-order electrooptic effect (Kerr effect)	$\Delta a_{ij} = r_{ijkl} E_k E_l$	p
Magnetoelectrooptic effect	$\Delta a_{ij} = n_{ijkl} H_k E_l$	a
2nd-order piezoelectricity	$P_i = \ldots + d_{ijklm} \sigma_{jk} \sigma_{lm}$	p

Table 12.3 Electric and magnetic effects (continued from Table 12.2).

Quantity	Magnetic phenomena	Type
Pyromagnetism	$\Delta B_i = \beta_i \Delta T$	a
Magnetic susceptibility	$B_i = \mu_{ij} H_j = \mu_0(\delta_{ij} + \kappa_{ij}) H_j$	p
Electromagnetic susceptibility	$B_i = \beta_{ij} E_j$	a
Piezomagnetism	$B_i = q_{ijk} \sigma_{jk}$	a
1st-order magnetostriction	$\varepsilon_{ij} = f_{ijk} H_k$	a
Nonlinear magnetic susceptibility	$B_i = \mu_{ijk} H_j H_k$	a
1st-order magnetooptic effect	$\Delta a_{ij} = m_{ijk} H_k$	a
Faraday effect	$D_i = \epsilon_{ij} E_j + z_{ijk} E_j H_k$ $(z_{ijk} = -z_{jik})$	a
1st-order electromagnetic effect	$B_i = \eta_{ijk} E_j H_k$	p
1st-order magnetic resistance change	$I_i = s_{ij} E_j + t_{ijk} E_j H_k$ $\Delta s_{ij} = t_{ijk} H_k$	a
2nd-order electromagnetic susceptibility	$B_i = \beta_{ijk} E_j E_k$	a
2nd-order magnetostriction	$\varepsilon_{ij} = f_{ijkl} H_k H_l$	p
Magnetoelectric resistance change	$\Delta s_{ij} = w_{ijkl} H_k E_l$	a
2nd-order magnetic resistance change	$\Delta S_{ij} = u_{ijkl} H_k H_l$	p
2nd-order nonlinear magnetic susceptibility	$B_i = \mu_{ijkl} H_j H_k H_l$	a
higher-order electromagnetic effects	$B_i = \eta_{ijkl} E_j H_k H_l$	a
higher-order electromagnetic effects	$B_i = \zeta_{ijkl} E_j E_k H_l$	p
higher-order electromagnetic effects	$B_i = \beta_{ijkl} E_j E_k E_l$	a
Magnetically induced piezomagnetism	$B_i =^m q_{ijkl} H_j \sigma_{kl}$	p
Magnetically induced optical activity (electrogyration)	$D_i =^m g_{ijkl} H_j \frac{\partial E_k}{\partial x_l}$	a
Electrically induced piezomagnetism	$B_i =^e q_{ijkl} E_j \sigma_{kl}$	a
2nd-order magnetooptical effect (Cotton-Mouton-Effekt)	$\Delta a_{ij} = m_{ijkl} H_k H_l$	p
2nd-order piezomagnetism	$B_i = \ldots + q_{ijklm} \sigma_{jk} \sigma_{lm}$	a

12.5
Tables of Standard Values

Data for more crystals can be obtained from the tables in "Landolt–Börnstein". Digits in parentheses indicate error intervals for the last digit given. All values are for approximately 293 K and standard pressure. Where the PSG is not indicated, the crystal is cubic.

Table 12.4 Density and specific heat.

	Density ρ [$g\,cm^{-3}$]	Specific heat C_p [$J\,g^{-1}K^{-1}$]
Diamond	3.5150(3)	0.511
Si	2.3283(2)	0.712
LiF	2.6406(2)	1.616
NaCl	2.1644(2)	0.870
CaF_2	3.1804(4)	0.912
$KAl(SO_4)_2 \cdot 12H_2O$	1.7530(3)	1.47
$CsAl(SO_4)_2 \cdot 12H_2O$	1.9995(3)	1.09
α-Quartz (32)	2.6481(3)	0.744
Calcite ($CaCO_3$) ($\bar{3}$m)	2.7102(3)	0.829
Quartz glass (Spectrosil)	2.200(1)	0.73(2)
Air (dry)	0.00121(1)	1.02(1)

Table 12.5 Pyroelectric effect.

	π_i^σ [$\mu C\,m^{-2}K^{-1}$]
Turmaline (3m)	$\pi_3^\sigma = 4.3$
Li_2GeO_3 (mm2)	$\pi_3^\sigma = -21.4$
$(CH_2NH_2COOH)_3 \cdot H_2SO_4$, TGS (2)	$\pi_2^\sigma = 350$
$(CH_2NH_2COOH)_3 \cdot H_2SeO_4$ (2)	$\pi_2^\sigma = 4200$
$BaTiO_3$ (4m)	$\pi_3^\sigma = 280$

Table 12.6 Relative permittivities of non-ferroelectric crystals at about 10 MHz at constant mechanical stress.

	$\epsilon_{rel,11}$
LiF	9.036(4)
NaCl	5.895(2)
CaF_2	6.799(2)
$KAl(SO_4)_2 \cdot 12H_2O$	6.55(5)
$CsAl(SO_4)_2 \cdot 12H_2O$	5.50(5)
α-Quartz (32)	$\epsilon_{rel,11} = 4.520(2)$, $\epsilon_{rel,33} = 4.640(2)$

Table 12.7 Relative permittivities of ferroelectric crystals at about 1 kHz close to the Curie temperature T_c. The asterisk * denotes the direction of the ferroelectric polarization P_s.

	T_c [K]	$\epsilon_{rel,11}$	$\epsilon_{rel,22}$	$\epsilon_{rel,33}$	P_s [$10^4 \mu Cm^{-2}$]
$NaKC_4H_4O_6 \cdot 4H_2O$	297	$\approx 3 \cdot 10^{3*}$	≈ 8.8	≈ 9.4	0.25
(Seignette salt, $22 \rightarrow 2$ and $2 \rightarrow 22$)[1]	255	$\approx 3 \cdot 10^{3*}$	≈ 8.8	≈ 9.4	
$(CH_2NH_2COOH)_3 \cdot H_2SO_4$ (TGS, $2/m \rightarrow 2$)	322	≈ 9.0	$\approx 5 \cdot 10^{5*}$	≈ 8	2.8
$BaTiO_3$ ($4/m3 \rightarrow 4m$)	408	$\approx 10^4$	$\approx 10^4$	$\approx 10^{4*}$	26.0
KH_2PO_4 (KDP, $\bar{4}2 \rightarrow m2$)	123	≈ 50	≈ 50	$\approx 10^{5*}$	4.75

[1] Seignette salt is ferroelectric in the interval between the Curie temperatures 297 K and 255 K.

Table 12.8 Refractivities for $\lambda_0 \approx 589$ nm.

	$n_1 = n_2$	n_3
Diamond	2.4190(1)	
Silicon	4.21(2)	
LiF	1.3915(1)	
NaCl	1.5443(1)	
$KAl(SO_4)_2 \cdot 12H_2O$	1.4564(1)	
CaF_2	1.43385(5)	
$MgAl_2O_4$	1.7274(1)	
$Y_3Al_5O_{12}$ (YAG)	1.823(1)	
α-Quartz (32)	1.54426(3)	1.55337(3)
Calcite $CaCO_3$ ($\bar{3}m$)	1.65835(5)	1.48640(5)
$NaNO_3$ ($\bar{3}m$)	1.5848(2)	1.3360(2)
Air (dry)	$n = 1.000272(2)$; $dn/dT = -0.93(2) \cdot 10^{-6}\ K^{-1}$	

Table 12.9 Specific optic rotation for $\lambda_0 = 589$ nm.

	α [grad mm^{-1}]
$NaBrO_3$ (23)	2.12
$Na_3SbS_4 \cdot 9H_2O$ (23)	2.35
α-Quartz (32)	21.72
Benzil $C_6H_5CO \cdot COC_6H_5$ (32)	24.84
$LiIO_3$ (6)	99.3
$Al(IO_3)_3 \cdot 2HIO_3 \cdot 6H_2O$ (6)	34.9

Table 12.10 Coefficients of thermal expansion.

	α_{11} $[10^{-6}K^{-1}]$		
Diamond	0.87		
Si	3.08(5)		
Cu	16.9(2)		
LiF	32.0(3)		
NaCl	39.1(3)		
$KAl(SO_4)_2 \cdot 12H_2O$	13.6(2)		
$CsAl(SO_4)_2 \cdot 12H_2O$	27.3(3)		
CaF_2	19(1)		
α-Quartz (32)	13.3(3)	$\alpha_{33} = 7.07(7)$	
Calcite ($\bar{3}$m)	−3.7(1)	$\alpha_{33} = 24.7(3)$	
$NaNO_3$ ($\bar{3}$m)	8.8(3)	$\alpha_{33} = 107(3)$	
$Ca(HCOO)_2$ (mmm)	−16.5(5)	$\alpha_{22} = 68.6(5)$	$\alpha_{33} = 29.8(5)$
Quartz glass (Spectrosil, $\rho = 2.200$)	0.5(1)		

Table 12.11 Magnetic susceptibilities in SI units.

	κ_{11} $[10^{-6}]$	κ_{33}
Diamond	−21.6	
Si	−3.3	
LiF	−12.9	
NaCl	−14.0	
CaF_2	−14.3	
$KAl(SO_4)_2 \cdot 12H_2O$	−11.7	
$KCr(SO_4)_2 \cdot 12H_2O$	+280	
$NH_4Al(SO_4)_2 \cdot 12H_2O$	−11.1	
$NH_4Fe(SO_4)_2 \cdot 12H_2O$	+650	
α-Quartz (32)	−15.6	−15.3
Calcite ($\bar{3}$m)	−12.1	−13.5

Table 12.12 Coefficients of electric conductivities (a definitive indication of errors is not possible because of the large contributions from crystal defects).

	s_{11} $[\Omega^{-1}cm^{-1}]$	s_{33}
Al	$4 \cdot 10^5$	
Fe	$1.16 \cdot 10^5$	
Cu	$6.45 \cdot 10^5$	
Si (high-purity)	$5 \cdot 10^{-2}$	
NaCl	$2 \cdot 10^{-8}$	
Zn (6/mm)	$1.72 \cdot 10^5$	$1.62 \cdot 10^5$
$LiIO_3$ (50 Hz) (6)	10^{-9}	$0.5 \cdot 10^{-6}$
Bi ($\bar{3}$m)	$0.917 \cdot 10^4$	$0.725 \cdot 10^4$

Table 12.13 Coefficients of thermal conductivity.

	λ_{11} [$Jm^{-1}s^{-1}K^{-1}$]		
Diamond (type I)	900(50)		
NaCl	6.2(2)		
$KAl(SO_4)_2 \cdot 12H_2O$	0.605(6)		
$CsAl(SO_4)_2 \cdot 12H_2O$	0.677(6)		
KH_2PO_4 ($\bar{4}2$)	$\lambda_{11} = 1.78(5)$	$\lambda_{33} = 1.36(4)$	
α-Quartz (32)	$\lambda_{11} = 6.2(3)$	$\lambda_{33} = 10.4(5)$	
$NaNO_3$ ($\bar{3}$m)	$\lambda_{11} = 1.64(3)$	$\lambda_{33} = 2.14(7)$	
$Ca(HCOO)_2$ (mmm)	$\lambda_{11} = 1.35(2)$	$\lambda_{22} = 0.561(5)$	$\lambda_{33} = 0.99(1)$

Table 12.14 Adiabatic piezoelectric constants for low frequencies (about 10 kHz).

	d_{ijk} [pCN^{-1}] ($= [10^{-12}mV^{-1}]$)		
$NaBrO_3$ (23)	$d_{123} = 1.21(1)$		
KH_2PO_4 ($\bar{4}2$)	$d_{123} = 0.65(2)$,	$d_{312} = -10.5(2)$	
α-Quartz (32)	$d_{111} = 2.31(1)$,	$d_{123} = -0.37(1)$	
$LiNbO_3$ (3m)	$d_{113} = 37.0(3)$,	$d_{222} = 20.8(1)$	$d_{311} = -0.86(2)$
	$d_{333} = 16.2(2)$		

Table 12.15 Adiabatic electrooptic constants for low frequencies (about 10 kHz) and $\lambda = 633$ nm (except for GaAs).

	r_{ijk} [$pm\,V^{-1}$]	
$NaBrO_3$ (23)	$r_{123} = 0.57(3)$	
$NaClO_3$ (23)	$r_{123} = 0.40(2)$	
GaAs ($\bar{4}3$)	$r_{123} = 1.43(7)$ (1150 nm)	
Hexamethylenetetramine ($\bar{4}3$)	$r_{123} = 0.80(5)$	
KH_2PO_4 ($\bar{4}2$)	$r_{123} = 10.3(2)$	$r_{231} = 8.6(2)$
α-Quartz (32)	$r_{111} = 0.48(1)$	$r_{231} = 0.23(1)$
$LiNbO_3$ (3m)	$r_{113} = 9.6$	$r_{222} = 6.8$
	$r_{311} = 32.6$	$r_{333} = 30.9$

Table 12.16 Nonlinear optical coefficients $\epsilon^*_{ijk} = \epsilon_0^{-1}\epsilon_{ijk}$ for the vacuum wavelength of the primary wave $\lambda = 1064$ nm (Nd glass laser).

	ϵ^*_{ijk} [$10^{-12}V^{-1}m$]		
GaAs ($\bar{4}3$)	$\epsilon^*_{123} = 134$		
$LiIO_3$ (6)	$\epsilon^*_{311} = -7.1$;	$\epsilon^*_{333} = -7.0$;	$\epsilon^*_{123} = 0.31$
KH_2PO_4 ($\bar{4}2$)	$\epsilon^*_{312} = 0.63$		
$NH_4H_2PO_4$ ($\bar{4}2$)	$\epsilon^*_{312} = 0.76$		
$LiNbO_3$ (3m)	$\epsilon^*_{311} = 5.9$;	$\epsilon^*_{333} = -34$;	$\epsilon^*_{222} = 4.0$
α-Quartz (32)	$\epsilon^*_{111} = 0.50$;	$\epsilon^*_{123} = -0.005$	
α-HIO_3 (22)	$\epsilon^*_{123} = 8.3$		

Table 12.17 Elastic constants c_{ij} $[10^{10}\text{N m}^{-2}]$.

	c_{11}	c_{12}	$c_{44} = c_{66}$
Diamond	104	17	55
Si	16.5(1)	6.4(1)	7.92(3)
LiF	11.37(2)	4.76(3)	6.35(3)
NaCl	4.944(8)	1.29(2)	1.266(5)
γ-NaCN	2.534(2)	1.444(6)	0.033(1)
CsCl	3.64(1)	0.92(2)	0.80(1)
CaF_2	16.357(20)	4.401(16)	3.392(13)
$Y_3Ga_5O_{12}$	28.70(1)	11.60(6)	9.04(4)
$KAl(SO_4)_2 \cdot 12H_2O$	2.465(3)	1.021(8)	0.867(5)
$CsAl(SO_4)_2 \cdot 12H_2O$	3.118(5)	1.541(9)	0.840(5)
$NaBrO_3$	5.478(5)	1.628(6)	1.505(5)
KH_2PO_4 ($\bar{4}2$)	$c_{11} = 7.165(5)$	$c_{33} = 5.640(5)$	$c_{12} = -0.627(6)$
	$c_{13} = 1.494(8)$	$c_{66} = 0.621(4)$	$c_{44} = 1.248(6)$
α-Quartz (32)	$c_{11}^E = 8.674$	$c_{33}^E = 10.72$	$c_{12}^E = 0.698$
	$c_{13}^E = 1.19$	$c_{44}^E = 5.79$	$c_{14}^E = -1.79$
Ga (mmm)	$c_{11} = 10.16(1)$	$c_{22} = 9.156(10)$	$c_{33} = 13.64(1)$
	$c_{12} = 4.601(10)$	$c_{13} = -3.057(10)$	$c_{23} = 2.804(10)$
	$c_{66} = 4.079(10)$	$c_{55} = 4.155(10)$	$c_{44} = 3.499(10)$

Table 12.18 Thermoelastic constants $T_{ij} = d \log c_{ij}/dT$ $[10^{-3}\text{K}^{-1}]$.

	T_{11}	T_{12}	$T_{44} = T_{66}$
Diamond	−0.0137	−0.057	−0.0125
Si	−0.081	−0.11	−0.063
LiF	−0.66	+0.01	−0.28
NaCl	−0.80	+0.17	−0.266
γ-NaCN	+0.47	−1.38	+27
CsCl	−0.42	−0.93	−1.29
CaF_2	−0.205	−0.291	−0.343
$Y_3Ga_5O_{12}$	−0.113	−0.092	−0.092
$KAl(SO_4)_2 \cdot 12H_2O$	+0.108	+1.91	−0.95
$CsAl(SO_4)_2 \cdot 12H_2O$	−0.517	−0.69	−0.43
$NaBrO_3$	−0.72	−0.45	−0.58
KH_2PO_4 ($\bar{4}2$)	$T_{11} = -0.635$	$T_{33} = -0.49$	$T_{12} = -0.62$
	$T_{13} = -0.185$	$T_{66} = -0.53$	$T_{44} = -0.57$
α-Quartz (32)	$T_{11} = -0.0443$	$T_{33} = -0.160$	$T_{12} = -2.69$
	$T_{13} = -0.55$	$T_{44} = -0.175$	$T_{14} = +0.11$
Ga (mmm)	$T_{11} = -0.47$	$T_{22} = -0.42$	$T_{33} = -0.38$
	$T_{12} = -0.029$	$T_{13} = +0.05$	$T_{23} = +0.115$
	$T_{66} = -0.72$	$T_{55} = -0.43$	$T_{44} = -0.59$

Table 12.19 Piezoelastic constants $P_{ij} = dc_{ij}/dp$ (dimensionless).

	P_{11}	P_{12}	P_{44}	$\frac{dK^{-1}}{dp}$
Diamond	5.72	4.18	3.19	4.69
Si	4.32	4.22	0.80	4.26
LiF	9.86	2.76	1.40	5.12
NaCl	11.85	2.06	0.37	5.32
γ-NaCN	5.55	5.98	−0.36	5.84
CsCl	6.77	5.13	3.52	5.68
CaF_2	6.11	4.55	1.32	5.07
$KAl(SO_4)_2 \cdot 12H_2O$	6.73	7.15	2.77	7.01
$CsAl(SO_4)_2 \cdot 12H_2O$	5.83	6.05	1.17	5.98
$MgAl_2O_4$	5.08	4.93	0.86	4.98
$K_2Hg(CN)_4$	−6.51	−6.23	−1.22	−6.32
α-Quartz (32)	$P_{11} = 3.30$	$P_{33} = 10.93$	$P_{12} = 83.7$	$\frac{dK^{-1}}{dp} = 19.8$
	$P_{13} = 5.95$	$P_{44} = 2.66$	$P_{14} = 2$	

Table 12.20 Piezooptic constants q_{ij} $[10^{-12}\text{m}^2\text{N}^{-1}]$ and elastooptic constants q_{ij} (dimensionless) for $\lambda_0 = 589$ nm. Abbreviations: $q_{11} = q_{1111}$; $q_{12} = q_{1122}$; $q_{44} = 2q_{2323}$ etc.; $p_{11} = p_{1111}$; $p_{12} = p_{1122}$; $p_{44} = p_{2323}$ etc.

	q_{11}	q_{12}	q_{44}	p_{11}	p_{12}	p_{44}
LiF (4/m3)	−0.40	1.12	−0.83	0.061	0.161	−0.053
NaCl (4/m3)	1.27	2.58	−0.84	0.129	0.177	−0.011
RbI (4/m3)	9.23	4.99	−7.92	0.275	0.181	−0.022
CaF_2 (4/m3)	−0.33	1.10	0.74	0.038	0.226	0.0254
Hexamethylenetetramine ($\bar{4}3$)	11.95	8.42	7.56	0.259	0.227	0.039
$KAl(SO_4)_2 \cdot 12H_2O$ (m3)	2.43	6.66	−0.55	0.199	0.260	−0.005
	$q_{13} = 6.99$			$p_{13} = 0.265$		
$CsAl(SO_4)_2 \cdot 12H_2O$ (m3)	−0.63	4.25	−1.82	0.124	0.201	−0.015
	$q_{13} = 5.05$			$p_{13} = 0.214$		
α-Quartz (32)	$q_{11} = 1.25$			$p_{11} = 0.15$		
	$q_{12} = 2.60$			$p_{12} = 0.26$		
	$q_{13} = 1.95$			$p_{13} = 0.265$		
	$q_{14} = -0.10$			$p_{14} = -0.029$		
	$q_{31} = 2.94$			$p_{31} = 0.27$		
	$q_{33} = 0.12$			$p_{33} = 0.10$		
	$q_{41} = -0.33$			$p_{41} = -0.045$		
	$q_{44} = -1.07$			$p_{44} = -0.074$		

Table 12.21 Nonlinear elastic coefficients c_{ijk} $[10^{10} \mathrm{N\,m^{-2}}]$; definition according to Brugger (1964).

	c_{111}	c_{112}	c_{113}	c_{123}	c_{144}	c_{155}	c_{166}	c_{456}
Si (4/m3)	−81.5	−44.8	−44.8	−7.0	−31	−31	−31	−7.5
LiF (4/m3)	−142	−26.4	−26.4	15.6	8.5	−27.3	−27.3	9.4
NaCl (4/m3)	−84.3	−3.3	−3.3	−3.6	2.0	−6.0	−6.0	2.0
CaF_2 (4/m3)	−124.6	−40	−40	−25.4	−12.4	−21.4	−21.4	−7.5
$KAl(SO_4)_2 \cdot 12H_2O$ (m3)	−22.2	−7.1	−8.6	−13.4	−2.3	−8.0	−7.44	−2.0
$CsAl(SO_4)_2 \cdot 12H_2O$ (m3)	−21.2	−11.1	−12.6	−9.0	−2.7	−5.90	−5.36	−1.6
α-Quartz (32)	$c_{111} = -21.4$;	$c_{112} = -34.1$;	$c_{113} = +1.9$;	$c_{114} = -15.5$;				
	$c_{123} = 28.2$;	$c_{124} = -2.0$;	$c_{133} = -31.8$;	$c_{134} = +35$;				
	$c_{144} = -15.8$;	$c_{155} = -17.9$;	$c_{222} = -33.6$;	$c_{333} = -82.6$;				
	$c_{344} = -12.9$;	$c_{444} = -22.6$;						

12.6 Bibliography

12.6.1 Books

M. Berek Anleitung zu optischen Untersuchungen mit dem Polarisationsmikroskop (Rinne-Berek). E. Schweizerbart'sche Verlagsbuchhandlung, Stuttgart 1953.

S. Bhagavantam Crystal symmetry and physical properties. Academic Press, New York 1966.

S. Bhagavantam, T. Venkatarayudu Theory of groups and its application for physical problems. Academic Press, New York, London 1969.

A. R. Billings Tensor properties of materials. Wiley-Interscience, London, New York, Sydney, Toronto 1969.

R. R. Birss Symmetry and magnetism. North-Holland, Amsterdam 1964.

N. Bloembergen Nonlinear optics. Benjamin, New York 1965.

F. D. Bloss An introduction to the methods of optical crystallography. Holt, Rinehart and Winston, New York 1961.

M. Born, K. Huang The dynamical theory of crystal lattices. Clarendon Press, Oxford 1954.

M. Born, E. Wolf Principles of optics. Macmillan, New York 1959.

F. C. Brown The physics of solids. Benjamin, New York 1967.

E. Buchwald Einführung in die Kristalloptik. Walter de Gruyter, Berlin 1937.

J. J. Burkhardt Die Bewegungsgruppen in der Kristallographie. Birkhäuser, Basel 1947.

W. G. Cady Piezoelectricitiy. McGraw-Hill, New York 1946.

A. P. Cracknell Applied group theory. Pergamon Press, Oxford 1967.

J. P. Elliott, P. G. Dawber Symmetry in physics. Vols. 1 and 2. Macmillan, London 1979.

J. C. Erdmann Wärmeleitung in Kristallen, theoretische Grundlagen und fortgeschrittene experimentelle Methoden. Springer, Berlin, Heidelberg, New York 1969.

L. Eyraud Diélectriques solides anisotropes et ferroélectricité. Gauthiers-Villars, Paris 1967.

F. I. Fedorov Theory of elastic waves in crystals. Übers. aus d. Russ. Plenum Press, New York 1968.

S. Flügge (ed.) Handbuch der Physik. Vol. XXV/2a (L. Genzel ed.) Licht und Materie, Springer, Berlin 1967.

A. J. Freeman, H. Schmid (eds.) Magnetoelectric interaction phenomena in crystals. Gordon and Breach, London 1975.

H. Fröhlich Theory of dielectrics. Clarendon Press, Oxford 1958, 1968 (2nd ed.).

W. Gebhardt, U. Krey Phasenübergänge und kritische Phänomene. Vieweg, Braunschweig, Wiesbaden 1980.

H. Geiger, K. Scheel (eds.) Handbuch der Physik. Vol. XXIV, 2. Aufl., Aufbau der zusammenhängenden Materie (A. Smekal ed.). Springer, Berlin 1933.

C. Giacovazzo (ed.) Fundamentals of crystallography. International Union of Crystallography and Oxford University Press, New York 1992.

P. Groth Elemente der physikalischen und chemischen Kristallogrophie. R. Oldenbourg, München, Berlin 1921.

C. Hamann, J. Heim, H. Burghardt Organische Leiter, Halbleiter und Photoleiter. Fr. Vieweg & Sohn, Braunschweig 1981.

W. A. Harrison Solid state theory. McGraw-Hill, New York 1970.

R. F.S. Hearmon Applied anisotropic elasticity. University Press, Oxford 1961.

K.-H. Hellwege Einführung in die Festkörperphysik, 2. Aufl. Springer, Berlin 1981.

R. Houwink Elastizität-Plastizität und Struktur der Materie, 3. Aufl. Theodor Steinkopff, Dresden, Leipzig 1958.

H. B. Huntington The elastic constants of crystals. Academic Press, New York, London 1958.

H. Jaffe Crystal chemistry and refractivity. Cambridge University Press, Cambridge 1988.

H. Jagodzinski Kristallographie. In: Handbuch der Physik (S. Flügge ed.). II, 1. Springer, Berlin 1955.

F. Jona, G. Shirane Ferroelectric crystals. Pergamon Press, Oxford, London 1962.

H. J. Juretschke Crystal physics. Benjamin, London, Amsterdam 1974.

C. Kittel Introduction to solid state physics. 4. Aufl. Wiley, New York 1971.

W. Kleber Angewandte Gitterphysik, 3. Aufl. Walter de Gruyter, Berlin 1960.

W. Kleber Einführung in die Kristallographie, 13. Aufl. VEB Verlag Technik, Berlin 1977.

W. Kleber, K. Meyer, W. Schoenborn Einführung in die Kristallphysik. Akademie-Verlag, Berlin 1968.

R. S. Knox, A. Gold Symmetry in the solid state. Benjamin, New York, Amsterdam 1964.

E. Kordes Optische Daten zur Bestimmung anorganischer Substanzen. Verlag Chemie, Weinheim 1960.

R. S. Krishnan (ed.) Progress in crystal physics. Vol. 1: thermal, elastic and optical properties. Viswanathan, Madras 1958.

R. S. Krishnan, R. Srinivasan, S. Devanarayanan Thermal expansion of crystals. Pergamon Press, Oxford, New York 1979.

E. Kröner Elastizität und Plastizität der Vielkristalle. In: Mechanische Anisotropie p. 149 (H. P. Stüwe, ed.). Springer, Wien, New York 1974.

W. D. Kusnezow Einfluß der Oberflächenenergie auf das Verhalten fester Körper. Akademie-Verlag, Berlin 1961.

L. D. Landau, E. M. Lifshitz Statistical physics (Vol. 5 Course of theoretical physics). Übers. aus d. Russ. Pergamon Press, Oxford, New York 1968. (2., verbess. Aufl.).

G. Leibfried Gittertheorie der mechanischen und thermischen Eigenschaften der Kristalle. In: Handbuch der Physik VII/1, pp. 104–324. Springer, Berlin 1955.

G. Leibfried, W. Ludwig Theory of anharmonic effects in crystals. In: Solid State Physics, Vol. 12, pp. 275–444 (F. Seitz, D. Turnbull eds.). Academic Press, New York, London 1961.

M. Levy, H. E. Bass, R. R. Stern Handbook of elastic properties of solids, liquids, and gases. Vols. I–IV. Academic Press, San Diego 2001.

Th. Liebisch Physikalische Kristallographie. Veit, Leipzig 1891.

M. E. Lines, A. M. Glass Principles and applications of ferroelectrics and related materials. Clarendon Press, Oxford 1977.

G. J. Ljubarski Anwendungen der Gruppentheorie in der Physik. VEB Deutscher Verlag der Wissenschaften, Berlin 1962.

A. E.H. Love A treatise of the mathematical theory of elasticity. Dover, New York 1944.

W. Ludwig Recent developments in lattice theory. Springer tracts in modern physics, Vol. 43 (G. Höhler ed.). Springer, Berlin 1967.

W. Ludwig Festkörperphysik. 2nd ed. Akademische Verlagsgesellschaft, Wiesbaden 1978.

B. Lüthi Interaction of magnetic ions with phonons. In: Dynamical Properties of Solids. Vol. 3, pp. 245–292 (G. K. Horton, A. A. Maradudin eds.). North-Holland, Amsterdam 1980.

B. Lüthi, W. Rehwald Ultrasonic studies near structural phase transitions. In: Structural phase transitions I, pp. 131–184 (K. A. Müller, H. Thomas eds.). Springer, Berlin 1981.

O. Madelung Festkörpertheorie I & II. Springer, Berlin 1972.

A. A. Maradudin, E. W. Montroll, G. H. Weiss Theory of lattice dynamics in the harmonic approximation. Solid State Physics, Suppl. 3 (F. Seitz, D. Turnbull eds.). Academic Press, New York, London 1963.

A. S. Marfunin Physics of minerals and inorganic materials. Übers. aus d. Russ. Springer, Berlin 1979.

H. J. Martin Die Ferroelektrika. Tech. phys. Monographien. Vol. 15. Akad. Verlagsgesellschaft Geest & Portig, Leipzig 1964.

W. P. Mason Piezoelectric crystals and their application to ultrasonics. Van Nostrand, Princeton, NJ 1950.

W. P. Mason (ed.) Physical acoustics, Vols. I–XII. Academic Press, New York, London 1964–1976.

W. P. Mason Crystal physics of interaction processes. Academic Press, New York 1966.

H. D. Megaw Ferroelectricity in crystals. Methuen, London 1957.

A. Migliori, J. L. Sarrao Resonant ultrasound spectroscopy. Wiley, New York 1997.

B. W. Mott Die Mikrohärteprüfung. Berliner Union, Stuttgart 1957.

M. J.P. Musgrave Crystal acoustics. Holden-Day, San Francisco 1970.

T. S. Narasimhamurty Photoelastic and electro-optic properties of crystals. Plenum Press, New York, London 1981.

R. E. Newnham Structure-property relations. Springer, Berlin 1975.

A. Nowick Crystal properties via group theory. Cambridge University Press, Cambridge 1995.

J. F. Nye Physical properties of crystals. Oxford University Press, London, New York 1957.

P. Paufler Phasendiagramme. Akademie-Verlag, Berlin 1981.

P. Paufler Physikalische Kristallographie. Akademie-Verlag, Berlin 1986.

P. Paufler, G. E.R. Schulze Physikalische Grundlagen mechanischer Festkörpereigenschaften I u. II. Vieweg, Braunschweig 1978.

F. Pockels Lehrbuch der Kristalloptik. Teubner, Leipzig 1906.

H. Poulet, J. P. Mathieu Vibration spectra and symmetry of crystals. Gordon and Breach, New York 1976.

H. Rabin, C. L. Tang (eds.) Quantum electronics: a treatise. Vol. I, Nonlinear Optics, Parts A and B. Academic Press, New York 1975.

G. N. Ramachandran, S. Ramaseshan Crystal optics. In: Handbuch der Physik (S. Flügge ed.). Vol. XXV/1. Springer, Berlin 1967.

R. Rath Theoretische Grundlagen der allgemeinen Kristalldiagnose im durchfallenden Licht. Springer, Berlin 1969.

Lord Rayleigh The theory of sound. Dover, New York 1945.

C. Z. Rosen, B. V. Hiremath, R. Newnham (eds.) Key papers in physics: Piezoelectricity. American Institute of Physics, New York 1992.

A. Seeger Kristallplastizität. In: Handbuch der Physik (S. Flügge ed.) Vol. VII/2 Kristallphysik II, pp. 1–210. Springer, Berlin 1958.

J. N. Sherwood (ed.) The plastically crystalline state. Wiley, Chichester 1979.

L. A. Shuvalov (ed.) Modern Crystallography IV (Physical properties of crystals). Übers. aus d. Russ. Springer, Berlin 1983.

G. A. Smolenskij, N. N. Krajnik Ferroelektrika und Antiferroelektrika. Teubner, Leipzig 1972.

U. Stille Messen und Rechnen in der Physik. Fr. Vieweg & Sohn, Braunschweig 1955.

G. Szivessy Kristalloptik. In: Handbuch der Physik. Vol. XX. Springer, Berlin 1928.

H. Tertsch Die Festigkeitserscheinungen der Kristalle. Springer, Wien 1947.

J. Tichý, G. Gautschi Piezoelektrische Meßtechnik. Springer, Berlin 1980.

H. F. Tiersten Linear piezoelectric plate vibrations. Plenum Press, New York 1969.

R. Truell, C. Elbaum, B. B. Chick Ultrasonic methods in solid state physics. Academic Press, New York, London 1969.

W. Voigt Lehrbuch der Kristallphysik. B. Teubner, Leipzig 1910 (2. Aufl. 1928).

D. C. Wallace Thermoelastic theory of stressed crystals and higher-order elastic constants. In: Solid State Physics, Vol. 25, pp. 301–404. Academic Press, New York, London 1970.

D. C. Wallace Thermodynamics of crystals. Wiley, New York 1972.

L. M. Wasiliev Physikalische Kristallographie (in Russian). Metallurgia, Moscow 1981.

W. A. Wooster A text-book on crystal physics. Cambridge University Press, Cambridge 1949.

W. A. Wooster Experimental crystal physics. Oxford University Press, New York 1957.

W. A. Wooster Diffuse X-ray reflections from crystals. Clarendon Press, Oxford 1962.

W. A. Wooster Tensors and group theory for the physical properties of crystals. University Press, Oxford 1973.

G. S. Zhdanov Crystal physics. Oliver & Boyd, Edinburgh, London 1965.

I. S. Zheludev Physics of crystalline dielectrics, I & II. Plenum Press, New York, London 1971.

J. M. Ziman Prinzipien der Festkörpertheorie. Harri Deutsch, Zürich, Frankfurt a.M., 1975.

12.6.2 Articles

K. Aizu Possible species of "ferroelastic crystals" and of simultaneously ferroelectric and ferroelastic crystals. J. Phys. Soc. Japan **27** (1969) 387–396.

C. Andeen, J. Fontanella, D. Schuele Low-frequency dielectric constants of the alkaline earth fluorides by the method of substitution. J. Appl. Phys. **42** (1971) 2216–2219.

R. Bechmann Contour modes of square plates excited piezoelectrically and determination of elastic and piezoelectric coefficients. Proc. Phys. Soc. (London) **B64** (1951) 323–337.

P. S. Bechthold Quantitative Interpretation der Maker-Interferenzen bei der Frequenzverdopplung in optisch zweiachsigen Kristallen. Diss. University Köln 1976.

L. Bohatý, S. Haussühl Quadratic electrostrictive effects in LiF, NaF, NaCl, NaBr, KCl, KBr, KI, RbCl, RbBr and RbI. Acta Crystallogr. **A33** (1977) 114–118.

K. Brugger Thermodynamic definition of higher order elastic coefficients. Phys. Rev. **A133** (1964) 1611–1612.

M. J. Buerger Crystallographic aspects of phase transformations. In: Phase Transformations in Solids, pp. 183–209. Wiley, New York 1951.

M. J. Buerger Phase transformations. Sov. Phys. Crystallography **16** (1972) 959–968.

C. F. Buhrer, L. Ho, J. Zucker Electrooptic effect in optically active crystals. Appl. Optics **3** (1964) 517–521.

S. Chandra, M. P. Hemkar Evaluation of the elastic constants of 1,3,5-triphenylbenzene from thermal diffuse scattering of X-rays. Acta Cryst. **A29** (1973) 25–28.

H. H. Demarest Cube-resonance method to determine the elastic constants of solids. J. Acoust. Soc. Am. **49** (1969) 768–775.

R. W. Dixon Acoustic diffraction of light in anisotropic media. IEEE J. Quant. Electron. **QE-3** (1967) 85–93.

H. J. Eichler Forced light scattering at laser induced gratings–a method for investigation of optically excited solids. Adv. Solid State Phys. **XVIII** (1978) 241–263.

W.v. Engelhardt, S. Haussühl Schleiffestigkeit und spezifische freie Grenzflächenenergie der Alkalihalogenide vom NaCl-Typus. Kolloid Zeitschr. **173** (1960) 20–35.

W.v. Engelhardt, S. Haussühl Festigkeit und Härte von Kristallen. Fortschr. Miner. **42** (1965) 5–49.

P. P. Ewald, C. Hermann Strukturbericht 1913–1928. Z. Kristallogr. Ergänzungsband 1931. Akad. Verlagsgesellschaft, Leipzig 1931.

R. L. Forgacs Improvements in the sing-around technique for ultrasonic velocity measurements. J. Acoust. Soc. Amer. **32** (1960) 1697–1698.

J. Gahm Ein neuer Mikrohärteprüfer. Zeiss Informationen 14. Jahrgang (1966) 120–127.

M. W. Guinan, A. D. Ritchie Evaluation of third-order elastic constants for cubic crystals. J. Appl. Phys. **41** (1970) 2256–2258.

U. Hauser, V. Oestreich, M. D. Rohrweck On optical dispersion in transparent molecular systems. Z. Physik **A280** (1977) 17–25, 125–130 et ibid. **A284** (1978) 9–19.

S. Haussühl Physikalische Kristallographie von Calcium- und Cadmiumformiat. Z. Kristallogr. **118** (1963) 33–50.

S. Haussühl Die Abweichungen von den Cauchy-Relationen. Phys. kondens. Materie **6** (1967) 181–192.

S. Haussühl Elastische und thermoelastische Konstanten von Benzil $C_6H_5COCOC_6H_5$, gemessen mit dem Schaefer-Bergmann-Verfahren. Acta Cryst. **23** (1967) 666–667.

S. Haussühl Elastic and thermoelastic properties of 1,3,5-triphenyl-benzene derived from ultrasonic wave velocities. Acta Cryst. **A30** (1974) 106–107.

S. Haussühl Ferroelasticity and phase transformation in $Rb_2Hg(CN)_4$ spinel. Acta Cryst. **A34** (1978) 965–968.

S. Haussühl Pyroelectric, dielectric, piezoelectric, elastic, and thermoelastic properties of triclinic lithium hydrogen oxalate hydrate, $LiHC_2O_4 \cdot H_2O$. VII. Intern. Conf. on Crystal Growth, Stuttgart 1983.

S. Haussühl, W. Müller Zur Auflösungsgeschwindigkeit von Kristallen in Wasser. Kristall und Technik **7** (1972) 533–554.

S. Haussühl, H. Siegert Bestimmung des Elastizitätstensors trikliner Kristalle: Beispiel $CuSO_4 \cdot 5H_2O$. Z. Kristallogr. **129** (1969) 142–246.

S. Haussühl, K. Uhl Die Bestimmung des piezooptischen Tensors monokliner Kristalle: Beispiel Taurin. Z. Kristallogr. **128** (1969) 418–423.

M. V. Hobden Phase-matched second-harmonic generation in biaxial crystals. J. Appl. Phys. **38** (1967) 4365–4372.

A. R. Hutson, D. L. White Elastic wave propagation in piezoelectric semiconductors. J. Appl. Phys. **33** (1962) 40–47.

J. Jerphagnon A contribution to the study of certain non-linear optical phenomena. Ann. Telecommunications **23** (1968) 203–232.

J. Jerphagnon, S. K. Kurtz Maker Fringes: a detailed comparison of theory and experiment for isotropic and uniaxial crystals. J. Appl. Phys. **41** (1970) 1667–1681.

R. C. Jones A new calculus for the treatment of optical systems. VII. Properties of the N-matrices. J. Opt. Soc. Amer. **38** (1948) 671–685.

S. K. Joshi, B. M.S. Kashyap Determination of elastic constants of benzalazine from thermal diffuse scattering of X-rays. Acta Cryst. **17** (1964) 629–632.

D. A. Kleinman Theory of second harmonic generation of light. Phys. Rev. **128** (1962) 1761–1775.

E. Kröner Calculation of the elastic constants of a polycrystalline material from the elastic constants of the single crystal. Z. Physik **151** (1958) 504–518.

H. Küppers Die Messung von Schallgschwindigkeiten in optisch anisotropen Medien mit dem Schaefer-Bergmann-Verfahren. Acta Cryst. **A27** (1971) 316–322.

J. J. Kyame Wave propagation in piezoelectric crystals. J. Acoust. Soc. Amer. **21** (1949) 159–167.

R. G. Leisure, F. A. Willis Resonant ultrasound spectroscopy. J. Phys. Condens. Matter **9** (1997) 6001–6029.

P. D. Maker, E. W. Terhune, M. Nisenoff, C. M. Savage Effects of dispersion and focussing on the production of optical harmonics. Phys. Rev. Letts. **8** (1962) 21–22.

H. J. McSkimin Pulse superposition method for measuring ultrasonic wave velocities in solids. J. Acoust. Soc. Amer. **33** (1961) 12–16.

R. Meier, K. Schuster Zur Theorie der Schallausbreitung in piezoelektrischen Kristallen. Ann. Physik **11** (1953) 397–406.

H. Mueller Determination of elasto-optical constants with supersonic waves. Z. Kristallogr. **99** (1938) 122–141.

D. F. Nelson, M. Lax New symmetry for acousto-optic scattering. Phys. Rev. Letters **24** (1970) 379–380.

D. F. Nelson, M. Lax Theory of the photoelastic ineraction. Phys. Rev. **B3** (1971) 2778–2794.

R. Nitsche, V. Krämer, M. Schuhmacher, A. Bussmann Preparation and properties of ferroelastic/ferroelectric polytypes of antimony(III) oxide iodide Sb_5O_7I. J. Cryst. Growth **42** (1977) 549–559.

I. Ohno Free vibration of a rectangular parallelepiped crystal and its application to determination of elastic constants of orthorhombic crystals. J. Phys. Earth **24** (1976) 355-379.

M. Onoe, H. F. Tiersten, A. H. Meitzler Shift in the location of resonant frequencies caused by large electromechanical coupling in thickness-mode resonators. J. Acoust. Soc. Amer. **35** (1963) 36–42.

P. Preu, S. Haussühl Dielectric properties and phase transition in $Li_2Ge_7O_{15}$. Solid State Comm. **41** (1982) 627–630.

P. Preu, S. Haussühl Quadratic electrostrictive effects in NaCl and $KAl(SO_4)_2 \cdot 12H_2O$ derived from stress dependence of dielectric constants. Solid State Comm. **45** (1983) 619–623.

N. W. Schubring, J. P. Nolta, R. A. Dork Ferroelectric hysteresis tracer featuring compensation and sample grounding. J. Sci. Instr. **35** (1964) 1517–1521.

A. Seeger, O. Buck Die experimentelle Ermittlung der elastischen Konstanten höherer Ordnung. Z. Naturforsch. **15a** (1960) 1056–1067.

L. A. Shuvalov Symmetry aspects of ferroelectricity. J. Phys. Soc. Japan **28** (1970) 38–51.

R. N. Thurston, K. Brugger Third-order elastic constants and the velocity of small amplitude elastic waves in homogenously stressed media. Phys. Rev. **A133** (1964) 1604–1610.

R. W. Whatmore New polar materials: their application to SAW (surface acoustic waves) and other devices. J. Cryst. Growth **48** (1980) 530–547.

H. Wondratschek Über Tensorsymmetrien in den einzelnen Kristallklassen. N. Jahrbuch Miner. Jahrgang 1952, 217–234.

I. S. Zheludev The change of complete symmetry of the crystals during the phase transitions in ferroelectrics and ferromagnetics. Proc. Ind. Acad. Sci. **LIX** (1964) 168–184.

12.6.3 Data Sources

Th.J. Ahrens (ed.) Mineral physics & crystallography: A handbook of physical constants. American Geophysical Union, Washington 1995.

S. L. Altmann, P. Herzig Point-group theory tables. Clarendon Press, Oxford 1994.

B. P. Belikov, K. S. Alexandrov, T. V. Ryshova Elastische Eigenschaften von gesteinsbildenden Mineralen und Gesteinen. In russ. Sprache. NAUK publishing, Moscow 1970.

M. G. Graham, H. E. Hagy (eds.) Thermal Expansion—1971. AIP Conf. Proceed. No. 3. American Institute of Physics, New York 1972.

P. Groth Chemische Krystallographie. Vol. I-V. V. W. Engelmann, Leipzig 1906–1919.

Landolt-Börnstein, 6. Aufl. Springer, Berlin 1959–1971

Vol. II/1 : Mechanisch-thermische Zustandsgrößen (1971).
Vol. II/4 : Kalorische Zustandsgrößen (1961).
Vol. II/6 : Elektrische Eigenschaften I (1959).
Vol. II/8 : Optische Konstanten (1962).
Vol. II/9 : Magnetische Eiegnschaften I (1962).
Vol. II/10 : Magnetische Eigenschaften II (1967).

Landolt-Börnstein, New Series. Springer, Berlin 1966–2000

Vol. III/1	: Elastic, piezoelectric, piezooptic and electrooptic constants of crystals (1966).
Vol. III/2	: Supplement and extension to III/1 (1969).
Vol. III/3	: Ferro and antiferroelectric substances 1969).
Vol. III/4a,b	: Magnetic and other properties of oxides and related compounds (1970).
Vol. III/9	: Ferro and antiferroelectric substances (1975).
Vol. III/11	: Elastic, piezoelectric, pyroelectric, piezooptic, electrooptic constants, and nonlinear dielectric susceptibilities of crystals (1979).
Vol. III/12a,b,c	: Magnetic and other properties of oxides and related compounds (1978-1981).
Vol. III/16a,b	: Ferroelectrics and related substances (1981, 1982).
Vol. III/17a,b,c	: Semiconductors (1981, 1982).
Vol. III/18a,b	: Ferroelectrics and related substances (1990).
Vol. III/29a	: Low frequency properties of dielectric crystals: Second and higher order elastic constants (1992).
Vol. III/29b	: Low frequency properties of dielectric crystals: Piezoelectric, pyroelectric and related constants (1993).
Vol. III/30a	: High frequency properties of dielectric crystals: Piezooptic and electrooptic coefficients and nonlinear dielectric susceptibilities (1996).
Vol. III/30b	: High frequency properties of dielectric crystals: Nonlinear dielectric susceptibilities (2000).
Vol. IV/4	: High-pressure properties of matter (1980).
Vol. V/1	: Physical properties of minerals and rocks (1981).

R. E. Taylor, G. L. Denman (eds.) Thermal expansion—1973. AIP Conf. Proceed. No. 17. American Institute of Physics, New York 1974.

Y. S. Touloukian, C. Y. Ho (eds.) Thermophysical properties of matter. Vols. 1–6, 12 and 13. The TPRC Data Series. IFI/Plenum, New York, Washington 1970 and 1972.

A. N. Winchell The optical properties of organic compounds. Academic Press, New York 1954.

A. N. Winchell, H. Winchell The microscopic character of artificial inorganic solid substances. Academic Press, New York 1964.

12.6.4 Journals

Acta Crystallographica
Crystal Research and Technology
Journal of Applied Crystallography
Journal of Crystal Growth
Kristallografiya (Soviet Physics Crystallography)
physica status solidi
Solid State Communications
Zeitschrift für Kristallographie

In addition, almost all Journals in physics and physical chemistry contain articles on crystal physics.

Index

α Quartz, 178, 271

abrasive hardness, 318
absolute measurement, 272
acoustical activity, 289
acute bisectrix, 134
adiabatic temperature increase, 301
adsorption energy, 321
angular constancy, 3
anisotropic crystals, 334
anisotropy, 334
antiferroelectricity, 108
antiferromagnetism, 110
axial angle, 133

basic law of crystaloptics, 117
basis vector, 2
beam bending, 219
black–white groups, 48
boundary surface energy, 318, 320
boundary surface energy, relative, 320
Bragg condition, 235
Bragg method, 49
Bravais indices, 47
Bravais types, 42, 44
– multiple-primitive, 43
– primitive, 43
Brillouin scattering, 231, 260
– stimulated, 262

capacitive dilatometer, 160, 219
Cartesian coordinate system, 5
Cauchy relations, 217, 342
character, 360
Christoffel determinant, 224
class, 359
Clausius–Clapeyron relation, 350
Clausius–Mossotti formula, 106, 149, 335
color groups, 48
combination elastic wave, 225
conjugated element, 359
converse piezoelectric effect, 182
correlation of properties, 329
coupling factors, 228
crystal
– ideal, 1
– real, 1
crystal class, 24
crystal systems, 32
crystallographic reference system, 2
cubic crystal system, 31
Curie–Weiss law, 108
cyclic groups, 366

Debye–Waller factor, 265
deformation tensor, 155
deformation tensor, mechanoelectric, 301
determination of orientation, 48

Physical Properties of Crystals. Siegfried Haussühl.

ISBN: 978-3-527-40543-5

deviation from Ohm's law, 194
diamagnetism, 110
dielectric tensor, 100, 113
– pressure dependence, 301
dipyramid, 36
direct matrix product, 364
dislocations, 314
dispersion, 148
disphenoid, 36
displacement tensor, 155
dissolution, 323
dome, 36
dynamic elasticity, 226
dynamic measurement methods, 231

elastic constants
– measurement strategy, 266
elastic properties, 332, 341
elastic stability, 267
elasticity tensor, 214
elastodynamic basic equations, 223
elastodynamics, 222
elastooptical tensor, 271, 273
elastostatics, 217
electrical conductivity tensor, 150
electroacoustical effects, 288
electrooptic effect
– second-order, 285
electrooptic tensor, 185
elektrogyration, 286
equations of state, 297
equivalent matrix, 360
Ewald construction, 275

Faraday effect, 204, 335
ferrimagnetism, 110
ferroelastics, 311
ferroelectricity, 108
ferromagnetism, 110
first basic law of crystallography, 3
first-order electrostriction, 182
first-order transformations, 351, 352
forced vibrations of piezoelectric crystals, 244
frequency doubling, 196
Fresnel formula, first, 123
Fresnel formula, second, 123

Gibbs potential, 298
Gibbs' condition, 4
glide directions, 311
glide lines, 311
glide mirror planes, 45
glide plane, 46, 311
glide system, 311
Grüneisen constant, 333
Grüneisen relation, 333
Grüneisen tensor, 402
group, 357
– isomorphic, 359
– representations, 358
group algebra, 385
group table, 358
group theory, 357
gyration vector, 142

habit, 4
Hall effect, 205
hardening, 314
hardness, 310
Hermann-Mauguin Symbole, 25
higher-rank tensors, 288
Hobden classes, 197
Hooke's law, 214, 290
hysteresis, 352

ideal crystal, 1
idempotents, 386
immersion method, 101
indentation hardness, 315
index equation, 115
index surface, 202

indicatrix, 116
inductive dilatometer, 159
inner Kronecker product, 364
interpretation of properties, 329
invariant linear subspace, 371
irreducible representation, 361
irreducible representation, construction, 364

Jamin interferometer, 187

Kleinman's rule, 200, 301
Kusnezow method, 318

Lagrangian deformation tensor, 156
λ transformation, 354
Lambert's absorption law, 139
Landau theory, 351
Laplace equation, 379
lateral contraction, 219
lattice energy, 336
Laue groups, 409
Laue method, 49
law of angular constancy, 3
law of rational indices, 8, 9
law of reflection, 119, 122
law of refraction, 122
Levi-Civitá tensor, 217, 342
$LiHSeO_3$ (PSG 222), 243
line groups, 47
linear compressibility, 220
linear compressibility tensor, 164
linear subspace, 387
linear vector space, 371
list of symbols, 405
lithium iodate, 179
lithium niobate, 178
longitudinal elastic wave, 225
Lorentz–Lorenz formula, 149, 284, 335

magnetic permeability tensor, 110, 113
magnetic point symmetry groups, 305
magnetic susceptibility, 335
magnetic susceptibility tensor, 109
Maker interferences, 197, 198
mass attenuation coefficient, 336
mass conductivity, 153
mass permeability tensor, 154
Maxwell relations, 115
– first-order, 300
– second–order, 300
Maxwell's equations, 113
mechanical stress tensor, 164
meroedries, 34
metric, 2
Michelson interferometer, 182
micro hardness tester, 317
Miller indices, 10
Mohs hardness, 323
mole polarization, 335
mole volume, 336
morphological symmetry, 42

Neutron scattering, 231, 263
non-tensorial properties, 309
nonadditive properties, 338
nonlinear dielectric susceptibility, 195
nonlinear elasticity, 290
nonlinear electrical conductivity, 194

obtuse bisectrix, 134
Onsager principle, 150
Onsager's principle, 152
optic axes, 118, 133
optic character, 134
optical absorption, 139
optical activity, 141

optical interference dilatometer, 160
optically biaxial, 118
optically isotropic, 118
optically negative, 134
optically positive, 134
optically uniaxial, 118
optorotation, 277
order parameter, 351
order–disorder transformations, 352

paramagnetism, 110
parametric oscillator, 201
pedion, 36
phase matching, 196, 201
phase transformation, 313
phase transformations, 347, 348
phonon dispersion curve, 263
photochromatic effect, 355
physical properties, 332
piezoacoustical effect, 290
piezoconductivity, 288
piezodielectric tensor, 284
piezoelastic constants, 270
piezoelastic properties, 344
piezoelectric effect, 173
piezoelectric effect, extreme values, 180
piezoelectric effect, second-order, 301
piezooptical tensor, 271
pinacoid, 36
plane groups, 47
plane principal axis transformation, 98
Poinsot construction, 94
point symmetry group, 24, 29, 30, 408
– determination, 42
Poisson's ratio, 220
polarization constants, 116
potassium bromate, 178
precision measurement of lattice constants, 161
principal axes' transformation, 95
principal refractive indices, 115
prism, 36
projection operators, 385, 388
pulse–echo methods, 231
pyramid, 36
pyroelectric effect, 86
pyroelectric groups, 86

quasiadditive properties, 331

real crystals, 1
reciprocal system, 15
refractive index, 115
regular representation, 360
resonances of plates and rods, 231, 239
resonant ultrasound spectroscopy (RUS), 255
Reuss case, 400
rotation, 21
rotation tensor, 155
rotoinversion, 21

sawing velocity, 324
scalar product, 13
scalar triple product, 15
scalenohedron, 36
Schaefer–Bergmann method, 231, 234
scratch hardness, 323
screw axes, 45
second basic law of crystallography, 9
second harmonic generation, 196
second-order electrostriction, 285, 301
second-order optical activity, 148
second-order transformations, 351, 354

Sellmeier equation, 149
sing-around-method, 234
skalares Produkt, 13
Snellius' law of refraction, 119
soft modes, 270
space groups, 42, 46
specific heat, 84
specific weight, 83
spectroscopic properties, 326
sphenoid, 36
spherical harmonics, 379
static and quasistatic methods of measurement, 174
stereographic projection, 4–6
strain gauge, 161, 219
strength, 317
strength properties, 309
streptohedron, 36
structure and properties, 329
substitution method, 102
superconductivity, 339
symmetry reduction
– first-rank tensors, 85
– fourth-rank tensors, 207
– second-rank tensors, 89
– tensors of nonlinear elasticity, 291
– third-rank tensors, 168

tensor components under auxiliary conditions, 301
tensor representation, 370
theory of forms, 32
thermal conductivity, 152
thermal conductometer, 153
thermal expansion, 159, 339
thermal scattering, 231
thermoelastic constants, 270
thermoelastic properties, 344
thermoelectrical effect, 307
thickness resonator, 232
thiourea, 354
third basic law of crystallography, 25
time reversal, 305
torsion modulus, 221
total reflection, 124
trace, 359
tracht, 4
transformation, dilative, 351
transformation, displacive, 351
transformation, reconstructive, 351
transformation, rotative, 351
translation, 20
transverse elastic wave, 225
twin formation, 47, 311

unitary matrix, 360
units, 407

vector addition, 10
vector product, 14
vector transformations, 18
vector triple product, 17
vectors, 10
Vegard's rule, 337
Verdet constant, 335
Vickers hardness, 316
Voigt case, 400
Voigt notation, 216
volume compressibility, 220, 336

wedge method, 197, 198
whiskers, 222, 317
Wulff's theorem, 4

X-cut, 232
X-ray scattering, 231, 264

Y-cut, 232
Young's modulus, 218

zero-cut, 271
zinc blende, 179